mostly finite geometries

PURE AND APPLIED MATHEMATICS

A Program of Monographs, Textbooks, and Lecture Notes

LECTURE NOTES IN PURE AND APPLIED MATHEMATICS

1. *N. Jacobson,* Exceptional Lie Algebras
2. *L.-Å. Lindahl and F. Poulsen,* Thin Sets in Harmonic Analysis
3. *I. Satake,* Classification Theory of Semi-Simple Algebraic Groups
4. *F. Hirzebruch et al., Differentiable Manifolds and Quadratic Forms*
5. *I. Chavel,* Riemannian Symmetric Spaces of Rank One
6. *R. B. Burckel,* Characterization of C(X) Among Its Subalgebras
7. *B. R. McDonald et al.,* Ring Theory
8. *Y.-T. Siu,* Techniques of Extension on Analytic Objects
9. *S. R. Caradus et al.,* Calkin Algebras and Algebras of Operators on Banach Spaces
10. *E. O. Roxin et al.,* Differential Games and Control Theory
11. *M. Orzech and C. Small,* The Brauer Group of Commutative Rings
12. *S. Thomier,* Topology and Its Applications
13. *J. M. Lopez and K. A. Ross,* Sidon Sets
14. *W. W. Comfort and S. Negrepontis,* Continuous Pseudometrics
15. *K. McKennon and J. M. Robertson,* Locally Convex Spaces
16. *M. Carmeli and S. Malin,* Representations of the Rotation and Lorentz Groups
17. *G. B. Seligman,* Rational Methods in Lie Algebras
18. *D. G. de Figueiredo,* Functional Analysis
19. *L. Cesari et al.,* Nonlinear Functional Analysis and Differential Equations
20. *J. J. Schäffer,* Geometry of Spheres in Normed Spaces
21. *K. Yano and M. Kon,* Anti-Invariant Submanifolds
22. *W. V. Vasconcelos,* The Rings of Dimension Two
23. *R. E. Chandler,* Hausdorff Compactifications
24. *S. P. Franklin and B. V. S. Thomas,* Topology
25. *S. K. Jain,* Ring Theory
26. *B. R. McDonald and R. A. Morris,* Ring Theory II
27. *R. B. Mura and A. Rhemtulla,* Orderable Groups
28. *J. R. Graef,* Stability of Dynamical Systems
29. *H.-C. Wang,* Homogeneous Branch Algebras
30. *E. O. Roxin et al.,* Differential Games and Control Theory II
31. *R. D. Porter,* Introduction to Fibre Bundles
32. *M. Altman,* Contractors and Contractor Directions Theory and Applications
33. *J. S. Golan,* Decomposition and Dimension in Module Categories
34. *G. Fairweather,* Finite Element Galerkin Methods for Differential Equations
35. *J. D. Sally,* Numbers of Generators of Ideals in Local Rings
36. *S. S. Miller,* Complex Analysis
37. *R. Gordon,* Representation Theory of Algebras
38. *M. Goto and F. D. Grosshans,* Semisimple Lie Algebras
39. *A. I. Arruda et al.,* Mathematical Logic
40. *F. Van Oystaeyen,* Ring Theory
41. *F. Van Oystaeyen and A. Verschoren,* Reflectors and Localization
42. *M. Satyanarayana,* Positively Ordered Semigroups
43. *D. L Russell,* Mathematics of Finite-Dimensional Control Systems
44. *P.-T. Liu and E. Roxin,* Differential Games and Control Theory III
45. *A. Geramita and J. Seberry,* Orthogonal Designs
46. *J. Cigler, V. Losert, and P. Michor,* Banach Modules and Functors on Categories of Banach Spaces
47. *P.-T. Liu and J. G. Sutinen,* Control Theory in Mathematical Economics
48. *C. Byrnes,* Partial Differential Equations and Geometry
49. *G. Klambauer,* Problems and Propositions in Analysis
50. *J. Knopfmacher,* Analytic Arithmetic of Algebraic Function Fields
51. *F. Van Oystaeyen,* Ring Theory
52. *B. Kadem,* Binary Time Series
53. *J. Barros-Neto and R. A. Artino,* Hypoelliptic Boundary-Value Problems
54. *R. L. Sternberg et al.,* Nonlinear Partial Differential Equations in Engineering and Applied Science
55. *B. R. McDonald,* Ring Theory and Algebra III
56. *J. S. Golan,* Structure Sheaves Over a Noncommutative Ring
57. *T. V. Narayana et al.,* Combinatorics, Representation Theory and Statistical Methods in Groups

58. *T. A. Burton,* Modeling and Differential Equations in Biology
59. *K. H. Kim and F. W. Roush,* Introduction to Mathematical Consensus Theory
60. *J. Banas and K. Goebel,* Measures of Noncompactness in Banach Spaces
61. *O. A. Nielson,* Direct Integral Theory
62. *J. E. Smith et al.,* Ordered Groups
63. *J. Cronin,* Mathematics of Cell Electrophysiology
64. *J. W. Brewer,* Power Series Over Commutative Rings
65. *P. K. Kamthan and M. Gupta,* Sequence Spaces and Series
66. *T. G. McLaughlin,* Regressive Sets and the Theory of Isols
67. *T. L. Herdman et al.,* Integral and Functional Differential Equations
68. *R. Draper,* Commutative Algebra
69. *W. G. McKay and J. Patera,* Tables of Dimensions, Indices, and Branching Rules for Representations of Simple Lie Algebras
70. *R. L. Devaney and Z. H. Nitecki,* Classical Mechanics and Dynamical Systems
71. *J. Van Geel,* Places and Valuations in Noncommutative Ring Theory
72. *C. Faith,* Injective Modules and Injective Quotient Rings
73. *A. Fiacco,* Mathematical Programming with Data Perturbations I
74. *P. Schultz et al.,* Algebraic Structures and Applications
75. *L Bican et al.,* Rings, Modules, and Preradicals
76. *D. C. Kay and M. Breen,* Convexity and Related Combinatorial Geometry
77. *P. Fletcher and W. F. Lindgren,* Quasi-Uniform Spaces
78. *C.-C. Yang,* Factorization Theory of Meromorphic Functions
79. *O. Taussky,* Ternary Quadratic Forms and Norms
80. *S. P. Singh and J. H. Burry,* Nonlinear Analysis and Applications
81. *K. B. Hannsgen et al.,* Volterra and Functional Differential Equations
82. *N. L. Johnson et al.,* Finite Geometries
83. *G. I. Zapata,* Functional Analysis, Holomorphy, and Approximation Theory
84. *S. Greco and G. Valla,* Commutative Algebra
85. *A. V. Fiacco,* Mathematical Programming with Data Perturbations II
86. *J.-B. Hiriart-Urruty et al.,* Optimization
87. *A. Figa Talamanca and M. A. Picardello,* Harmonic Analysis on Free Groups
88. *M. Harada,* Factor Categories with Applications to Direct Decomposition of Modules
89. *V. I. Istrăţescu,* Strict Convexity and Complex Strict Convexity
90. *V. Lakshmikantham,* Trends in Theory and Practice of Nonlinear Differential Equations
91. *H. L. Manocha and J. B. Srivastava,* Algebra and Its Applications
92. *D. V. Chudnovsky and G. V. Chudnovsky,* Classical and Quantum Models and Arithmetic Problems
93. *J. W. Longley,* Least Squares Computations Using Orthogonalization Methods
94. *L. P. de Alcantara,* Mathematical Logic and Formal Systems
95. *C. E. Aull,* Rings of Continuous Functions
96. *R. Chuaqui,* Analysis, Geometry, and Probability
97. *L. Fuchs and L. Salce,* Modules Over Valuation Domains
98. *P. Fischer and W. R. Smith,* Chaos, Fractals, and Dynamics
99. *W. B. Powell and C. Tsinakis,* Ordered Algebraic Structures
100. *G. M. Rassias and T. M. Rassias,* Differential Geometry, Calculus of Variations, and Their Applications
101. *R.-E. Hoffmann and K. H. Hofmann,* Continuous Lattices and Their Applications
102. *J. H. Lightbourne III and S. M. Rankin III,* Physical Mathematics and Nonlinear Partial Differential Equations
103. *C. A. Baker and L. M. Batten,* Finite Geometrics
104. *J. W. Brewer et al.,* Linear Systems Over Commutative Rings
105. *C. McCrory and T. Shifrin,* Geometry and Topology
106. *D. W. Kueke et al.,* Mathematical Logic and Theoretical Computer Science
107. *B.-L. Lin and S. Simons,* Nonlinear and Convex Analysis
108. *S. J. Lee,* Operator Methods for Optimal Control Problems
109. *V. Lakshmikantham,* Nonlinear Analysis and Applications
110. *S. F. McCormick,* Multigrid Methods
111. *M. C. Tangora,* Computers in Algebra
112. *D. V. Chudnovsky and G. V. Chudnovsky,* Search Theory
113. *D. V. Chudnovsky and R. D. Jenks,* Computer Algebra
114. *M. C. Tangora,* Computers in Geometry and Topology
115. *P. Nelson et al.,* Transport Theory, Invariant Imbedding, and Integral Equations
116. *P. Clément et al.,* Semigroup Theory and Applications

117. *J. Vinuesa,* Orthogonal Polynomials and Their Applications
118. *C. M. Dafermos et al.,* Differential Equations
119. *E. O. Roxin,* Modern Optimal Control
120. *J. C. Díaz,* Mathematics for Large Scale Computing
121. *P. S. Milojevič,* Nonlinear Functional Analysis
122. *C. Sadosky,* Analysis and Partial Differential Equations
123. *R. M. Shortt,* General Topology and Applications
124. *R. Wong,* Asymptotic and Computational Analysis
125. *D. V. Chudnovsky and R. D. Jenks,* Computers in Mathematics
126. *W. D. Wallis et al.,* Combinatorial Designs and Applications
127. *S. Elaydi,* Differential Equations
128. *G. Chen et al.,* Distributed Parameter Control Systems
129. *W. N. Everitt,* Inequalities
130. *H. G. Kaper and M. Garbey,* Asymptotic Analysis and the Numerical Solution of Partial Differential Equations
131. *O. Arino et al.,* Mathematical Population Dynamics
132. *S. Coen,* Geometry and Complex Variables
133. *J. A. Goldstein et al.,* Differential Equations with Applications in Biology, Physics, and Engineering
134. *S. J. Andima et al.,* General Topology and Applications
135. *P Clément et al.,* Semigroup Theory and Evolution Equations
136. *K. Jarosz,* Function Spaces
137. *J. M. Bayod et al.,* *p*-adic Functional Analysis
138. *G. A. Anastassiou,* Approximation Theory
139. *R. S. Rees,* Graphs, Matrices, and Designs
140. *G. Abrams et al.,* Methods in Module Theory
141. *G. L. Mullen and P. J.-S. Shiue,* Finite Fields, Coding Theory, and Advances in Communications and Computing
142. *M. C. Joshi and A. V. Balakrishnan,* Mathematical Theory of Control
143. *G. Komatsu and Y. Sakane,* Complex Geometry
144. *I. J. Bakelman,* Geometric Analysis and Nonlinear Partial Differential Equations
145. *T. Mabuchi and S. Mukai,* Einstein Metrics and Yang–Mills Connections
146. *L. Fuchs and R. Göbel,* Abelian Groups
147. *A. D. Pollington and W. Moran,* Number Theory with an Emphasis on the Markoff Spectrum
148. *G. Dore et al.,* Differential Equations in Banach Spaces
149. *T. West,* Continuum Theory and Dynamical Systems
150. *K. D. Bierstedt et al.,* Functional Analysis
151. *K. G. Fischer et al.,* Computational Algebra
152. *K. D. Elworthy et al.,* Differential Equations, Dynamical Systems, and Control Science
153. *P.-J. Cahen, et al.,* Commutative Ring Theory
154. *S. C. Cooper and W. J. Thron,* Continued Fractions and Orthogonal Functions
155. *P. Clément and G. Lumer,* Evolution Equations, Control Theory, and Biomathematics
156. *M. Gyllenberg and L. Persson,* Analysis, Algebra, and Computers in Mathematical Research
157. *W. O. Bray et al.,* Fourier Analysis
158. *J. Bergen and S. Montgomery,* Advances in Hopf Algebras
159. *A. R. Magid,* Rings, Extensions, and Cohomology
160. *N. H. Pavel,* Optimal Control of Differential Equations
161. *M. Ikawa,* Spectral and Scattering Theory
162. *X. Liu and D. Siegel,* Comparison Methods and Stability Theory
163. *J.-P. Zolésio,* Boundary Control and Variation
164. *M. Křížek et al.,* Finite Element Methods
165. *G. Da Prato and L. Tubaro,* Control of Partial Differential Equations
166. *E. Ballico,* Projective Geometry with Applications
167. *M. Costabel et al.,* Boundary Value Problems and Integral Equations in Nonsmooth Domains
168. *G. Ferreyra, G. R. Goldstein, and F. Neubrander,* Evolution Equations
169. *S. Huggett,* Twistor Theory
170. *H. Cook et al.,* Continua
171. *D. F. Anderson and D. E. Dobbs,* Zero-Dimensional Commutative Rings
172. *K. Jarosz,* Function Spaces
173. *V. Ancona et al.,* Complex Analysis and Geometry
174. *E. Casas,* Control of Partial Differential Equations and Applications
175. *N. Kalton et al.,* Interaction Between Functional Analysis, Harmonic Analysis, and Probability

176. *Z. Deng et al.*, Differential Equations and Control Theory
177. *P. Marcellini et al.* Partial Differential Equations and Applications
178. *A. Kartsatos*, Theory and Applications of Nonlinear Operators of Accretive and Monotone Type
179. *M. Maruyama*, Moduli of Vector Bundles
180. *A. Ursini and P. Aglianò*, Logic and Algebra
181. *X. H. Cao et al.*, Rings, Groups, and Algebras
182. *D. Arnold and R. M. Rangaswamy*, Abelian Groups and Modules
183. *S. R. Chakravarthy and A. S. Alfa*, Matrix-Analytic Methods in Stochastic Models
184. *J. E. Andersen et al.*, Geometry and Physics
185. *P.-J. Cahen et al.*, Commutative Ring Theory
186. *J. A. Goldstein et al..*, Stochastic Processes and Functional Analysis
187. *A. Sorbi*, Complexity, Logic, and Recursion Theory
188. *G. Da Prato and J.-P. Zolésio*, Partial Differential Equation Methods in Control and Shape Analysis
189. *D. D. Anderson*, Factorization in Integral Domains
190. *N. L. Johnson*, Mostly Finite Geometries
191. *D. Hinton and P. W. Schaefer*, Spectal Theory and Computational Methods of Sturm–Liouville Problems
192. *W. H. Schikhof et al.*, *p*-adic Functional Analysis

Additional Volumes in Preparation

mostly finite geometries

in celebration of T. G. Ostrom's 80th birthday

edited by

Norman L. Johnson

University of Iowa
Iowa City, Iowa

MARCEL DEKKER, INC. NEW YORK • BASEL • HONG KONG

Library of Congress Cataloging-in-Publication Data

Mostly finite geometries: in celebration of T. G. Ostrom's 80th
birthday / edited by Norman L. Johnson.
p. cm. — (Lecture notes in pure and applied mathematics; v. 190)
Proceedings of TEDFEST '96, held Mar. 1996 at the University of Iowa.
Includes index.
ISBN 0-8247-0035-X (pbk. : alk. paper)
1. Finite geometries—Congresses. I. Ostrom, T. G. (Theodore
Gleason). II. Johnson, Norman L. (Norman Lloyd).
III. TEDFEST '96 (1996: University of Iowa) IV. Series.
QA167.2.M67 1997
516'.13—dc21

97-1028
CIP

The publisher offers discounts on this book when ordered in bulk quantities. For more information, write to Special Sales/Professional Marketing at the address below.

This book is printed on acid-free paper.

MARCEL DEKKER, INC.
270 Madison Avenue, New York, New York 10016
http://www.dekker.com

Current printing (last digit):
10 9 8 7 6 5 4 3 2 1

PRINTED IN THE UNITED STATES OF AMERICA

Preface

In March of 1996, a conference on “Mostly Finite Geometries” was held at the University of Iowa to honor Professor T.G. Ostrom and to celebrate his eightieth birthday. Mathematicians world-wide came to participate in TEDFEST ’96. Most, if not all, of these mathematicians have been influenced by the work of Ted Ostrom in varying degrees and in different manners.

So, it is appropriate here to revisit some of Ted’s mathematical creations and ideas. I wrote a similar piece on the geometry of Ostrom in 1981 in commemoration of his retirement from Washington State University where he had taught since 1960 – it is difficult to believe that it has been sixteen years (“The Geometry of Ostrom; The contributions to Geometry of T.G. Ostrom.” **Finite Geometries.** Lecture notes in pure and applied mathematics, Vol.82, Marcel Dekker, Inc., 1983, pp 1-22.) (**G**).

In the above mentioned article, I wrote of the great originality of Ted’s work. The essence of his work can best be given by describing some of his ideas which are simple, deep, and fundamental simultaneously. In fact, it is difficult to find virtually any area of finite geometry that does not have some thread back to an idea of Ted. As I detailed many of these ideas before in **G**, let me pull back my favorite thread and mention one of the most stunning.

Perhaps the most powerful and simple method for the construction of geometries deals with the idea of “covering” a subgeometry by another in such a way so as to construct another related geometry.

Let me describe Ted’s conception of “derivation”. Let π be an affine plane of order n^2. An affine subplane π_o of order n has n^2 points; as does any line; and $n+1$ parallel classes. Suppose for every pair of points P and Q which are joined by a line in one of the parallel classes of π_o, there is an affine subplane of order n, $\pi_{P,Q}$, which contains P and Q and also shares the $n + 1$ parallel classes of π_o. If we define a new incidence structure whereby the lines of the parallel classes of π_o are traded for the subplanes of order n with the properties above, a new affine plane is created. The new plane is called the plane “derived” from π.

This, of course, seems now like a natural idea, as the subplanes have the same size as lines so perhaps we can replace some of the lines by subplanes and create a new affine plane. But, derivation was a revolutionary concept, a simple idea and one which promised to be extremely deep, producing new planes by the handfuls where previously there was a dearth of examples.

Mathematicians have been trying to determine the nature of derivation for the last thirty-five years or so since Ted thought this up in 1960. Generalizations abound as one can consider the derivation of infinite affine planes or consider what the structure theory should be merely for the substructure involved in the construction process, the derivable net. A derivable net is then a net which is covered by Baer subplanes (the infinite analogue of a subplane of order n in an affine plane of order n^2) in the manner specified above. One could further consider "subplane covered" nets where the term "Baer subplane" is replaced simply by the word "subplane".

It is now known that subplane covered nets are intimately connected to projective spaces. In fact, if Σ is any projective space, finite or infinite, of dimension > 2 and N is any projective subspace of codimension two, consider the combinatorial structure of "lines" L as points of $\Sigma - N$, "points" P as lines of Σ skew to N, "parallel classes" C as the hyperplanes containing N and the "subplanes" B as the projective subplanes of Σ which intersect N in a point. Let I denote the incidence inherited from that of Σ . Then (P, L, C, B, I) is a subplane covered net. It turns out that any subplane covered net can be realized in this way.

So, the idea of derivation is simple, deep and fundamental. Derivable nets and planes and their collineation groups are then intrinsically connected with projective geometry and projective groups respectively.

I had the privilege to work with Ted on derivable nets recently where we considered direct products of affine planes (N.L. Johnson and T.G. Ostrom, Direct products of affine partial linear spaces. Journal of Combinatorial Theory (A), vol. 75, no. 1 (1996), 99-140). The germ of the idea for this paper was, of course, Ted's and I was intrigued by the application of the projective space corresponding to a derivable net so it was decided to consider when a direct product of two or more affine planes produces a derivable net and generalize all of this to the consideration and construction of subplane covered nets by products or sums. To give an inkling of what was done in this work, I might mention the following result:

A net is subplane covered if and only if there exists a set ρ such that the net is isomorphic to a ρ–fold sum of a Desarguesian affine plane.

A net is derivable if and only if it is the regular direct product of two isomorphic Desarguesian affine planes.

It was a joy to work with this material as the structure fits together so well - Ted's ideas just keep working in very general situations whether describing subplane covered nets in terms of combinatorial incidence structures in projective spaces or considering derivable nets as direct products. The great geometer B. Segre classified geometries which somehow relate back to or are classified by projective spaces as the "beautiful" geometries. It seems natural and fitting that many of Ted's beautiful ideas lead to beautiful geometries.

However, Ted would probably still say "well, all of this is fine, but I still think it is easier to see this **my** way".

For all of us who participated in "Tedfest '96", for those of us who were and are touched and inspired by Ted's ideas and mathematics, for all of us who love Ted, it is my distinct honor and pleasure to dedicate this volume, "Mostly Finite Geometries", to Professor T.G. Ostrom on the occasion of his eightieth birthday.

This conference and proceedings grew out of a special session of a regional meeting of the American Mathematical Society. So, I would like to thank the society for providing the foundation for this book.

The University of Iowa celebrated its sesquicentennial in 1996 and the Department of Mathematics, in turn, responded in kind by hosting eight special sessions, conferences and workshops during and around the regional meeting of the AMS. One of these activities became Tedfest '96 and I am indebted to the University, in general, and the Department of Mathematics, in particular, for their support–financial, logistic, and collegial.

I would also like to thank my students, Changmin Ding, Dean Draayer and Xiaobin Liu, who made sure that the conference ran smoothly. Also, thanks goes to Brian Treadway for invaluable help with the typing and styling of various of the papers.

To all of the participants of the conference and contributors to the proceedings, I owe a great debt of gratitude for making this event truly memorable.

Finally, I thank my wife Bonnie L. Hemenover for her continual encouragement and immeasurable 'behind the scenes' help and assistance with Tedfest '96.

Norman L. Johnson

Contents

Contributors

Michael A. Abramson Mathematics Department, Bowling Green State University, Bowling Green, Ohio 43402

Dieter Betten Mathematisches Seminar, Universität Kiel, Ludewig-Meyn-Str. 4, 24098 Kiel, Germany

Mauro Biliotti Dipartimento di Matematica, Università di Lecce, Via Arnesano, 73100 Lecce, Italy

A. Bonisoli Dipartimento di Matematica, Università della Basilicata, via N. Sauro 85, 85100 Potenza, Italy

Francis Buekenhout C.P. 216, Université Libre de Bruxelles, Bd. du Triomphe, B-1050 Bruxelles, Belgium

Phillippe Cara Departement Wiskunde, Vrije Universiteit Brussel, Pleinlaan 2, B-1050 Brussel, Belgium

Minerva Cordero Department of Mathematics, Texas Tech University, Lubbock, Texas 79409

Michel Dehon C.P. 216, Université Libre de Bruxelles, Bd. du Triomphe, B-1050 Bruxelles, Belgium

Jeremy M. Dover Department of Mathematics, Moravian College, 1200 Main Street, Bethlehem, Pennsylvania 18018

G. L. Ebert Mathematical Sciences Department, 510 Ewing Hall, University of Delaware, Newark, Delaware 19716

Raul F. Figueroa Department of Mathematics, University of Puerto Rico, PO Box 23355, Rio Piedras, Puerto Rico 00931

John Fuelberth Division of Mathematics and Sciences, Wayne State College, Wayne, Nebraska 68787

Athula Gunawardena Division of Mathematics and Sciences, Wayne State College, Wayne, Nebraska 68787

Christoph Hering Universitaet Tuebingen, Mathematisches Institut, Auf der Morgenstelle 10, D-72076 Tuebingen, Germany

Yutaka Hiramine Department of Mathematics, Faculty of Education, Kumamoto University, Kurokami, Kumamoto, Japan

Chat Yin Ho Department of Mathematics, 358 Little Hall, University of Florida, Gainesville, Florida 32607

Vikram Jha Mathematics Department, Glasgow-Caledonian University, Cowcaddens Road, Glasgow, Scotland

Norman L. Johnson Mathematics Department, University of Iowa, Iowa City, Iowa 52242

Anna Kasikova Department of Mathematics, Kansas State University, Manhattan, Kansas 66502

Robert A. Liebler Department of Mathematics, Colorado State University, Fort Collins, Colorado 80523

Xiaobin Liu Department of Mathematics, University of Iowa, Iowa City, Iowa 52242

Geoffrey Mason Mathematics Department, University of California, Santa Cruz, California

G. Eric Moorhouse Department of Mathematics, University of Wyoming, Laramie, Wyoming 82071-3036

T. G. Ostrom Department of Pure and Applied Mathematics, Washington State University, Pullman, Washington 99164-3113

S. E. Payne Department of Mathematics, Campus Box 170, University of Colorado at Denver, Denver, Colorado 80217-3364

Tim Penttila Department of Mathematics, University of Western Australia, Nedlands 6907, Western Australia, Australia

Rolando Pomareda Department of Mathematics, Universidad de Chile, Casilla 653, Santiago, Chile

Alan R. Prince Department of Mathematics, Heriot-Watt University, Edinburgh, Scotland EH14 4AS

Gordon F. Royle Department of Computer Science, University of Western Australia, Nedlands 6907, Western Australia, Australia

Hans-Jörg Schaeffer Universitaet Tuebingen, Mathematisches Institut, Auf der Morgenstelle 10, D-72076 Tuebingen, Germany

F. A. Sherk Department of Mathematics, University of Toronto, Toronto, Ontario, Canada M5S 1A1

Ernest Shult Department of Mathematics, Kansas State University, Manhattan, Kansas 66502

Thomas Q. Sibley Mathematics Department, St. John's University, Collegeville, Minnesota 56321

Leo Storme Department of Pure Mathematics and Department of Computer Algebra, University of Gent, Galglaan 2, Gent, Belgium

Chiro Suetake Amagasake-minami High School, Amagasaki, Hyogo 660, Japan

G. P. Wene Department of Mathematics, University of Texas at San Antonio, San Antonio, Texas

T. G. OSTROM

Conference Participants

M. Abramson, Bowling Green State University, Bowling Green, Ohio
C. Baker, University of Manitoba, Winnipeg, Canada
D. Betten, University of Kiel, Kiel, Germany
M. Biliotti, University of Lecce, Lecce, Italy
J. Brown, York University, Downsview, Ontario, Canada
A. Bonisoli, University of Basilicata, Potenza, Italy
F. Buekenhout, Free University of Brussels, Brussels, Belgium
P. Cara, Free University, Brussels, Belgium
W. Cherowitzo, University of Colorado at Denver, Denver, Colorado
M. Cordero, Texas Tech University, Lubbock, Texas
M. Dehon, Free University of Brussels, Brussels, Belgium
C. Ding, University of Iowa, Iowa City, IA
J. Dover, University of Delaware, Newark, DE
D. Draayer, University of Iowa, Iowa City, IA
G. Ebert, University of Delaware, Newark, DE
R. Figueroa, University of Puerto Rico, Río Piedras, Puerto Rico
J. Fink, Kalamazoo College, Kalamazoo, Michigan
J. Fuelberth, Wayne State College, Wayne, NE
C. Garner, Carleton College, Canada
A. Gunawardena, Wayne State College, Wayne, NE
C. Hering, University of Tübingen, Germany
Y. Hiramine, Kumamoto University, Kurokami, Kumamoto, Japan
H. Klein, University of Kiel, Kiel, Germany
E. Kleinfeld, University of Iowa, Iowa City, IA
M. Kleinfeld, University of Iowa, Iowa City, IA
C.Y. Ho, University of Florida, Gainesville, Florida
S. Hobart, University of Wyoming, Laramie, WY
V. Jha, Glasgow-Caledonian University, Glasgow, Scotland
N.L. Johnson, University of Iowa, Iowa City, IA
P. Johnson, University of London Queen Mary-Westfield, London, England

M. Kallaher, Washington State University, Pullman, Washington
W. Kantor, University of Oregon, Eugene, Oregon
A. Kasikova, Kansas State University, Manhattan, Kansas
R.A. Liebler, Colorado State University, Fort Collins, Colorado
X. Liu, University of Iowa, Iowa, City, IA
M. Lorimer, University of Toronto, Toronto, Canada
G. Mason, University of California, Santa Cruz, CA
M. Miller, University of Colorado at Denver, Denver, Colorado
G.E. Moorhouse, University of Wyoming, Laramie, WY
T.G. Ostrom, Washington State University, Pullman, WA
S.E. Payne, University of Colorado at Denver, Denver, Colorado
T. Penttila, University of Western Australia, Nedlands, Australia
R. Pomareda, University of Chile, Santiago, Chile
A.R. Prince, Heriot-Watt University, Edinburgh, Scotland
M. de Resmini, University of Rome, Italy
G.F. Royle, University of Western Australia, Nedlands, Australia
H-J. Schaeffer, University of Tübingen, Germany
F.A. Sherk, University of Toronto, Toronto, Canada
E. Shult, Kansas State University, Manhattan, Kansas
T.Q. Sibley, St. John's University, Collegeville, MN
L. Storme, University of Ghent, Ghent, Belgium
C. Suetake, Amagasaki-minami High School, Amagasaki, Hyogo, Japan
J. Swaby, University of Wyoming, Laramie, Wyoming
V. Tonchev, Michigan Tech. University, Houghton, MI
M. Vincent, Colorado State University, Fort Collins, Colorado
G.P. Wene, University of Texas at San Antonio, San Antonio, Texas
F. Wilke, University of Missouri at St. Louis, Missouri

Introduction

This book is entitled "Mostly Finite Geometries" for two reasons. First of all most of the geometries encountered herein are finite. However, there are the important topological geometries which are necessarily infinite and there is a growing appreciation of the ideas of generalized quadrangles and flocks of geometries in the infinite case which originate with ideas of finite geometric objects. Also, the geometries are also mostly finite in the sense that so many of what one might call "interesting" infinite geometries were conceptually born of finite ancestry.

There is an extensive relatively new body of knowledge concerning "flocks" and their associated geometries. Flocks seem to be connected to an ever increasing number of diverse geometric and combinatorial structures such as sharply transitive sets, spreads, generalized quadrangles, esoteric projective planes, families of ovals, partial spreads to name a few.

The theory of generalized quadrangles has become a major player in many geometric arenas and this will become clear as the reader considers the variety of articles written which somehow touch on or connect with a generalized quadrangle.

T.G. Ostrom was a pioneer in the study and construction of finite projective and affine planes and their building blocks, finite nets. In particular, Ostrom was interested in the possible finite groups that can act on finite translation planes. And, of course, his construction techniques such as "derivation" and more generally "net replacement" are still sources for on-going research.

There are two major schools in the study of translation planes: André and Bruck-Bose. Although equivalent at least for finite planes, the difference between the models is mainly that the André translation planes sit in a vector space whereas the Bruck-Bose translation planes lie in a projective space. So, mathematicians interested in representation theory would work with the André model and others primarily interested in projective geometry might feel more at home working with the Bruck-Bose construct.

Of course, many of the geometries considered may be more generally thought of in terms of diagram geometries or in other more abstract forms. So, there is a movement toward the study of general geometries by various methods.

To describe the contents of this book, I have decided to collect the papers herein into the sets of articles mostly connected with flocks, generalized quadrangles, projective planes, projective geometries, and general geometries. Of course, the classes are not disjoint so inclusion in one set is somewhat arbitrary.

FLOCKS

By flocks here is meant flocks of quadratic cones, although there are some generalizations which the various papers consider.

Johnson and Payne provide a general survey of the geometries which are somehow connected to or may be constructed from a flock. In the finite case, one such connected geometry is a type of generalized quadrangle that may be constructed via Kantor's theory of coset geometries. Actually, such generalized quadrangles, it may be argued, are central to the subject. For example, Payne, Penttila and Royle construct some new finite flocks which can only be described by consideration of the associated generalized quadrangles.

In the infinite case, a flock need not give rise to an associated generalized quadrangle but there is an associated spread so it is often convenient to work in a related translation plane. Johnson and Liu survey the known infinite flocks and flock spreads and show that the Kantor-Knuth flocks behave in the infinite case very much like they do in the finite. Biliotti and Johnson consider a method of Dembowski to produce flocks of elliptic quadrics which when properly interpreted can produce infinite flocks of quadratic cones.

A partial flock corresponds to a partial spread which is the union of reguli that share a common line. Johnson and Storme consider whether such partial spreads can be embeddable in spreads in $PG(3,q)$. If approximately $3q/4$ reguli are involved in the union and there is an embedding of the partial spread into a spread then the partial flock may be uniquely extended to a flock.

GENERALIZED QUADRANGLES

Of course, all of the papers mentioned under the previous heading could be included here. Furthermore, Kasikova and Shult construct a new infinite class of "extended" generalized quadrangles. And, Mason uses representation theory and the action of $SL(2,q)$ for the study of generalized quadrangles of characteristic two. Abramson constructs generalized quadrangles from blueprints. In particular, there are generalized quadrangles of orders (q,q^2) and (q^2,q^3).

PROJECTIVE PLANES
Group Theory and projective planes

As was mentioned, Ostrom was fascinated by finite translation planes that admit certain "interesting" groups either intrinsically or by their action. Previously, Ostrom and Mason collaborated on the construction of translation planes admitting A_5 acting in a particular manner. However, only finitely many of these have been previously constructed. Moorhouse shows how to construct various infinite classes of translation planes with the properties mentioned. More generally, Moorhouse gives root lattice constructions of ovoids in $PG(5,q)$ which project to spreads in $PG(3,q)$ via the Klein quadric.

Ostrom has shown that any Klein 4-group all of whose involutions are Baer and which acts on a translation plane of square order forces the spread into at least a 7-dimensional projective space. Hering and Schaeffer construct a new translation plane of order 3^4 admitting a Baer-Klein 4-group.

Bonisoli studies finite projective planes which contain an oval upon which acts a doubly transitive collineation group. A theorem of Hering is used to restrict the possible group actions. Cordero and Figueroa also consider doubly transitive group actions on a projective plane. In particular, they study semifield planes which admit an autotopism group acting transitively on the nonvextex points of one side of the triangle and provide various classifications of such planes. Although not quite connected to group theory but rather to semifield planes, Wene reviews some classical constructions of semifield planes, noting how to generalize some results of Dickson.

Ho considers the geometry of Singer groups and analyzes arc subgroups of planar groups. There is a necessary and sufficient condition given for such subgroups based on the injectivity of functions associated with the difference sets.

Betten surveys the classification of 4-dimensional flexible projective planes. These are topological planes which admit a collineation group of dimension at least 7. In particular, two of the classes of 4-dimensional projective planes with group of dimension 8 also correspond to flocks of infinite quadratic cones.

Further consideration of infinite translation planes is undertaken by Jha and Johnson who try to make sense in the arbitrary or infinite case of the fundamental theorem of André on finite groups generated by homology groups . Also, there is a general structure theory available for affine Baer subplanes and their point-Baer and line-Baer generalizations admitting certain groups which mirrors that of the configuration of axis and coaxis of central collineations.

Construction Techniques

The concept of "derivation" is studied in various ways. Recall that a regulus in $PG(3,q)$ is the classical example of a partial spread which produces a derivable net. Derivation replaces the regulus with its opposite regulus and its derivable net by its derived net.

An important class of flocks may be obtained using certain translation planes which are constructed via nest replacement. A t-nest is a set of t reguli in $PG(3,q)$ whose union forms a partial spread of cardinality $t(q+1)$. Nest replacement is a process of replacing exactly half of the lines of each regulus of a set of t reguli while producing a corresponding translation net of the same degree. Ebert gives a survey of the known translation planes which may be constructed using t-nests of reguli in Desarguesian spreads for $t = (q+3)/2, q-1, q, 2(q-1)$.

More generally, Liebler extends the ideas of derivation by the consideration of the $Z-$span Λ of the lines of pairs of opposite reguli. Roughly speaking, two spreads S and S^* are "virtually" derived from each other if their difference is in Λ. This process is very general and, in fact, there are no spreads in $PG(3,q)$ which are not known to be virtually derived from a regular (Desarguesian) spread.

Buekenhout has given previously constructions of types of unitals which sit in every translation plane of order q^2 with spread in $PG(3,q)$. Using derivation of a regular spread, Dover shows that there exist non-Buekenhout unitals in the Hall planes.

A different approach is taken by Sherk who considers various spreads and corresponding translation planes using indicator sets among other ideas and giving particular attention to weighted configurations associated with certain spreads.

PROJECTIVE SPACES

Again, many, if not all, of the articles on projective planes may be placed within this category.

Pomareda considers a generalization called a "hyper-regulus" of the notion of a regulus in projective spaces. In particular, he shows that the so-called André hyper-reguli have unique opposite hyper-reguli in $PG(5,q)$ for suitable q thereby answering a previously posed question of Ostrom.

Moorhouse determines the $p-$ranks of various submatrices of the incidence matrix of points versus hyperplanes in $PG(3,q)$ thus generalizing results on quadrics among others. And, Fuelberth and Gunawardena study the orbits of $4-$caps in orthogonal spaces giving various formulae in specific situations.

Prince determines certain of the parallelisms (q^2+q+1 spreads covering the lines of $PG(3,q)$) in PG(3,3). In particular it is shown that there exist exactly seven mutually nonisomorphic parallelisms invariant under a group of order 5.

GENERAL GEOMETRIES

In this category, I am placing papers on designs, diagrams, linear spaces and coset geometries other than generalized quadrangles again somewhat arbitrarily.

Hiramine and Suetake consider biplanes of order $k-2$ (symmetric designs with parameters $(v, k, 2)$). There are exactly two known biplanes of order 11, which are those of Aschbacher. The authors show that if there are other biplanes of order 11, then the order of the automorphism group is $2^a \cdot 3^b$ for nonnegative integers a and b.

Buekenhout, Dehon and Cara classify all pairs (Γ, G) which are inductively minimal where Γ is a firm, finite geometry with a connected digon diagram and G is a flag-transitive automorphism group.

Sibley considers various geometrical structures which may be constructed as coset geometries and relate somewhat to certain linear spaces.

Finally, Ostrom, taking a departure from affine and projective planes, discusses finite linear spaces (systems) with parallelism on lines. Examples, collineations, replacement problems and techniques are introduced. In particular, it is shown that such structures may be constructed from affine 3-spaces by the replacement of affine planes embedded in the affine space.

mostly finite geometries

Remarks Concerning Linear Systems with Parallelism

T.G. OSTROM Department of Pure and Applied Mathematics, Washington State University, Pullman, WA 99164-3113

T.G. OSTROM Department of Pure and Applied Mathematics, Washington State University, Pullman, WA 99164-3113

DEFINITION 1. A linear system with parallelism is a set of "points" together with certain designated point sets called "lines" such that

1. For some integer $n > 1$ (to be called the *order* of the system) each line contains exactly n points.
2. Each pair of distinct points belongs to exactly one line.
3. There is an equivalence relation (*parallelism*) on the lines such that each point belongs to exactly one line of each equivalence class (*parallel class*).

DEFINITION 2. Let h be the number of lines in a parallel class, let k be the number of parallel classes, and let N be the total number of points.

LEMMA 1. *$N = hn$, $N = 1 + k(n-1)$, and there exists an integer r such that $h = 1 + r(n-1)$ and $k = 1 + nr$. The parameter h is independent of the parallel class; each two of the parameters N, h, k, n, r determines the order.*

PROOF. The union of the lines in a parallel class is equal to the total number of points, so $N = hn$.

The union of the k lines through a given point P is also equal to N. These lines through P have no other points in common, so $N = 1 + k(n-1)$. Since $hn = N = 1+k(n-1)$, $k \equiv 1 \mod n$. Putting $k = 1+nr$, we get $h = 1+(n-1)r$. The rest of the proof is left to the reader. ∎

DEFINITION 3. The linear system with parallelism with parameters as in Definition 2 will be denoted by $I(n, r)$.

EXAMPLE . An affine plane of order n is an $I(n, 1)$. The points and lines of an affine space of order n and dimension d form a system $I\left(n, n^{d-2} + n^{d-1} + \cdots + 1\right)$. Here

$$k = 1 + rn = n^{d-1} + n^{d-2} + \cdots + n + 1,$$
$$h = 1 + r(n-1) = 1 + \left(n^{d-2} + \cdots + 1\right)(n-1) = n^{d-1}, \quad \text{and}$$
$$N = n^d.$$

REMARK . A system $I(2,r)$ can be interpreted as a complete graph on $N = 2h = 2(1+r)$ vertices with the edges of the graph being taken as the lines. Conversely, Beth, Jungnickel, and Lenz [1, p. 42] have shown that a complete graph on an even number N of vertices admits a parallelism. Since each line contains 2 points, each parallel class will contain exactly $\frac{1}{2}N$ lines. If we take $h = 1 + r$, the number of parallel classes will be the number of lines through a point, i.e., $N - 1 = 2(1+r) - 1 = 2r + 1 = k$.

There is another way of looking at some of the systems $I(2,r)$: A hyperoval $\mathcal{H}$ in a projective plane of even order q is a set of $q+2$ points, no three of which are collinear. Now choose some line l of the projective plane which does not intersect $\mathcal{H}$ and convert the projective plane into an affine plane by taking l as the "line at infinity". The lines of $I(2,r)$ will be the points of intersection of secants of $\mathcal{H}$ with $\mathcal{H}$—i.e., each line consists of 2 points. Two lines of $I(2,r)$ are parallel iff the corresponding secants are parallel—i.e., intersect each other in a point of l_∞. We then get a system where $N = q+2$, $h = \frac{1}{2}(q+2)$, $k = q+1$, $r = k - h = (2r+1) - (r+1) = \frac{1}{2}q$, and $n = 2$. This situation is of interest for two reasons. First, $I(2,r)$ is embedded in a natural way in $I(q,1)$—i.e., the affine plane of order q. Secondly, by choosing a different l_∞, we can get a different parallelism on the "lines" of $\mathcal{H}$.

It may be worth noting that an oval in a plane of odd order q gives a parallel system in which condition (3) of Definition 1 is violated: these are, so to speak, two different values for h. We have a set of $q+1$ points, no three of which are collinear in a projective plane of order q. If we take an exterior line as our "line at infinity", then there are two kinds of parallel classes with respect to the oval—i.e., one parallel class has $\frac{1}{2}(n+1)$ members while the other has $\frac{1}{2}(n-1)$ members.

Another example: The authors of [1] show, in effect, that the Kirkman schoolgirl problem is an $I(3,2)$. Thus $n = 3$ and $r = 2$. The other parameters are $h = 5$, $k = 7$.

DEFINITION 4. A system $I(n_1, r_1)$ will be said to be embedded in $I(n,r)$ if the following hold:

1. The points of $I(n_1, r_1)$ form a proper subset of the points of $I(n,r)$.
2. For each line l_1 of $I(n_1, r_1)$ there is a unique line l such that l includes l_1 (as point sets).
3. Two lines l_1 and m_1 of $I(n_1, r_1)$ are parallel iff the corresponding lines l and m of $I(n,r)$ are parallel.

LEMMA 2. *If $I(n_1, r_1)$ (with $h_1 = 1 + (n_1 - 1) r_1$, $N_1 = h_1 n_1$ and $k_1 = 1 + n_1 r_1$) is embedded in $I(n,r)$, then $N > N_1$, $n \geq n_1$, $h \geq h_1$, and $k \geq k_1$. If equality holds in two of the cases, then we get the contradiction that $N = N_1$. If $k = k_1$ then $n > n_1$ and $r < r_1$.*

PROOF. The proof is left to the reader. Note that $k = k_1$ implies $nr = n_1 r_1$. ∎

REMARK . In the example of a hyperoval with $q+2$ points mentioned previously, the hyperoval was embedded in an affine plane of even order q. Here we have $I(2, \frac{1}{2}q)$ embedded in $I(q,1)$. That is, $r = 1$ and $r_1 = \frac{1}{2}q$. Here $k = k_1 = q+1$.

THEOREM 1. *If $I(n_1, r_1)$ is embedded in $I(n,r)$ and $k > k_1$ then $nr > n_1 r_1$ and $(n-1) r \geq (n_1 - 1)(n_1 r_1 + 1)$.*

PROOF. If $k > k_1$, $I(n,r)$ has a parallel class none of whose $h = 1+(n-1)r$ lines contains a line of $I(n_1,r_1)$ and hence do not contain two points of $I(n_1,r_1)$. Thus $1+(n-1)r \geq n_1 h_1 = n_1[1+(n_1-1)r_1]$ and $(n-1)r \geq (n_1-1)+n_1(n_1-1)r_1 = (n_1-1)(1+n_1r_1)$. ∎

COROLLARY 1. *If $I(n,r_1)$ is embedded in $I(n,r)$, then $r \geq nr_1+1$.*

PROOF. If $n = n_1$, Theorem 1 gives $(n-1)r \geq (n-1)(nr_1+1)$. ∎

COROLLARY 2. *If $I(n_1,r)$ is embedded in $I(n,r)$, then $n \geq 1+(n_1-1)\left(n_1+\frac{1}{r}\right)$.*

PROOF. Putting $r = r_1$ we get

$$(n-1)r \geq (n_1-1)(n_1r+1),$$
$$r[(n-1)-n_1(n_1-1)] \geq n_1-1.$$

Hence $n-1 \geq \frac{n_1-1}{r} + n_1(n_1-1) = (n_1-1)\left(n_1+\frac{1}{r}\right)$. ∎

REMARK . A simple example of Corollary 1 is the case where an affine space of order n is embedded in an affine space of order n but of higher dimension. In particular, if $I(n,r)$ is an affine 3-space of order n, then $k = 1+n+n^2 = 1+n(n+1)$ so $r = n+1$. If $I(n,r_1)$ is an affine plane of order n then $r_1 = 1$, so we get a case where equality holds in Corollary 1. If we take $r = 1$ in Corollary 2, we get $n \geq n_1^2$. Bruck's famous result for subplanes of projective planes gives an improvement: either $n = n_1^2$ or $n \geq n_1^2 + n_1$.

For another common class of examples, let Π be an affine Desarguesian plane coordinatized by $\mathcal{GF}(q^d) = K$. Recall that the additive group of $\mathcal{GF}(q^d)$ is a vector space of dimension d over $\mathcal{GF}(q) = F$. Also recall that given a vector space of dimension d over F, there is an affine space, also of dimension d, over F. The points of the affine space are the elements of the vector space. The affine subspaces are the translates of the vector spaces. Now consider the set of points (x,y) such that $y = x^q$. Call this set M. Note that M is a vector space of dimension d over F, as is $\{(x,y) \mid y = xm,\ m \in K\}$. Now $(c_1, cm) \in M \cap (y = xm)$ iff $c^q = cm$ or $m = c^{q-1}$. It follows that dimension $[M \cap (y = xm)] = 1$ if $m^{\frac{q^d-1}{q-1}} = 1$; otherwise the intersection is the point $(0,0)$. From this, we get that the affine space over F defined by M is embedded in Π. In our notation, we have a case of $I\left(q, 1+q+\cdots+q^{d-2}\right)$ is embedded in $I\left(q^d, 1\right)$—i.e., $n = q^d$, $r = 1$, $n_1 = q$, $r_1 = 1+\cdots+q^{d-2}$.

This generalizes as follows: Let V be a vector space of dimension $2d$ over $F = \mathcal{GF}(q)$. Let S be a spread on V—i.e., a set of d-dimensional subspaces (the components of S) such that each non-zero element of V belongs to exactly one component of S.

Then it is well known that there is a translation (affine) plane whose points are the elements of V and whose lines are the translates of the components. Then it frequently happens (see [2]) that V has a subspace V_0 of dimension d such that the dimension of the intersection of V with a component of S is either 1 or 0. Then V_0 defines an affine space of dimension d embedded in the translation plane. The lines of the affine space V_0 are subsets of translates of those components which intersect V_0 non-trivially. Thus $n = q^d$, $r = 1$, $n_1 = q$, $k = q^d+1$, $k_1 = \frac{q^d-1}{q-1} = q^{d-1}+\cdots+1$, $r = 1$, $r_1 = 1+\cdots+q^{d-2}$.

It may be interesting to note that if we set $r = 1$ and $r_1 = \frac{n_1^e - 1}{n_1 - 1}$ in the inequality $(n-1)\,r \geq (n_1 - 1)\,(n_1 r + 1)$ we get that $n - 1 \geq \frac{n_1^{e+1} - 1}{n-1}$.

DEFINITION 5. A collineation of $I\,(n,r)$ is a permutation of the points which carries lines into lines. A translation is a collineation which fixes no points but leaves each parallel class invariant. Exception: The identity map is a translation.

DEFINITION 6. If P and Q are two points, PQ will denote the line containing P and Q.

LEMMA 3. *If τ is a translation and $Q = P\tau$ for some point P then the line $l = PQ$ is invariant under τ.*

PROOF. $l\tau$ is a line through P parallel to l, so $l\tau = l$. ∎

REMARK . In the cases known to us, if a translation fixes a line l, it fixes all lines parallel to l (and no other lines. However, it is not clear that this must always be the case. The question is: are there translations with less than h fixed lines?)

DEFINITION 7. A translation τ is a strong translation if the set of lines fixed by τ is the set of h lines of a parallel class.

DEFINITION 8. A regular parallel system will be called a translation system if it admits a group of strong translations which is transitive on the points.

DEFINITION 9. In a translation system, let O be an arbitrary reference point and let τ_Q be the translation which carries O into Q. Then $P\tau_Q$ will be written as $P + Q$.

THEOREM 2. *The translation group of a translation system is elementary abelian. For some prime p, the parameters N, n, and h are powers of p.*

PROOF. The proof is essentially the same as the well known one for translation planes. Let P, Q be two points not collinear with O. Let P_∞, Q_∞ respectively denote the parallel classes to which OP and OQ belong. Since $P \in OP_\infty \cap PQ_\infty$, τ_Q fixes PQ_∞ while $(OP_\infty)\,\tau = QP_\infty$, it follows that $P\tau_Q = P+Q = PQ_\infty \cap QP_\infty$. Similarly $Q + P = PQ_\infty \cap QP_\infty$. Hence $P + Q = Q + P$ if P, Q, O are not collinear. If P_1, P_2, O are collinear while Q is not on the line OP_1, $(P_1 + P_2) + Q = P_1 + (P_2 + Q) = P_1 + (Q + P_2) = (Q + P_2) + P_1 = (P_2 + P_1) + Q$. So $P_1 + P_2 = P_2 + P_1$ if O, P_1, P_2 are collinear. Finally, let P_i denote $P + \cdots + P$ i times. Then $(P+Q)_i = P_i + Q_i$. Moreover, P, P_i, O are collinear as are Q, Q_i, O and $P + Q_1 P_i + Q_i O$. Suppose that $P_i = O$ for some i. Then either $Q_i = O$ or $OP + Q$ is the same line as OQ. Hence P and Q have the same additive order if P, Q, O are not collinear. From this it follows that all elements other than O have the same additive order and hence that all nontrivial translations have the same order. ∎

REMARK . There are many known non-isomorphic affine planes of the same order n. However, affine spaces of the same dimension (greater than 2) and order are isomorphic translation systems. This raises the question as to whether there can be non-isomorphic translation systems other than the affine plane case.

DEFINITION 10. Let $I\,(n,r)$ and $\bar{I}\,(n,r)$ be translation systems. Suppose that

1. The two systems have exactly the same points.

2. Not every line of $\bar{I}(n,r)$ is a line of $I(n,r)$.
3. If T is group of strong translations acting on $I(n,r)$, the permutation of points induced by T also acts as a group of strong translations of $\bar{I}(n,r)$.
4. $I(n,r)$ and $\bar{I}(n,r)$ are not isomorphic.

Then $\bar{I}(n,r)$ will be said to be a translation consistent replacement for $I(n,r)$. This last expression will usually be abbreviated to "$\bar{I}(n,r)$ is a replacement for $I(n,r)$".

THEOREM 3. *Suppose that $I(n,s)$ is embedded in the translation system $I(n,r)$. Suppose that the group T of (strong) translations has a subgroup T_1 which acts as a group of translations on $I(n,s)$. Suppose further that $I(n,s)$ has a translation consistent replacement $\bar{I}(n,s)$. Then $I(n,r)$ has a replacement $\bar{I}(n,r)$ which contains $\bar{I}(n,s)$.*

PROOF. Case: Pet P, Q be two points of $I(n,r)$ such that the line $l = PQ$ is not parallel to any line of $I(n,s)$. Then l is the line of $\bar{I}(n,r)$ which contains P and Q.

Case: Suppose that $l = PQ$ is parallel to a line of $I(n,s)$. Note that the transitivity of T on the points of $I(n,r)$ means that all lines of $I(n,r)$ in the same parallel class are in the same orbit under T. Suppose that $l\tau$ is a line of $I(n,s)$. It will contain $P\tau$ and $Q\tau$. Now let $\bar{l}$ denote the line of $\bar{I}(n,s)$ which contains $P\tau$ and $Q\tau$. Then the line of $\bar{I}(n,r)$ which contains P and Q is defined to be $\bar{l}\tau^{-1}$.

Thus each two points of $\bar{I}(n,r)$ is joined by a line of $\bar{I}(n,r)$. Just as the orbit of a line of $I(n,r)$ under T is the set of lines of a parallel class, the orbit of a line of $\bar{I}(n,r)$ under T is a parallel class of $\bar{I}(n,r)$. So that the number of lines of a parallel class of $\bar{I}(n,r)$ is equal to the number of lines of a parallel class of $I(n,r)$. Hence $\bar{I}(n,r)$ is a parallel system admitting T. Furthermore, T still acts as a group of strong translations on $\bar{I}(n,r)$.

In the second paragraph of this proof, we took a line l and a translation τ such that $l\tau$ is a line of $I(n,s)$. We must show that the construction is independent of the choice of T. If $l\tau$ is a line of $I(n,s)$, each line of $I(n,s)$ which is parallel to l and $l\tau$ and be written in the form $l\tau\tau_1$ where $\tau_1 \in T_1$. Then $\bar{l}_1$ is the line of $\bar{I}(n,s)$ which contains $P\tau\tau_1$ and $Q\tau\tau_1$. We must show that $\bar{l}\tau^{-1} = \bar{l}_1\tau^{-1}\tau_1^{-1}$. But the fact that the replacement $I(n,s) \to \bar{I}(n,s)$ is translation consistent implies that $\bar{l}_1 = \bar{l}\tau_1$ and hence that $\bar{l}\tau_1^{-1} = \bar{l}_1\tau^{-1}\tau_1^{-1}$. ∎

REMARK . It may be helpful to note that the points of $I(n,s)$ form an orbit under T_1. If $\tau^* \in T$ carries $I(n,s)$ into the isomorphic system $I(n,s)\tau^*$ then the points of $I(n,s)\tau^*$ form an orbit under the coset $T_1\tau^*$. The lines of $I(n,s)\tau^*$ are parallel to the lines of $I(n,s)$ and every line of $I(n,r)$ which is parallel to some line in $I(n,s)$ belongs to one of the $I(n,s)\tau^*$ for some τ^*.

COROLLARY 3. *There exist translation systems $I(n,n+1)$ which are not isomorphic to affine three-space if n is a prime power but not a prime and is larger than eight.*

PROOF. Recall that affine 3-space of order n is a system $I(n,n+1)$ and contains systems $I(n,1)$—i.e., affine planes which in this case are Desarguesian. There are many known non-Desarguesian affine planes of prime-power order greater than 8 which can be obtained by translation consistent replacements. See [2]. ∎

There are certain "replacements" (ways of changing one system into the other that are analogous and/or generalizations of methods used on affine planes, especially translation planes).

Suppose $I(n_1, r_1)$ is embedded in $I(n, r)$ and $N_1 = n_1 h_1 = n_1 [1 + (n_1 - 1) r_1] = n$. Actually we will be interested in the situation where there are several copies of $I(n_1, r_1)$ in $I(n, r)$. We are interested in cases where $I(n, r)$ has a replacement $\bar{I}(n, r)$ where some of the lines of $\bar{I}(n, r)$ are isomorphic to $I(n_1, r_1)$.

Recall that if $I(n, r)$ is a translation system, then the translation group is elementary abelian. Every elementary abelian group can be thought of as the additive group of a vector space.

DEFINITION 11. Let V be a vector space of dimension st. A partition of V is a set of s-dimensional subspaces (the components of the partition) such that each non-zero vector belongs to exactly one component.

EXAMPLE . If $s = 1$, the components of the partition are the 1-spaces in V.

EXAMPLE . Let W be a vector space of dimension t over $F = \mathcal{G}F(q^s)$. Then the additive group of W is an elementary abelian group of order q^{st}. This group is also the additive group of a vector space V of dimension st over F; the 1-spaces over F are subspaces of V with dimension s.

If V is a vector space of dimension st with a partition into s-spaces, the system whose points are the elements of V and whose lines are the translates of the components of the partition.

LEMMA 4. *Let V be a vector space of dimension st which a partition into a set $\mathcal{P}$ as s-dimensional subspaces. Let $\mathcal{P}_0$ be a subset of $\mathcal{P}$. Thus $\mathcal{P}_0$ is a set of s-dimensional subspaces such that each member of $\mathcal{P}_0$ is a member of $\mathcal{P}$. Suppose that there is a set $\bar{\mathcal{P}}_0$ of s-dimensional subspaces such that*

1. *The union of the components of $\bar{\mathcal{P}}_0$ is equal to the union of the components of $\mathcal{P}_0$;*
2. *If $\bar{l}_1$ and $\bar{l}_2$ are members of $\bar{\mathcal{P}}_0$, then $\bar{l}_1 \cap \bar{l}_2$ is the zero vector.*

Then, with $\tilde{\mathcal{P}}_0$ denoting the component of $\mathcal{P}_0$ in $\mathcal{P}$, $\bar{\mathcal{P}}_0 \cup \tilde{\mathcal{P}}_0$ is a partition of V. If $\bar{\mathcal{P}}_0 \neq \mathcal{P}_0$, the translation systems defined by $\bar{\mathcal{P}}_0 \cup \tilde{\mathcal{P}}_0$ is a replacement for the system defined by $\mathcal{P}$.

Let W be a vector space of dimension t over $K = \mathcal{G}F(q^s)$. Let V be the vector space of dimension st over $F = \mathcal{G}F(q)$ such that the additive group of V is the same as the additive group of W. Let $I(n, r)$ be the affine space of dimension t over K whose lines the translates of the 1-spaces of W = translates of the components of the partition of V into s-spaces. (Here a component of the partition on V is the s-dimensional space in V which is also a 1-space on W). Choose a basis so that each element of W (point of the affine space) is represented by a t-tuple $(x_1, \ldots, x_t)$, $x_i \in K$. If $a \neq 0$ belongs to K then the mapping $(x_1, \ldots, x_t) \to (ax_1, \ldots, ax_t)$ is a collineation of the affine space; in particular each component of the partition is preserved. If $a \in F$, each 1-space over F (1-space of V) is invariant. If $a \neq F$ the image of a 1-space over F is disjoint (except for $\mathcal{O}$) from its image.

THEOREM 4. *In the context of the above discussion, let $\bar{l}$ be an r-dimensional subspace of V such that either $l \cap \bar{l} = \mathcal{O}$ or $l \cap \bar{l}$ is a 1-dimensional subspace of V.*

Let $\mathcal{P}$ be the original partition and let $\bar{\mathcal{P}}_0$ be the set of distinct images of $\bar{l}$ under the mappings $(x_1, \ldots, x_t) \to (x_1 a, \ldots, x_t a)$.

Then $\bar{\mathcal{P}}_0$ satisfies the condition of Lemma 4. The lines of the linear system with parallelism are of two types:

1. *translates of components of $\tilde{\mathcal{P}}_0$;*
2. *translates of components of $\bar{\mathcal{P}}_0$.*

Note that the lines of $\bar{\mathcal{P}}_0$ are images of $\bar{l}$ under collineations of the original affine space.

EXAMPLE . Choose a basis so that each point is a t-tuple $(x_1, \ldots, x_t)$, $x_i \in K$. If $s \leq t$ let $\bar{l}$ be the set of points such that (a) $x_i \in F$ if $i \leq s$; (b) $x_i = 0$ if $i > s$.

REMARK . If $s = t = 2$ we have an example of "derivation".

REFERENCES

[1] T. Beth, D. Jungnickel, and H. Lenz, *Design Theory,* Cambridge University Press, New York, 1986.

[2] T.G. Ostrom, Replaceable nets, net collineations, and net extensions, *Canad. J. Math.* **18** (1966), 666–672.

On the Classification of 4-Dimensional Flexible Projective Planes

Dieter Betten Mathematisches Seminar, Universität Kiel, Ludewig-Meyn-Str. 4, 24098 Kiel, Germany

Abstract

A 4-dimensional plane is a topological projective plane which is, roughly speaking, homeomorphic to the plane over $\mathbb{C}$. We call the plane flexible if the collineation group has an open orbit in the space of flags. All 4-dimensional flexible planes have been classified. We give an outline of this classification.

1991 Mathematical Subject Classification: 51H10, 51H20, 51A35.

1 Introduction

A projective plane $(P, \mathcal{L})$ is called *topological* if P and $\mathcal{L}$ are topological spaces such that the geometric operations of joining points and intersecting lines are continuous. The classical examples are the planes over $\mathbb{R}$, $\mathbb{C}$, $\mathbb{H}$ and $\mathbb{O}$.

We call a plane *4-dimensional* if it is homeomorphic to the plane over $\mathbb{C}$. By this we mean that the point space and the line space of the plane are homeomorphic to those of the complex plane and all lines and all points (viewed as pencils of lines) are 2-spheres. The group of continuous collineations is a Lie group of dimension ≤ 16 and if we assume *flexibility*, i.e., if the collineation group has an open orbit in the space of flags, then the dimension of the group is ≥ 6.

All 4-dimensional flexible projective planes have been classified in the last 25 years. Our aim is to give a survey of this classification and to describe the various types of planes which were obtained. We do not give a full description of all these planes. For some of them we outline only the lines and the conditions on the parameters. For some of the planes we display only their collineation group in order to give an impression of the situation. For exact references the reader is referred to the original articles which are cited in the bibliography. Some of the material concerning translation planes and shift planes can also be found in the book [43].

2 Foundations

The foundations were laid by H. Salzmann in the years 1969–1973. Denoting by Σ the component of the identity in the group of all continuous collineations, we summarize some results from the fundamental papers [39], [40] and [42] with the following

THEOREM:

i) Σ is a Lie group with $\dim \Sigma \leq 16$,

ii) if $\dim \Sigma \geq 9$ then the plane is classical, i.e., isomorphic to the plane over $\mathbb{C}$,

iii) if $\dim \Sigma = 8$ then the plane is a translation plane up to duality.

Comments: With i) the rich theory of Lie groups and Lie transformation groups may be applied. The dimension of the group is a convenient measure for the homogeneity of the plane (large dimension $\approx$ homogeneous, low dimension $\approx$ rigid). The deep theorem iii) implies that the non-desarguesian planes with highest dimension of the collineation group will be translation planes.

H. Salzmann developed also the main tools which are needed in the study of 4-dimensional planes: for instance

Lemma on quadrangles: If a collineation $\sigma \in \Sigma$ fixes the points of a quadrangle, then $\sigma = id$.

We mention also the *method of homologies and elations* first used in [41] and refined by H. Hähl in [18].

3 Translation planes

In [1] the following construction principle was established. Let $\tau : \mathbb{R}^2 \to \mathbb{R}^2, (\alpha, \beta) \mapsto (f(\alpha, \beta), g(\alpha, \beta))$, be a homeomorphism, which is *transversal.* This means: for each line L in $\mathbb{R}^2$ (more precisely, L a one-dimensional affine subspace of $\mathbb{R}^2$), the curve L^τ intersects any parallel L' of L exactly once.

THEOREM: The system of 2-dimensional subspaces of $\mathbb{R}^4$

$$S \cup \left\{ \begin{pmatrix} \alpha & \beta \\ f(\alpha,\beta) & g(\alpha,\beta) \end{pmatrix} ;\ \alpha, \beta \in \mathbb{R} \right\}$$

is a topological spread in $\mathbb{R}^4$: we describe 2-dimensional subspaces by matrices and add the vertical line S. The 4-dimensional translation plane generated by this spread is desarguesian if and only if τ is an affine map.

3.1 Translation planes with $\dim \Sigma = 8$.

It turns out that for $\dim \Sigma = 8$ there exist exactly a one-parameter family and two single translation planes. They can be characterized by: a) the translation complement $GL_2\mathbb{R}$ acts reducibly on the translation group $\mathbb{R}^4$, b) it acts irreducibly, c) the collineation group is solvable.

a) The family, for which $GL_2\mathbb{R}$ acts reducibly on $\mathbb{R}^4$ consists of the following spreads:

$$\mathcal{S}_w = \{S\} \cup \left\{ \begin{pmatrix} s & -t \\ t & s \end{pmatrix} ;\ s, t \in \mathbb{R}, t \geq 0 \right\} \cup \left\{ \begin{pmatrix} s & -t \\ wt & s \end{pmatrix} ;\ s, t \in \mathbb{R}, t \leq 0 \right\},$$

where $w \geq 1$. The plane defined by $\mathcal{S}_w$ is desarguesian if and only if $w = 1$, and different parameters w and w' define non-isomorphic planes [1, Satz 5].

Remark: These spreads are distorted versions of the classical spread, obtained by modifying one half of the spread using the parameter $w \geq 1$.

b) The single plane with $GL_2\mathbb{R}$ irreducible on $\mathbb{R}^4$ is generated by the spread

$$\mathcal{S} = \{S\} \cup \left\{ \begin{pmatrix} -3s^2 + t & 2s \\ -2s^3 - 2st & s^2 + t \end{pmatrix} ;\ s, t \in \mathbb{R},\ t \geq 0 \right\}$$

$$\cup \left\{ \begin{pmatrix} -3s^2 + 3t & 2s \\ -2s^3 + 2st & s^2 + \frac{t}{3} \end{pmatrix} ;\ s, t \in \mathbb{R},\ t \leq 0 \right\}.$$

This single spread [3] has finite analogues; see [19].

c) The single plane with a solvable collineation group is generated by the following spread:

$$\mathcal{S} = \{S\} \cup \left\{ \begin{pmatrix} t & s \\ -\frac{s^3}{3} & s^2 + t \end{pmatrix} ;\ s, t \in \mathbb{R} \right\}.$$

Remark: This spread [2] also has finite analogues found by M. Walker, [48]. For this reason the spread is sometimes called the Betten-Walker spread.

3.2 Translation planes with $\dim \Sigma = 7$.

Each 4-dimensional translation plane has a 5-dimensional group of dilatations, generated by the group of translations and the real homotheties. This group acts trivially on the line at infinity. In order to get a 2-dimensional orbit there, we need two more dimensions for the collineation group. Therefore, if we want to have a flexible translation plane, we need $\dim \Sigma \geq 7$.

THEOREM [1]–[6]: A 4-dimensional translation plane is flexible if and only if $\dim \Sigma \geq 7$. All such planes are known.

Let us give a rough survey of the translation planes with a 7-dimensional collineation group. The connected component of the collineation group is solvable since it is an extension of the (solvable) dilatation group by a 2-dimensional (solvable) group. Therefore, by a simple group representation argument, there is a fixed point ∞ on the translation axis W. Let this point correspond to the parallel class of all vertical lines.

Now we form subcases according to the action of Σ on $W \setminus \{\infty\}$: either this action is transitive, or it has a one-dimensional orbit (but no fixed point) or it fixes some point (together with $\{\infty\}$ we then have two fixed points on the axis W). If we denote by μ the minimal dimension of orbits on $W \setminus \{\infty\}$, then these cases may be described by $\mu = 2$, 1 and 0, respectively.

Σ acts transitively on $W \setminus \{\infty\}$, ($\mu = 2$):

In this case the translation complement is commutative and we get a single plane and a one-parameter family of planes.

The single plane is generated by the spread

$$\mathcal{S} = \{S\} \cup \left\{ \begin{pmatrix} t & s \\ -\frac{s^3}{3} + s & s^2 + t \end{pmatrix} ; \; s, t \in \mathbb{R} \right\},$$

see [43, 73.10].

The planes of the one-parameter family were first derived in [6]. By [27, Prop. 5.8] the three real parameters occurring there may be reduced to $w = 0$, $c = 2$ and $0 < n \le 2$. After this reduction the translation complement is generated by the real homotheties and the groups

$$B = \left\{ \begin{pmatrix} \cos s & \sin s & & \\ -\sin s & \cos s & & \\ 2\,s\cos s & 2\,s\sin s & \cos s & \sin s \\ -2\,s\sin s & 2\,s\cos s & -\sin s & \cos s \end{pmatrix} ; \; s \in \mathbb{R} \right\},$$

$$A = \left\{ \begin{pmatrix} 1 & & & \\ & 1 & & \\ & t & 1 & \\ -t & & & 1 \end{pmatrix} ; \; t \in \mathbb{R} \right\}.$$

The translation complement fixes the vertical line S through the origin but no one-dimensional subspace in S. Therefore the collineation group has no one-dimensional normal subgroups.

The planes are generated by the following spreads, $0 < n \le 2$:

$$\mathcal{S} = \{S\} \cup \left\{ \begin{pmatrix} n\sin^2 s + 2s & n\sin s\cos s + t \\ n\sin s\cos s - t & n\cos^2 s + 2s \end{pmatrix} ; \; s, t \in \mathbb{R} \right\}.$$

One-dimensional orbit of Σ on $W \setminus \{\infty\}$, ($\mu = 1$):

These planes are classified in [4]. We shall not describe these planes in detail here, but instead write down the translation complement C of the translation group. Suppose first that this complement contains the group

$$\left\{ \begin{pmatrix} 1 & & & \\ & 1 & & \\ t & & 1 & \\ & t & & 1 \end{pmatrix} ; \; t \in \mathbb{R} \right\}$$

of shears with axis S and center ∞. Then C is generated by this group of shears, the group of real homotheties and the following one-parameter group:

either

$$B = \left\{ \begin{pmatrix} e^{as} & & & \\ & e^{ds} & & \\ & & e^{(a+1)s} & \\ & & & e^{(d+1)s} \end{pmatrix} ; \ s \in \mathbb{R} \right\}, \ a + d + 1 = 0$$

[4, Satz 1], two 1-dimensional subspaces of S are fixed,
or

$$B = \left\{ \begin{pmatrix} e^{-s/2} & & & \\ se^{-s/2} & e^{-s/2} & & \\ & & e^{s/2} & \\ & & se^{s/2} & e^{s/2} \end{pmatrix} ; \ s \in \mathbb{R} \right\}$$

[4, Satz 2], one 1-dimensional subspace of S is fixed,
or

$$B = \left\{ \begin{pmatrix} e^{-s/2}\cos s & -e^{-s/2}\sin s & & \\ e^{-s/2}\sin s & e^{-s/2}\cos s & & \\ & & e^{s/2}\cos s & -e^{s/2}\sin s \\ & & e^{s/2}\sin s & e^{s/2}\cos s \end{pmatrix} ; \ s \in \mathbb{R} \right\}$$

[4, Satz 3], no 1-dimensional subspace of S is fixed. These three cases may also be described by: the collineation group has 2, 1 or no 1-dimensional normal subgroup.

Now suppose that the group of shears with respect to S is 0-dimensional, then the translation complement is generated by the real homotheties and the following two one-parameter groups:

$$A = \left\{ \begin{pmatrix} 1 & & & \\ t & 1 & & \\ t^2/2 & t & 1 & \\ t^3/6 & t^2/2 & t & 1 \end{pmatrix} ; \ t \in \mathbb{R} \right\}$$

$$B = \left\{ \begin{pmatrix} e^{-3s/2} & & & \\ & e^{-s/2} & & \\ & & e^{s/2} & \\ & & & e^{3s/2} \end{pmatrix} ; \ s \in \mathbb{R} \right\}.$$

These planes are studied in [4, Satz 5].

Fixed point on $W \setminus \{\infty\}$, $(\mu = 0)$:

Here the collineation group has two fixed points on the line at infinity and the translation complement C fixes two lines through the origin, say $S = \{(0,0,u,v)|u,v \in \mathbb{R}\}$ and $X = \{(x,y,0,0)|x,y \in \mathbb{R}\}$. In each of these 2-dimensional subspaces the group C either acts transitively on the space of one-dimensional subspaces or fixes (at least) one 1-dimensional subspace. If in both we had a 1-dimensional fixed space, this would give a contradiction to the lemma on quadrangles. Therefore we may suppose that C is transitive on the space of one-dimensional subspaces of , say, X. Let us give an idea of the planes by writing down the translation complement C, i.e., two one-parameter groups besides the group of real homotheties:

The planes of [5, Satz 1] have the following shift complement:

$$\Delta_w = \left\{ \begin{pmatrix} r\cos\varphi & -r\sin\varphi & & \\ r\sin\varphi & r\cos\varphi & & \\ & & rs & \\ & & & rs^w \end{pmatrix} ; \begin{matrix} r>0 \\ s>0 \\ \varphi\in\mathbb{R} \end{matrix} \right\}, \ 0<|w|<1.$$

Here, the group C fixes two 1-dimensional subspaces of S. As H. Hähl noticed, these geometries are topological only for $p=0$. For the same reason we set $p=0$ also in the following case.
Next, we have the group

$$\Delta = \left\{ \begin{pmatrix} r\cos\varphi & -r\sin\varphi & & \\ r\sin\varphi & r\cos\varphi & & \\ & & re^t & \\ & & rte^t & re^t \end{pmatrix} ; \ \varphi, t\in\mathbb{R}, r>0 \right\}$$

[5, Satz 2], C fixes one 1-dimensional subspace of S.

If C acts transitively on the space of one-dimensional subspaces of S there are two further families of translation planes with full translation complement

$$\Delta_{p,q} = \left\{ \begin{pmatrix} ae^{ps}\cos s & ae^{ps}\sin s & & \\ -ae^{ps}\sin s & ae^{ps}\cos s & & \\ & & ae^{qt}\cos t & ae^{qt}\sin t \\ & & -ae^{qt}\sin t & ae^{qt}\cos t \end{pmatrix} ; \begin{matrix} a>0 \\ s,t\in\mathbb{R} \end{matrix} \right\}$$

where p and q are real parameters with $(p,q)\neq(0,0)$, [5, Satz 3], and

$$\Delta_{m,n} = \left\{ \begin{pmatrix} a\cos m\varphi & a\sin m\varphi & & \\ -a\sin m\varphi & a\cos m\varphi & & \\ & & b\cos n\varphi & b\sin n\varphi \\ & & -b\sin n\varphi & b\cos n\varphi \end{pmatrix} ; \begin{matrix} a,\ b>0 \\ 0\le\varphi<2\pi \end{matrix} \right\}$$

with integers $0<m\le n$, [5, Satz 4]. The space of one-dimensional subspaces in the two fixed spaces is a circle, the product of both is a 2-torus. Then the last two families of planes can be distinguished in the following way: for the planes of Theorem 3 the group C is transitive on this torus; for the planes in Theorem 4 this action is not transitive.

Remark: Though all flexible 4-dimensional translation planes are hereby classified, one may also consider planes with a smaller group, for instance planes which have besides the group of homotheties a group isomorphic to $\mathbb{R}$ or to $SO_2(\mathbb{R})$ in the translation complement; see [6, p.140]. Going on, one may even ask if there are 4-dimensional translation planes which have other than the obvious group of dilatations no further collineations. Planes of this type, so-called "rigid" translation planes, have been constructed by R. Riesinger, [37].

Remark: Part of the classification can now be found in [43, sections 71-73]. There the theoretical background is developed and the planes with an 8-dimensional collineation group are treated at length.

Another construction method for 4-dimensional translation planes is given by N. Knarr [27]: Let $\varphi:\mathbb{C}\mapsto\mathbb{C}$ be a *contraction*, that is, a map satisfying the condition
(i) for $z\neq w\in\mathbb{C}$, $|\varphi(z)-\varphi(w)|<|z-w|$.
If moreover
(ii) $\big||z|-|\varphi(z)|\big|\to\infty$ for $z\to\infty$,

then a compact spread on $\mathbb{C}^2$ is obtained by defining

$$\mathcal{S} = \{L_z \ ; \ z \in \mathbb{C} \cup \{\infty\}\},$$

where $L_\infty = \{0\} \times \mathbb{C}$ and $L_z = \{(x, zx + \varphi(z)\overline{x}) \ ; \ x \in \mathbb{C}\}$.
By this method some of the translation planes get very simple descriptions, for instance
Planes of [6]: $\varphi(a+bi) = ce^{ia}, c \in \mathbb{R}, 0 < |c| \leq 1$.
Planes [4] Theorem 3:

$$\varphi(a+bi) = \begin{cases} ube^{2ilog|b|} & \text{if } b < 0 \\ 0 & \text{if } b = 0 \\ vbe^{2ilog|b|} & \text{if } b > 0 \end{cases}$$

where $(u,v) \in \mathbb{C}^2 \setminus \{(0,0)\}, 5|u|^2 \leq 1$, and $5|v|^2 \leq 1$.

Planes [5] Theorems 3 and 4:

$$\varphi(z) = u\frac{z^k}{|z|^{k-1}}e^{ivlog|z|}$$

where $k \in \mathbb{N} \cup \{0\}$, and $u, v \in \mathbb{R}$ with the conditions $0 < u < 1$, $0 \leq v$ and $u^2v^2 \leq (1-u^2)(1-k^2u^2)$.

In [27] a survey about various construction methods for translation planes is also given. In particular the method of transversal homeomorphisms is compared with the construction of finite translation planes by T. G. Ostrom [34], [35].

4 3-dimensional Translation Group

After all flexible translation planes had been classified, it was not clear how to go on. It took some years till the right idea came up, namely to consider planes which have a 3-dimensional translation group, [8]. Let us describe the setup for this situation.

Assumption: There exists some line W such that the translation group $T = \Sigma_{[W,W]}$ with respect to W is 3-dimensional and $T \cong \mathbb{R}^3$. We have coordinates on the affine part of the plane

$$P \setminus W = \{(x,y,u,v); \ x,y,u,v \in \mathbb{R}\},$$

in such a way that the action of T is:

$$(x,y,u,v) \mapsto (x, y+y_1, u+u_1, v+v_1), \quad y_1, u_1, v_1 \in \mathbb{R}.$$

There are the following "vertical lines"

$$L_{(x,y)} = \{(x,y,u,v); \ u,v \in \mathbb{R}\}, \quad x,y \in \mathbb{R},$$

which all pass through a fixed point $\infty \in W$. The translation group T is partitioned in the following way; c.f. the *center map* [40, 4.5], [43, 61.7]: There is a 2-dimensional group T_∞ of translations in the direction of $\infty \in W$

$$(x,y,u,v) \mapsto (x, y, u+u_1, v+v_1), u_1, v_1 \in \mathbb{R}.$$

and a set of one-dimensional subgroups

$$(x, y, u, v) \mapsto (x, y + t, u + \xi t, v + \eta t), t \in \mathbb{R}$$

of T which define translation groups in the direction of the points of $W \setminus \{\infty\}$. In this way these points may be identified with one-parameter subgroups of T and we have natural coordinates $[\xi, \eta], \xi, \eta \in \mathbb{R}$, on $W \setminus \{\infty\}$. The action of Σ on W can be seen by letting Σ act by conjugation on T (and therefore on the set of one-dimensional subspaces of T). In particular, the induced action of Σ on

$$W \setminus \{\infty\} = \{[\xi, \eta];\ \xi, \eta \in \mathbb{R}\}$$

is an affine action.

Let $\Xi = \mathbb{R}^4/T$ be the space of orbits in $\mathbb{R}^4$ under T, coordinatized by the x-coordinate. Then two cases may occur:

A) one orbit is fixed under T: "fixed orbit", or
B) Σ acts transitively on Ξ: "Transitivity" .

Under additional assumptions both situations lead to projective planes of a new type; planes which are defined by gluing two halves of translation planes and the so-called shift planes.

5 Glued Translation Planes

Take a plane with 3-dimensional translation group and case A): fixed orbit. Assume further that the plane admits all real homotheties

$$H = \{(x, y, u, v) \mapsto (rx, ry, ru, rv);\ r > 0\}.$$

Then a line L can be constructed on the region $x > 0$ by applying H to one of its points followed by the one-parameter translation group in direction $L \cap W$. This leads to a part of a two-dimensional affine subspace of $\mathbb{R}^4$ and the same is true for the part of L on the region $x < 0$. Those two parts of L are glued along some one-dimensional affine subspace of the orbit $\mathbb{R}^3$ at $x = 0$. Since this is true for all non-vertical lines, the whole (affine) plane consists of two halves of translation planes which are glued along the fixed translation group orbit in a suitable way.
One has to look at the list of 4-dimensional translation planes in order to find halves of translation planes which may be put together in this way. Since $x = 0$ is fixed we lose one dimension of the translation group in the translation planes we use. Moreover we need a fixed point on the line at infinity that is incident with all vertical lines $L_{(x,y)}$. Either the translation planes have a fixed point on W or we take that point from a one-dimensional orbit on W: this again reduces the collineation group by one dimension. Thus, when we are interested in getting flexible planes there are strong restrictions on the translation planes to begin with. The process of finding halves of translation planes and gluing them was begun in [8] and continued in [46]. As a result we have the following

THEOREM: All flexible glued translation planes are classified and known explicitly.

We give here only two examples of families of planes which are constructed in this way.

EXAMPLE: Gluing the translation planes with reducible translation complement $GL_2(\mathbb{R})$; see [8, Satz 2]:

On the affine point-space $\mathbb{R}^4 = \{(x, y, u, v);\ x, y, u, v \in \mathbb{R}\}$ we take all vertical lines $L_{(x,y)}$, $x, y \in \mathbb{R}$, and the lines defined by the following equations:

$$\begin{aligned} u &= sx - ty + u_0, \quad u_0, s, t \in \mathbb{R}, t \geq 0 \\ v &= w(x)tx + sy + v_0, \quad v_0, s, t \in \mathbb{R}, t \leq 0 \end{aligned}$$

where $w(x) = w_- > 0$ for $x < 0$ and $w(x) = w_+ > 0$ for $x > 0$ is chosen suitably. Thus, the parameter w which modifies one half of the desarguesian spread is taken differently for $x > 0$ and for $x < 0$.

Remark: It is allowed that one of the parameters w_- or w_+ is $= 1$. Then we have on one side half of the plane over $\mathbb{C}$. This gluing process can be considered as a generalization of the Moulton construction in dimension 2.

EXAMPLE: Gluing the translation planes of [2, Satz 2]; see [8, Satz 3]: In [43, 73.10] it is shown that the parameters f and g can be reduced to $g = 1$ and $f = h \in \{0, -1\}$. Therefore this family of translation planes consists of two single planes, one with an 8-dimensional collineation group, the so-called Betten-Walker plane ($h = 0$) and one with a 7-dimensional collineation group ($h = -1$). For the gluing process we allow the parameter h to take arbitrary values ≤ 0 again and "split" it. Then we get glued planes whose non-vertical lines are described by the following equations:

$$\begin{aligned} u &= tx + sy + u_0, \quad u_0, s, t \in \mathbb{R}, \\ v &= (-\frac{s^3}{3} + h(x)s)x + (s^2 + t)y + v_0, \quad v_0, s, t \in \mathbb{R}, \end{aligned}$$

where $h(x) = h_- \leq 0$ for $x < 0$ and $h(x) = h_+ \leq 0$ for $x > 0$.

6 Shift Planes

Next we consider planes with the property: Σ acts transitively on $P \setminus W \approx \mathbb{R}^4$. We assume in addition that there is a subgroup $\cong \mathbb{R}^4$ of Σ which acts regularly on $P \setminus W$. This situation leads to planes of a new type, the so-called shift planes:

DEFINITION: A map $f : \mathbb{R}^2 \to \mathbb{R}^2 : (x, y) \mapsto (g(x, y), h(x, y))$ is called *planar*, if the set $L = \{(x, y, g(x, y), h(x, y));\ x, y \in \mathbb{R}\}$ generates an affine plane, i.e., the system $L^{\mathbb{R}^4}$ together with all verticals defines the lines of an affine plane on $\mathbb{R}^4$.

We call such a plane a *shift plane* and assume that f is continuous in order to get a topological 4-dimensional shift plane. Most interesting is the following situation: the planar map is differentiable and the system of all differentials defines a topological spread of $\mathbb{R}^4$.

EXAMPLE (First single shift plane of N. Knarr 1983): A generating line is $L = \{(x, y, xy - \frac{1}{3}x^3, \frac{1}{2}y^2 - \frac{1}{12}x^4) | x, y \in \mathbb{R}\}$.
Here, differentiation gives the following Jacobian matrices:

$$\begin{pmatrix} \frac{\partial f}{\partial x} & \frac{\partial f}{\partial y} \\ \frac{\partial g}{\partial x} & \frac{\partial g}{\partial y} \end{pmatrix} = \begin{pmatrix} y - x^2 & x \\ -\frac{1}{3}x^3 & y \end{pmatrix}, \quad x, y \in \mathbb{R}.$$

Substituting $y_1 = y - x^2, y = y_1 + x^2$ we see that this is the "Betten-Walker-spread" with solvable collineation group. Hence we can get spreads by differentiating differentiable planar maps. Conversely one may try to integrate the known spreads in order to get planar functions. This was first carried out in [7]. A thorough study of non-desarguesian shift planes was made by N. Knarr in [23] who proved among other things the following:

The shift group

$$\mathbb{R}^4 = \{(x, y, u, v) \mapsto (x + x_0, y + y_0, u + u_0, v + v_0); x_0, y_0, u_0, v_0 \in \mathbb{R}\}$$

is a normal subgroup of the collineation group which fixes the line at infinity. Therefore the collineation group is a semidirect product of the shift group and the stabilizer of the origin, the so-called shift complement. All 4-dimensional shift planes are self-dual and admit ovals.

THEOREM [7], [23], [13], [24]: All flexible 4-dimensional shift planes are classified and they are explicitly known.

Let us give a rough sketch of the results. There is only one plane with a 7-dimensional collineation group, the "first single shift plane" by N. Knarr cited above. All other flexible shift planes have a 2-dimensional shift complement. It can be shown that the torus group $SO_2(\mathbb{R}) \times SO_2(\mathbb{R})$ is not possible. Hence there remain three cases:

$$\mathbb{R}^2, \quad \mathbb{R} \times SO_2(\mathbb{R}), \text{ or } L_2.$$

a) $\Sigma_0 \cong \mathbb{R}^2$.
This group leads to the "second single shift plane" by N. Knarr, [23], which is generated by the line

$$L = \{(x, y, xy - \frac{1}{3}x^3 + x, \frac{1}{2}(y^2 + x^2) - \frac{1}{12}x^4); \ x, y \in \mathbb{R}\}.$$

By differentiation we see that the set of Jacobians is the spread of the single translation plane [43, 73.10] with a 7-dimensional group.

b) $\Sigma \cong \mathbb{R} \times SO_2(\mathbb{R})$, [13]:
Here one gets a 2-parameter family of planes, generated by the following multiplicative homomorphism of $\mathbb{C}$:

$$f_{[w]} = z^2 |z|^{w-2}, \quad w = (u + iv) \in \mathbb{C}.$$

This map is planar and generates a topological shift plane if the parameters u and v satisfy the conditions $1 < u \leq 4$ and

$$v^6 + 2(u^2+4)v^4 + (u^4 + 16u^3 - 40u^2 + 16u + 16)v^2 + 16u(u-1)^3(u-4) \leq 0.$$

The shift complement is the group

$$\{(x,y) \mapsto (cx, f_{[w]}(c)y);\ c \in \mathbb{C}^*\}, x, y \in \mathbb{C}.$$

The parameter $w = 2$ defines the complex plane.

The derivatives of these shift planes are among the translation planes [5], Theorems 3 and 4.

c) Non-commutative shift complement L_2, [24]:
In this situation there exist four families of planes.

c1) The skew parabola planes [24, Satz 4.1] are generated by the following planar maps

$$f : \mathbb{R}^2 \to \mathbb{R}^2 : (x,y) \mapsto (\frac{1}{2}x^2 + u(y)y^2, xy + v(y)y^2).$$

Here $u(y) = \begin{cases} -\frac{1}{2} & \text{for } y \geq 0 \\ u_- & \text{for } y < 0 \end{cases}, \quad v(y) = \begin{cases} 0 & \text{for } y \geq 0 \\ v_- & \text{for } y < 0 \end{cases},$

with $v_- \geq 0$, $2u_- + v_-^2 \leq -1$ and $(u_-, v_-) \in \mathbb{R}^2$. The derivatives of these shift planes are the translation planes with reducible action of $GL_2(\mathbb{R})$. Compare also [7] where the derivation-integration process was established. Note that this process is not a 1-1 correspondence: the spreads form a one-parameter family, but the integrated shift planes depend on two parameters.

In the following three families there are functions $u(w)$ and $v(w)$ involved with $u(w) = u_+ \in \mathbb{R}$ for $w \geq 0$ and $u(w) = u_- \in \mathbb{R}$ for $w < 0$ and a similar "splitting" for $v(w)$. We shall write down the generating planar maps and the conditions for the real constants u_+, u_-, v_+, v_- in order to give a rough idea of the planes.

c2) The planes [24, Satz 4.2] are generated by

$$f : \mathbb{R} \to \mathbb{R} : (x,y) \mapsto (\frac{1}{2}x^2 + u(y)|y|^{2p}, xy + v(y)|y|^{(1+p)}).$$

Conditions:

$$\begin{aligned} &\tfrac{1}{2} < p,\ p \neq 1,\ u_+ < 0,\ u_- < 0, \\ &u_+ \leq -\frac{p(p+1)^2}{8(2p-1)}v_+^2,\quad u_- \leq -\frac{p(p+1)^2}{8(2p-1)}v_-^2, \\ &(p, u_+, u_-, v_+, v_-) \neq (2, -\tfrac{3}{4}m^2, -\tfrac{3}{4}m^2, m, -m),\quad m \neq 0. \end{aligned}$$

The derivatives belong to the spreads of the planes [4, Satz 1].

c3) The planes [24, Satz 4.3] are generated by

$$f:\mathbb{R}^2\to\mathbb{R}^2:(x,y)\mapsto(u(x)x^2+xy-x^2\log|x|,$$
$$\frac{1}{2}y^2+v(x)x^2+u(x)x^2\log|x|-\frac{1}{2}x^2(log|x|)^2)$$

with $u_+^2+2v_++\frac{5}{4}\le 0$ and $u_-^2+2v_-+\frac{5}{4}\le 0$. These planes correspond to the spreads of [4, Satz 2].

c4) The planes [24, Satz 5.1] are generated by the planar maps

$$f:\mathbb{R}^2\to\mathbb{R}^2:(x,y)\to\left(u\left(y-\frac{1}{2}x^2\right)|\,y-\frac{1}{2}x^2\,|^{3/2}+x\left(y-\frac{1}{2}x^2\right)+\frac{1}{6}x^3,\right.$$
$$v\left(y-\frac{1}{2}x^2\right)\left(y-\frac{1}{2}x^2\right)^2+u\left(y-\frac{1}{2}x^2\right)x\,|\,y-\frac{1}{2}x^2\,|^{3/2}+$$
$$\left.+\frac{1}{2}x^2\left(y-\frac{1}{2}x^2\right)+\frac{1}{24}x^4\right)$$

with the following conditions for the parameters:

$$2v_+-\tfrac{9}{8}u_+^2>0,\qquad 2v_-+\tfrac{9}{8}u_-^2>0,$$

$$\left(1+\tfrac{9}{8}u_+^2\right)\left(v_++\tfrac{1}{2}\right)^2\le\left(3v_+-\tfrac{1}{2}-\tfrac{9}{4}u_+^2\right)^2,$$

$$\left(3v_--\tfrac{1}{2}+\tfrac{9}{4}u_-^2\right)^2\le\left(1-\tfrac{9}{8}u_-^2\right)\left(v_-+\tfrac{1}{2}\right)^2,$$

$$(u_+,u_-,v_+,v_-)\ne\left(0,0,\tfrac{1}{2},\tfrac{1}{2}\right).$$

These planes correspond to the spreads of [4, Satz 5].

Remark: The collineation groups of these shift planes are rather similar to those of the associated translation planes. Only the group of real homotheties is now missing.

7 Large Commutative Subgroup

The 4-dimensional translation planes and the 4-dimensional shift planes both have a subgroup $\mathbb{R}^4$ which acts regularly on the affine point set: in particular the collineation group has a large commutative subgroup. In [26] the case that Σ contains a 4-dimensional connected commutative subgroup Δ is studied in detail. It is proved that five cases may occur:

1) $\Delta\cong\mathbb{R}^2\times(SO_2)^2$.

In this case Δ fixes three non-collinear points defining three homology groups isomorphic to $\mathbb{R}\times SO_2$. The group Δ is the direct product of two of these groups and the plane is of Lenz-Barlotti type at least I.4. It is not known if such a topological plane exists, but it cannot be flexible.

3)-5) $\Delta\cong\mathbb{R}^4$ and the plane is a shift plane, a translation plane or a dual translation plane.

2) $\Delta \cong \mathbb{R}^3 \times SO_2(\mathbb{R})$.

THEOREM [25]: In the situation 2) there exists a one-parameter family $\mathcal{A}_p, 0 < p \leq 2$ of planes.

Let us list some properties of these planes:

a) The planes have Lenz-Barlotti type II.2: exactly the central groups $\Sigma_{[\infty,W]}$ and $\Sigma_{[u,S]}$ are transitive.

b) The collineation group Σ fixes exactly two points (∞ and u) and exactly two lines (W and S).

c) Δ is a maximal abelian subgroup of Σ. This group is isomorphic to $\mathbb{R}^3 \times SO_2(\mathbb{R})$.

d) Define on $\mathbb{C}$ a new multiplication by

$$z \circ w = \begin{cases} zw^{ip \log|z| \log|w|} & \text{if} \quad zw \neq 0 \\ 0 & \text{if} \quad zw = 0 \end{cases}.$$

Then $(\mathbb{C}, +, \circ)$ is a topological ternary field (if $0 < |p| \leq 2$) defining the plane $\mathcal{A}_p$.

COROLLARY: If a 4-dimensional flexible projective plane has a collineation group whose Lie algebra has a 4-dimensional commutative subalgebra, then the plane must be either a plane with two fixed points and two fixed lines, or a translation plane (up to duality) or a shift plane. Since all flexible translation planes and shift planes have already been classified, we may from now on exclude a 4-dimensional commutative subalgebra.

8 7-dimensional collineation group

Let us resume the classification program according to the size of the collineation group. By Salzmann's results we know that planes with $\dim \Sigma \geq 9$ are isomorphic to the complex plane and planes with 8-dimensional collineation groups are translation planes (up to duality). These planes are all classified, and therefore the case $\dim \Sigma \geq 8$ is finished. Since we have met several other planes up to now, we might ask if the case $\dim \Sigma = 7$ has been already settled. The following could be proved in 1990:

THEOREM [9], [28]: Each 4-dimensional topological projective plane with $\dim \Sigma = 7$ is either isomorphic (or dual) to a translation plane or it is the first single shift plane of N. Knarr.

In the proof essential use is made of the classification of all transitive actions of connected Lie groups on 2-manifolds; [31] . In [28] it is assumed that the collineation group is not solvable and the solvable case is settled in [9].

Since we are interested in flexible planes, and these planes have $\dim \Sigma \geq 6$, the only dimension left is 6.

9 $\dim \Sigma = 6$, Σ Solvable

We now suppose that the collineation group is 6-dimensional. If the collineation group is not solvable, then by [28] we have $\Sigma \cong \mathbb{R}^2 \rtimes GL_2(\mathbb{R})^+$. This group leads to new planes which are being worked out by N. Knarr, R. Löwen and H. Klein.

Besides this special situation we may assume that the collineation group is solvable from now on.

THEOREM: Σ fixes some flag $\infty \in W$.

This was proved in [11]. We may think of W as the line at infinity and take the point ∞ as the parallel class of all vertical lines. With respect to these fixed elements we may define subcases in several ways.

Geometric assumptions:

We look for central collineations with respect to the fixed elements W and ∞. Thus we have the natural invariants (σ, τ), where $\tau = \dim \Sigma_{[W,W]}$, $\sigma = \dim \Sigma_{[\infty,\infty]}$.

Since all flexible translation planes are already classified we may suppose that

$$(\sigma, \tau) \leq (3, 3)$$

in the following.

Transformation group assumptions:

The action of Σ on $W \setminus \{\infty\}$ may have a fixed point, or a one-dimensional orbit (but no fixed point), or it may be transitive. In these cases the minimal dimension μ_W of orbits on $W \setminus \{\infty\}$ is 2, 1 or 0. Similarly we have for the action of Σ on $\mathcal{L}_\infty \setminus \{W\}$ the three cases $\mu_\infty = 2, 1$ or 0. Thus we have the invariant (μ_W, μ_∞).

By dualizing, if necessary, this leads to the six *orbit types*

$$(\mu_W, \mu_\infty) = (0,0), (0,1), (0,2), (1,1), (1,2), (2,2).$$

Algebraic structure of the collineation group:

The Lie algebra of the collineation group Σ is a 6-dimensional real solvable algebra which has a well defined *nilradical* N, namely the maximal nilpotent ideal. Most Lie groups will be simply connected in the following, and so we need not be too careful when switching from the group to its Lie algebra and vice-versa. We shall also call the corresponding connected normal subgroup of Σ the nilradical. Fortunately all 6-dimensional real solvable Lie algebras are classified. All 6-dimensional nilpotent algebras can be found in [30]; see also [47]. All 6-dimensional real Lie algebras with a 5-dimensional nilradical are classified in [33], and the case with a 4-dimensional nilradical was settled only a few years ago, [45].
For algebraic reasons, $\dim N \geq \frac{1}{2} \dim \Sigma$, [32]: therefore we have to consider only the cases $\dim N = 3, 4, 5, 6$.

10 6- and 5-dimensional Nilradical

A 6-dimensional nilradical means that Σ itself is nilpotent. In [10] it has been proved that this case leads to exactly one plane, the second singular shift plane of Knarr 1983, which is generated by

$$L = \{(x, y, xy - \frac{1}{3}x^3 + x, \frac{1}{2}(y^2 + x^2) - \frac{1}{12}x^4);\ x, y \in \mathbb{R}\}.$$

Note: Both singular shift planes found by N. Knarr 1983 are remarkable: the first one is the only 4-dimensional projective plane with a 7-dimensional collineation group which is not a translation plane. The second one is the only 4-dimensional flexible projective plane which has a nilpotent collineation group.

Now let us suppose that the group has a 5-dimensional nilradical; see [12]. The list of all 5-dimensional nilpotent real Lie algebras ([30] or [17]) is not too long. We may ignore those algebras which have a subalgebra $\cong \mathbb{R}^4$ since these lead to known planes, c.f. the section on large commutative subgroups. It turns out that only two algebras have to be considered:

$$\mathcal{N}_1 = \{< \mathbf{y}, \mathbf{u}, \mathbf{v}, \mathbf{s}, \mathbf{n} > |[\mathbf{s}, \mathbf{y}] = \mathbf{u}, [\mathbf{s}, \mathbf{u}] = \mathbf{v}, [\mathbf{n}, \mathbf{y}] = \mathbf{v}\},$$

$$\mathcal{N}_2 = \{< \mathbf{u}, \mathbf{v}, \mathbf{x}, \mathbf{y}, \mathbf{s} > |[\mathbf{x}, \mathbf{y}] = \mathbf{v}, [\mathbf{s}, \mathbf{x}] = \mathbf{y}, [\mathbf{s}, \mathbf{y}] = \mathbf{u}, [\mathbf{s}, \mathbf{u}] = \mathbf{v}\}.$$

The first algebra is a split extension of $\mathbb{R}^3 =< \mathbf{y}, \mathbf{u}, \mathbf{v} >$ by the commutative algebra generated by

$$\mathbf{s} = \begin{pmatrix} 0 & & \\ 1 & 0 & \\ 0 & 1 & 0 \end{pmatrix}, \qquad \mathbf{n} = \begin{pmatrix} 0 & & \\ 0 & 0 & \\ 1 & 0 & 0 \end{pmatrix}.$$

The second algebra is an extension of $\mathbb{R}^3$ by the endomorphisms $\mathbf{s}$ und $\mathbf{x}$. Since $[\mathbf{s}, \mathbf{x}] = \mathbf{y}$ this extension does not split.

Case A: $\mathcal{N} = \mathcal{N}_2$, $(\sigma, \tau) = (2, 3)$, orbit type: $(2, 2)$

As a full Lie algebra we get the very last algebra $\mathcal{G}_{6,99}$ of Mubarakzjanov's list: $\mathcal{L}(\Sigma) =< \mathcal{N}_2, \mathbf{a} >=< \mathbf{x}, \mathbf{y}, \mathbf{u}, \mathbf{v}, \mathbf{s}, \mathbf{a} > |[\mathbf{a}, \mathbf{v}] = 5\mathbf{v}$, $[\mathbf{a}, \mathbf{u}] = 4\mathbf{u}, [\mathbf{a}, \mathbf{y}] = 3\mathbf{y}, [\mathbf{a}, \mathbf{x}] = 2\mathbf{x}, [\mathbf{a}, \mathbf{s}] = \mathbf{s}$.

The action of Σ on $P \setminus W$ has a regular nilpotent normal subgroup

$$(x, y, u, v)^{(x_1, y_1, u_1, v_1)} = (x + x_1, y + y_1, u + u_1, v + v_1 - x_1 y),$$

and a stabilizer L_2 which acts on the normal subgroup (by conjugation) in the following way:

$(x, y, u, v)^{(a,s)} = (a^2 x, a^3 y + sa^2 x, a^4 u + \frac{1}{2}a^2 xs^2 + a^3 ys,$
$a^5 v + \frac{1}{6}a^2 xs^3 + \frac{1}{2}a^3 ys^2 + a^4 us - \frac{1}{2}a^4 x^2 s).$

Equations for a typical line:

$$L = \{(x, y, f(x), g(x));\ x, y \in \mathbb{R}\}, \text{ where}$$

$$f(x) = \begin{cases} w_-(-x)^2 & : x \le 0 \\ w_+x^2 & : x \ge 0 \end{cases}, \qquad g(x) = \begin{cases} z_-(-x)^{\frac{5}{2}} & : x \le 0 \\ z_+x^{\frac{5}{2}} & : x \ge 0 \end{cases}.$$

Necessary conditions for w_-, z_-, w_+, z_+:

$$z_+^2 \le -\left(\frac{8}{15}\right)^2 2w_+(2w_+ - 1), \qquad z_-^2 \le \left(\frac{8}{15}\right)^2 2w_-(2w_- - 1).$$

Case C: $\mathcal{N}_1$, $(\sigma, \tau) = (3,3)$, orbit type: (1,2).

In this case there are two possibilities to extend the 5-dimensional nilpotent algebra to a 6-dimensional solvable one.
$\mathcal{L}(\Sigma) = \mathcal{G}_{6,94}$:

We get the following collineation group with fixed orbit $x = 0$: it is a semidirect product of $\mathbb{R}^3$ by the following one-parameter groups:

$$\left\{\begin{pmatrix} 1 & & & \\ s & 1 & & \\ \frac{s^2}{2} & s & 1 & \\ \frac{s^3}{6} & \frac{s^2}{2} & s & 1 \end{pmatrix}; s \in \mathbb{R}\right\}, \quad \left\{\begin{pmatrix} 1 & & & \\ & 1 & & \\ t & & 1 & \\ & t & & 1 \end{pmatrix}; t \in \mathbb{R}\right\},$$

$$\left\{\begin{pmatrix} e^r & & & \\ & e^{(1+k)r} & & \\ & & e^{(1+2k)r} & \\ & & & e^{(1+3k)r} \end{pmatrix}; r \in \mathbb{R}\right\}.$$

A typical line is

$$L^{(0,0)} = \{(x, y, (\epsilon x)^{1+2k} w_\epsilon, (\epsilon x)^{1+3k} z_\epsilon);\ x, y \in \mathbb{R}\},$$

and the conditions for the existence of a plane are

$k > 0,\ w_-^2 + z_-^2 + w_+^2 + z_+^2 > 0,\ w_- \ge 0,\ w_+ \le 0.$

$\mathcal{L}(\Sigma) = \mathcal{G}_{6,98}$:

Here, the full Lie algebra is the semidirect product of
$\mathbb{R}^3 = \{(0, y, u, v) | y, u, v \in \mathbb{R}\}$ with the endomorphisms

$$\mathbf{s} = \begin{pmatrix} 0 & & & \\ 0 & 0 & & \\ 0 & 1 & 0 & \\ 0 & 0 & 1 & 0 \end{pmatrix}, \quad \mathbf{n} = \begin{pmatrix} 0 & & & \\ 0 & 0 & & \\ 0 & 0 & 0 & \\ 0 & 1 & 0 & 0 \end{pmatrix}, \quad \mathbf{a} = \begin{pmatrix} 0 & & & \\ 0 & 1 & & \\ 0 & 0 & 1 & \\ 0 & 0 & h & 1 \end{pmatrix}, h \in \mathbb{R}.$$

A typical line is $(h > 0, \xi \neq 0)$:

$L^{(\xi,\eta)} = \{(s, n, e^a(w_\epsilon s - \epsilon s^2 + \epsilon(n - 2has)) + \eta s,$
$hae^a(w_\epsilon s - \epsilon s^2 - \epsilon has) + e^a(z_\epsilon s + w_\epsilon \frac{s^2}{2} - \epsilon\frac{s^3}{3}) + \eta n) | s, n \in \mathbb{R}\},$

where $\xi = e^a \epsilon$ and $\epsilon = +1, -1$.

Conditions for the existence of planes:

$$\begin{aligned} \epsilon = +1 &: \quad w_+^2 + 4h^2 + 4z_+ \leq 0 \\ \epsilon = -1 &: \quad w_-^2 + 4h^2 - 4z_- \leq 0. \end{aligned}$$

Case B: $\mathcal{N}_1, (\sigma, \tau) = (3,3)$, orbit type: $(2,2)$

The full Lie algebra is isomorphic to $\mathcal{G}_{6,94}$ with $\lambda = 2$, in our notation: $\mathcal{L}(\Delta) = < \mathcal{N}, \mathbf{a} >$ with

$$\mathcal{N} = \{< \mathbf{y}, \mathbf{u}, \mathbf{v}, \mathbf{s}, \mathbf{n} > | [\mathbf{s}, \mathbf{y}] = \mathbf{u}, [\mathbf{s}, \mathbf{u}] = \mathbf{v}, [\mathbf{n}, \mathbf{y}] = \mathbf{v}\},$$

$$[\mathbf{a}, \mathbf{y}] = 2\mathbf{y}, [\mathbf{a}, \mathbf{u}] = 3\mathbf{u}, [\mathbf{a}, \mathbf{v}] = 4\mathbf{v}, [\mathbf{a}, \mathbf{s}] = \mathbf{s}, [\mathbf{a}, \mathbf{n}] = 2\mathbf{n}.$$

This Lie algebra is a semidirect product of $\mathbb{R}^3 \cong < \mathbf{y}, \mathbf{u}, \mathbf{v} >$ with the endomorphisms

$$\mathbf{s} = \begin{pmatrix} 0 & & \\ 1 & 0 & \\ 0 & 1 & 0 \end{pmatrix}, \quad \mathbf{n} = \begin{pmatrix} 0 & & \\ 0 & 0 & \\ 1 & 0 & 0 \end{pmatrix}, \quad \mathbf{a} = \begin{pmatrix} 2 & & \\ & 3 & \\ & & 4 \end{pmatrix}.$$

A typical line has the form

$$L^{0,0} = \{(x, y, w_{\epsilon(x)} x^3, z_{\epsilon(x)} x^4) | x, y \in \mathbb{R}\},$$

where $w_{\epsilon(x)} = w_+$, $z_{\epsilon(x)} = z_+$ for $x > 0$,
$w_{\epsilon(x)} = w_-$, $z_{\epsilon(x)} = z_-$ for $x < 0$ and
$w_{\epsilon(x)} = z_{\epsilon(x)} = 0$ for $x = 0$.

Conditions for existence:

$$\left(w_- - \frac{1}{3}\right)^2 + \frac{4}{3}\left(z_- - \frac{1}{16}\right) \leq 0, \qquad \left(w_+ - \frac{1}{3}\right)^2 + \frac{4}{3}\left(z_+ - \frac{1}{16}\right) \leq 0.$$

11 4-dimensional Nilradical

There exist 3 nilpotent real Lie algebras of dimension 4, namely

$$\mathbb{R}^4, \quad \mathbb{R}^3 \rtimes \mathbb{R} \begin{pmatrix} 0 & & \\ 0 & 0 & \\ 0 & 1 & 0 \end{pmatrix}, \quad \mathbb{R}^3 \rtimes \mathbb{R} \begin{pmatrix} 0 & & \\ 1 & 0 & \\ 0 & 1 & 0 \end{pmatrix}.$$

The commutative algebra $\mathbb{R}^4$ only leads to planes already known. The third algebra has a unique extension to a 6-dimensional solvable algebra, the algebra $\mathcal{N}_{6,28}$ in [45]. But H. Klein [20] has proved that this algebra does not lead to a plane.

So, we are left with the second Lie algebra, which also may be written as

$$N \cong \mathrm{nil} \times \mathbb{R}, \quad \mathrm{nil} = \{< x, y, z >;\ [x, y] = z\}.$$

If $\dim N = 3$ then the structure of the collineation group or its Lie algebra is very restricted; there are only two groups: $L_2 \times L_2 \times L_2$ and $L_2(\mathbb{C}) \times L_2$. In [21] H. Klein proved that these groups do not lead to planes.

As a consequence we may assume that the nilradical of Σ is $N \cong nil \times \mathbb{R}$. We take this as a general assumption in the rest of the article.

12 Planes with Two Fixed Points

Let us now turn to the orbit type cases. Orbit type (0,0) means that there is a fixed point on W different from ∞ and a fixed line through ∞ different from W. In other words, we have a plane with exactly two fixed points and two fixed lines. Thus by [25] it must be one of the planes of Lenz-Barlotti type II.2 considered earlier.

In [11] it was shown that the orbit type (0,1) also leads to planes already known, namely some glued translation planes. Thus we come next to the orbit type (0,2); see [14].

In this case, there are two fixed points on the line W, and we shall call these points ∞ and p. The nilradical N of Σ is the simply connected Lie group with Lie algebra $\mathrm{nil}{\times}\mathbb{R}$. It can be shown that N acts regularly on the affine point space $P\backslash W$. In particular, the affine point space can be written as $P\backslash W = \{(x, y, u, v); x, y, u, v \in \mathbb{R}\}$ such that the stabilizer Σ_o acts as a two-dimensional automorphism group of N. Since $\Sigma' = N$, the group Σ_o is abelian. By changing the roles of ∞ and p, if necessary, we can assume that $\dim N_{[\infty]} = 2$ and $\dim N_{[p]} = 1$. The group $\Omega := N_{[\infty]} \cdot N_{[p]}$ is a three-dimensional elation group with axis W, i.e. $\tau = 3$ and $\sigma \geq 2$.

As above, we use this elation group to introduce coordinates ξ, η on the line W such that the point p has the coordinates $[0, 0]$. Then the action of the stabilizer Σ_o on W has the simple form $[\xi, \eta] \mapsto [e^t \xi, e^s \eta]$. Using this result it is simple to see that the action of Σ_o on $P\backslash W$ is

$$(x, y, u, v) \mapsto (e^{s-t}x, e^{bs-2t}y, e^{bs-t}u, e^{(b+1)s-2t}v)$$

with a real parameter $b > 1$: in particular the group Σ depends only on one parameter. The line joining o with the point $[\xi, \eta]$ at infinity is given by the equations

$$\begin{aligned} u &= \xi y + \frac{\alpha(\xi x - \eta) - \alpha(-\eta)}{\xi} \\ v &= \eta y + \frac{bx}{b+1}\frac{\alpha(\xi x - \eta)}{\xi} + \frac{\eta}{b+1}\frac{\alpha(\xi x - \eta) - \alpha(-\eta)}{\xi^2} \end{aligned},$$

where $\alpha(x) = x^b$ if $x \geq 0$ and $\alpha(x) = c(-x)^b$ if $x < 0$, with a constant $c \geq 1$.

It can be proved that the above equations define a topological affine plane; and this plane is neither a translation plane nor a dual translation plane if $(b, c) \neq (2, 1)$.

Finally, we may determine the value of $\sigma = \dim \Sigma_{[\infty]}$. If $b \neq 2$, then $\sigma = 2$, but if $b = 2$, then the subgroup of Σ_o given by $s = t$ is a homology group with center ∞ and axis $o \vee p$, i.e. $\sigma = 3$.

13 Planes with Orbit Type (1,1)

These planes are classified in [16]. It turns out that they have a 3-dimensional translation group and the collineation group Σ fixes at least one orbit of this translation group. Hence we have the situation "fixed orbit" described in the section on 3-dimensional translation groups and we can introduce (x, y, u, v)-coordinates on the affine point space and $[\xi, \eta]$-coordinates on the line W at infinity. The two one-dimensional orbits of the "orbit type (1,1)" correspond to $x = 0$ and to $\xi = 0$. The nilradical is 4-dimensional and has the following form:

$$N = \left\{ \begin{pmatrix} 1 & & & & \\ 0 & 1 & & & \\ y & 0 & 1 & & \\ u & n & 0 & 1 & \\ v & 0 & n & 0 & 1 \end{pmatrix} ; \; y, u, v, n \in \mathbb{R} \right\}.$$

It comprises the translation group $\{(y, u, v) | y, u, v \in \mathbb{R}\} \cong \mathbb{R}^3$ with respect to the axis W and the dual translation group $\{(n, u, v) | n, u, v \in \mathbb{R}\} \cong \mathbb{R}^3$ with respect to the center ∞. For the full 6-dimensional collineation group one finds three candidates

A)

$$N \rtimes \left\{ \begin{pmatrix} 1 & & & & \\ & e^s & & & \\ & & e^t & & \\ & & & e^{as+bt} & \\ & & & & e^{(a-1)s+(b+1)t} \end{pmatrix} ; \; s, t \in \mathbb{R} \right\},$$

where $a \neq 0, 1$ and $b \neq -1, 0, 1$.

B)

$$N \rtimes \left\{ \begin{pmatrix} 1 & & & & \\ & e^t & & & \\ & se^t & e^t & & \\ & & & e^s & \\ & & & se^s & e^s \end{pmatrix} ; \; s, t \in \mathbb{R} \right\},$$

C)

$$N \rtimes \left\{ \begin{pmatrix} 1 & & & & \\ & e^{s+at} & & & \\ & se^{s+at} & e^{s+at} & & \\ & & & e^t & \\ & & & se^t & e^t \end{pmatrix} ; \; s, t \in \mathbb{R} \right\},$$

with $a \neq 0$.

In the following a superscript $+$, $(-)$ always refers to the right (left) half-plane of the coordinatized part of the line at infinity, i.e., that part of this line for which $\xi > 0$ $(\xi < 0)$. A subscript $+$, $(-)$ refers to the right (left) half-space of the affine part of the plane, i.e., the right $(x > 0)$, left $(x < 0)$ half-space of the $\mathbb{R}^4$ under consideration. We sketch how the possible line set $\mathcal{L}$ for the planes with collineation group A) can be constructed.

a) All verticals $S_{x,y}$, where $x, y \in \mathbb{R}$, have to belong to $\mathcal{L}$.

b) We construct the line $L^{[\kappa 1,0]}$, $\kappa \in \{+,-\}$ through the origin that has slope $[\kappa 1, 0] (\in W)$.

For this we set $A = \frac{a}{1-b}$ and $C = \frac{2a+b-1}{1-b} (= 2A - 1)$ and denote the intersection points of $L^{[\kappa 1,0]}$ with the vertical lines $S_{(\epsilon 1,0)}, \epsilon = +,-$, by $(\epsilon 1, 0, w_\epsilon^\kappa, z_\epsilon^\kappa)$. Applying a stabilizing one-parameter group to these points followed by the translation group in direction of $[\kappa 1, 0]$ we get the equations of the two special lines.

c) The other lines through the origin can then be constructed by using the stabilizer of the origin. As a result one gets the following pencil of lines through the origin:

$$L^{(\xi,\eta,m,n)} = \{(x, y, w_\epsilon^\kappa |\xi|^b |x|^A + |\xi|^{(b-1)} \kappa y + \eta x + m, z_\epsilon^\kappa |\xi|^{b+1} |x|^C + \eta y + n);\ x, y \in \mathbb{R}\},$$

$\xi, \eta, m, n \in \mathbb{R}$. The missing lines of the one-dimensional fixed orbit $\xi = 0$ can be determined by a limit process since we want to have a topological plane. We find

$$\{(x, y, \eta x + m, \eta y + n);\ x, y \in \mathbb{R}\}, \eta, m, n \in \mathbb{R}.$$

d) By homogeneity one has to construct only two further pencils, namely the lines through the points $(1,0,0,0)$ and $(-1,0,0,0)$. Afterwards one has to check the incidence axioms. Though the lines look very complicated one is able to find a very simple discriminant condition for the existence of planes. More accurately, there are four conditions:

THEOREM: The collineation group A) leads to a 4-dimensional compact projective plane of orbit-type $(1,1)$ if the w's, the z's and the parameters a and b are chosen such that the following conditions are satisfied.

$$a \neq 0, 1; b > 1; \quad C = \frac{2a+b-1}{1-b} (= 2A - 1) > 0 \qquad (\Longleftrightarrow A > 1/2)$$

$$z_+^\kappa z_-^\kappa < 0; \quad z_\epsilon^+ z_\epsilon^- < 0$$

$$b^2 (w_\epsilon^\kappa)^2 + 4(b^2 - 1)\epsilon\kappa z_\epsilon^\kappa \leq 0, \quad \epsilon, \kappa = +, -.$$

In a similar way the collineation groups B) and C) also give rise to compact planes.

Remark: Four of the six orbit-type cases are settled so far. Only the two cases

$$(\mu_\infty, \mu_W) \geq (1, 2)$$

have to be considered in the following.

14 Planes with a 3-dimensional Elation Group

THEOREM (with J. - H. Im [15]): All planes which have, up to duality, a 3-dimensional translation group occur in the list of planes we know so far.

Way of proof: We use the list of P. Turkowski [45] of all 6-dimensional solvable Lie algebras with a 4-dimensional nilradical. A combination of algebraic and geometric considerations leads to only few transformation groups which might be the collineation group of a plane. Then one has to try to construct the plane. Only after very strenuous and lenghty calculations one gets a contradiction (or a known plane).

Note: Therefore we may assume $(\sigma, \tau) \leq (2, 2)$ from now on.

15 Planes with a 2-transitive Fixed Pencil

The cases which are still open have been narrowed down to
a) central collineations: $(\sigma, \tau) \leq (2, 2)$,
b) orbit types: $(\mu_\infty, \mu_W) \geq (1, 2)$
c) nilradical: $\dim N = 4$, $N = \mathrm{Nil} \times \mathbb{R}$.

In this final situation,[22], we may assume that Σ contains no 3-dimensional group of elations with a common center or axis, and that Σ acts transitively on the pencil $L_\infty \backslash \{W\}$ and fixes no point in the set $W \backslash \{p\}$, i.e. $\mu_\infty = 2$ and $\mu_W \geq 1$. The nilradical N of Σ is isomorphic to $\mathrm{Nil} \times \mathbb{R}$ and it acts regularly on the affine point set $P \backslash W$. It can be proved that the case $\mu_W = 1$ is impossible, i.e. Σ acts transitively on $W \backslash \{\infty\}$. In particular, this situation is self-dual. It turns out that $\sigma = \tau = 2$; more precisely we have that $\Sigma_{[W]} = \Sigma_{[\infty]} = Z(N)$ is the 2-dimensional center of N.

In order to understand the full group Σ, we have to determine the action of the group Σ_o on the affine point set $P \backslash W = N$ and the action of Σ_o on W. The group $N/Z(N)$ acts regularly on $W \backslash \{\infty\}$ and this implies that Σ_o acts on W as a two-dimensional abelian affine group. Using this information, it can be shown that Σ_o fixes exactly one point $b \in W \backslash \{\infty\}$. If Y denotes the line containing o and ∞, the stabilizer Σ_Y equals $\Sigma_o \times Z(N)$ and since $Z(N)$ fixes all of W, the full group Σ_Y fixes b.

Hence, the stabilizer on a line $Y \neq W$ through ∞ fixes a point on $W \backslash \{p\}$. Since the stabilizer on a quadrangle is trivial, this fact has a dramatic consequence: Σ acts doubly transitively on $L_\infty \backslash \{W\}$ and on $W \backslash \{\infty\}$. In particular, the one-point stabilizer Σ_o has to be a cylinder group. The one-dimensional torus group in Σ_o is a group of homologies with center b and axis Y. In coordinates on $P \backslash W = \{(x, y, u, v) : x, y, u, v \in \mathbb{R}\}$ this torus group is given by the formula:

$$(x, y, u, v) \xrightarrow{S(s)} (x \cos s - y \sin s, x \sin s + y \cos s, u, v - xy \sin^2 s + \\ + \frac{1}{4}(x^2 - y^2) \sin(2s)).$$

There are exactly two possibilities for the action of a second one-parameter group in Σ_o. These are given by

$$(x, y, u, v) \xrightarrow{H(t)} (e^{\frac{1}{2}t}x, e^{\frac{1}{2}t}y, e^{at}u, e^{t}v)$$

with a parameter $a > 0$ and

$$(x, y, u, v) \xrightarrow{H(t)} (e^{\frac{1}{2}t}x, e^{\frac{1}{2}t}y, e^{t}u, e^{t}v + te^{t}u).$$

Both cases for the group Σ_o lead to a series of four-dimensional compact projective planes with one geometrical parameter, i.e. there are two parameters in total for the first series and one parameter for the second series.

In order to describe these planes it is sufficient to define one line not through ∞ since the remaining affine lines are simply the images of this line under the action of the nilradical N. In the first case one line is given by

$$\begin{aligned} u &= (x^2 + y^2)^a \\ v &= (x^2 + y^2)\, v_0 + \frac{1}{2}xy, \end{aligned}$$

with a geometrical parameter $v_0 \geq 0$. The values of v_0 that define a projective plane are explicitly known. The second case is given by

$$\begin{aligned} u &= (x^2 + y^2)u_0 \\ v &= \frac{1}{2}xy + u_0\omega(x^2 + y^2), \end{aligned}$$

with a parameter $u_0 > 0$, where ω is defined by $\omega(0) = 0$, and $\omega(t) = t\log(t)$ for $t > 0$. This defines a projective plane if u_0 is in some open interval $(0, U)$ $(U > 0)$.

16 Summary

All 4-dimensional flexible projective planes are now classified. Let us list the various types of planes appearing in this classification:

1) Translation planes
2) Glued translation planes
3) Shift planes
4) Planes of Lenz-Barlotti type II.2
5) Planes with a 5-dimensional nilradical
6) Planes with two fixed points
7) Planes with a 2-transitive fixed pencil
8) Planes with collineation group $\mathbb{R}^2 \rtimes GL_2(\mathbb{R})^+$.

References

[1] Betten, D., 4-dimensionale Translationsebenen, Math. Z. 128, 129-151 (1972).

[2] Betten, D., 4-dimensionale Translationsebenen mit 8-dimensionaler Kollineationsgruppe, Geom. Ded. 2, 327-339 (1973).

[3] Betten, D., 4-dimensionale Translationsebenen mit irreduzibler Kollineationsgruppe, Arch. Math. 24, 552-560 (1973). Math. Z. 132, 249-259 (1973).

[4] Betten, D., 4-dimensionale Translationsebenen mit genau einer Fixrichtung, Geom. Ded. 3, 405-440 (1975).

[5] Betten, D., 4-dimensionale Translationsebenen mit 7-dimensionaler Kollineationsgruppe, J. Reine Angew. Math. 285, 126-148 (1976).

[6] Betten, D., 4-dimensionale Translationsebenen mit kommutativer Standgruppe, Math. Z. 154, 125-141 (1977).

[7] Betten, D., Komplexe Schiefparabel-Ebenen, Abh. Math. Sem. Univ. Hamburg 48, 76-88 (1979).

[8] Betten, D., 4-dimensionale projektive Ebenen mit 3-dimensionaler Translationsgruppe, Geom. Ded. 16, 179-193 (1984).

[9] Betten, D., 4-dimensional compact projective planes with a 7-dimensional collineation group, Geom. Ded. 36, 151-170 (1990).

[10] Betten, D., 4-dimensional compact projective planes with a nilpotent collineation group, Mitt. Math. Ges. Hamburg 12, 741-747 (1991).

[11] Betten, D., Orbits in 4-dimensional compact projective planes, Journal of Geometry 42, 30-40 (1991).

[12] Betten, D., 4-dimensional compact projective planes with a 5-dimensional nilradical, Geom. Ded.58, 259-289 (1995).

[13] Betten, D. und Knarr, N., Rotationsflächen-Ebenen, Abh. Math. Sem. Univ. Hamburg 57, 227-234 (1988).

[14] Betten, D. and Klein, H., Four-dimensional compact projective planes with two fixed points, J. of Geom. 55, 31-56 (1996).

[15] Betten, D. and Im, J.-H., 4-dimensional compact projective planes with a 3-dimensional elation group, Note di Mathematica

[16] Betten, D. and Polster, B., 4-dimensional projective planes with orbit type (1,1).

[17] Dixmier, J., Sur les représentations unitaires des groupes de Lie nilpotentes III, Can. J. Math. 10, 321-348 (1958).

[18] Hähl, H., Homologies and elations in compact, connected projective planes, Topology Appl. 12, 49-63 (1981).

[19] Hering, C., A new class of quasifields, Math. Z. 118, 56-57 (1970).

[20] Klein, H., 4-Dimensional compact projective planes with collineation group $N_{6,28}$, Results Math. 28, 100-116 (1995).

[21] Klein, H., 4-Dimensional Compact Projective Planes with Small Nilradical, Geom. Ded. 58, 53-62 (1995).

[22] Klein, H., 4-Dimensional compact projective planes with doubly transitive action on the fixed pencil, Geom. Ded. 61, 227-255 (1996).

[23] Knarr, N., Topologische Differenzenflächenebenen, Kiel 1983.

[24] Knarr, N., Topologische Differenzenflächenebenen mit nichtkommutativer Standgruppe, Dissertation Kiel 1986.

[25] Knarr, N., 4-dimensionale projektive Ebenen vom Lenz-Barlotti-Typ II.2, Resultate Math. 12, 134-147 (1987).

[26] Knarr, N., 4-dimensionale projektive Ebenen mit grosser abelscher Kollineationsgruppe, J. Geom. 31, 114-124 (1988).

[27] Knarr, N., Translation Planes, Springer Lecture Note 1611, 1995.

[28] Löwen, R., Four-dimensional compact projective planes with a nonsolvable automorphism group, Geom. Ded. 36, 225-234 (1990).

[29] Lüneburg, M., Kompakte, 4-dimensionale projektive Ebenen mit 8-dimensionaler Kollineationsgruppe, Diplomarbeit Tübingen 1988.

[30] Morozov, V.V., Klassifikation nilpotenter Lie Algebren der 6. Ordnung, Izv. Vyss. Uchebn. Zaved. Matematika no. 4(5), 161-171 (1958).

[31] Mostow, G.D., The extensibility of local Lie groups of transformations and groups on surfaces, Ann. of Math. 52, 606-636 (1950).

[32] Mubarakzjanov, G.M., On solvable Lie-algebras, Izv. Vyssh. Uchebn. Zaved. Mathematica 1, 114-123 (1963).

[33] Mubarakzjanov, G.M., Classification of solvable Lie algebras of sixth order with one non-nilpotent basis element, Izv. Vyssh. Uchebn. Zaved. Matematica 4, 104-116 (1963).

[34] Ostrom, T.G., Derivable nets, Can. Math. Bull. 8, 601-613 (1965).

[35] Ostrom, T.G., Finite Translation Planes, Springer, Berlin, 1970.

[36] Plaumann, P. - Strambach, K., Zusammenhängende Quasikörper mit Zentrum, Arch. Math. 21, 455-465 (1970).

[37] Riesinger, R., Beispiele starrer, topologischer Faserungen des reellen 3-dimensionalen projektiven Raumes, Geom. Ded. 40, 145-163 (1991).

[38] Salzmann, H.R., Topological Planes, Adv. Math. 2, 1-60 (1967).

[39] Salzmann, H.R., Kollineationsgruppen kompakter vier-dimensionaler Ebenen, Math. Z. 117, 112-124 (1970).

[40] Salzmann, H.R., Kollineationsgruppen kompakter 4-dimensionaler Ebenen II, Math. Z. 121, 104-110 (1971).

[41] Salzmann, H.R., Elations in four-dimensional planes, Topology Appl. 3, 121-124 (1973).

[42] Salzmann, H.R., Kompakte, vier-dimensionale projektive Ebenen mit 8-dimensionaler Kollineationsgruppe, Math. Z. 130, 235-247 (1973).

[43] Salzmann, H.R. - Betten, D. - Grundhöfer, T. - Hähl, H. - Löwen, R. - Stroppel, M., Compact Projective Planes, de Gruyter, Berlin 1995.

[44] Strambach, K., Vierdimensionale affine Ebenen, Arch. Math. 23, 342-345 (1972).

[45] Turkowski, P., Solvable Lie algebras of dimension six, J. Math. Phys. 31, 1344-1350 (1990).

[46] Weigand, C., Konstruktion topologischer projektiver Ebenen, die keine Translationsebenen sind, Mitt. Math. Sem. Giessen 177, (1987).

[47] Vergne, M., Variétés des algèbres de Lie nilpotentes, Thèse Paris 1966.

[48] Walker, M., A class of translation planes, Geom. Ded. 5, 135-146 (1976).

Replaceable Nests

G. L. Ebert
Mathematical Sciences Department
University of Delaware
Newark, DE 19716

Dedicated to T. G. Ostrom on the occasion of his eightieth birthday

1 INTRODUCTION

In [31] T. G. Ostrom described a general method called "net replacement" for creating new affine planes from old ones. Basically, the idea was to remove some collection of parallel classes of lines from the given plane, and then find a replacement set of parallel classes that would complete the resulting partial plane to a new plane which hopefully was nonisomorphic to the original plane. The best known example of this idea is Ostrom's technique of "derivation" [30], which has been studied and used by many authors over the years. In this paper we restrict ourselves to the construction of two-dimensional translation planes by working with certain net replacements in the Desarguesian plane which we call "nests". We discuss several infinite families of translation planes so obtained, including a discussion of their collineation groups and some related geometrical properties, and then survey recent attempts at characterizing these planes in terms of the collineation groups admitted. Some open problems and new directions for research will also be mentioned.

2 PRELIMINARY RESULTS

It is well known that every finite translation plane arises from a "spread" of $PG(2n-1, q)$ of some appropriate integer $n \geq 2$ (see [2] or [9]). In particular, when $n = 2$ a *spread* of $\Sigma = PG(3, q)$ is a collection of $q^2 + 1$ mutually skew lines which necessarily partition the points of Σ. A *regulus* of Σ is any collection R of $q + 1$ mutually skew lines in Σ with the property that any line meeting three lines of R necessarily meets all lines of R. It is easy to see that the transversals to a regulus R form another regulus R^{opp}, called the *opposite regulus* to R, which covers the same set of points as does R. The set of $(q + 1)^2$ points in Σ covered by R (or R^{opp}) is a doubly-ruled or hyperbolic quadric. Moreover, one easily shows that any three mutually skew lines of

This research was partially supported by NSA grant MDA 904-94-H-2033.

Σ uniquely determine a regulus containing them, and a spread of Σ is called *regular* if the regulus determined by any three of its lines is completely contained in the spread. The translation plane $\pi(S)$ associated with the spread S will be Desarguesian if and only if S is regular (see [10]). It should be noted that $\pi(S)$ has order q^2, and the lines of S represent the "points at infinity" in $\pi(S)$. From an algebraic point of view, $\pi(S)$ can be coordinatized by a quasifield of order q^2 with $GF(q)$ in its kernel (see [9]).

Let Ω be some regular spread of $\Sigma = PG(3,q)$, and thus $\pi(\Omega) \simeq PG(2,q^2)$ as discussed above. In this setting net replacement amounts to replacing some subset U of Ω by another set V of mutually skew lines in Σ which cover exactly the same points as did U. As long as the resulting spread $S = (\Omega \setminus U) \cup V$ is not regular, we will have constructed a nondesarguesian translation plane $\pi(S)$ of order q^2 with kernel $GF(q)$. Of course, the simplest replacement would be replacing a regulus R of Ω by its opposite regulus R^{opp}. The resulting translation plane would be a Hall plane (see [9], for instance). In Ostrom's terminology the Hall plane is obtained by *deriving* the Desarguesian plane. To see that these two constructions are equivalent, one must carefully look at the 4-space model $\pi(\Omega)$ for $PG(2,q^2)$. In this model we are replacing planes of $PG(4,q)$ that meet Σ (treated as a hyperplane) in lines of R by planes that meet Σ in lines of R^{opp}. The former planes represent lines of $PG(2,q^2)$ with slope points in R, while the latter planes represent Baer subplanes of $PG(2,q^2)$ with $q+1$ points on the line at infinity (namely, the elements of R). This is precisely the notion of derivation. Similarly, if we simultaneously replace a "linear" set of mutually disjoint reguli in Ω by their opposite reguli, we obtain an André plane (see [8]).

What are some other replaceable subsets of the regular spread Ω? In [11] A. Bruen suggested looking at a set of $(q+3)/2$ reguli in Ω such that any two reguli of the set have two lines in common but no three reguli of the set have a common line. Such a collection of reguli in Ω is now typically called a *Bruen chain.* The very definition of chain forces every line lying in some regulus of the chain to lie in exactly two reguli of the chain. Hence a chain of reguli contains exactly $(q+1)(q+3)/4$ distinct lines of Ω, and a possible replacement set consists of $(q+1)/2$ lines carefully chosen from the opposite regulus to each regulus in the chain so that the resulting collection of lines is mutually skew. In [11] Bruen showed that a chain of reguli does exist when $q = 5$ and found a replacement set consisting of 4 "opposite half-reguli" as described above. Moreover, he showed the resulting translation plane of order 25 is nondesarguesian. It is also shown in [11] that any replacement set for a Bruen chain must consist of "opposite half-reguli".

In the last 15 years a number of authors have constructed Bruen chains for various small values of q, all of which turned out to be replaceable. It is now known that replaceable Bruen chains exist for $q = 5, 7, 9, 11, 13, 17, 19, 23$, and 31. A recent computer search (see[22]) has shown that up to projective equivalence there are precisely two Bruen chains for $q = 7, 11, 13, 17$, and 19 and exactly one for $q = 5$ and $q = 23$. Moreover, in [22] O. Heden proved that all Bruen chains must be replaceable; in fact, they are replaceable using "opposite half-reguli" which are orbits under a cyclic group. Unfortunately, it is still not known if Bruen chains exist for infinitely many values of q. It should also be mentioned that while most translation planes obtained from Bruen chains have solvable collineation groups, there is one example for $q = 7$ in which $SL(2,5)$ is contained in the translation complement (see [28]).

It thus seems natural to relax somewhat the definition of chain in hopes of proving the existence of infinite families of replaceable subsets of Ω. In [3] Baker and Ebert defined a *nest* of reguli in a regular spread Ω to be a set N of reguli such that every line of Ω is contained in precisely 0 or 2 reguli of N. That is, the lines in the reguli of N are "doubly covered" by N. If N contains t reguli, then N is called a t-nest and there are precisely $(q+1)t/2$ distinct lines of Ω contained in the reguli of N. Simple counting shows that necessarily $\frac{q+3}{2} \leq t \leq 2(q-1)$. Hence Bruen chains are t-nests for the smallest possible value of t. The combinatorics of a nest N again suggests that if U denotes the set of lines in Ω contained in the reguli of N, a possible replacement set V for U could be obtained by carefully choosing "opposite half-reguli" so that no two chosen lines meet. Of course, there is not guarantee that t-nests, if found, will be replaceable.

To search for nests it seems judicious to use the one-to-one correspondence of Bruck [8] between the lines and reguli of Ω and the points and circles of a miquelian inversive plane $M(q)$ of order q. Namely, computations in $M(q)$ are much simpler than computations in Ω. We will thus search for nests of circles in $M(q)$, and then pull back via the Bruck correspondence to a nest of reguli in Ω to see if we obtain a replaceable subset of lines in our regular spread. The model we use for $M(q)$ will be described in the next section.

The following result, found in [37], is a generalization of some of Bruen's work on chains and will serve to focus our search for replaceable t-nests, at least when $t \leq q$.

THEOREM 2.1 : *Let N be a replaceable t-nest with $t \leq q$, and let V be a replacement partial spread for the lines U contained in the reguli of N. Let $S = (\Omega \setminus U) \cup V$ denote the resulting spread of Σ. Then*

(i) V consists of $(q+1)/2$ lines from the opposite regulus to each regulus of N,

(ii) V contains no reguli,

(iii) U contains no reguli other than those of N,

(iv) $Aut(S)$ is inherited from $Aut(\Omega)$ for $q > 7$, and

(v) any regulus of S is contained in $\Omega \setminus U$ for $t < q-1$. ∎

3 SOME NEST CONSTRUCTIONS

Let $q \geq 5$ be an odd prime power. We will use a model for $M(q)$ that is carefully developed in [29], which we now briefly describe. The points of $M(q)$ are taken to be the elements of $GF(q^2) \cup \{\infty\}$, and the circles of $M(q)$ are represented by matrices of the form $C = \begin{pmatrix} \alpha & a \\ b & -\alpha^q \end{pmatrix}$, where $\alpha \in GF(q^2)$ and $a, b \in GF(q)$ with $\alpha^{q+1} + ab \neq 0$. Two such matrices represent the same circle if and only if one is a $GF(q)$-scalar multiple of the other. The $q+1$ points incident with a given circle C (as above) are those points of $M(q)$ fixed by inversion ϕ_C with respect to that circle, where

$$\phi_C : z \longrightarrow \frac{\alpha z^q + a}{bz^q - \alpha^q}$$

and we take the usual conventions on ∞.

Let $A = \begin{pmatrix} \alpha_1 & \alpha_2 \\ \alpha_3 & \alpha_4 \end{pmatrix}$, where $\alpha_1, \alpha_2, \alpha_3, \alpha_4 \in GF(q^2)$ with $\alpha_1\alpha_4 - \alpha_2\alpha_3 \neq 0$. Then

$$\theta_A : z \longrightarrow \frac{\alpha_1 z + \alpha_2}{\alpha_3 z + \alpha_4} \text{ and } \psi_A : z \longrightarrow \frac{\alpha_1 z^q + \alpha_2}{\alpha_3 z^q + \alpha_4}$$

are automorphisms of $M(q)$ whose action on circles is easily described. Namely, if C represents a circle as above, then $\theta_A : C \longrightarrow ACA'^*$ and $\psi_A : C \longrightarrow AC'A'^*$, where A' denotes the matrix obtained by raising each entry of A to the qth power and A^* denotes the classical adjoint of A. Finally, if C and D are matrices representing circles of $M(q)$, we define

$$\begin{aligned} h(C,D) &= det(C+D) - det(C) - det(D), \text{ and} \\ C^*D &= (\tfrac{1}{2}h(C,D))^2 - det(C)det(D). \end{aligned} \qquad (\dagger)$$

Then, as shown in [29], C and D represent circles that are disjoint, tangent, or secant accordingly as C^*D in a nonzero square, zero, or nonsquare in $GF(q)$. Throughout the paper we will let $\square_q$ and $\not\square_q$ denote the nonzero squares and nonsquares of $GF(q)$, respectively, and we will let $F^{\#}$ denote the nonzero elements of the field F.

Let β denote a primitive element of $GF(q^2)$, and thus $w = \beta^{q+1}$ is a primitive element of the subfield $GF(q)$. The bundle of $q+1$ circles passing through 0 and ∞ is easily seen to be

$$J(0,\infty) = \left\{ \begin{pmatrix} \lambda & 0 \\ 0 & -\lambda^q \end{pmatrix}; \lambda = \beta^0, \beta^1, \beta^2, \cdots, \beta^q \right\},$$

while the (linear) flock of $q-1$ mutually disjoint circles with carriers 0 and ∞ is

$$K(0,\infty) = \left\{ \begin{pmatrix} 0 & a \\ 1 & 0 \end{pmatrix} : a \in GF(q)^{\#} \right\}.$$

Similarly, it is easily verified that

$$L(\infty) = \left\{ \begin{pmatrix} 1 & 0 \\ 0 & -1 \end{pmatrix} \right\} \cup \left\{ \begin{pmatrix} a & 1 \\ 0 & -a \end{pmatrix} : a \in GF(q)^{\#} \right\}$$

is a pencil of q circles mutually tangent at ∞. We will now construct four infinite families of nests, all of which turn out to be replaceable.

We begin by looking at the pencil $L(\infty)$ of q circles mutually tangent at ∞ described above. Let $\epsilon = \beta^{(q+1)/2}$, so that $\epsilon^2 = w$ and $\epsilon^q = -\epsilon$. Suppose first that $q \equiv 3 \pmod 4$. Then $D_0 = \begin{pmatrix} \epsilon & w-1 \\ 1 & \epsilon \end{pmatrix}$ represents a circle of $M(q)$. Clearly D_0 does not contain ∞, and it is easily seen from (†) that D_0 is not tangent to any circle of $L(\infty)$ since $q \equiv 3 \pmod 4$. As the circles of $L(\infty)$ cover all points of $M(q)$, D_0 is necessarily secant to $\frac{q+1}{2}$ circles of $L(\infty)$ and disjoint from the rest.

Let G denote the subgroup of $\mathrm{Aut}(M(q))$ consisting of all mappings of the form $z \longrightarrow z + a\epsilon$ as a varies over $GF(q)$. Thus G is an elementary abelian p-group of order q, where p is the characteristic of $GF(q)$. Moreover, it is easily seen that G fixes ∞ and leaves invariant each circle of $L(\infty)$. In fact, G acts regularly on the points other than ∞ in each such circle. Hence the orbit $\tilde{N}_q$ of D_0 under G is a q-nest of circles in $M(q)$ doubly covering the points other than ∞ that lie on the $\frac{q+1}{2}$ circles of $L(\infty)$ which are secant to D_0. If $q \equiv 1 \pmod 4$, picking $D_0 = \begin{pmatrix} \epsilon & 0 \\ 1 & \epsilon \end{pmatrix}$ as the "purely secant base circle" to the pencil $L(\infty)$ and again taking the orbit of D_0 under G will yield a q-nest of circles in this case.

Next look at the bundle $J(0, \infty)$ of $q+1$ circles passing through 0 and ∞. Choose $b \in GF(q)$ so that $b(b+1)$ and $\frac{-1}{bw}$ are both nonzero squares in $GF(q)$. The fact that b can thus be chosen for $q \leq 5$ follows from elementary cyclotomy (see [15], for instance). Then $D_0 = \begin{pmatrix} 1 & 1 \\ b & -1 \end{pmatrix}$ represents a circle of $M(q)$ as $b \neq -1$. Just as above, one now shows that D_0 contains neither 0 nor ∞, and D_0 is tangent to no circle in $J(0, \infty)$ (see [4] for details). Hence D_0 is secant to $\frac{q+1}{2}$ circles in $J(0, \infty)$ and disjoint from the rest. Let G denote the cyclic subgroup of order $q-1$ in $\mathrm{Aut}(M(q))$ generated by the mapping $z \to wz$, where $w = \beta^{q+1}$ as above. One easily computes that G fixes 0 and ∞, and G leaves invariant each circle of $J(0, \infty)$ by acting regularly on the points other than 0 and ∞ which lie on any such circle. Hence the orbit $\tilde{N}_{q-1}$ of D_0 under G is a $(q-1)$-nest of circles in $M(q)$ doubly covering the points other than 0 and ∞ which lie on the $\frac{q+1}{2}$ circles in $J(0, \infty)$ that are secant to D_0. It should also be noted that $D_0' = \begin{pmatrix} \epsilon & (b+1)w \\ 1 & \epsilon \end{pmatrix}$ is a circle in $M(q)$ that is secant to the other $\frac{q+1}{2}$ circles in $J(0, \infty)$, and again D_0' passes through neither 0 nor ∞ (see [16]). Hence the orbit of D_0' under G is a companion $(q-1)$-nest to $\tilde{N}_{q-1}$, doubly covering the remaining points of $M(q)$ other than 0 and ∞. Together one obtains a $2(q-1)$-nest, achieving the previously mentioned upper bound on the size of a nest.

Finally, look at the flock $K(0, \infty)$ of $q-1$ circles with carriers 0 and ∞. Hence these circles partition the points of $M(q)$ other than 0 and ∞. Choose $b \in GF(q)$ so that $b+1 \in \not\Box_q$. Then $D_0 = \begin{pmatrix} 1 & 1 \\ b & -1 \end{pmatrix}$ represents a circle of $M(q)$ that contains neither 0 nor ∞ and is tangent to no circle in $K(0, \infty)$ (see [17] for details). Hence D_0 is secant to $\frac{q+1}{2}$ circles in $K(0, \infty)$ and disjoint from the rest. Let $\alpha = \beta^{q-1}$, and consider the cyclic subgroup of order $q+1$ in $\mathrm{Aut}(M(q))$ generated by the mapping $z \to \alpha z$. It is easy to see that G fixes 0 and ∞, and G leaves invariant each circle of $K(0, \infty)$ by regularly acting on the points of any such circle. Thus the orbit $\tilde{N}_{q+1}$ of D_0 under G is a $(q+1)$-nest of circles in $M(q)$, doubly covering the points on the $\frac{q+1}{2}$ circles in $K(0, \infty)$ which are secant to D_0.

To see if any of these nests are replaceable, we pull back to a regular spread of $\Sigma = PG(3, q)$ using the previously mentioned Bruck correspondence. More explicitly, using homogeneous coordinates for Σ, we let $l_\infty = \langle (1,0,0,0), (0,1,0,0) \rangle$ and $l_{(x,y)} = \langle (x, wy, 1, 0), (y, x, 0, 1) \rangle$ for any $x, y \in GF(q)$. Then $\Omega = \{l_\infty\} \cup \{l_{(x,y)} : x, y \in GF(q)\}$ is a regular spread of Σ (see [8]), and we identify l_∞ with the point ∞ and $l_{(x,y)}$ with the point $z = x + y\epsilon$ in $M(q)$. We can now explicitly coordinatize the reguli of Ω corresponding to the circles in the various nests constructed above. For instance, the regulus R_0 corresponding to the base circle $D_0 = \begin{pmatrix} 1 & 1 \\ b & -1 \end{pmatrix}$ in our $(q-1)$-nest $\tilde{N}_{q-1}$ of circles, where $b \in GF(q)$ with $b(b+1), \frac{-1}{bw} \in \Box_q$, is easily seen to be $R_0 = \{l_{(x,y)} : wby^2 = bx^2 - 2x - 1\}$. Then straightforward computations show that the unique line of R_0^{opp} passing through $<(x, wy, 1, 0)>$, when $wby^2 = bx^2 - 2x - 1$, is

$$L = \begin{cases} <(x, wy, 1, 0),\ (x+1, wy, bx-1, wby)> & \text{if } y \neq 0 \\ <(x, wy, 1, 0),\ (abx, -1, ab, bx)> & \text{if } y = 0. \end{cases}$$

Here a denotes an element of $GF(q)$ such that $a^2 = \frac{-1}{wb}$.

Let T_w denote the collineation of Σ induced by the matrix

$$\begin{bmatrix} w & 0 & 0 & 0 \\ 0 & w & 0 & 0 \\ 0 & 0 & 1 & 0 \\ 0 & 0 & 0 & 1 \end{bmatrix},$$

and let T_{ker} denote the collineation of Σ induced by the matrix

$$\begin{bmatrix} s & w^{-1}t & 0 & 0 \\ t & s & 0 & 0 \\ 0 & 0 & s & w^{-1}t \\ 0 & 0 & t & s \end{bmatrix}$$

acting on column vectors, where $f(x) = (x-s)^2 - w^{-1}t^2$ is the minimal polynomial of β over $GF(q)$. Then T_w is a pre-image of the mapping $z \to wz$ under the natural homomorphism $\mathcal{B}$: $\text{Aut}(\Omega) \to \text{Aut}\ (M(q))$, and T_{ker} is a generator of the (cyclic) kernel of this homomorphism. That is, T_{ker} is a collineation of order $q+1$ that leaves invariant each line of Ω by cyclically permuting the points on such a line. Of course, the orbit of R_0 (above) under the cyclic group generated by T_w is the $(q-1)$-nest N_{q-1} corresponding to $\tilde{N}_{q-1}$.

It is easy to see that T_w and T_{ker} commute, and the order of the collineation $T = T_w T_{\text{ker}}$ is $\frac{1}{2}(q^2-1)$. In fact, it is shown in [4] that the orbit V of the line $L \in R_0^{\text{opp}}$ described above under the cyclic group generated by T is a replacement set for the subset U of lines contained in the reguli of N_{q-1}. Hence N_{q-1} is a replaceable $(q-1)$-nest, and the resulting spread $\mathcal{S} = (\Omega \setminus U) \cup V$ determines a nondesarguesian translation plane of order q^2 for any odd prime power $q \geq 5$.

One similarly shows that the q-nests, $(q+1)$-nests, and $2(q-1)$-nests of circles constructed above have corresponding t-nests of reguli in Ω that are replaceable. Although the replacement sets are not always cyclically generated, as was true for the $(q-1)$-nests, the replacement sets found always consist of $\frac{q+1}{2}$ lines carefully chosen from the opposite regulus to each regulus of the nest. In fact, each of these "opposite half-reguli" is an orbit under the cyclic group generated by T is a replacement set for the subset U of lines contained in the reguli of N_{q-1}. Hence N_{q-1} is a replaceable $(q-1)$-nest, and the resulting spread $\mathcal{S} = (\Omega \setminus U) \cup V$ determines a nondesarguesian translation plane of order q^2 for any odd prime power $q \geq 5$.

One similarly shows that the q-nests, $(q+1)$-nests, and $2(q-)$-nests of circles constructed above have corresponding t-nests of reguli in Ω that are replaceable. Although the replacements sets are not always cyclically generated, as was true for the $(q-1)$-nests, the replacement sets found always consist of $\frac{q+1}{2}$ lines carefully chosen from the opposite regulus to each regulus of the nest. In fact, each of these "opposite half-reguli" is an orbit under the cyclic group generated by T_{ker}^2. The following theorem summarizes results found in [3], [4], [16], [17] and [33].

THEOREM 3.1 : *For any odd prime power $q \geq 5$ there exist replaceable t-nests for $t = q,\ q-1, 2(q-1)$, and $q+1$. The resulting spreads determine nondesarguesian translation planes of order q^2.* ■

A reasonable question to ask at this stage is whether or not the planes constructed in Theorem 3.1 are "new". S. E. Payne [33] showed that the planes obtained from replaceable q-nests are isomorphic to the planes obtained from the Fisher flocks of a quadratic cone (see [20], for instance). It is also not hard to see that the planes obtained from replaceable $q+1$-nests are isomorphic to those constructed by Pabst and Sherk [32] using indicator sets. This isomorphism will become increasingly clearer when we discuss the properties of these planes in the next section. The planes arising from replaceable $(q-1)$-nests and $2(q-1)$-nests appear to be new. Moreover, one typically obtains nonisomorphic planes by starting with different "base reguli" R_0. The number of mutually nonisomorphic translation planes of a given order that can be obtained from $(q-1)$-nests (or t-nests for other values of t) remains unknown.

4 Algebraic and Geometric Properties

With the coordinatization established in the previous section it is now fairly straightforward to compute collineation groups of the spreads constructed from replaceable nests, to determine the reguli contained in such spreads, and so on. As an example, we again work with the $(q-1)$-nests constructed above. Let U denote the subset of Ω consisting of the lines contained in the reguli of N_{q-1}, and let V denote the previously described replacement set for U. That is, V is the orbit of $L \in R_0^{opp}$ under the cyclic group of order $\frac{1}{2}(q^2-1)$ generated by $T = T_w T_{ker}$, as discussed in Section 3. Let $S = (\Omega \setminus U) \cup V$ be the resulting spread, and let $\pi = \pi(S)$ be the (affine) translation plane of order q^2 determined by S. It is well known that $Aut(\pi)$ is the semidirect product of the translation group (of order q^4) by the translation complement, where the translation complement is the group of collineations of π fixing the origin. In fact, the translation complement consist of those nonsingular semilinear transformations of a 4-dimensional vector space over $GF(q)$ that induce collineations of $\Sigma = PG(3,q)$ leaving S invariant. Hence, to determine $\text{Aut}(\pi)$, it suffices to determine $\text{Aut}(S)$.

Since $t = q-1 \leq q$, we know from Theorem 2.1 that $\text{Aut}(S)$ is inherited from $\text{Aut}(\Omega)$ as long as $q > 7$. It is easy to see that $T : \Omega \setminus U \to \Omega \setminus U$ and $T : V \to V$, thereby implying $T \in \text{Aut}(S)$. It is also clear from comments made just prior to Theorem 3.1 that $T_{ker}^2 \in \text{Aut}(S)$. Using these two collineations, S is partitioned into two orbits of length 1 ($l_{(0,0)}$ and l_∞), $\frac{1}{2}(q+1)$ orbits of length $q-1$ (reguli of the bundle that are disjoint from R_0, with $l_{(0,0)}$ and l_∞ deleted from each such regulus), and one orbit of size $\frac{1}{2}(q^2-1)$ (namely, V). In [4] two involutions of $\text{Aut}(S)$ are constructed that combine some of the above orbits, the coalescence being slightly different for $q \equiv 1 \pmod 4$ and $q \equiv 3 \pmod 4$. The general form of these involutions was determined after looking at several examples for small values of q using the software package CAYLEY [12], which was the predecessor of currently available package MAGMA [13]. The following result is found in [4].

THEOREM 4.1 : *Let $q \geq 5$ be an odd prime power, and let S be a spread of Σ obtained by replacing a $(q-1)$-nest in the regular spread Ω as described above. Then*

Aut(S) contains a linear collineation group H of order $2(q^2 - 1)$ *with the following orbit structure on S:*

(i) one orbit of length 2, one orbit of length $q-1$, $\frac{1}{4}(q-1)$, *orbits of length* $2(q-1)$, *and one orbit of length* $\frac{1}{2}(q^2-1)$ *whenever* $q \equiv 1 \pmod 4$, *or*

(ii) one orbit of length 2, $\frac{1}{4}(q+1)$ *orbits of length* $2(q-1)$, *and one orbit of length* $\frac{1}{2}(q^2-1)$ *whenever* $q \equiv 3 \pmod 4$.

In particular, $\pi(S)$ *admits a cyclic collineation group of order* $q^2 - 1$ *in its linear translation complement which has an orbit of components of length* $q - 1$. ∎

It should be noted that using CAYLEY it was determined that $H = \mathrm{Aut}(S) \cap \mathrm{Aut}(\Omega)$ for $q = 7$ and $H = \mathrm{Aut}(S)$ for $q = 11$. It seems likely that H is the full linear collineation group stabilizing S for $q \geq 7$. For $q = 5$ any spread obtained by replacing a $(q-1)$-nest as described above will be projectively equivalent to the spread found by Bruen in [11], and hence will have an automorphism group of order 144 with orbits of size 12, 8, and 6. The existence of a larger automorphism group in that case is no doubt related to the fact that a $(q-1)$-nest is a chain if and only if $q = 5$.

Similar results have been obtained for the automorphism groups of spreads obtained by replacing q-nests and $(q+1)$-nests. The following results are found in [3] and [17].

THEOREM 4.2 : *Let* $q \geq 5$ *be an odd prime power, and let S be a spread of* Σ *obtained by replacing a q-nest in the regular spread* Ω. *Then Aut(S) contains a linear collineation group H of order* $q(q+1)$ *with the following orbit structure on S:*

(i) one orbit of length 1, $\frac{1}{4}(q-1)$ *orbits of length* $2q$, *and one orbit of length* $\frac{1}{2}q(q+1)$ *whenever* $q \equiv 1 \pmod 4$, *or*

(ii) one orbit of length 1, one orbit of length q, $\frac{1}{4}(q-3)$ *orbits of length* $2q$, *and one orbit of length* $\frac{1}{2}q(q+1)$, *whenever* $q \equiv 3 \pmod 4$.

In particular, $\pi(S)$ *admits an abelian collineation group of order* $q(q+1)$ *with a Sylow subgroup of order q that leaves some (affine) line invariant without fixing every point of the line.* ∎

THEOREM 4.3 : *Let* $q \geq 5$ *be an odd prime power, and let S be a spread of* Σ *obtained by replacing a* $(q+1)$*-nest in the regular spread* Ω. *Then Aut(S) contains a linear collineation group H of order* $2(q+1)^2$.

(i) If $q \equiv 1 \pmod 4$, *then the orbit structure on S induced by H consists of one orbit of size 2, one orbit of size* $q+1$, $\frac{1}{4}(q-5)$ *orbits of size* $2(q+1)$, *and one orbit of size* $\frac{1}{2}(q+1)^2$.

(ii) If $q \equiv 3 \pmod 4$, *then the orbit structure on S induced by H either consists of one orbit of size 2, two orbits of size* $q+1$, $\frac{1}{4}(q-7)$ *orbits of size* $2(q+1)$ *and one orbit of size* $\frac{1}{2}(q+1)^2$ *or else one orbit of size 2,* $\frac{1}{4}(q-3)$ *orbits of size* $2(q+1)$ *and one orbit of size* $\frac{1}{2}(q+1)^2$.

In particular, $\pi(S)$ admits a collineation group of order $(q+1)^2$ in its linear translation complement which is the direct product of two cyclic homology groups of order $q+1$ with affine axes. ∎

A reasonable question to ask at this stage is whether the planes described in Theorems 4.1, 4.2, and 4.3 may be characterized by the action of some collineation group. In fact, the final statement in each of those theorems essentially gives such a characterization. The following results are due to N. Johnson and his colleagues, proofs of which may be found in [23], [24] and [26].

THEOREM 4.4 (Johnson) : *Let π be a translation plane of order q^2 with $GF(q)$ contained in its kernel. If π admits a cyclic collineation group of order q^2-1 in the linear translation complement which has an orbit of components of length $q-1$, then π is one of the following;*

(i) a Desarguesian plane,

(ii) a Hall plane,

(iii) a regular nearfield plane or a derived regular nearfield plane,

(iv) a plane constructed from a Desarguesian plane by replacing a $(q-1)$-nest, or

(v) a plane derived from a plane of type (iv). ∎

THEOREM 4.5 (Johnson and Pomareda) : *Let π be a translation plane of order q^2 with $GF(q)$ in its kernel. Assume that π admits a collineation group in its linear translation complement that is the direct product of two cyclic homology groups of order $q+1$ with affine axes. Then one of the following holds:*

(i) π is an André plane,

(ii) q is odd and π is obtained from a Desarguesian plane by replacing a $(q+1)$-nest, or

(iii) q is odd and π is obtained from a Desarguesian plane by replacing a $(q+1)$-nest and an set of mutually disjoint reguli. ∎

THEOREM 4.6 (Jha and Johnson) : *Let π be a translation plane of odd order q^2 with $GF(q)$ in its kernel. Assume $q \equiv 1 \pmod 4$. If π admits a collineation group of order $q(q+1)$ in its linear translation complement which has a Sylow subgroup of order q leaving some line invariant but not fixed pointwise, then π is one of the following:*

(i) a plane constructed from a Desarguesian plane by replacing a q-nest,

(ii) a plane derived from one of type (i), or

(iii) a plane constructed from a Desarguesian plane by replacing a certain kind of $(q+2)$-nest. ∎

It should be noted that in [18] it is shown that possibility (iii) never arises in Theorem 4.6 and hence the given group action characterizes q-nest planes, at least for $q \equiv 1 \pmod 4$. It seems likely that the characterization also holds for $q \equiv 3 \pmod 4$, although the proof would certainly need to be modified.

Since several of the above results discuss derivation of planes obtained from replaceable t-nests for $t = q-1$, $q+1$ and q, it seems reasonable to address the issue of reguli contained in the spreads described in Theorems 4.1, 4.2, and 4.3. We first look at spreads $S = (\Omega \setminus U) \cup V$ obtained by replacing a $(q-1)$-nest. From our earlier work we know that there is a partial bundle of $\frac{1}{2}(q+1)$ reguli through $l_{(0,0)}$ and l_∞ which comprise the lines of $\Omega \setminus U$. We also know there is a set of $(q-1)$ reguli in $\Omega \setminus U$ which form a complementary nest to the given $(q-1)$-nest. These $\frac{1}{2}(3q-1)$ reguli, all of which are contained in $\Omega \setminus U$, are the only known reguli of S. In fact, it is conjectured that S contains no other reguli, a fact that has been verified for small values of q by computer.

We now turn our attention to spreads $S = (\Omega \setminus U) \cup V$ obtained by replacing $(q+1)$-nests. It is clear from the construction of such spreads that $\Omega \setminus U$ can be partitioned into $\{l_{(0,0)}, l_\infty\}$ and $\frac{1}{2}(q-3)$ mutually disjoint reguli. Moreover, it is shown in [17] that the replacement set V may be partitioned in two different ways as a union of $\frac{1}{2}(q+1)$ mutually disjoint reguli. Thus S contains at least $\frac{1}{2}(3q-1)$ reguli, and again it is conjectured these are the only reguli of S. One should note that S admits two different "regular elliptic covers". That is, the spread S contains two different sets of $q-1$ mutually disjoint reguli, necessarily partitioning all but two lines of S. These are the only non-André spreads known to the author which admit a regular elliptic cover.

Finally, we turn our attention to the geometry of the spreads S obtained by replacing a q-nest. The construction of these spreads implies that there are $\frac{1}{2}(q-1)$ reguli in $\Omega \setminus U$ which share a common line, namely l_∞, and partition the remaining lines of $\Omega \setminus U$. In [3] it is shown that the lines of the replacement set V are partitioned into $\frac{1}{2}(q+1)$ sets of size q such that each set together with l_∞ forms a regulus. Thus the spread S is a union of q reguli which pairwise meet in a common line. That is, S admits a "regular conical cover" and hence is projectively equivalent to some spread arising from the flock of a quadratic ocne (see [20]). In fact, as previously mentioned, this flock must be the Fisher [19] flock. Moreover, the q reguli of S forming the conical cover are the only reguli contained in S (see [21]).

In any case we have seen that spreads obtained by replacing a t-nest for $t = q-1, q,$ or $q+1$ have numerous reguli contained in them, and thus the resulting translation planes are certainly derivable (and usually multiply derivable). On the other hand, the spread S obtained by replacing a $2(q-1)$-nest as described above (the union of two disjoint $(q-1)$-nests) appears to be regulus-free, a fact that has been verified by computer for small values of q. Of course, in the latter case all but two lines of the regular spread Ω get replaced in forming S.

5 Robustness of the Technique

We have now shown the existence of infinite families of replaceable t-nests for $t = q-1, q, q+1$ and $2(q-1)$. Some of the resulting translation planes can be obtained by other methods, and some appear to be new. But how robust is this technique?

What about other values of t in the range of possible sizes for nests? These questions are addressed in this section.

We begin as we did in the construction of our $(q+1)$-nest; that is, we begin with a linear flock $K(0,\infty)$ of $q-1$ mutually disjoint circles in $M(q)$ with carriers 0 and ∞. Once again choose $b \in GF(q)$ so that $b+1 \in \not\square_q$. Then $D_0 \begin{pmatrix} 1 & 1 \\ b & -1 \end{pmatrix}$ represents a circle in $M(q)$ that is secant to $\frac{1}{2}(q+1)$ circles in $K(0,\infty)$ and disjoint from the rest. Just as before, let $\alpha = \beta^{q-1}$ and consider the automorphism of $M(q)$ given by $\theta : z \longrightarrow \alpha z$. Then θ fixes 0 and ∞, and θ leaves invariant each circle of $K(0,\infty)$ by cyclically permuting the $q+1$ points of any such circle. However, instead of working with the cyclic group generated by θ as we did for $(q+1)$-nests, we now work with the cyclic group G of order $\frac{1}{2}(q+1)$ generated by θ^2. Thus the points of each circle in $K(0,\infty)$ are partitioned into two sets of size $\frac{1}{2}(q+1)$ by the action of G.

Of the $\frac{1}{2}(q+1)$ circles in $K(0,\infty)$ that are secant to D_0, suppose i of them meet D_0 in (two) points from different orbits under G. The points in these circles of $K(0,\infty)$ will only be covered once in the orbit of D_0 under G. Adding these i circles to the orbit D_0^G will thus generate a t-nest $\tilde{N}$ of circles in $M(q)$ of size $t = \frac{1}{2}(q+1)+i$.

As shown in [6] the circle $C_a = \begin{pmatrix} 0 & a \\ 1 & 0 \end{pmatrix}$, $a \in GF(q)^*$, of $K(0,\infty)$ will be in the nest $\tilde{N}$ if and only if $(ab-1)^2 - 4a \in \not\square_q$ and $a \in \not\square_q$. Thus determining the number of possibilities for a (and hence computing i) is a problem in cyclotomy. In the special case of $q \equiv 3 \pmod 4$ and $b = -2$ (so that $b+1 = -1 \in \not\square_q$), it can be shown that $i = \frac{1}{4}(q+1)$ (see [6]). In general using a result of Schmidt [36] it is shown in [6] that $\frac{1}{4}(q+1) - \frac{1}{2}\sqrt{q} \le i \le \frac{1}{4}(q+1) + \frac{1}{2}\sqrt{q}$, and hence the t-nests $\tilde{N}$ constructed above have size t with

$$\frac{3}{4}(q+1) - \frac{1}{2}\sqrt{q} \le t \le \frac{3}{4}(q+1) + \frac{1}{2}\sqrt{q}.$$

In practice these bounds appear to be quite sharp. For instance, when $q = 37$ the above bounds imply $26 \le t \le 31$. In fact, by varying the choice of b and hence the base circle in our construction, nests of size 26, 27, 28, 29, 30 and 31 were constructed for $q = 37$. Moreover, it should be noted that for any $(\frac{1}{2}(q+1)+i)$-nest constructed above, there is a complementary $(q+1-i)$-nest and hence there is a symmetry about the midpoint in the interval of possible values for t (see [6]).

Pulling back any such nest $\tilde{N}$ of circles to a nest N of reguli in Ω via the Bruck correspondence, a rather messy coordinte argument shows that N is always replaceable. Just as in the previous section, one can then determine reguli contained in the resulting spread S and compute automorphism groups of S. The following results are found in [6].

THEOREM 5.1 *Let $q \ge 5$ be an odd prime power, and let $S = (\Omega \setminus U) \cup V$ be a spread of Σ obtained by replacing a $(\frac{1}{2}(q+1)+i)$-nest in the regular spread Ω as described above. Then S contains at least $\frac{1}{2}(q-3)$ reguli, all of which must be contained in $\Omega \setminus U$. In general different base regli will generate inequivalent spreads. Moreover, Aut(S) contains a linear collineation group $H \cong (\mathbb{Z}_{\frac{1}{2}(q+1)} \times \mathbb{Z}_{\frac{1}{2}(q+1)}) \rtimes K_4$ with the following orbit structure on S:*

(i) one orbit of size 2, one orbit of size $\frac{1}{2}(q+1), \frac{1}{4}(q-5)+i$ orbits of size $q+$

$1, \frac{1}{4}(q-1) - \frac{i}{2}$ *orbits of size* $2(q+1)$, *and one orbit of size* $\frac{1}{4}(q+1)^2$ *whenever* $q \equiv 1$ *(mod 4) and* $b \in \Box_q$,

(ii) one orbit of size 2, one orbit of size $\frac{1}{2}(q+1), \frac{1}{4}(q-1)+i$ *orbits of size* $q+1, \frac{1}{4}(q-1) - \frac{1}{2}(i+1)$ *orbits of size* $2(q+1)$, *and one orbit of size* $\frac{1}{4}(q+1)^2$ *whenever* $q \equiv 1$ *(mod 4) and* $b \in \not\Box_q$,

(iii) one orbit of size 2, two orbits of size $\frac{1}{2}(q+1), \frac{1}{4}(q+1)+i-2$ *orbits of size* $q+1, \frac{1}{4}(q+1) - \frac{1}{2}(i+1)$ *orbits of size* $2(q+1)$, *and one orbit of size* $\frac{1}{4}(q+1)^2$ *whenever* $q \equiv 3$ *(mod 4) and* $b \in \Box_q$,

(iv) one orbit of size 2, two orbits of size $\frac{1}{2}(q+1), \frac{1}{4}(q-3)+i$ *orbits of size* $q+1, \frac{1}{4}(q-3) - \frac{i}{2}$ *orbits of size* $2(q+1)$, *and one orbit of size* $\frac{1}{4}(q+1)^2$ *whenever* $q \equiv 3$ *(mod 4) and* $b \in \not\Box_q$.

In particular, $\pi(S)$ *admits a collineation group in its linear translation complement which is the direct product of two cyclic homology groups of order* $\frac{1}{2}(q+1)$ *with affine axes.* ■

The final statement in the above theorem is the same as the final statement of Theorem 5.1 on $(q+1)$-nests except that the commuting homology groups now have order $\frac{1}{2}(q+1)$ instead of order $q+1$. One might hope that the existence of such homology groups would characterize the planes obtained from $(\frac{1}{2}(q+i)+i)$-nests, just as Theorem 4.5 characterized $(q+1)$-nest planes. However, it seems that an extra assumption is needed. The following result is found in [27], where the term "i- mixed nest" is used instead of the term "$(\frac{1}{2}(q+1)+i)$-nest." The authors make this terminology change because all t-nests in this construction contain a mixture of regulus nets obtained via a base regulus and certain André nets.

THEOREM 5.2 (Johnson and Pomareda) : *Let* π *be a translation plane of odd order* $q^2 \neq 3^2, 7^2$ *with* $GF(q)$ *contained in its kernel. Suppose* π *admits two commuting cyclic affine homology groups of order* $\frac{1}{2}(q+1)$, *say* H_1 *and* H_2. *Let* $e = gcd(2, \frac{1}{2}(q-1))$, *and suppose* π *admits an abelian collineation group* G *of order* $\frac{1}{2e}(q+1)^2$ *in the linear translation complement which contains* $H_1 \times H_2$. *If there is exactly one component orbit of length larger than* $\frac{1}{e}(q+1)$ *under the action of* G, *then* π *is one of the following;*

(i) a plane constructed from a Desarguesian plane by replacing a mixed i-nest, or

(ii) a plane constructed from a Desarguesian plane by replacing a mixed i-nest and a set of André nets whose carriers are the axes of H_1 *and* H_2. ■

6 Concluding Remarks

It is now clear that generalizing the idea of a Bruen chain to that of a nest produces a method for constructing many infinite families of two-dimensional translation planes. In fact, the number of mutually nonisomorphic planes of order q^2 produced by this method appears to approach infinity as $q \to \infty$, although this has not been proven. In any case the method is certainly quite robust and encompasses several previously known families of planes. It should also be noted that several sporadic nests for various values of t and q have been constructed in [18]. These sporadic nests were all obtained as the orbit under some cyclic group. Most of these nests were replaceable, but not all of them.

The non-replaceable t-nests mentioned in the above paragraph all had $t > q$. It is conjectured that all t-nests with $t \leq q$ are replaceable. In fact, it is conjectured that all such t-nests are "naturally" replaceable in the sense that the opposite half-reguli comprising the replacement set (see Theorem 2.1) are orbits under $T_{\mathbf{ker}}^2$, where $T_{\mathbf{ker}}$ generates the kernel of the Bruck homomorphism as previously described. This result would be completely analogous to Heden's result [22] on Bruen chains.

Finally, it should be mentioned that all this work has recently been extended by N. Johnson [25] to certain infinite fields. In particular, let K be any field of characteristic different from two whose multiplicative subgroup of nonzero squares is an index two subgroup. Such a field K is called a *full* field. Moreover, assume K has a full field quadratic extension $K(\sqrt{\gamma})$, where γ is a nonsquare of K. In [25] infinite analogues of t-nests for $t = q-1, q, q+1$, and $2(q-1)$ in $PG(3, K)$ are constructed, as are infintie analogues of "mixed" nests. The basic idea is to work in the circle geometry over $K(\sqrt{\gamma})$, and then rewrite each circle as a regulus net in the associated Pappian affine plane coordinatized by $K(\sqrt{\gamma})$.

Acknowledgment: The idea of nesting was hatched as a result of countless stimulating conversations with R. D. Baker about a decade ago. The author is grateful not only for those conversations but also for the opportunity to conduct the ensuing joint work, some of which has been surveyed in this article.

References

[1] V. Abatangelo, A translation plane of order 81 and its full collineation group, *Bull. Austral. Math. Soc. 29* (1984), 19-34.

[2] J. André, Uber nicht-Desarguesche Ebenen mit transitiver Translationsgruppe, *Math. Z. 60* (1954), 156-186.

[3] R. D. Baker and G. L. Ebert, A new class of translation planes, *Annals of Disc. Math. 37* (1988), 7-20.

[4] R. D. Baker and G. L. Ebert, Nests of size $(q-1)$ and another family of translation planes, *J. London Math. Soc. 38* (1988), 341-355.

[5] R. D. Baker and G. L. Ebert, A Bruen chain for $q = 19$, *Designs, Codes and Crypt. 4* (1994), 307-312.

[6] R. D. Baker and G. L. Ebert, Filling the nest gaps, *Finite Fields and Their Appl. 2* (1996), 42-61.

[7] R. D. Baker, G. L. Ebert, and R. Weida, Another look at Bruen Chains, *J. Combin. Theory Ser. A 48* (1988), 77-90.

[8] R. H. Bruck, Construction problems of finite projective planes, *Combinatorial Mathematics and Its Applications*, North Carolina Press, 1969, 426-514.

[9] R. H. Bruck and R. C. Bose, The construction of translation planes from projective spaces, *J. Algebra 1* (1964), 85-102.

[10] R. H. Bruck and R. C. Bose, Linear representations of projective planes in projective spaces, *J. Algebra 4* (1966), 117-172.

[11] A. A. Bruen, Inversive geometry and some translation planes, I., *Geom. Dedicata 7* (1978), 81-98.

[12] J. Cannon, *A Language for Group Theory*, Univ. of Sydney, Sydney, Australia, 1982.

[13] J. Cannon and C. Playoust, *An Introduction to Magma*, Univ. of Sydney, Sydney, Australia, 1994.

[14] M. Capursi, A translation plane of order 11^2, *J. Combin. Theory Ser. A 35* (1983), 289-300.

[15] L. E. Dickson, *Linear Groups*, Dover Publications (reprint), New York, 1958.

[16] G. L. Ebert, Nests, covers and translation planes, *Ars. Combin. 25c* (1988), 213-233.

[17] G. L. Ebert, Spreads admitting regular elliptic covers, *Europ. J. Combin. 10* (1989), 319-330.

[18] G. L. Ebert, Some nonreplaceable nets, *Combinatorics '88*, Research and Lecture Notes in Mathematics, Vol. 1, Mediterranean Press, pp. 353-372, 1991.

[19] J. C. Fisher, Geometry according to Euclid, *Amer. Math Monthly 86* (1979), 260-270.

[20] H. Gevaert and N. L. Johnson, Flocks of quadratic cones, generalized quadrangles and translation planes, *Geom. Dedicata 27* (1988), 301-317.

[21] H. Gevaert, N. L. Johnson and J. A. Thas, Spreads covered by reguli, *Simon Stevin 62* (1988), 51-62.

[22] O. Heden, On Bruen chains, *Disc. Math. 146* (1995), 69-96.

[23] V. Jha and N. L. Johnson, Nests of reguli and flocks of quadratic cones, *Simon Stevin 63* (1989), 311-338.

[24] N. L. Johnson, Nest replaceable translation planes, *J. Geom. 36* (1989), 49-62.

[25] N. L. Johnson, Infintie nests of reguli, submitted to *Geom. Dedicata.*

[26] N. L. Johnson and R. Pomareda, André planes and nests of reguli, *Geom. Dedicata 31* (1989), 245-260.

[27] N. L. Johnson and R. Pomareda, Mixed nests, to appear in *J. Geom.*

[28] G. Korchmaros, A translation plane of order 49 with non-solvable collineation group, *J. Geom. 24* (1985), 18-30.

[29] W. R. Orr, *The miquelian inversive plane $IP(q)$ and the associated projective planes*, Ph.D. Thesis, Univ. of Wisconsin, Madison, 1973.

[30] T. G. Ostrom, Semi-translation planes, *Trans. Amer. Math. Soc. 111* (1964), 1-18.

[31] T. G. Ostrom, Replaceable nets, net collineations, and net extensions, *Canad. J. Math. 18* (1966), 666-672.

[32] G. Pabst and F. A. Sherk, Indicator sets, reguli and a new class of spreads, *Canad. J. Math. 29* (1977), 132-154.

[33] S. E. Payne, Spreads, flocks and generalized quadrangles, *J. Geom. 33* (1988), 113-128.

[34] G. Pellegrino and G. Korchmaros, Translation planes of order 11^2, *Annals of Disc. Math. 14* (1981), 249-264.

[35] G. Raguso and B. Larato , Piano di traslazione di ordine 13^2, *Riv. Mat. Univ. Parma 10* (1984), 223-233.

[36] W. M. Schmidt, *Equations over Finite Fields, An Elementary Approach*, Lecture Notes in Mathematics, Vol. 536, Springer-Verlag, New York, 1976.

[37] R. A. Weida, An extension of Bruen chains, *Geom. Dedicata 30* (1989), 11–21.

Flocks of Laguerre Planes and Associated Geometries

Norman L. Johnson Mathematics Department, University of Iowa, Iowa City, Iowa 52242

S. E. Payne Department of Mathematics, Campus Box 170, University of Colorado at Denver, Denver, Colorado 80217-3364

This article is dedicated to T.G. Ostrom on the occasion of his eightieth birthday.

1. Introduction

In 1984 the monograph [72] by S. E. Payne and J. A. Thas appeared giving a rather complete treatment of what was known concerning finite generalized quadrangles. As soon as that monograph was completed, W.M Kantor [56] and S.E. Payne [61] found other new families of generalized quadrangles with parameters (q^2, q) using sets of 2×2 matrices over $GF(q)$ that came to be known as $q-$clans and were soon shown by J.A. Thas [76] to be equivalent flocks of a quadratic cone in $PG(3, q)$. The connection between flocks of cones and linespreads of $PG(3, q)$ had already been discovered by J. A. Thas and independently by M. Walker. Almost immediately N.L. Johnson explained how to recognize which spreads were associated with conical flocks, and there began an extensive study of projective planes associated with this general family of geometries.

A large part of the theory concerning finite generalized quadrangles that was developed between 1984 and 1993 has been surveyed in the HANDBOOK OF INCIDENCE GEOMETRY [13] (see especially chapters 7 and 9), but we now have a better understanding of the connections with other geometries, partial structures, and certain infinite generalizations. Moreover, the Subiaco examples that exist for $q = 2^e$, $e \geq 4$, did not quite make it into the HANDBOOK except as a kind of footnote. And the number of articles on the topics just mentioned has grown at a dizzying rate even since the appearance of the HANDBOOK.

This article attempts to bring the reader further up to date and surveys (mostly without proofs) the connections between flocks of quadratic cones and a variety of other geometric structures that includes q-clans, generalized quadrangles, linespreads of 3-dimensional projective space and various associated projective planes. When the underlying field is finite with characteristic 2 there are also herds of ovals in the associated desarguesian planes. In particular there are the recently constructed Subiaco q-clans and their geometries. Partial flocks, infinite flocks (i.e., over infinite fields) and a variety of related structures are also considered. Several recently discovered families of examples are presented.

Recently the theory of flocks of cones over an arbitrary oval in a Desarguesian plane of characteristic 2 first introducted by J.A. Thas [76], has been developed further by W. Cherowitzo [15] and by V. Jha and N.L. Johnson ([30] and [37]). In 1984, Cherowitzo used a computer to produce a few new examples of ovals in $PG(3, 2^e)$, e odd. In [15], W. Cherowitzo uses the theory of flocks of a cone over translation ovals to prove that his ovals belong to an infinite family. So flocks of oval cones seem to have earned a place in the general theory. Furthermore, certain of the construction techniques and classification results are valid over infinite fields of characteristic two. More generally, it is possible to construct flocks of oval cones which are not translation oval cones at least in the infinite case. We shall discuss this more generally in the section on examples.

It can be argued that the central object that ties all these geometries together is the flock of the cone (quadratic or oval) and in some cases a partial flock. Since there are some fascinating differences between the finite and infinite cases, wherever possible, we formulate our definitions without the assumption that the structures are finite.

Hence, we begin with a general definition of a partial flock of an oval cone over a skewfield. Actually, we may generalize further and define partial flocks of arbitrary Laguerre planes. We include a section due to Knarr which describes geometrically how to connect partial flocks and partial general quadrangles. However, this is valid only in finite projective space $PG(3, q)$ where q is odd.

More generally, but for flocks only and in the finite case, we consider $q-$clans, flocks of quadratic cones and spreads in $PG(3, q)$ and show how these are connected both algebraically and by the Klein quadric (the Thas-Walker construction).

We then turn to flocks and spreads in projective spaces $PG(3, K)$ but where K is an arbitrary field. The field assumption is necessary to ensure that reguli and quadratic cones exist.

Given any translation plane π with spread in $PG(3, q)$, then any affine dual of the projective extension of π is derivable. If there is an elation group E of π which leaves a $GF(q)-$Baer subplane π_o (and subspace) invariant and acts as the full elation group of π_o with a fixed axis, then by dualizing so that the center of E becomes the line at infinity for the dual plane, there is a derivable net such that derivation produces a projective plane admitting at least one incident point-line transitivity. Furthermore, if π is not a semifield plane then this derived plane admits exactly one such transitivity and so is of Lenz-Barlotti class II-1. In this article, we consider the more general situation of a spread in $PG(3, K)$ for K a field but for those spreads which correspond to flocks of quadratic cones. It follows that the dual planes are, in fact, derivable and there is a corresponding elation group E of the type mentioned above.

It has been realized previously that the set of finite partial flocks of $(q-1)$ conics of a quadratic cone in $PG(3, q)$ (that is, deficiency one partial flocks) is equivalent to the set of translation planes with spreads in $PG(3, q)$ that admit a Baer collineation group of order q. Furthermore, it is possible to uniquely extend such partial flocks. We more generally consider partial flocks of deficiency one over infinite or arbitrary quadratic cones and note that there is a set of corresponding translation planes admitting Baer collineation groups B which act regularly on the nonfixed $1-$ dimensional subspaces which lie on components of $FixB$. And, we consider the existing extension theory of partial flocks.

For even order flocks, we have noted that there are associated sets of ovals called "herds". These are introduced and used as tools for extension theory and for the analysis of "additive" flocks.

Our longest section is devoted to the Fundamental Theorem of $q-$clan Geometry which has applications to finite generalized quadrangles and provides motivation for recoordinatizations of arbitrary translation oval flocks.

We include a listing of the known finite flocks and $q-$clans and corresponding translation planes. We pay particular attention to the most recent constructions of the so-called "Subiaco" flocks and their associated oval herds.

It turns out that there are no nonclassical even order translation generalized quadrangles which correspond to flocks of quadratic cones, or equivalently, there are no nonlinear semifield spreads of even order which correspond to flocks of quadratic cones. However, there are odd order examples and for these there is a related construction called the "translation dual" which produces potentially new semifield spreads. In fact, the Ganley semifield spreads may be applied to produce new semifield spreads.

We note the existing examples of translation oval flocks and give some characterizations of translation oval flocks. In particular, consideration is given to flocks of oval cones no four planes of which share a point. When the oval cone is a quadratic cone, J.A. Thas has completely classified these flocks. We note that there is a more general result for oval cones which may be deduced.

In the infinite case, it is possible to construct maximal partial flocks of quadratic cones which have no finite analog. In fact, it is possible for an infinite partial flock of a quadratic cone to cover all of the points of a particular line of the cone but not cover all of the points of a cone. There is a somewhat natural coordinate system for the corresponding translation net which we call a quasi-quasifield. The associated geometric partial spreads are called quasifibrations. We include some general theory of quasifibrations and associated geometries.

There are myriad examples of infinite flocks of quadratic cones particularly over the field of real numbers. We mention only a few of these leaving the more complete listing to another article in this volume.

J.A. Thas has completely classified finite flocks whose planes share a point (the so-called "star" flocks). Every star flock admits a linear automorphism group which fixes each plane of the flock. We call such groups "rigid" groups. More generally, a star partial flock also admits such a locally rigid group. We include a section on "rigidity" and show how to use this theory for the extension of the result of Thas mentioned above and for the construction of certain maximal partial flocks.

There are several sections devoted to the analysis of "transitive" partial flocks. For example, we consider a subsection on the construction of Bader-Lunadon-Thas called "derivation", by which any flock in $PG(3,q)$ produces or corresponds to $(q+1)$ flocks. More generally, any partial flock of $t-$conics produces a set of $t+1$ partial flocks. There are corresponding sets of translation planes which are called the skeletons of the planes. If all of the planes of the skeleton are isomorphic then there is a corresponding transitive group. If the initial plane is a semifield plane, there is a doubly transitive group. Similarly, it is possible to classify the flocks which admit a doubly transitive group on the associated planes. And, we now have a complete classification of all transitive even order partial flocks of quadratic cones of deficiency one.

We may use the theory of rigidity to analyze maximal partial spreads of $q^2 - q + 2$ lines in $PG(3,q)$. There is a well known construction due to Bruen which when applied to spreads corresponding to flocks of quadratic cones gives an interesting characterization.

Finally, we note that there are generalizations of many of these ideas and theories in $PG(n,q)$.

2. Partial Flocks of Laguerre Planes

We recall some definitions for convenience.

Definition 2.1. *A Laguerre plane (P, C, E) is a set of points P, circles C and an equivalence relation E called parallelism on points with the following properties:*

(i) For any three mutually non parallel points, there is a unique circle incident with these points.

(ii) Given a flag (x, c) where x is a point incident with a circle c then for any point y which is not parallel to x there is a unique circle incident with x and y which is tangent to c(intersects c in exactly one point x),

(iii) Each circle is incident with at least three points and not all points are on the same circle.

Definition 2.2. *A partial flock of a Laguerre plane is a mutually disjoint set of circles.*

A flock of a Laguerre plane is a partial flock which covers the points of the plane.

2.1. Partial flocks of oval cones.

Definition 2.3. *Let P be a projective plane. An oval in P is a set of points no three of which are collinear and such that there is a unique tangent line at each point.*

We note that ovals in Desarguesian planes can be quite diverse. Of course, any conic defines an oval but also the union of two semi-ellipses in the Pappian plane over the reals can produce an oval which is not a conic. There are ovals in the quaternion planes of charactistic two which are not translation ovals.

Definition 2.4. *Let K be a skewfield. And let O denote an oval in a Desarguesian plane π over K. Embed the plane in $\Sigma \simeq PG(3,K)$ and let v be a point of $\Sigma - \pi$. The oval cone C_O is defined to be the set of points on lines αv where α is a point of O. We shall call v the vertex of the oval cone.*

Remark 1. *(1) The known finite Laguerre planes are oval Laguerre planes in the sense that they are constructed via ovals in three dimensional projective planes. The points of the Laguerre plane are the non-vertex points of an oval cone and the circles are the ovals of intersection by planes which do not contain the vertex. Two points* *are parallel if and only if they are on the same line of the oval cone. However, there are a variety of infinite non-oval Laguerre planes (e.g. [80]).*

(2) Given any point p of a Laguerre plane, the residue of p is the set of points which are not parallel to p and lines the circles which are incident with p. The residue is an affine plane. When the Laguerre plane is an oval Laguerre plane the residues are Desarguesian affine planes.

(3) So, flocks of oval cones are equivalent to flocks of oval Laguerre planes. However, there are some interesting problems which may be formed for affine planes which may be best formed thinking of the residues of oval Laguerre planes.

Definition 2.5. *Let C_O be an oval cone in $PG(3, K)$.*

(1) A partial flock of C_O is a set of planes of Σ not incident with v whose pairwise intersections are disjoint from C_O.

(2) A maximal partial flock of C_O is a partial flock F such that there is no plane ρ such that $F \cup \{\rho\}$ is a partial flock.

(3) A maximal partial flock is said to be a class-covered partial flock if and only if some line of the oval cone is completely covered by the planes of the set and $\{v\}$.

(4) A maximal partial flock which covers $C_O - \{v\}$ is called a flock of the oval cone.

(5) A partial flock such that the planes share a common point is called a star partial flock.

(6) A partial flock such that the planes share a common line is called a linear partial flock.

(7) A partial flock such that each two of the planes share one of n lines is called an n-linear partial flock.

Definition 2.6. *Let L denote an arbitrary Laguerre plane. A partial flock F is said to have deficiency t if and only if for each point equivalence class c then*

$$t = card\{c - c \cap F\}$$

where $c \cap F$ denotes the set of points of c on planes of F.

When t is finite, this means that the partial flock covers all but exactly t points on any point equivalence class.

When L is an oval Laguerre plane then the partial flock covers all but t nonvertex points on any line of the oval cone.

Remark 2. *(1) If K is an algebraically closed field then any plane not on the vertex defines a maximal partial flock of a quadratic cone.*

(2) There are examples of proper class-covered partial flocks of quadratic cones.

(3) Later in the section, we shall indicate some structure theory that is available for star flocks and certain partial flocks. And, we shall be particularly interested in partial flocks of deficiency one.

Proof: To see that (1) is valid, assume that there are two planes that intersect the cone in conics which are disjoint and let M denote the line of intersection. Choose homogeneous coordinates (x, y, z, w) and represent the quadratic cone by the equation

$xy = z^2$. Represent one of the planes in question by $w = 0$ and the other by $ax + by + cz + w = 0$. We may assume that $ab \neq 0$ to avoid an intersection on the cone with $w = 0$ of the form $(1, 0, 0, 0)$ or $(0, 1, 0, 0)$. Multiply the equation by y and divide by y^2 to obtain the required condition for common intersection as $a(z/y)^2 + b + c(z/y) = 0$. Since there is always a solution if K is algebraically closed, (1) follows.

As mentioned above, there are certain fundamental geometric structures associated with flocks and partial flocks of Laguerre planes. In this section, we introduce the major connecting geometries and explicate the associations in later sections.

2.2. Partial flocks and partial spreads.

We first consider flocks and partial flocks of quadratic cones. That is, the oval defining the cone is a conic. Thus, we assume that there is an associated field (commutative) K and an associated quadraic cone in $PG(3, K)$. We point out that Thas [76] and Walker [81] showed that the following construction is valid for finite flocks. It is not difficult to generalize to partial flocks over arbitrary fields. The reader might also consult De Clerck and Van Maldeghem [21] and Jha, Johnson [33]. For the converse theorems, see Gevaert-Johnson-Thas [25].

Theorem 2.7. *(1) Let F be a partial flock of a quadratic cone C in $PG(3, K)$ and let $\{\pi_i$ for i in $\lambda\}$ denote the set of planes containing the conics of the flock. Embed $PG(3, K)$ in $PG(5, K)$ so that C is contained in the Klein quadric Q. Form the polar planes $\pi_i^\perp$ where $\perp$ denotes the polarity associated with Q.*

Then $\{\pi_i^\perp \cap Q$ for i in $\lambda\}$ is a partial ovoid of Q that consists of a set of conics that share a common point.

(2) Applying the Klein correspondence, $\{\pi_i^\perp \cap Q$ for i in $\lambda\}$ yields a partial spread in $PG(3, K)$ consisting of a set of reguli that share a common line. Equivalently, the set is a partial spread of 2-dimensional K-subspaces that generate a set of regulus nets that share a common component (line) (note that we call the lines of the associated net containing the zero vector "components" of the net).

(3) If F is a flock then the partial spread in $PG(3, K)$ is a spread.

(4) Conversely, if S is any partial spread in $PG(3, K)$ which is a union of reguli that mutually share a line then S corresponds to a partial flock of a quadratic cone in $PG(3, K)$.

Actually, the partial spreads corresponding to the partial flocks may be characterized by certain intrinsic automorphism groups.

Theorem 2.8. *(Jha-Johnson [28])*

Let S be any partial spread of 2-dimensional K-subspaces and let $N(S)$ denote the associated translation net.

If there is a collineation group E which fixes a line L of S pointwise and some E-orbit of a component union L is a regulus, then S is a union of reguli that mutually share L, and $N(S)$ admits E as an elation group.

There is a more satisfying way of describing the translation nets that correspond to partial flocks which we shall come to in due course.

2.3. Partial flocks and partial generalized quadrangles.

In order to see how generalized quadrangles come into the picture, we now consider the work of Knarr [57] generalized to partial flocks. We note that partial generalized quadrangles have not been of particular general interest but the connections to such objects and partial flocks are easily described.

Definition 2.9. *A partial generalized quadrangle is a set (P, L, I) of points P , lines L, and incidence I such that*

(i) every two distinct points (lines) are incident with ≤ 1 lines (points) respectively,

(ii) given any point-line pair (x, M) where x is not incident with M , there exists at most one point-line pair (y, N) such that that y is incident with M and x is incident with N, and

(iii) each point (line) is incident with at least two lines (points) respectively.

We note that a generalized quadrangle is a partial generalized quadrangle with the wording "there exists at most one ..." in (ii) replaced by "there exists exactly one point-line pair . . .".

3. Knarr Theory

In this section, K shall always denote a field of characteristic 0 or odd.

Definition 3.1. *Let Q be a generalized quadrangle. A partial$-(0,2)$ set S of points is a set of points such that every point x of $Q-S$ is incident with either $0,1$ or 2 points of S. When each point x is always incident with either 0 or 2 points then the partial $-$ $(0,2)-$set is called simply a $(0,2)-$ set of points.*

Dually, we may consider partial $-$ $(0,2)-$ sets of lines.

Remark 3. *The set of totally isotropic points and totally isotropic lines with respect to a quadric Q_4 in $PG(4,K)$ forms a generalized quadrangle $Q(4,K)$. A partial $-(0,2)$ $-$set of points in this case is called a partial BLT set after L.Bader, G.Lunardon, and J.A. Thas [4] who first considered such sets in the contruction of finite flocks of odd order.*

To see the significance of partial BLT sets, we note the following theorem first proved by Knarr for BLT sets.

Theorem 3.2. *(See Knarr [57]). (Also see Bader-Lunardon-Thas [4] and Johnson [44]).*

Let Q_4 denote a quadric in $PG(4,K)$ and let S be a set of points of $Q(4,K)$.

Then S is a partial BLT set if and only if for each point x of S the set

$$\{x^{\perp} \cap y^{\perp}\ \forall\ y \in S-x\} \tag{3.1}$$

is a partial flock of planes of the quadratic cone $x^{\perp} \cap Q_4$ with vertex x.

Proof: Note that the conics in the projective 3-space $x^\perp$ are the sets $x^\perp \cap y^\perp \cap Q_4$ and there is an intersection with say $x^\perp \cap z^\perp \cap Q_4$ and $x^\perp \cap y^\perp \cap Q_4$ provided there is a point w in $x^\perp \cap y^\perp \cap z^\perp$. Hence, partial flocks and partial BLT sets are equivalent objects in this context.

Now to see how this is related to partial generalized quadrangles, we first observe the following connection to $Q(4, K)$.

Theorem 3.3. *Let $W(5, K)$ denote the polar space of the totally isotropic points and totally isotropic lines of a sympectic polarity in $\Sigma \simeq PG(5, K)$.*

Definition 3.4. *Define the set of "points" and "lines" of the residue R_x with respect to a point x of $W(5, K)$ to be the set of lines and planes (totally isotropic) of $W(5, K)$ that are incident with x.*

(1) The residue R_x is a generalized quadrangle isomorphic to $W(3, K)$ (totally isotropic points and totally isotropic lines of a symplectic polarity in $PG(3, K)$).

(2) Choose y to be a point which is not collinear with x. Define a generalized quadrangle $A(x, y)$ of "point sets" and "line sets" to be the set of points $W(5, K)$ collinear (incident) with both x and y and lines all of whose points are collinear with both x and y.

Then R_x is isomorphic to $A(x, y)$ under the mapping which takes a line z of $W(5, K)$ incident with x to the unique point in R_x of z which is collinear with both x and y and which takes a plane π incident with x to the intersection line $\pi \cap y^\perp$ (which is a line all of whose points are collinear with both x and y).

(3) The dual of $W(3, K)$ is isomorphic to $Q(4, K)$.

Thus, we see that a partial BLT set of points in $Q(4, K)$ produces a partial $-(0, 2)-$set of lines in $W(3, K)$. Using partial $-$ $(0, 2)$ sets of lines, henceforth partial BLT sets of lines in $W(3, K)$, a partial generalized quadrangle can be constructed which is then intimately connected with a partial flock of a quadratic cone.

3.1. Knarr's construction

Consider the polar space $W(5, K)$ and assume in some residue $R_x \simeq W(3, K)$ that S is a partial BLT set of lines of R_x. Note that S is a set of planes of $W(5, K)$ that lie on x.

Define S_y for y a point of $W(5, K) - x^\perp$,where $\perp$ refers to the symplectic polarity, to be the set of intersections of lines of S with the 4-dimensional projective space $y^\perp$, $\{M \cap y^\perp$ for all M in $S\}$. Notice from the above theorem part (2) that this set is a partial BLT set of lines of the generalized quadrangle $A(x, y) \simeq R_x$.

Let E be a plane of $W(5, K)$ which is not a plane of R_x. Define $E_x = E \cap x^\perp$. If z is a point of E not collinear to x, then E_x is a line of $A(x, z)$.

Now define an incidence structure $P(S)$:

Points:

(i) points of $W(5, K) - x^\perp$,

(ii) lines of S_y for all points y of $W(5, K) - x^\perp$ (the partial BLT set of lines in $A(x, y)$)

(iii) the point x.

Lines:

(a) all planes E of $W(5,K)$ which are not collinear with x such that the planes $< E_x, x >$ are lines of S. (lines of R_x of S), and

(b) the elements of S.

Theorem 3.5. *(Knarr [57]) $P(S)$ is a partial generalized quadrangle. Furthermore, the structure is a generalized quadrangle if and only if S is a BLT - set of lines.*

Remark 4. *In the finite case, $W(3,q)$ is a generalized quadrangle of order (q,q) and Knarr notes that a partial BLT set of lines of cardinality $(q+1)$ actually becomes a BLT set of lines.*

Remark 5. *(1) Note that in the above structure, the point x is incident with all lines of the partial BLT set.*

(2) Furthermore, a plane E which is not collinear with x of type (a) above has q^2 points which are not in $x^{\perp}$ and exactly one point of type

(ii) $E \cap x^{\perp} =< E \cap x^{\perp}, x > \cap y^{\perp}$ where y is a point of $E - E \cap x^{\perp}$. Thus, the associated generalized quadrangles are of order (q^2, q).

(3) When S is a partial BLT set of lines ,the collineation group of $W(5,K)$ which fixes x linewise induces a collineation group of $P(S)$. Knarr notes that since the planes containing x are also fixed then the group is transitive on the points not in $x^{\perp}$ so that $P(S)$ is a partial elation generalized quadrangle.

Actually, Knarr proved the above results more generally when S is a $(0,2)-$set of lines in the residue of a polar space of rank three.

Knarr's proof actually shows that $P(S)$ is a partial generalized quadrangle in the situation under consideration.

We note that it might be possible to produce partial $-(0,2)-$sets of lines in various classical generalized quadrangles even though it proved by Knarr that three of the classical cases do not contain $(0,2)-$ sets of lines ($Q(4,q)$, $H(3,q^2), H(4,q^2)$ (the residues of the polar spaces associated with unitary polarities of a 5- or 6-dimensional projective space over $GF(q^2)$ respectively)).

Hence, certain families of flocks of quadratic cones in $PG(3,K)$ are equivalent to partial BLT sets in $Q(4,K)$ which are, in turn, equivalent to partial BLT sets of lines in $W(3,K)$ (which is considered within $W(5,K)$ as a residue) and thus yields a partial generalized quadrangle.

Bader-Lunadon-Thas show that given a finite flock of a quadratic cone in $PG(3,q)$ for q odd then there is a corresponding BLT set, so there is a family of $(q+1)$ $-$flocks and thus a corresponding generalized quadrangle.

Furthermore, the above fact is also true when q is even (using algebraic techniques), but it is not true universally in $PG(3,K)$ for K an arbitrary field as Biliotti and Johnson [9] provide some examples of flocks in $PG(3,K)$ for which there is no corresponding generalized quadrangle.

We shall turn to this in the next section. But before we do so, we shall close this chapter with some problems based on consideration of the residue affine planes of a Laguerre plane.

Remark 6. *Let π be a Laguerre plane and let x be any point of π. Then the residue R_x at x is the point-line geometry of points not parallel to x and lines as circles containing x and point parallel classes not containing x . This structure becomes an affine plane π_x. Each circle not on x becomes an oval of π_x which shares a point on the line at infinity.*

Any flock of an Laguerre plane induces a covering of π_x by a set of ovals which share an infinite point and a line of π_x.

Problem. Let ρ be an affine plane. Suppose that for each line M of ρ, there corresponds a set of mutually disjoint ovals O_M which share the parallel class of M as a point and $M \cup O_M$ is a disjoint cover of the points of ρ. Show that there is a Laguerre plane such that ρ is a residue.

If we call such sets of ovals *flocks* of affine planes, then for each line of an affine plane there is a flock of ovals,what can be said of the plane.

Problem. Let ρ be a finite affine plane of order q. Assume there exists a set of $q^3 - q^2$ ovals of ρ which share a common infinite point. Show that there is a Laguerre plane such that ρ is a residue.

4. q-Clans and Associated Geometries

In this section, we begin with a finite version of some of the constructions of generalized quadrangles as in the previous section, except that here it is not required that the field be of odd order.

Let q be an arbitrary prime power, and let $F = GF(q)$ be the Galois field with q elements. Let

$$A = \begin{pmatrix} x & y \\ w & z \end{pmatrix} \text{ and } A' = \begin{pmatrix} x' & y' \\ w' & z' \end{pmatrix}$$

be arbitrary 2×2 matrices over F.

Definition 4.1. *We say A and A' are equivalent and write $A \equiv A'$ provided $x = x'$, $z = z'$, and $y + w = y' + w'$.*

This definition is chosen so that $\alpha A \alpha^T = \alpha A' \alpha^T$ for all $\alpha \in F^2$ iff $A \equiv A'$.

Definition 4.2. *The matrix A is said to be* anisotropic *provided $\alpha A \alpha^T = 0$ iff $\alpha = (0, 0)$.*

Using the quadratic formula when q is odd and the usual trace condition when $q = 2^e$, the following result is easily obtained.

Theorem 4.3. *The matrix $A = \begin{pmatrix} x & y \\ w & z \end{pmatrix}$ is anisotropic iff*

1. *$-det(A + A^T) = (y + w)^2 - 4xz$ is a nonsquare in F, if q is odd;*
2. *$xz/(y + w)^2$ has absolute trace equal to 1, if $q = 2^e$.* ■

When q is odd, $\begin{pmatrix} x & y \\ w & z \end{pmatrix} \equiv \begin{pmatrix} x & \frac{y+w}{2} \\ \frac{y+w}{2} & z \end{pmatrix}$; when $q = 2^e$, $\begin{pmatrix} x & y \\ w & z \end{pmatrix} \equiv \begin{pmatrix} x & y+w \\ 0 & z \end{pmatrix}$. Hence in what follows, when we deal with an anisotropic matrix A, for q odd we may write $A \equiv \begin{pmatrix} x & y/2 \\ y/2 & z \end{pmatrix}$, and for q even or odd we may write $A \equiv \begin{pmatrix} x & y \\ 0 & z \end{pmatrix}$.

Definition 4.4. *A q-clan is a set $\mathcal{C} = \left\{ A_t \equiv \begin{pmatrix} x_t & y_t \\ 0 & z_t \end{pmatrix} : t \in F \right\}$ of q (equivalence classes of) 2×2 matrices over F for which all pairwise differences $A_s - A_t$ $(s, t \in F,\ s \neq t)$ are anisotropic.*

Theorem 4.3 has an obvious interpretation for q-clans.

Theorem 4.5. *Let $X : t \mapsto x_t$, $Y : t \mapsto y_t$, $Z : t \mapsto z_t$ be three functions from F to F. Put $\mathcal{C} = \left\{ A_t \equiv \begin{pmatrix} x_t & y_t \\ 0 & z_t \end{pmatrix} : t \in F \right\}$. Then $\mathcal{C}$ is a q-clan iff the following holds:*

(i) $-det\left((A_t + A_t^T) - (A_s + A_s^T)\right) = (y_t - y_s)^2 - 4(x_t - x_s)(z_t - z_s)$ is a nonsquare in F for distinct $t, s \in F$, if q is odd.

(ii) $tr\left(\frac{(x_s+x_t)(z_s+z_t)}{(y_s+y_t)^2}\right) = 1$ for distinct $s, t \in F$, if $q = 2^e$. ■

For q odd, we assume in general that $A_t \in \mathcal{C}$ is symmetric and write $A_t = \begin{pmatrix} x_t & y_t/2 \\ y_t/2 & z_t \end{pmatrix}$. For $q = 2^e$ we assume in general that $A_t \in \mathcal{C}$ is upper triangular and write $A_t = \begin{pmatrix} x_t & y_t \\ 0 & z_t \end{pmatrix}$.

So for both q even and q odd we have $\mathcal{C} = \left\{ A_t \equiv \begin{pmatrix} x_t & y_t \\ 0 & z_t \end{pmatrix} : t \in F \right\}$, where $X : t \mapsto x_t$, $Y : t \mapsto y_t$, $Z : t \mapsto z_t$ are three functions from F to F. It is easy to check that if $\mathcal{C}$ is a q-clan, then X and Z must be one-to-one, and for $q = 2^e$ Y is also one-to-one. Moreover, if $0 \neq \mu \in F$, $\sigma \in Aut(F)$, $B = \begin{pmatrix} a & b \\ c & d \end{pmatrix} \in GL(2, q)$, and if x, y, z are any fixed elements of F, replacing $A_t \in \mathcal{C}$ with $A_t' \equiv \mu B A_t^\sigma B^T + \begin{pmatrix} x & y \\ 0 & z \end{pmatrix}$ gives a new q-clan $\mathcal{C}' = \{A_t' : t \in F\}$. Hence it follows that the definition below of equivalence for q-clans is a reasonable one.

Definition 4.6. *Let $\mathcal{C} = \{A_t : t \in F\}$ and $\mathcal{C}' = \{A_t' : t \in F\}$ be q-clans. We say $\mathcal{C}$ and $\mathcal{C}'$ are equivalent and write $\mathcal{C} \sim \mathcal{C}'$ provided there exist the following: $0 \neq \mu \in F$, $B \in GL(2, q)$, $\sigma \in Aut(F)$, M a 2×2 matrix over F, and a permutation $\pi : t \mapsto \bar{t}$ on F for which the following holds:*

$$A_{\bar{t}}' \equiv \mu B A_t^\sigma B^T + M, \text{ for all } t \in F. \tag{4.1}$$

4.1. Flocks of a Quadratic Cone

Let $K = \{(x_0, x_1, x_2, x_3) \in PG(3,q) : x_1^2 = x_0x_2\}$. So K is a quadratic cone in $PG(3,q)$ with vertex $V = (0,0,0,1)$.

As mentioned previously, a *flock* of K is a partition $\mathcal{F} = \{C_t : t \in F\}$ of $K \setminus \{V\}$ into q, pairwise disjoint conics. Each conic $C_t \in \mathcal{F}$ is a plane intersection $C_t = \pi_t \cap K$, where $\pi_t = [x_t, y_t, z_t, 1]^T$ is a plane not containing the vertex V. We also speak of π_t as a *plane of the flock* $\mathcal{F}$. The following important connection between flocks and q-clans was first discovered by J. A. Thas [76].

Theorem 4.7. *Let $X : t \mapsto x_t$, $Y : t \mapsto y_t$, $Z : t \mapsto z_t$ be three functions on F. For $t \in F$, put $\pi_t = [x_t, y_t, z_t, 1]^T$, $C_t = \pi_t \cap K$, $\mathcal{F} = \{C_t : t \in F\}$. Also put $A_t \equiv \begin{pmatrix} x_t & y_t \\ 0 & z_t \end{pmatrix}$ and $\mathcal{C} = \{A_t : t \in F\}$. Then $\mathcal{C}$ is a q-clan iff $\mathcal{F}$ is a flock.*

4.2. Elation GQ from q-Clans

Here we give only the most basic concepts pertaining to elation generalized quadrangles (EGQ). For a fuller treatment see the monograph [72].

Let $\mathcal{S} = (P, B, I)$ be a generalized quadrangle (GQ) of order (s,t), $s > 1, t > 1$. Let $\mathcal{G}$ be a group of collineations of $\mathcal{S}$ that leaves invariant each line through a fixed point p and acts regularly on the set $P \setminus p^\perp$ of points not collinear with p. Then $|\mathcal{G}| = s^2t$. Fix $y \in (P \setminus p^\perp)$, and suppose that $pIL_iIz_iIM_iIy$, $0 \le i \le t$. Put $A_i = \{g \in \mathcal{G} : M_i^g = M_i\}$, $A_i^* = \{g \in \mathcal{G} : z_i^g = z_i\}$. Then $\{e\} < A_i < A_i^* < \mathcal{G}$, $|A_i| = s$, and $|A_i^*| = st$. Moreover, the family $\mathcal{J} = \{A_i : 0 \le i \le t\}$ together with $\mathcal{J}^* = \{A_i^* : 0 \le i \le t\}$ satisfy the following two properties first identified by W. M. Kantor [55]:

K1. $AB \bigcap C = \{e\}$ for distinct $A, B, C \in \mathcal{J}$;

K2. $A^* \bigcap B = \{e\}$ for distinct $A, B \in \mathcal{J}$.

W. M. Kantor [55]was also the first to observe the following converse. Let $\mathcal{G}$ be a finite group with $|\mathcal{G}| = s^2t$, $s > 1, t > 1$. Let $\mathcal{J}$ be a family of $1 + t$ subgroups of $\mathcal{G}$, each with order s. For each $A \in \mathcal{J}$, let A^* be a subgroup of $\mathcal{G}$ containing A and with $|A^*| = st$. We say that $\mathcal{J}$ is a *4-gonal family for* $\mathcal{G}$ provided $\mathcal{J}$ and $\mathcal{J}^* = \{A^* : A \in \mathcal{J}\}$ satisfy K1 and K2. (For each $A \in \mathcal{J}$, A^* is uniquely determined as the union of A with the collection of all right cosets of A disjoint from all members of $\mathcal{J}$. However, it seems necessary to include in the hypothesis that A^* be a subgroup.) Define as follows a point-line incidence geometry $\mathcal{S} = (P, B, I)$ with pointset P, lineset B, and incidence relation I.

The points of $\mathcal{S}$ are of three types:

(*i*) The elements $g \in \mathcal{G}$,
(*ii*) The right cosets A^*g, $g \in \mathcal{G}$, $A \in \mathcal{J}$,
(*iii*) A symbol (∞).

The lines of $\mathcal{S}$ are of two types:

(*a*) The right cosets Ag, $g \in \mathcal{G}, A \in \mathcal{J}$,
(*b*) Symbols $[A]$, $A \in \mathcal{J}$.

A point g of type (i) is incident with each coset Ag, $A \in \mathcal{J}$. A point A^*g of type (ii) is incident with each coset Ahg contained in A^*g, $g \in \mathcal{G}$, $A \in \mathcal{J}$, and with the line $[A]$. The point (∞) is incident with each line $[A]$ of type (b). There are no other incidences. Then $\mathcal{S}$ is a GQ of order (s,t) precisely because $(\mathcal{G}, \mathcal{J}, \mathcal{J}^*)$ satisfies K1 and K2. We often write $\mathcal{S} = (\mathcal{G}, \mathcal{J})$.

It is clear that right multiplication by an element g of $\mathcal{G}$ induces an *elation about* (∞), i.e., a collineation of $\mathcal{S}$ that fixes each line through (∞) and acts semiregularly on the points not collinear with (∞). In this way $\mathcal{G}$ act regularly on the points not collinear with (∞). We say $\mathcal{G}$ is a group of *elations about* (∞) and express this by writing $\mathcal{S} = (\mathcal{S}^{(\infty)}, \mathcal{G})$ is an EGQ (elation generalized quadrangle). See Chapters 8 - 10 of the monograph [72] for many additional results concerning EGQ.

The reason for introducing 4-gonal families and the corresponding construction of GQ at this point is that q-clans provide 4-gonal families for a special group with $s = q^2$ and $t = q$. This construction is as follows.

Let $\mathcal{C}$ be a given q-clan normalized so that A_0 is the zero matrix. And $P = \begin{pmatrix} 0 & 1 \\ -1 & 0 \end{pmatrix}$.

For $\alpha = (a,b)$, $\beta = (c,d) \in F^2$, put $\alpha \cdot \beta = ac + bd = \alpha(\beta)^T$. Then we may define a group $\mathcal{K} = \{(\alpha, c, \beta) \in F^2 \times F \times F^2 : \alpha, \beta \in F^2, c \in F\}$ with binary operation

$$(\alpha, c, \beta) \circ (\alpha', c', \beta') = (\alpha + \alpha', c + c' + \beta \cdot \alpha', \beta + \beta'). \tag{4.2}$$

This makes $\mathcal{K}$ into a group of order q^5 with $(\alpha, c, \beta)^{-1} = (-\alpha, \alpha \cdot \beta - c, -\beta)$. Starting with the q-clan $\mathcal{C}$ put $K_t = A_t + A_t^T$ for all $A_t \in \mathcal{C}$. Also put $g_t(\alpha) = \alpha A_t \alpha^T$. It is then easy to check that the following holds.

$$g_t(\alpha + \beta) = g_t(\alpha) + g_t(\beta) + \alpha K_t \beta^T. \tag{4.3}$$

Then we may define the following subgroups of $\mathcal{K}$.

$A(\infty) = \{(\underline{0}, 0, \beta) \in \mathcal{K} : \beta \in F^2\}$;

$A^*(\infty) = \{(\underline{0}, c, \beta) \in \mathcal{K} : c \in F, \beta \in F^2\}$;

and for all $t \in F$

$A(t) = \{(\alpha, g_t(\alpha), \alpha K_t) \in \mathcal{K} : \alpha \in F^2\}$;

$A^*(t) = \{(\alpha, c, \alpha K_t) \in \mathcal{K} : \alpha \in F^2, c \in F\}$.

Put $\tilde{F} = F \cup \{\infty\}$, $\mathcal{J}(\mathcal{C}) = \{A(t) : t \in \tilde{F}\}$, $\mathcal{J}^*(\mathcal{C}) = \{A^*(t) : t \in \tilde{F}\}$.

Theorem 4.8. *The triple $(\mathcal{K}, \mathcal{J}(\mathcal{C}), \mathcal{J}^*(\mathcal{C}))$ satisfies the conditions K1 and K2. The resulting GQ is denoted $GQ(\mathcal{C})$ and has parameters (q^2, q).*

4.3. q-Clans, Flocks & Spreads of PG(3,q)

Starting with a q-clan $\mathcal{C}$ we show how to construct a line spread of $PG(3,q)$ (similar to a construction of Gevaert and Johnson [24]). Also we start with the associated flock $\mathcal{F}(\mathcal{C})$ and use the Klein correspondence, as discovered independently by Thas and Walker [76, 81], to construct a spread of $PG(3,q)$. These two approaches are equivalent, and each has its special advantages (see (2.7)). There is a standard

(Bose-Barlotti) construction of a translation plane from a spread of $PG(3,q)$, with isomorphic translation planes corresponding to projectively equivalent spreads. The Bose-Barlotti construction uses a 4–dimensional projective space. Equivalently, one may use the André construction via 4–dimensional vector space. We shall compare both methods a little later.

With $P = \begin{pmatrix} 0 & 1 \\ -1 & 0 \end{pmatrix}$, let A, B be any 2×2 matrices over F. Then $A \equiv B$ iff $B = A + \mu P$ for some $\mu \in F$. A line of $PG(3,q)$ will be given as the row space of a 2×4 matrix with rank 2. For example, if O (resp., I) is the 2×2 zero (resp., identity) matrix over F, put

$$L_\infty = \langle (O, I) \rangle . \tag{4.4}$$

Then if $u \in F$ and A is any 2×2 matrix over F, put

$$L_{(A,u)} = \langle (I, A + uP) \rangle . \tag{4.5}$$

Lemma 4.9. *$\mathcal{R}_A = \{L_\infty\} \cup \{L_{(A,u)} : u \in F\}$ is a regulus.*

For a line L, L^* denotes the set of points incident with L. And for a matrix A, $\mathcal{R}_A^*$ denotes the set of points incident with a line of the regulus $\mathcal{R}_A$.

Lemma 4.10. *Let A, B be two 2×2 matrices over F.*
(*i*) $\langle (I, A) \rangle = \langle (I, B) \rangle$ *iff* $A = B$.
(*ii*) $\langle (I, A) \rangle \in \mathcal{R}_B$ *iff* $A \equiv B$ *iff* $\mathcal{R}_A = \mathcal{R}_B$.
(*iii*) $\mathcal{R}_A^* \cap \mathcal{R}_B^* \supseteq L_\infty^*$, *with equality iff* $A - B$ *is anisotropic.*

As an immediate corollary we have the following theorem.

Theorem 4.11. *Let $X : t \mapsto x_t$; $Y : t \mapsto y_t$; $Z : t \mapsto z_t$ be three functions from F to F. For $t \in F$ put $A_t = \begin{pmatrix} x_t & y_t \\ 0 & z_t \end{pmatrix}$. Then put $\mathcal{C} = \{A_t : t \in F\}$. With the notation as above, $\mathcal{R}_t = \mathcal{R}_{A_t}$ is a regulus for each $t \in F$. Put $\mathcal{S}(\mathcal{C}) = \cup\{\mathcal{R}_t : t \in F\}$. Then $\mathcal{C}$ is a q-clan iff $\mathcal{S}(\mathcal{C})$ is a spread of $PG(3,q)$, in which case X and Z are both one-to-one.*

It follows that with each q-clan $\mathcal{C}$ there is a corresponding line spread $\mathcal{S}(\mathcal{C})$ of $PG(3,q)$, from which a standard construction provides a translation plane $\pi(\mathcal{C})$. It turns out that two q-clans $\mathcal{C}$, $\mathcal{C}'$ are equivalent iff their associated spreads $\mathcal{S}(\mathcal{C})$, $\mathcal{S}(\mathcal{C}')$ are projectively equivalent.

We now use the Klein map from lines of $PG(3,q)$ to points of the Klein quadric to give the Thas-Walker construction of a spread of $PG(3,q)$ starting with a flock of a quadratic cone. (Their construction is actually more general. See J. A. Thas [76], and M. Walker [81]). We have previously seen this construction in (2.7). Now we shall consider only the finite case and add a few details.

4.4. The Thas-Walker construction in the finite case.

Let $\mathcal{F} = \{C_1, \ldots, C_q\}$ be a flock of the cone K (as in Section 4.1). Embed K in the hyperbolic (or Klein) quadric $\mathcal{H}$ of $PG(5,q)$. Denote the plane of C_i by π_i. So π_i intersects $\mathcal{H}$ in a conic. The polarity $\pi_i \mapsto \pi_i^\perp$ defined by $\mathcal{H}$ takes planes of this type to planes of this type. (In algebraic notation: $\mathcal{H} = O^+(6,q) = O(3,q) \perp O(3,q)$.) Denote $\pi_i^\perp \cap \mathcal{H}$ by $C_i^\perp$, and let $\mathcal{O} = \cup\{C_i^\perp : 1 \le i \le q\}$. The cone K lies in a solid Σ whose perp is a tangent line L (since Σ meets $\mathcal{H}$ in a cone). The point of tangency is the vertex V of the cone. Thus all $\pi_i^\perp$ contain L and hence contain V. Moreover, since the planes π_i in $\mathcal{F}$ are distinct, their perps are distinct, so they pairwise intersect exactly in L. Other than V, each $\pi_i^\perp$ has q points of $\mathcal{H}$, so $|\mathcal{O}| = 1 + q^2$. And $\mathcal{O}$ is the union of q conics pairwise intersecting in V.

We claim that no two points of $\mathcal{O}$ are on a common line of $\mathcal{H}$. If p and q are two points of $\mathcal{O}$ lying in the same $\pi_i^\perp$, they lie on a conic of $\mathcal{H}$ and are not on a common line of $\mathcal{H}$. So suppose they lie in $\pi_i^\perp$ and $\pi_j^\perp$, respectively, $i \ne j$. Then $\pi_i \cap \pi_j$ is a line M external to $\mathcal{H}$. Hence $M^\perp = \langle \pi_i^\perp, \pi_j^\perp \rangle$ is a 3-space that intersects $\mathcal{H}$ in an elliptic quadric, since $O^+(6,q) = O^-(2,q) \perp O^-(4,q)$. Hence no two points of $M^\perp \cap \mathcal{H}$, and hence of $\mathcal{O}$, are on a common line of $\mathcal{H}$.

A subset $\mathcal{O}$ of size $1+q^2$ of $\mathcal{H}$ with no two points on a line of $\mathcal{H}$ is an *ovoid* of $\mathcal{H}$. Equivalently, $\mathcal{O}$ is a subset of $\mathcal{H}$ intersecting each plane of $\mathcal{H}$ in exactly one point.

Now consider the Klein correspondence between points of the Klein quadric $\mathcal{H}$ and lines of $PG(3,q)$. The points of $\mathcal{O}$ correspond to lines of a spread $S(\mathcal{F})$. And the points of each conic $C_i^\perp$ correspond to the lines of a regulus. Hence $S(\mathcal{F})$ is the union of q reguli pairwise meeting in a line L_∞ that Klein-corresponds to the vertex V. This completes the Thas-Walker construction.

Starting with a q-clan $\mathcal{C} = \{A_t \equiv \begin{pmatrix} x_t & y_t \\ 0 & z_t \end{pmatrix} : t \in F\}$, we have the flock $\mathcal{F}(\mathcal{C}) = \{C_t : t \in F\}$, where $C_t = \pi_t \cap K$, $\pi_t = [x_t, y_t, z_t, 1]^T$. We now have two constuctions of a spread of $PG(3,q)$ starting with $\mathcal{C}$, one going directly, algebraically, from the q-clan, the other one using the associated conical flock and using the Klein correspondence.

Lemma 4.12. *Our two constructions of spreads from a q-clan are equivalent. (The proof is a variation on a treatment in Gevaert and Johnson [24].)*

Proof. What we wish to establish is that starting with a q-clan $\mathcal{C} = \left\{A_t = \begin{pmatrix} x_t & y_t \\ 0 & z_t \end{pmatrix} : t \in F\right\}$, the lines of the regulus $\mathcal{R}_t$ Klein-correspond to to the points of a conic lying in a plane $\pi_t^\perp$, where π_t is the usual plane $[x_t, y_t, z_t, 1]^T$ of the associated flock. We know that the points $p(L_\infty)$ and $p(L_{t,u})$ for $u \in F$ form a conic in some plane π_t^T. So the plane π_t will be the intersection of $p(L_\infty)^\perp$, $p(L_{t,0})^\perp$, and $p(L_{t,1})^\perp$.

Given a line $L = \left\langle\begin{pmatrix} x_0, x_1, x_2, x_3 \\ y_0, y_1, y_2, y_3 \end{pmatrix}\right\rangle$, put $p_{ij} = \begin{vmatrix} x_i & x_j \\ y_i & y_j \end{vmatrix}$, and use the specific Klein map

$$p(L) = (p_{01}, p_{02}, p_{03}, p_{12}, -p_{13}, p_{23}).$$

So $p(L_\infty) = (0,0,0,0,0,1)$. And $L_{(t,u)} = \left\langle\left(\begin{smallmatrix} 1, & 0, & x_t, & y_t + u \\ 0, & 1, & -u, & z_t \end{smallmatrix}\right)\right\rangle$, so $p(L_{(t,u)}) =$ $(1, -u, z_t, -x_t, y_t + u, u^2 + y_t u + x_t z_t)$. Here $\mathcal{H} = \{(a_1, \ldots, a_6)$ $\in PG(5,q) : a_1a_6 + a_2a_5 + a_3a_4 = 0\}$.

Put $(a_i) \circ (b_j) = \sum_{i=1}^{6} a_i b_{7-i}$. So $(a_i) \perp (b_j)$ iff $(a_i) \circ (b_j) = 0$. Then $\mathrm{p}(\mathrm{L}_\infty)^\perp = \{(b_1, \ldots, b_6) : b_1 = 0\}, p(L_{(t,0)})^\perp = (1, 0, z_t, -x_t, y_t, x_t z_t)^\perp =$ $\{(\mathrm{b}_i) : b_6 + z_t b_4 - x_t b_3 + y_t b_2 + x_t z_t b_1 = 0\}$.

And $p(L_{(t,1)})^\perp = (1, -1, z_t, -x_t, y_t + 1, 1 + y_t + x_t z_t)^\perp =$ $\{(b_i) : b_6 - b_5 + z_t b_4 - x_t b_3 + (y_t + 1)b_2 + (1 + y + t + x_t z_t)b_1 = 0\}$

The intersection of these three perps is $\{(0, b_2, b_3, b_4, b_2, b_6) : (-b_3)x_t + b_2 y_t + b_4 z_t + b_6 = 0\}$. We identify $\{(0, b_2, b_3, b_4, b_2, b_6)$ (which is on $\mathcal{H}$ iff $b_2^2 = (-b_3)b_4$) with $(-b_3, b_2, b_4, b_6) = (x_0, x_1, x_2, x_3) \in \Sigma$. So $\Sigma \cap \mathcal{H} = K : x_1^2 = x_0 x_2$. And $(-b_3)x_t + b_2 y_t + b_4 z_t + b_6 = 0$ becomes

$$(x_0, x_1, x_2, x_3) \begin{bmatrix} x_t \\ y_t \\ z_t \\ 1 \end{bmatrix} = 0.$$

This says that the regulus $\mathcal{R}_t$ Klein - corresponds with $C_t^\perp$ where $C_t = \pi_t \cap K$, $\pi_t = [x_t, y_t, z_t, 1]^T$. Hence the two spread constructions are essentially the same. ■

Lemma 4.13. *The transversal* $L = \left\langle\left(\begin{smallmatrix} a & b & ax & ay + bz \\ 0 & 0 & -b & a \end{smallmatrix}\right)\right\rangle$ *to* $\mathcal{R}_A$ *where* $A = \left(\begin{smallmatrix} x & y \\ 0 & z \end{smallmatrix}\right)$ *that meets* L_∞ *at* $(0,0,-b,a)$ *Klein- corresponds to the point of* Σ *given by* $(-a^2, -ab, -b^2, a^2x + aby + b^2z)$. *This implies that the* q *transversals (of the reguli* $\mathcal{R}_A$*) through the point* $(0,0,-b,a)$ *of* L_∞*, which all lie in a plane* $[b, -a, 0, 0]^T$ *containing* L_∞*, must Klein-correspond to points on the generator of* K *through the point* $(a^2, ab, b^2, 0)$. *Hence the* $q+1$ *planes through* L_∞ *Klein-correspond to the* $q+1$ *generators of the cone* K.

Proof. The proof is a routine application of the Klein correspondence and subsequent identification with points of Σ as given in the proof of Lemma 4.12. ■

Not every line spread of $PG(3,q)$ can arise from a flock of a quadratic cone. The following result of Gevaert, Johnson and Thas [25] gives precisely the right characterization of such spreads.

Theorem 4.14. *A flock of a quadratic cone in* $PG(3,q)$ *is equivalent to a line spread* S *of* $PG(3,q)$*, where* S *is the union of* q *reguli* $\mathcal{R}_i$*,* $1 \leq i \leq q$*, which pairwise intersect in the same line* L.

We have seen in the previous section that the above theorem is actually valid for arbitrary fields K.

Moreover, we recall that it sometimes useful to have the corresponding characterization by elation groups.

Theorem 4.15. *A flock of a quadratic cone in $PG(3,q)$ is equivalent to a line spread in $PG(3,q)$ whose corresponding translation plane admits an elation group E whose component orbits union the axis L of the group are reguli.*

4.5. Coordinates for the spreads.

In matrix form, the spread coming from the $q-$clan may be represently on the 4−dimensional vector space (x,y) where x,y are 2−vectors in the following form
$x = 0, y = x\begin{bmatrix} x_t & u+y_t \\ -u & z_t \end{bmatrix}$ for all u,t in $GF(q)$.

Let $x_t = t$ and change bases by $T = \begin{bmatrix} 0 & -1 & 0 & 0 \\ 1 & 0 & 0 & 0 \\ 0 & 0 & 1 & 0 \\ 0 & 0 & 0 & 1 \end{bmatrix}$ to produce the spread in the form
$x = 0, y = x\begin{bmatrix} u & -zt \\ t & u+y_t \end{bmatrix}$ for all u,t in $GF(q)$. Let $z_t = -f(t)$ and $y_t = g(t)$ for two functions f and g on $GF(q)$. Then the spread has the form:

$$x = 0, y = x\begin{bmatrix} u+g(t) & f(t) \\ t & u \end{bmatrix} \text{ for all } u,t \text{ in } GF(q).$$

We call this form the **standard form**.

Also note that if f and g are functions which produce a spread as above then also f and $-g$ are functions which produce a spread. It is not difficult to prove that the two spreads are isomorphic.

To see why this transformation is useful, consider a spread in $PG(3,q)$ (considered as a 4−dimensional vector space V_4 over $GF(q)$) as a set of mutually disjoint 2−dimensional vector subspaces whose union is a coverof V_4. Representing $V_4 = W \oplus W$ where W is a 2−dimensional $GF(q)-$vector space, then the translation plane may be defined as follows: The points of the plane are the vectors and the lines are the translates of the spread lines considered as 2−dimensional vector subspaces. This is the André construction of a translation plane corresponding to a spread in $PG(3,q)$. There is also the Bruck-Bose construction which produces equivalent translation planes and which we shall discuss briefly somewhat later.

Given a translation plane as above with points (x,y), a regulus which contains $W \oplus 0 \equiv (y=0)$ and $0 \oplus W \equiv (x=0)$ and $y = x$ may be represented as follows: $x = 0, y = x\begin{bmatrix} u & 0 \\ 0 & u \end{bmatrix}$ for all u in $GF(q)$. Moreover, any other regulus that contains $x = 0$ has the form $x = 0, y = x\begin{bmatrix} u+g(t) & f(t) \\ t & u \end{bmatrix}$ for t fixed, g and f functions on K and for all u in $GF(q)$. If the spread consists of a union of q reguli that share a fixed line, we may choose this line to be represented as $x = 0$ so we obtain the following result.

Theorem 4.16. *A translation plane π with spread in $PG(3,q)$ corresponds to a flock* *of a quadratic cone if and only if there exist functions f and g on $GF(q)$ such that the spread for π may be represented in the following form:*

$x=0, y=x\begin{bmatrix} u+g(t) & f(t) \\ t & u \end{bmatrix}$ *for g and f functions on $GF(q)$ and for all t, u in $GF(q)$.*

In this case, there is an elation group

$$E=<\begin{bmatrix} 1 & 0 & u & 0 \\ 0 & 1 & 0 & u \\ 0 & 0 & 1 & 0 \\ 0 & 0 & 0 & 1 \end{bmatrix} \forall\, u \in GF(q)>.$$

We note that the corresponding $q-$clan matrices now have the form $\begin{bmatrix} t & g(t) \\ 0 & -f(t) \end{bmatrix}$.

5. Flocks and spreads in $PG(3,K)$ for K an arbitrary field

We have mentioned that it is not always that case that corresponding to a flock of a quadratic cone in $PG(3,K)$ for K a field, there is always a generalized quadrangle. However, flocks and spreads related as in the previous theorem but over K instead of $GF(q)$ do correspond. The following result may be found in Jha-Johnson [33]

Theorem 5.1. *For every translation plane with spread in $PG(3,K)$, for K an arbitrary field, of the form*

$x=0, y=x\begin{bmatrix} u+g(t) & f(t) \\ t & u \end{bmatrix}$ *for g and f functions on K and for all t, u in K, there is an associated flock of a quadratic cone defined as follows: Let (x_0,x_1,x_2,x_3) denote homogeneous coordinates for $PG(3,K)$, and consider the conic C_o with equation $x_ox_1=x_2^2$ in the plane $x_3=0$. Let $v_o=(0,0,0,1)$ and form the cone v_oP for all P in C_o. Then the following planes define, by intersection, a flock of the cone:*

$\pi_t: tx_o-f(t)x_1+g(t)x_2+x_3=0.$

Conversely, a flock of a cone in $PG(3,K)$ with equation $x_ox_1=x_2^2$ and vertex $(0,0,0,1)$ has plane equations of the form given above and defines a translation plane with spread in $PG(3,K)$ with components defined as above.

We note previously that there is also the corresponding Klein quadric construction connecting flocks of quadratic cones in $PG(3,K)$ with spreads in $PG(3,K)$ which are unions of reguli sharing a common line.

Since we shall shortly be concerned with finer connections between the points of the cone and the associated translation plane, we note the following result.

Theorem 5.2. *Let F be a flock of a quadratic cone in $PG(3,K)$ where K is a field and let π_F denote the associated translation plane whose spread is a union of reguli which share exactly one line L. Then the corresponding nets of reguli(called base regulus nets) sharing a common component have the following correspondence:*

(1)(a) The non vertex points on the quadratic cone correspond to the Baer subplanes incident with the zero vector of the base regulus nets.

(b) The points of a conic of the flock are the set of all Baer subplanes incident with the zero vector of a particular base regulus net.

(2) The lines M of the quadratic cone correspond to the 1-dimensional K-subspaces on L. Note that for each 1-dimensional K-subspace X on L there is exactly one Baer subplane of each base regulus net that contains X. The set of these Baer subplanes as points of the quadratic cone consists of the nonvertex points of a line M of the quadratic cone.

(3) Let Y be a 1−dimensional K-subspace which is not on L. Let E denote the elation group noted in the previous result.

(a) Then $<YE>$ is a 2−dimensional K -subspace which defines a Baer subplane of one of the base regulus nets.

(b) Let (∞) denote the parallel class containing L and let Y_∞ denote the set of lines $(\infty)P$ for all points P of Y where Y is a 1-dimensional K subspace which is not on L. Let Z be any 1-dimensional K-subspace which is not on L such that $Z_\infty = Y_\infty$.

Then $<YE> \cap L =< ZE> \cap L$.

Proof: Most of the above result follows more-or-less directly from the properties of the Klein quadric. To see that $(3)(b)$ is valid, let L be $x = 0$ where the points of the vector space are represented by (x, y) and x,y are 2−vectors. Let (c, m) be a point of Y and (c, n) be a point of Z. Then E has the form

$$\langle \begin{bmatrix} 1 & 0 & u & 0 \\ 0 & 1 & 0 & u \\ 0 & 0 & 1 & 0 \\ 0 & 0 & 0 & 1 \end{bmatrix} \textit{for all } u \textit{ in } K \rangle.$$

Then $<YE> \cap L = \{(c, cu) \text{ for all } u \text{ in } K\} =< ZE> \cap L$.

Corollary 5.3. *Let F be a flock of a quadratic cone in $PG(3, K)$ for K a field and let π_K denote the corresponding translation plane with spread in $PG(3, K)$ which is the union of a set of reguli sharing a line L (component) on the parallel class (∞) . Let M be any other component of π_K and for each 1-dimensional K space Y on M denote the set of lines incident with a point of Y and (∞) by Y_∞.*

Then the permutation group induced on the lines of the quadratic cone is permutation isomorphic to the permutation group induced by the stabilizer of L on the set $\{Y_\infty$ for all Y in $M\}$.

Theorem 5.4. *Let F be a flock of a quadratic cone in $PG(3, K)$ and π_F the associated translation plane with spread in $PG(3, K)$. Let G denote the collineation group of π_F in $\Gamma L(4, K)$ which permutes the base nets and leaves invariant the axis of the elation group E defining the reguli of the spread. Let K^* denote the kernel homology collineation group which fixes each parallel class which is induced from $K - \{0\}$.*

Then the full automorphism group of the flock F is G/EK^*.

Proof: The group of the flock must leave the cone invariant and permute the conics of the flock. The conics of the flock are the Baer subplanes of the base reguli. The automorphism group which fixes each Baer subplane of each base reguli may be shown to be EK^*. It follows that the automorphism group of the flock in $P\Gamma L(4, K)$ is exactly G/EK^*.

We refer to the group G of the translation plane corresponding to a flock above which normalizes the group E as the group inherited from the flock or the inherited group. If the plane is finite and not Desarguesian then Johnson [40] has shown

Theorem 5.5. *Let π be a finite non-Desarguesian translation plane that corresponds to a flock of a quadratic cone. Then the full collineation group of π is the inherited group.*

If the plane is infinite then Biliotti and Johnson [9] provide examples of non-Pappian planes corresponding to infinite flocks such that the inherited group is not the full collineation group. The planes of the flocks share one of two lines and are thus called "bilinear".

That the bilinear flocks or bilinear planes play the part of Desarguesian planes in the finite case is illustrated by the following result of Biliotti and Johnson [9].

Theorem 5.6. *Let K be an ordered field such that all positive elements are squares and F is a flock of a quadratic cone in $PG(3, K)$ and let π denote the associated translation plane. Furthermore, let the full collineation group of the flock be denoted by G_F and let the full collineation group of the associated translation plane be denoted by G_π.*

Then one of the following holds:

Corollary 5.7. (1) *the plane is Pappian,*

(2) *the plane is a bilinear flock plane or*

(3) *G_π/EK^* is isomorphic to G_F where E is the regulus inducing elation group induced from the regulus cover and K^* denotes the group of homologies induced from the nonzero scalars in K.*

6. Semi-translation planes derived from dual translation planes.

In this section, we shall be interested in planes obtained by the derivation of the duals of translation planes with spreads in $PG(3, K)$ for K certain skewfields. We note that although every dual translation plane of a plane with spread in $PG(3, q)$ must be derivable, it is possible to construct nonderivable dual translation planes but which contain derivable nets.

Definition 6.1. *Recall that a spread S in $PG(2k-1, K)$, for K a skewfield, and k a positive integer > 1 is a set of skew $(k-1)$-dimensional subspaces such that each point of $PG(2k-1, K)$ is incident with one of the subspaces of S.*

A dual spread S^* is a set of skew $(k-1)$-dimensional subspaces such that each hyperplane of $PG(2k-1, K)$ contains a line of S^*.

When K is finite any spread is a dual spread and conversely. However, for any infinite skewfield, there are spreads which are not dual spreads and dual spreads which are not spreads. (All of these examples are in $PG(3,K)$) [14](p.41, 4.1.6). Actually, the spreads mentioned in Cameron are generalizations of those given in Bruen and Fisher [12] for spreads over countable fields. Also, see Barlotti [5] and Bernardi [7]).

There is an intimate connection between spreads which are dual spreads and Baer subplanes.

Definition 6.2. *Let π be a projective plane and π_o a subplane of π.*

π_o shall be said to be a point-Baer subplane if and only if every point of the plane is incident with a line of π_o.

π_o shall be said to be a line-Baer subplane if and only if every line of the plane is incident with a point of π_o.

When the projective plane is finite then every point-Baer subplane is also a line-Baer subplane. However, in the infinite case, this is no longer the case as is shown by Barlotti [5]. Barlotti shows that there exist line-Baer subplanes which are not point Baer. Actually, Barlotti uses the Bruck-Bose model to prove this statement whereas the following treatment uses the André approach.

There are situations when one can always guarantee the line-Baer case.

Theorem 6.3. *Let S be a spread in $PG(2k-1,K)$. Let π_S denote the associated affine translation plane obtained by taking vectors in the $2k$- dimensional vector space over K as points and translates of the elements of S (taken as vector subspaces) as lines.*

Let π_o be a $k-$dimensional $K-$ vector subspace of π_S. Assume that k is even and assume that nontrivial intersections with the spread elements with π_o are always $k/2$-dimensional K subspaces of π_o.

(1) Then π_o is an affine plane which is a line-Baer subplane when considered projectively.

(2) When $k=2$ then a spread S is a dual spread if and only if any 2-dimensional K-subspace which is not a line of π_S is a Baer subplane of π_S.

Using this, Johnson proved the following (see Johnson [46]):

Theorem 6.4. *Let S be a spread of $PG(3,K)$ for K a skewfield. Let Σ be a particular plane. Let π_0 be a line of Σ but not a line of S and write $\pi_o = X_o \oplus Y_o$ where X_o and Y_o are 1-dimensional K subspaces.*

(1) Then π_o is a line-Baer subplane of the associated translation plane π_S.

Denote the parallel classes of π_S containing the lines containing X_o and Y_o respectively by (∞) and (0). Form the lines $(\infty)\alpha$ for all α in Y_o. Dualize the translation plane, sending (∞) to the line at infinity of $D(\pi_o^+)$ and the line at infinity of the translation plane to $(\infty)^*$.

(2) The affine net D with infinite points $(\infty)^*, (\infty)\alpha$ for all α in Y_o is a derivable net.

(3) If S is a dual spread then the affine dual translation plane containing D obtained by deleting (∞) is derivable by derivation of D.

(4) If S is not a dual spread and Σ is a plane which does not contain a line of S then the affine dual translation plane containing D obtained by deleting (∞) is not derivable by derivation of D.

Using the algebraic representation of a flock of a quadratic cone, it is possible to show that every corresponding spread is a dual spread. J.A. Thas pointed out to one of the authors that it is possible to use the Klein quadric to observe the same result. (See Johnson [46].)

Theorem 6.5. *Let S be a spread in $PG(3,K)$ for K a field. Then S is a dual spread in either of the following two situations:*

(1) *S is a union of reguli that share exactly one line.*

(2) *S is a union of reguli that mutually share two lines.*

Hence, we obtain

Theorem 6.6. *Let π be any translation plane with spread in $PG(3,K)$, for K an arbitrary field. Assume there exists an elation group E such that some component orbit union the axis is a regulus in $PG(3,K)$.*

Choose a component M distinct from the axis of E. Let L denote the axis of E of the parallel class (∞). Choose a 1-dimensional K subspace Y in M and let Y_∞ denote the set of lines parallel to L which contain a point of Y.

(1) The subplane defined by the points $< YE >$ is a Baer subplane of π which has Y_∞ as its lines on the parallel class (∞).

(2) The dual subplane of $< YE >$ defines a derivable net in the affine restriction by deletion of (∞) of the dual of the projective extension of π.

(3) There is an associated flock F of a quadratic cone in $PG(3,K)$. Let G denote the collineation group of π which is inherited from the flock. Then G_{Y_∞} induces a collineation group of the dual translation plane. Furthermore, $G_{Y_\infty}EK^*/EK^*$ is the full automorphism group of the flock F which fixes the line L_{Y_∞} corresponding to 1-dimensional K space $< YE > \cap L$.

Remark 7. *In the finite case, every spread is a dual spread. It follows from Johnson [39] that if the plane is not Desarguesian and of order > 16 then the full collineation group of the dual translation plane is G_{Y_∞}. And, we have mentioned previously that the full collineation group of the flock is G/EK^*.*

Hence, in the finite case, if the flock is not linear then the full collineation group of the dual translation plane is an extension of the stabilizer of a line of the cone in the automorphism group of the flock.

More generally, the isomorphism classes of the finite planes introduced in this section (each of which is derived from the point-line dual of a translation plane already associated with a flock of a cone) are in one-to-one correspondence with the orbits on the generators of the cone under the semi-linear group preserving the cone K and the given flock. Hence this increases the number of geometries related to a q-clan and also motivates the determination of the orbits on the generators of the cone under the appropriate group.

We have viewed the translation planes associated with flocks of quadratic cones above using an associated vector space. For those interested in the Bruck-Bose approach, we consider the equivalent representation within projective space. This construction is valid for spreads in $PG(3,K)$ which are also dual spreads and is equivalent to the construction given above.

Let $P = PG(4, K)$ and let Σ be a hyperplane of P. If S is a spread of Σ which is a dual spread, we will denote by $\pi = \pi(P, \Sigma, S)$ the translation plane associated with S, whose points and lines will be called, respectively, π-points and π-lines. The π-points are the points of $P \setminus \Sigma$ and the lines of S. The π-lines are the planes of P which intersect Σ at a line of S, and Σ. The incidence relation is inclusion.

Now let π^d denote the dual plane of π. Let Σ_1 be a hyperplane of P different from Σ. As $\Sigma \cap \Sigma_1$ is a plane of Σ, there is exactly one line L of S contained in the plane $\Sigma \cap \Sigma_1$. Following A. Barlotti and R. C. Bose [6] we can construct a projective plane $\Delta = \Delta(P, \Sigma, \Sigma_1, S)$ using P, Σ, Σ_1 and S. The points and the lines of Δ will be called, respectively, Δ-points and Δ-lines.

The Δ-points are of three types. The Δ-points of type (1) are the planes of P which intersect Σ at a line of S different from L. The Δ-points of type (2) are the planes of P which intersect Σ at L and are not contained in Σ_1. The Δ-points of type (3) are the points of L.

The Δ-lines are of three types. The Δ-lines of type (1) are the points of P which do not belong to Σ or to Σ_1. The Δ-lines of type (2) are the planes of Σ_1 which do not contain L. The line L is the unique Δ-line of type (3).

A Δ-line α of type (2) is incident with a Δ-point β of type (2) iff $\alpha \cap \beta$ is a line. In all other cases, a Δ-line and a Δ-point are incident iff they are incident in P.

Let $R(\Sigma_1)$ be the set of all π-lines α such that α is either a plane of Σ_1 containing L or $\alpha = \Sigma$. In π^d, $R(\Sigma_1)$ defines a derivable net and Δ is the derived plane of π^d obtained by replacing the derivable net $R(\Sigma_1)$.

To see how the two approaches connect, note that considering (6.4) and (6.6), notice that the infinite points of a derivable net in the dual translation plane may be defined by a component L and $1-$dimensional space Y_∞ union the line at infinity L_∞. The $3-$dimensional space $< L, Y_\infty >$ union L_∞ corresponds to $R(\Sigma_1)$.

In the finite case, there is some general theory concerning the collineation groups of derived dual translation planes. There has been essentially no accompanying theory in the infinite case. Hence, we now restrict ourselves to the finite case.

The following theorem on Δ-planes appears in N. L. Johnson [41].

Theorem 6.7. *For finite planes, the planes $\Delta(P, \Sigma, \Sigma_1, S)$ and $\Delta(P, \Sigma, \Sigma_2, S)$ are isomorphic iff there is a collineation σ of Σ such that*
(1)$\sigma : S \mapsto S$
(2) if $\alpha_i = \Sigma \cap \Sigma_i$ for $i = 1, 2$ then $\sigma : \alpha_1 \mapsto \alpha_2$.

Using the fact that an equivalence between q-clans corresponds in a precise way to a projective equivalence between their corresponding spreads, as well as Lemma 4.13 and Theorem 6.7, we obtain the following corollary:

Starting with a given q-clan $\mathcal{C}$, the number of pairwise distinct Δ-planes derived from the plane of the spread $\mathcal{S}(\mathcal{C})$ is the same as the number of orbits on the generators of the cone K under the group of semilinear collineations preserving the flock $\mathcal{F}(\mathcal{C})$.

The following result appears in Johnson [41] in the finite case. Also see Johnson and Lunardon [50]. It is an open question whether the analogous statement remains valid in the infinite case.

Theorem 6.8. *Let π be a translation plane in $PG(3,q)$ and correspond to a flock of a quadratic cone. Let $D(\pi)$ be an affine restriction of the dual of the projective extension of π and let N denote an associated derivable net in $D(\pi)$ constructed as above and let π_N denote the derived plane.*

If π is not a semifield plane then the projective extension of π_N is a plane which admits exactly one incident point-line transitivity. That is, π_N is of Lenz-Barlotti class II-1.

7. Baer Groups and Extension Theory.

7.1. Baer groups

It turns out that the structure of a Baer group in a translation plane provides a link to translation planes corresponding to flocks of quadratic cones. It is also possible to study point-Baer and/or line-Baer subplanes and groups which fix such subplanes pointwise. In particular, we are interested in the analysis of translation planes that contain point-Baer subplanes which are fixed pointwise by certain maximal point-Baer groups. Moreover, the reader is referred to the papers of Jha-Johnson([36], [34], [35]) which more generally concern point-Baer collineation groups, the generalized André theorem and infinite Baer nets..

In particular,

Theorem 7.1. *(Jha-Johnson [36]). Let π be a translation plane with spread in $PG(3,K)$ where K is a field. Let B denote a Baer collineation group in $\Gamma L(4,K)$ which acts regularly on the 1-dimensional K subspaces not in $FixB$ which lie on any component of $FixB$. (1) Then any nontrivial orbit of components of π union $FixB$ is regulus in $PG(3,K)$.*

(2) The reguli of part (1) correspond to a partial flock of deficiency one of a quadratic cone.

Conversely,

Theorem 7.2. *(Jha-Johnson [36]). Let P be a partial flock of deficiency one of a quadratic cone in $PG(3,K)$, for K a field.*

(1) Then there is an associated translation plane π_P with spread in $PG(3,K)$ which admits a Baer group B that acts regularly on the 1-dimensional K subspaces not in $FixB$ which lie on any component of $FixB$.

(2) The partial flock P may be extended uniquely to a flock if and only if the net defined by $FixB$ is a $K-$regulus net.

The idea of the proof is that corresponding to a partial flock of deficiency one is a translation net which is the union of a set of regulus nets sharing a component L. It is then shown that there is a corresponding elation group B with axis L which leaves each regulus net invariant. However, the natural setting is to consider L as a potential Baer subplane of a putative translation plane admitting B as a Baer group. The deficiency one assumption allows that indeed this is the case and conversely such translation planes admitting such Baer groups correspond to partial flocks of deficiency one.

In the finite case, Payne and Thas [73] have shown

Theorem 7.3. *A partial flock of deficiency one of a quadratic cone in $PG(3,q)$ may be uniquely extended to a flock.*

Remark 8. *(1) The proofs for q even or odd are quite different. The case, q odd depends ultimately on the associated partial BLT set and corresponding set of partial flocks in $PG(4,q)$. The even order case depends on properties of conics in even order Desarguesian planes.*

(2) In the infinite case, since there is not necessarily an associated BLT set, the corresponding analogous attempt for odd characteristic is not available. In general, there is nothing known about possible extensions of deficiency one partial flocks in the infinite case.

It might also be mentioned that there are connections with flocks of hyperbolic quadrics and translation planes admitting certain Baer groups which are considered by Jha-Johnson in [36]

In the next sections, we shall see that for q even, there is a much stronger extension result for partial flocks.

8. Herds of k-Arcs and Partial Flocks; q-Even

In this section we assume that $q = 2^e$. Let K be the usual cone $x_0x_2 = x_1^2$ with vertex $V = (0,0,0,1)$. A partial flock $\mathcal{F}$ of K is a set of disjoint (irreducible) conics on K. The partial flock $\mathcal{F}$ is complete if it is not contained in a larger partial flock. Suppose that the k planes $\pi_i = [x_i, y_i, z_i, 1]^T$, $0 \le i \le k-1$, define the k conics of $\mathcal{F}$. Then the k conics $C_i = \pi_i \cap K$ form a partial flock of K precisely because

$$tr\left(\frac{(x_i + x_j)(z_i + z_j)}{(y_i + y_j)^2}\right) = 1 \text{ for } 0 \le i \ne j \le k-1.$$

Without loss of generality we may assume that $\pi_0 : x_3 = 0$, i.e., $x_0 = y_0 = z_0 = 0$.

Theorem 8.1. *Let I be a subset of F containing k elements including 0. The k-planes $\pi_t = [x_t, y_t, z_t, 1]^T$, $t \in I$, with $\pi_0 = [0,0,0,1]^T$, define a partial flock $\mathcal{F}$ of the cone $K : x_0x_2 = x_1^2$ of $PG(3,q)$ if and only if for $(0,0) \ne (a_1, a_2) \in F^2$, the set*

$$K_{(a_1,a_2)} = \{(1, \sqrt{a_1^2x_t + a_1a_2y_t + a_2^2z_t}, t) : t \in I\} \cup \{(0,1,0), (0,0,1)\}$$

is a $(k+2)$-arc of $PG(2,q)$.

Such a set of $q^2 - 1$ $(k+2)$- arcs of $PG(2,q)$ is called a *herd* of arcs. In this way to each flock of K corresponds a herd of hyperovals of $PG(2,q)$, and conversely.

The existence of the herd of hyperovals starting with a flock is implicit in S. E. Payne [61], where it follows from the theory of generalized quadrangles. That a herd of hyperovals implies the existence of a flock first appears in W. Cherowitzo et al. [17]. Theorem 8.1 as given here first appears in L. Storme and J. A. Thas [75], where it is credited to S. E. Payne and T. Penttila by Private Communication.

As mentioned previously, S. E. Payne and J. A. Thas [73] proved that a partial flock of $q-1$ conics always is contained in a flock. By using the relationship between partial flocks and herds, L. Storme and J. A. Thas [75] were able to improve this considerably.

Theorem 8.2. *(L. Storme and J. A. Thas) Let $\mathcal{F}$ be a partial flock of K consisting of k conics in $PG(3,q)$. If*

(a) $k > q - \sqrt{q} - 1$ when q is a square, and

(b) $k > q - \sqrt{2q}$ when q is a nonsquare, $q > 2$,

then $\mathcal{F}$ can be extended uniquely to a flock.

9. Partial flocks containing linear subflocks

Any partial linear flock of an oval cone certainly may be extended to a linear flock. But, suppose one has a partial flock of an oval cone which is not linear but which has a reasonable large subflock which is linear. In the general case, it is not clear what might be meant by a large linear subflock. In the finite case for flocks of oval cones in $PG(3,q)$, we note the following result.

Remark 9. *Let P be any nonlinear partial flock of an oval cone in $PG(3,q)$. Then the maximal cardinality of a linear subflock is $(q-1)/2$ if q is odd or $q/2-1$ if q is even.*

Proof: Let F denote the linear flock containing the linear subflock of P all of whose plane contains the line L. Let π_o denote a plane of P which does not contain L. Then the oval of intersection in π_o can share at most two points from each of the planes of the partial flock $F-P$. Thus, at most $(q+1)/2$ for q odd or $q/2+1$ planes of $F-P$ are required to accomplish this.

Thus, we have the following question:

If a partial flock P of an oval cone in $PG(3,q)$ contains a linear subflock of cardinality $(q-1)/2$ or $q/2-1$ respectively as q is odd or even, can P be extended to a flock of an oval cone?

Actually, we shall see in a later chapter that when q is even then any partial flock of a quadratic cone which contains a linear subflock of cardinality $q/2-1$ has the property that all of the planes of the partial flock share a common point. When q is odd, Payne and Thas [73] proved

Theorem 9.1. *(Payne and Thas). There is a unique nonlinear flock of a quadratic cone in $PG(3,q)$ for q odd, which contains a linear subflock of $(q-1)/2$ conics.*

Now all of this may be considered for partial flocks which properly contain a suitable linear sub-partial flock.

We shall state this in the most general manner.

Definition 9.2. *Let M be a linear partial flock of a quadratic cone in $PG(3, K)$ for K an arbitrary field with the property that any linear partial flock which properly contains M may be extended uniquely to a linear flock. Then M shall be called a maximal linear partial flock. When K is finite and isomorphic to $GF(q)$ then any linear partial flock of $(q-1)/2$ conics for q odd or $(q/2-1)$ conics for q even is a maximal linear partial flock.*

Assume that P is a nonlinear partial flock of a quadratic cone in $PG(3, K)$ for K an arbitrary field which proper contains a maximal linear partial flock. Can P be extended to a (necessarily nonlinear) flock?

In Johnson [45], the following extension result is proved.

Theorem 9.3. *Let P be any nonlinear partial flock which properly contains a linear subflock of $(q-1)/2$ conics for q odd. Then P may be uniquely extended to a flock of a quadratic cone in $PG(3, q)$.*

9.1. Flock partial spreads in spreads.

A partial spread corresponding to a flock of a quadratic cone consists of a set of reguli in $PG(3, K)$ for K a field , that share a line. Suppose that such a partial spread is contained within a spread in $PG(3, K)$. Is the containing partial spread a flock spread? This may be considered a question on extension of sorts. In an article in this volume, Johnson and Storme [53] consider this question and show roughly speaking that if the original partial flock spread contains at least $3q/4$ conics and is embedded in a spread then the spread is, in fact, a flock spread and the original partial flock may be extended. More generally, Johnson and Storme consider such questions for other sets of reguli that correspond to flocks of elliptic and hyperbolic quadrics.

10. The Fundamental Theorem of q-Clan Geometry

Throughout this chapter $\mathcal{C} = \{A_t : t \in F\}$ will be a fixed q-clan normalized so that $A_0 = \begin{pmatrix} 0 & 0 \\ 0 & 0 \end{pmatrix}$ and each A_t is symmetric or upper triangular according as q is odd or even. In addition, we assume without loss of generality that when q is odd the members of $\mathcal{C}$ are indexed so that $x_t = t$ for all $t \in F$. And if $q = 2^e$, so that $t \mapsto y_t$ is a permutation, we assume that $y_t = t$ for all $t \in F$. $GQ(\mathcal{C})$ is the GQ of order (q^2, q) constructed from $\mathcal{C}$. Also P is always the matrix $P = \begin{pmatrix} 0 & 1 \\ -1 & 0 \end{pmatrix}$.

Moreover, $\mathcal{C}' = \{A'_t : t \in F\}$ will be a second q-clan normalized in the same way. Basically, the Fundamental Theorem (FT) shows how to interpret each equivalence between $\mathcal{C}$ and $\mathcal{C}'$ as an explicit isomorphism between the corresponding geometries of whatever kind: generalized quadrangles, conical flocks, or spreads of $PG(3, q)$.

Theorem 10.1. *For any prime power q, let $\mathcal{C} = \{A_t \equiv \begin{pmatrix} x_t & y_t \\ 0 & z_t \end{pmatrix} : t \in F\}$ and $\mathcal{C}' = \{A'_t \equiv \begin{pmatrix} x'_t & y'_t \\ 0 & z'_t \end{pmatrix} : t \in F\}$ be two (not necessarily distinct) normalized q-clans. Then the following are equivalent:*

(i) $\mathcal{C} \sim \mathcal{C}'$;

(ii) The flocks $\mathcal{F}(\mathcal{C})$ and $\mathcal{F}(\mathcal{C}')$ are projectively equivalent;

(iii) $GQ(\mathcal{C})$ and $GQ(\mathcal{C}')$ are isomorphic by an isomorphism mapping (∞) to (∞), $[A(\infty)]$ to $[A'(\infty)]$, and $(\bar{0},0,\bar{0})$ to $(\bar{0},0,\bar{0})$;

(iv) The associated spreads $\mathcal{S}(\mathcal{C})$ and $\mathcal{S}(\mathcal{C}')$ are equivalent.

Moreover, these four conditions hold iff there exist $\sigma \in Aut(F)$, $B \in GL(2,q)$, $0 \neq u \in F$, and a permutation $\pi : t \mapsto \bar{t}$ on F for which

(v) $A'_{\bar{t}} \equiv uB^{-1}A_t^{\sigma}B^{-T} + A'_{\bar{0}}$ for all $t \in F$.

The corresponding isomorphism $\theta : GQ(\mathcal{C}) \mapsto GQ(\mathcal{C}')$ is given by

$$(vi)\quad \theta = \theta(u,B,\sigma,\pi) : (\alpha, c, \beta) \mapsto (\alpha^{\sigma}B, uc^{\sigma} + \alpha^{\sigma}BA'_{\bar{0}}B^{T}(\alpha^{\sigma})^{T}, u\beta^{\sigma}B^{-T} + \alpha^{\sigma}BK'_{\bar{0}}).$$

And the corresponding isomorphism between the associated translation planes is induced by the following projective equivalence of the associated linespreads of $PG(3,q)$.

$$(vii)\quad T : (x_0,x_1,x_2,x_3) \mapsto (x_0,x_1,x_2,x_3)\begin{pmatrix} u^{-1}B & u^{-1}BA'_{\bar{0}} \\ O & B^{-T} \end{pmatrix} : \mathcal{S}(\mathcal{C}) \mapsto \mathcal{S}(\mathcal{C}').$$

If $B^{-1} = \begin{pmatrix} a & b \\ c & d \end{pmatrix}$, the corresponding isomorphism $T_{\theta} : \mathcal{F}(\mathcal{C}) \mapsto \mathcal{F}(\mathcal{C}')$ is defined for planes of the flocks by

$$(viii)\qquad T_{\theta} : \begin{bmatrix} x_t \\ y_t \\ z_t \\ 1 \end{bmatrix} \mapsto \begin{bmatrix} x'_{\bar{t}} \\ y'_{\bar{t}} \\ z'_{\bar{t}} \\ 1 \end{bmatrix} = \begin{pmatrix} ua^2 & uab & ub^2 & x'_{\bar{0}} \\ 2uac & u(ad+bc) & 2ubd & y'_{\bar{0}} \\ uc^2 & ucd & ud^2 & z'_{\bar{0}} \\ 0 & 0 & 0 & 1 \end{pmatrix} \begin{bmatrix} x_t^{\sigma} \\ y_t^{\sigma} \\ z_t^{\sigma} \\ 1 \end{bmatrix}.$$

10.1. Applications to $GQ(\mathcal{C})$

In this section we consider the FT under the additional hypothesis that $\mathcal{C} = \mathcal{C}'$, or more specifically that $A_t = A'_t$, $t \in F$. Then for the normalized q-clan $\mathcal{C}$, we define the following groups of collineations of $GQ(\mathcal{C})$, where $\mathcal{G}$ denotes the group of all collineations of $GQ(\mathcal{C})$ that fix the point ∞ .

$$\begin{array}{llll} (i) & \mathcal{G}_0 & = & \{\theta \in \mathcal{G} : (\bar{0}, 0, \bar{0})^\theta = (\bar{0}, 0, \bar{0})\}; \\ (ii) & \mathcal{H} & = & \{\theta \in \mathcal{G}_0 : [A(\infty)]^\theta = [A(\infty)\}; \\ (iii) & \mathcal{M} & = & \{\theta \in \mathcal{H} : [A(0)]^\theta = [A(0)]\}; \\ (iv) & \mathcal{N} & = & \{\theta = \theta_a \in \mathcal{M} : \theta_a : (\alpha, c, \beta) \mapsto (a\alpha, a^2 c, a\beta), 0 \neq a \in F\}. \end{array} \tag{10.1}$$

The following observation is a trivial consequence of the FT.

$\mathcal{M}$ always contains the following subgroup:

$\mathcal{M}^* = \{\theta = \theta(a^2, aI, \sigma, \pi : t \mapsto t^\sigma) : (\alpha, c, \beta) \mapsto (a\alpha^\sigma, a^2 c^\sigma, a\beta^\sigma) : 0 \neq a \in F, \sigma \in Aut(F)$ with $A_{t^\sigma} = A_t^\sigma$ for $t \in F\}$.

With the q-clans normalized as they are it is easy to establish the following.

$$\text{For } \theta(\sigma, B, \mu, \pi) \in \mathcal{H} \text{ with } B^{-1} = \begin{pmatrix} a & b \\ c & d \end{pmatrix}, \tag{10.2}$$

$$\begin{array}{ll} \text{(i)} & t^\pi = \mu(a^2 t^\sigma + aby_t^\sigma + b^2 z_t^\sigma) + 0^\pi, \text{ if } q \text{ is odd.} \\ \text{(ii)} & t^\pi = \mu(ad + bc)t^\sigma + 0^\pi, \text{ if } q \text{ is even.} \end{array}$$

Now let $\theta_i = \theta(\mu_i, B_i, \sigma_i, \pi_i)$, $i = 1, 2$. We may compute $\theta = \theta_1 \circ \theta_2$, where the notation means do θ_1 first.

$$\theta(\mu_1, B_1, \sigma_1, \pi_1) \circ \theta(\mu_2, B_2, \sigma_2, \pi_2) = \theta(\mu_1^{\sigma_2}\mu_2, B_1^{\sigma_2}B_2, \sigma_1 \circ \sigma_2, \pi) \tag{10.3}$$

Using the above results it is possible to prove the following.

Theorem 10.2. *$T : \theta \mapsto T_\theta$ is a homomorphism from $\mathcal{H}$ onto the subgroup of $P\Gamma L(4, q)$ leaving invariant the cone K and the flock $\mathcal{F}(\mathcal{C})$. The kernel of T is $\mathcal{N}$.* ■

We define $\mathcal{N}$ to be the *kernel* of $GQ(\mathcal{C})$ This definition of kernel may not be as satisfactory as that for translation generalized quadrangles (TGQ) studied in [72] but for many nonclassical q-clans $\mathcal{C}$ the kernel of $GQ(\mathcal{C})$ plays a role similar to one played by the multiplicative group of the kernel of a TGQ. Note that the kernel $\mathcal{N}$ is a group of collineations of $GQ(\mathcal{C})$ fixing each line through the point (∞) for every q-clan $\mathcal{C}$.

10.2. Recoordinatization: The Shift-Flip

Any automorphism of $\mathcal{K}$ replaces one 4-gonal family for $\mathcal{K}$ with another. But in general we cannot expect that a 4-gonal family $\mathcal{J}(\mathcal{C})$ arising from a q-clan $\mathcal{C}$ will be replaced with one for which a q-clan is clearly present. In this section we study three basic automorphisms of $\mathcal{K}$ that even replace normalized q-clans with normalized q-clans. One goal is to develop a procedure to use the FT to describe all elements of $\mathcal{G}_0$. A second goal is to assign to each line through (∞) in some $GQ(\mathcal{C})$ an equivalence class of flocks (or q-clans) in such a way that two lines through (∞) are in the same

$\mathcal{G}_0$-orbit if and only if their flocks (or q-clans) are equivalent. This provides a very concrete algebraic approach to the concept of derivation first presented in [4] when q is odd and includes the case for $q = 2^e$ first worked out in [2] for a representation of $\mathcal{K}$ that works only in characteristic 2. That representation will be considered in a later section where q is restricted to be a power of 2.

Throughout this section we assume that $\mathcal{C}$ is a normalized q-clan with the specific indexing given above. Our first observation is just a special corollary of the FT.

Observation 1. *Let $B \in GL(2,q)$ and $\pi : F \to F : t \mapsto \overline{t}$ a permutation. For each $A_t \in \mathcal{C}$ set $A'_{\overline{t}} = B^{-1}A_tB^{-T}$. Then $\mathcal{C}' = \{A'_{\overline{t}} : t \in F\}$ is a q-clan equivalent to $\mathcal{C}$, and $\theta : GQ(\mathcal{C}) \to GQ(\mathcal{C}') : (\alpha, c, \beta) \mapsto (\alpha B, c, \beta B^{-T})$ is an isomorphism.* ■

Lemma 10.3. *Shift by s.* *Fix $s \in F$. Define $\tau_s : \mathcal{K} \to \mathcal{K} : (\alpha, c, \beta) \mapsto (\alpha, c - \alpha A_s \alpha^T, \beta - \alpha K_s)$. For $A_t \in \mathcal{C}$, put $A^{\tau_s}_{\overline{t}} = A_t - A_s$, so $A^{\tau_s}_x = A_{x+s} - A_s$. Then $\mathcal{C}^{\tau_s} = \{A^{\tau_s}_t : t \in F\}$ is a normalized q-clan equivalent to $\mathcal{C}$, and $\tau_s = \theta(1, I, id, \pi : t \mapsto t - s) : GQ(\mathcal{C}) \to GQ(\mathcal{C}^{\tau_s})$ is an isomorphism.*

Proof: This is immediate from the FT. For q odd, $x_t = t$, so $x^{\tau_s}_t = (t+s) - s = t$. For q even, $y_t = t$, so $y^{\tau_s}_t = (t+s) - s = t$. Hence $\mathcal{C}^{\tau_s}$ is normalized and $\overline{0} = -s$. ■

Lemma 10.4. *Scale by a.* *Let $0 \neq a \in F$. Define $\sigma_a : \mathcal{K} \to \mathcal{K} : (\alpha, c, \beta) \mapsto (\alpha, ac, a\beta)$. For $A_t \in \mathcal{C}$, put $A^{\sigma_a}_{\overline{t}} = aA_t$, so $A^{\sigma_a}_{at} = aA_t$, or $A^{\sigma_a}_t = aA_{t/a}$. Then $\mathcal{C}^{\sigma_a} = \{A^{\sigma_a}_t : t \in F\}$ is a normalized q-clan equivalent to $\mathcal{C}$, and $\sigma_a = \theta(a, I, id, \pi : t \mapsto at) : GQ(\mathcal{C}) \to GQ(\mathcal{C}^{\sigma_a})$ is an isomorphism.* ■

Before considering the third type of automorphism of $\mathcal{K}$, we collect some rather trivial results that are helpful in doing computations.

Lemma 10.5. *Put $P = \begin{pmatrix} 0 & 1 \\ -1 & 0 \end{pmatrix}$, and for $B \in GL(2,q)$, put $\Delta = \det(B)$.*

(i) $P^T = P^{-1} = -P$.

(ii) $P^TBP = \Delta B^{-T} = PBP^T \equiv \Delta B^{-1}$.

(iii) $B^TPB = \Delta P$; $B^TP^TB = \Delta P^T$.

(iv) *If q is odd and A is nonsingular and symmetric with $K = A + A^T = 2A$, then $P^TK^{-1}AKP = (\det(K))^{-1}A$.*

(v) *If $q = 2^e$ and A is upper triangular with $K = A + A^T = (\det(K))^{1/2}P$ nonsingular, then $P^TK^{-1}AKP \equiv \det(K))^{-1}A$.*

Lemma 10.6. *The flip.* *Define $\varphi : \mathcal{K} \to \mathcal{K} : (\alpha, c, \beta) \mapsto (\beta P, c - \alpha\beta^T, \alpha P^T)$. Then φ is an automorphism of $\mathcal{K}$ that interchanges $A(0)$ and $A(\infty)$. It maps a normalized q-clan $\mathcal{C} = \{A_t : t \in F\}$ to a q-clan $\mathcal{C}^\varphi = \{A^\varphi_{\overline{t}} \equiv P^TK_t^{-1}A_tK_t^{-1}P \equiv$-$(\det(K_t))^{-1}A_t : t \in F\}$. The permutation $\pi : t \mapsto \overline{t}$ is chosen so that $\mathcal{C}^\varphi$ will have a normalized indexing. If q is even with $y_t = t$, clearly $\pi : t \mapsto \overline{t} = t^{-1}$ for all $t \in \tilde{F}$. For q odd with $x_t = t$, the permutation $\pi : t \mapsto \overline{t}$ depends on the functions $t \mapsto y_t$ and $t \mapsto z_t$. Explicitly: $\pi : t \mapsto \overline{t} = t/(4x_tz_t - y_t^2)$.*

Proof: It is routine to check that φ is an automorphism of $\mathcal{K}$ interchanging $A(0)$ and $A(\infty)$. So $A^{\varphi}(0) = A(\infty)$, $A^{\varphi}(\infty) = A(0)$. And

$$\begin{aligned}\varphi : (\alpha, \alpha A_t\alpha^T, \alpha K_t) &\mapsto (\alpha K_t P, \alpha A_t\alpha^T - \alpha K_t\alpha^T, \alpha P^T)\\ &= (\alpha K_t P, \alpha K_t P(-P^T K_t^{-1} A_t K_t^{-1} P)(\alpha K_t P)^T, \alpha K_t P(P^T K_t^{-1} P^T))\\ &\doteq (\gamma, \gamma(-(\det(K_t))^{-1} A_t)\gamma^T, -(\det(K_t))^{-1}\gamma K_t),\\ &\quad \text{for } \gamma = \alpha K_t P,\ 0 \neq t \in F. \blacksquare\end{aligned}$$

Shifting by s and scaling by a are automorphisms of $\mathcal{K}$ that map normalized q-clans to equivalent normalized q-clans, and affect on $\tilde{F}$ (the index set for the lines through (∞)) the permutations $t \mapsto \overline{t} = t - s$ and $t \mapsto \overline{t} = at$, respectively. So the group generated by these permutations stabilizes ∞. On the other hand, the flip φ affects a permutation π on $\tilde{F}$ that interchanges 0 and ∞, and for $q = 2^e$ $\pi : t \mapsto t^{-1}$ for all $t \in \tilde{F}$. In general for q odd, π is determined by $t \mapsto y_t$ and $t \mapsto z_t$, and $\mathcal{C}$ and $\mathcal{C}^{\varphi}$ are not equivalent!

For $s \in F$ define the *shift-flip* i_s by

$$\begin{aligned}&\text{(i)} && i_s = \tau_s \circ \varphi,\ s \in F. && (10.4)\\ &\text{(ii)} && i_s : (\alpha, c, \beta) \mapsto ((\beta - \alpha K_s)P, c - \alpha\beta^T + \alpha A_s\alpha^T, \alpha P^T)\\ &\text{(iii)} && i_s^{-1} : (\alpha, c, \beta) \mapsto (\beta P, c - \alpha\beta^T + (\beta P)A_s(\beta P)^T, \alpha P^T + \beta P K_s).\end{aligned}$$

Compute the effect of i_s on an element of $A(t)$. $i_s : (\alpha, \alpha A_t\alpha^T, \alpha K_t) \mapsto (\alpha(K_t - K_s)P, -\alpha(A_t - A_s)^T, \alpha P^T)$. Put $\gamma = \alpha(K_t - K_s)P$, so for $t \neq s$, $\alpha = \gamma P(K_t - K_s)^{-1}$. Then the image is
$(\gamma, \gamma(-P^T(K_t - K_s)^{-1}(A_t - A_s)(K_t - K_s)^{-1}P)\gamma^T, -\gamma P^T(K_t - K_s)^{-1}P)$
$=$
$(\gamma, \gamma(-(\det(K_t - K_s))^{-1}(A_t - A_s))\gamma^T, \gamma(-(\det(K_t - K_s))^{-1})(K_t - K_s))$. So we note that

$$\begin{aligned}&\text{(i)} && i_s : \mathcal{C} \to \mathcal{C}^{i_s} : A_t \mapsto A_{\overline{t}}^{i_s} = && (10.5)\\ &&& -(\det(K_t - K_s))^{-1}(A_t - A_s),\\ &&& \overline{t} = -(t - s)(\det(K_t - K_s))^{-1}.\\ &\text{(ii)} && \text{If } q = 2^e,\ \overline{t} = (t + s)^{-1}.\end{aligned}$$

Also define $i_\infty = id : \mathcal{K} \to \mathcal{K}$. Then to each line of $GQ(\mathcal{C})$ through (∞) we are going to assign an equivalence class of q-clans and its associated class of flocks. Start with a normalized q-clan $\mathcal{C}$. For each $s \in \tilde{F}$, applying i_s to $\mathcal{K}$ yields a normalized q-clan $\mathcal{C}^{i_s}$. We *assign to the line* $[A(s)]$ *of* $GQ(\mathcal{C})$ *the class of* q*-clans equivalent to* $\mathcal{C}^{i_s}$, and also *the class of flocks projectively equivalent to* $\mathcal{F}(\mathcal{C}^{i_s})$. The following application of the FT makes this assignment natural and useful.

Theorem 10.7. *Let* $\mathcal{C}$ *be a normalized* q*-clan. Then for arbitrary* $s, t \in \tilde{F}$, *there is a collineation (an automorphism) of* $GQ(\mathcal{C})$ *mapping* $[A(s)]$ *to* $[A(t)]$ *if and only if* $\mathcal{F}(\mathcal{C}^{i_s})$ *and* $\mathcal{F}(\mathcal{C}^{i_t})$ *are projectively equivalent (i.e., if and only if* $\mathcal{C}^{i_s} \sim \mathcal{C}^{i_t}$*).*

Proof: Since τ_s always effects the permutation $t \mapsto t - s$ and φ always interchanges 0 and ∞, the shift-flip i_s always maps s to ∞, i.e., $i_s : GQ(\mathcal{C}) \to GQ(\mathcal{C}^{i_s}) : [A(s)] \mapsto [A^{i_s}(\infty)]$. Let θ be an automorphism of $GQ(\mathcal{C})$ mapping $[A(s)]$ to $[A(t)]$. Without loss of generality we may assume that θ fixes $(\overline{0}, 0, \overline{0})$ (and of course (∞)). Then $\overline{\theta} = i_s^{-1} \circ \theta \circ i_t : GQ(\mathcal{C}^{i_s}) \to GQ(\mathcal{C}^{i_t}) : [A^{i_s}(\infty)] \mapsto [A^{i_t}(\infty)]$. Since clearly $\overline{\theta} : (\infty) \mapsto (\infty)$ and $(\overline{0}, 0, \overline{0}) \mapsto (\overline{0}, 0, \overline{0})$, by the FT. $\mathcal{F}(\mathcal{C}^{i_s})$ and $\mathcal{F}(\mathcal{C}^{i_t})$ are projectively equivalent. Conversely, given that $\mathcal{F}(\mathcal{C}^{i_s})$ and $\mathcal{F}(\mathcal{C}^{i_t})$ are projectively equivalent, there is an appropriate $\overline{\theta} : GQ(\mathcal{C}^{i_s}) \to GQ(\mathcal{C}^{i_t}) : [A^{i_s}(\infty)] \mapsto [A^{i_t}(\infty)]$. Hence we may put $\theta = i_s \circ \overline{\theta} \circ i_t^{-1} : GQ(\mathcal{C}) \to GQ(\mathcal{C}) : [A(s)] \mapsto [A(t)]$. ■

10.3. A Tensor Product Notation in Case $q = 2^e$

For q an arbitrary prime power, let $F = GF(q)$ and put $P = \begin{pmatrix} 0 & 1 \\ -1 & 0 \end{pmatrix}$. The standard group $\mathcal{K}$ used for the q–clan geometry (at least for the generalized quadrangles) is

$$\mathcal{K} = \{(\alpha, c, \beta) \in F^2 \times F \times F^2 : \alpha, \beta \in F^2, c \in F\}, \tag{10.6}$$

with binary operation

$$(\alpha, c, \beta) \cdot (\alpha', c', \beta') = (\alpha + \alpha', c + c' + \beta(\alpha')^T, \beta + \beta'). \tag{10.7}$$

The map $\psi^{\otimes} : (\alpha, c, \beta) \mapsto ((\alpha, \beta P), c) \in F^2 \times F^2 \times F$ is an isomorphism from $\mathcal{K}$ to a group $G^{\otimes}$ given by

$$G^{\otimes} = \{((\alpha, \beta), c) \in F^2 \times F^2 \times F : \alpha, \beta \in F^2, c \in F\}, \tag{10.8}$$

with binary operation

$$((\alpha, \beta), c) \cdot ((\alpha', \beta'), c') = ((\alpha + \alpha', \beta + \beta'), c + c' + \beta \circ \alpha'), \tag{10.9}$$

where

$$(\alpha, \beta) \mapsto \alpha \circ \beta = -\alpha P \beta^T \tag{10.10}$$

is a nonsingular, alternating bilinear form. In particular, $\alpha \circ \beta = 0$ iff $\{\alpha, \beta\}$ is F-dependent

In the published literature on q-clan geometry with $q = 2^e$ there is a third representation of the underlying group. In this representation, which here we denote by G, the typical element is denoted (α, c, β) and the binary operation is $(\alpha, c, \beta) \circ (\alpha', c', \beta') = (\alpha + \alpha', c + c' + \sqrt{\beta \circ \alpha'}, \beta + \beta')$. However, in this article we prefer to write $((\alpha, \beta), c)$ in place of (α, c, β).
Then

$$\psi : G^{\otimes} \to G : ((\alpha, \beta), c) \mapsto ((\alpha, \beta), c^{\frac{1}{2}}) \tag{10.11}$$

is an isomorphism from $G^{\otimes}$ to the group G with binary operation

$$((\alpha, \beta), c) \circ ((\alpha', \beta'), c') = ((\alpha + \alpha', \beta + \beta'), c + c' + \sqrt{\beta \circ \alpha'}). \tag{10.12}$$

This latter representation only works when $q = 2^e$.

For the remainder of this section, we assume that $q = 2^e$.

There are two families of subgroups of $G^{\otimes}$, each subgroup having order q^3 and being elementary abelian, that play important roles in q-clan geometry.

$$\text{For } \underline{0} \neq \gamma \in F^2, \mathcal{L}_\gamma = \{(\gamma \otimes \alpha, c) \in G^{\otimes} : \alpha \in F^2, c \in F\}, \tag{10.13}$$

$$\text{For } \underline{0} \neq \alpha \in F^2, \mathcal{R}_\alpha = \{(\gamma \otimes \alpha, c) \in G^{\otimes} : \gamma \in F^2, c \in F\}. \tag{10.14}$$

For nonzero α, $\gamma \in F^2$, $\mathcal{L}_\gamma = \mathcal{L}_\alpha$ (resp., $\mathcal{R}_\gamma = \mathcal{R}_\alpha$) iff $\{\alpha, \gamma\}$ is F-dependent. Hence we may think of the $\mathcal{L}_\gamma$ (resp., $\mathcal{R}_\alpha$) as indexed by the points of $PG(1,q)$. The scalar multiplication $d((\gamma \otimes \alpha), c) = (d\gamma \otimes \alpha, d^2c) = (\gamma \otimes d\alpha, d^2c)$ makes each $\mathcal{L}_\gamma$ and $\mathcal{R}_\alpha$ into a 3-dimensional vector space over F. So there are associated projective planes $\bar{L}_\gamma \cong \bar{R}_\alpha \cong PG(2,q)$. Clearly the center $Z = \{((\underline{0},\underline{0}), c) \in G^{\otimes} : c \in F\}$ of $G^{\otimes}$ is contained in each L_γ and R_α.

Let $\sigma \in \mathrm{Aut}(F)$, $0 \neq x \in F$, and let H and S be 4×4 matrices over F. Define $\theta(\sigma, x, H, S) : G^{\otimes} \to G^{\otimes}$ by

$$\theta(\sigma, x, H, S) : ((\alpha, \beta), c) \mapsto ((\alpha^\sigma, \beta^\sigma)H, xc^\sigma + (\alpha^\sigma, \beta^\sigma)S(\alpha^\sigma, \beta^\sigma)^T). \tag{10.15}$$

A routine computation shows that

$$\begin{aligned} &\theta(\sigma_1, x_1, H_1, S_1) \cdot \theta(\sigma_2, x_2, H_2, S_2) = \\ &\theta(\sigma_1 \cdot \sigma_2, x_1^{\sigma_2}x_2, H_1^{\sigma_2}H_2, x_2 S_1^{\sigma_2} + H_1^{\sigma_2}S_2(H_1^{\sigma_2})^T). \end{aligned} \tag{10.16}$$

For a given $\theta = \theta(\sigma, x, H, S)$, we may write H and S in block form, with 2×2 blocks. Without loss of generality, we may assume that S is in block-upper triangular form. Say $H = \begin{pmatrix} A & K \\ B & M \end{pmatrix}$, and $S = \begin{pmatrix} C & D \\ 0 & E \end{pmatrix}$. Recall that $\begin{pmatrix} a & b \\ c & d \end{pmatrix} \equiv \begin{pmatrix} x & y \\ z & w \end{pmatrix}$ means $a = x, d = w, b + c = y + z$. Clearly, C (resp., E) may be replaced with any matrix $C' \equiv C$ (resp., $E' \equiv E$), so we usually take C and E to be upper triangular. Then a computation similar to that used in the proof of 10.5.2 of Gevaert and Johnson [24] shows that θ is an automorphism of $G^{\otimes}$ if and only if H is nonsingular and

$$C + C^T = KPA^T \tag{10.17}$$

$$D = KPB^T \tag{10.18}$$

$$D = APM^T + xP \tag{10.19}$$

$$E + E^T = MPB^T \tag{10.20}$$

For each automorphism of $G^{\otimes}$ of interest here there will be matrices $\bar{A} = \begin{pmatrix} d & b \\ c & a \end{pmatrix}$ and $\bar{B}$, both in $GL(2,q)$, for which $A = d\bar{B}$, $B = c\bar{B}$, $K = b\bar{B}$, $M = a\bar{B}$, i.e., $H = \bar{A} \otimes \bar{B}$. Put $\mu = \det(\bar{A})$, $\Delta = \det(\bar{B})$. Then we may rewrite Eq. 10.15 as

$$\begin{gathered} \theta = \theta(\sigma, x, \bar{A} \otimes \bar{B}, \begin{pmatrix} C & D \\ O & E \end{pmatrix}) : ((\alpha, \beta), c) \mapsto \\ ((\alpha^\sigma, \beta^\sigma)(\bar{A} \otimes \bar{B}), xc^\sigma + \alpha^\sigma C(\alpha^\sigma)^T + \alpha^\sigma D(\beta^\sigma)^T + \beta^\sigma E(\beta^\sigma)^T). \end{gathered} \tag{10.21}$$

And the conditions of Eqs. 10.17 through 10.20 are equivalent to the following.

$$C + C^T = \Delta bdP \tag{10.22}$$

$$D = \Delta bcP \tag{10.23}$$

$$x = \mu\Delta \tag{10.24}$$

$$E + E^T = \Delta acP. \tag{10.25}$$

This means that the symbols x and D may be suppressed in the notation for θ. So temporarily we write

$$\begin{gathered} \theta = \theta(\sigma, \bar{A} \otimes \bar{B}, C, E) : ((\alpha, \beta), c) \mapsto \\ ((\alpha^\sigma, \beta^\sigma)(\bar{A} \otimes \bar{B}),\ \mu\Delta c^\sigma + \alpha^\sigma C(\alpha^\sigma)^T + \Delta bc(\alpha^\sigma \circ \beta^\sigma) + \beta^\sigma E(\beta^\sigma)^T), \end{gathered} \tag{10.26}$$

where $\mu = \det(\bar{A})$, $\Delta = \det(\bar{B})$, $C + C^T = \Delta bdP$ and $E + E^T = \Delta acP$.

NOTE:

$$\theta(\sigma, \bar{A} \otimes \bar{B}, C, E) : \mathcal{L}_\gamma \to \mathcal{L}_{\gamma^\sigma \bar{A}}, \tag{10.27}$$

$$\theta(\sigma, \bar{A} \otimes \bar{B}, C, E) : \mathcal{R}_\alpha \to \mathcal{R}_{\alpha^\sigma \bar{B}}. \tag{10.28}$$

Identify elements of $\tilde{F} = F \cup \{\infty\}$ with points of $PG(1,q)$ in the following way. For $t \in F$, write $\gamma_t = (1,t)$, and $\gamma_\infty = (0,1)$. So for $\underline{0} \neq \gamma = (a,b)$, $\gamma \equiv \gamma_{b/a}$. For $t \in \tilde{F}$, we then write $\mathcal{L}_t = \mathcal{L}_{\gamma_t}$ and $\mathcal{R}_t = \mathcal{R}_{\gamma_t}$. If $\bar{A} = \begin{pmatrix} a_4 & a_2 \\ a_3 & a_1 \end{pmatrix}$ and $\bar{B} = \begin{pmatrix} b_4 & b_2 \\ b_3 & b_1 \end{pmatrix}$, then we have

$$\theta(\sigma, \bar{A} \otimes \bar{B}, C, E) : \mathcal{L}_t \to \mathcal{L}_{\bar{t}}, \text{where } \bar{t} = \frac{a_1 t^\sigma + a_2}{a_3 t^\sigma + a_4},\ t \in \tilde{F}. \tag{10.29}$$

$$\theta(\sigma, \bar{A} \otimes \bar{B}, C, E) : \mathcal{R}_t \to \mathcal{R}_{\bar{t}}, \text{where } \bar{t} = \frac{b_1 t^\sigma + b_2}{b_3 t^\sigma + b_4},\ t \in \tilde{F}. \tag{10.30}$$

A q-**clan** $\mathcal{C}$ is said to be $\tau-$**normalized** for some $\tau \in Aut(F)$ provided $A_t = \begin{pmatrix} x_t & t^\tau \\ 0 & z_t \end{pmatrix}$ for all $t \in F$, and $A_0 = \begin{pmatrix} 0 & 0 \\ 0 & 0 \end{pmatrix}$. If $\mathcal{C}$ is $\tau-$normalized for $\tau = id$, we usually just say that $\mathcal{C}$ is normalized. Also put $A_\infty = \begin{pmatrix} 0 & 0 \\ 0 & 0 \end{pmatrix}$, so that strictly speaking $y_\infty = 0$. But, put $\gamma_{y_\infty} = (0,1)$.

When the q-clan $\mathcal{C} = \{A_t \in F\}$ is understood to be given, we write $g_t(\alpha) = \alpha A_t \alpha^T$, for $t \in \tilde{F}$, $\alpha \in F^2$. Then the following is easily checked.

$$g_t(\alpha+\beta) \quad = \quad g_t(\alpha) + g_t(\beta) + y_t(\alpha \circ \beta). \tag{10.31}$$

If $\mathcal{C}' = \{A'_t : t \in F\}$ is a second (not necessarily distinct) q-clan, we write $g'_t(\alpha) = \alpha A'_t \alpha^T$.

Let $\mathcal{C} = \{A_t = \begin{pmatrix} x_t & y_t \\ 0 & z_t \end{pmatrix} : t \in F\}$ be a given q-clan. Then we may define subgroups of $G^{\otimes}$ as follows.

$$\text{For } t \in \tilde{F},\ A(t) = \{(\gamma_{y_t} \otimes \alpha, g_t(\alpha)) \in G^{\otimes} : \alpha \in F^2\}. \tag{10.32}$$

It follows that $A(t)$ is a subgroup of $\mathcal{L}_{y_t}$ having order q^2. And as usual, $\mathcal{J}(\mathcal{C}) = \{A(t) : t \in \tilde{F}\}$ is a 4-gonal family for $G^{\otimes}$ with $\mathcal{L}_{y_t}$ being the **tangent space** of $\mathcal{J}(\mathcal{C})$ at $A(t)$. (See [2] for details using the isomorphic group G with only superficially different notation.) The notation here has been chosen so that if $\mathcal{C}$ is normalized (with $\tau = id$), then for $\gamma = (a,b) \in PG(1,q)$,

$$A(\gamma) = A(b/a) = \{(\gamma \otimes \alpha, \alpha A_{b/a}\alpha^T) : \alpha \in F^2\}. \tag{10.33}$$

For some $\tau \in \mathrm{Aut}(F)$, let $\mathcal{C} = \{A_t : t \in \tilde{F}\}$ and $\mathcal{C}' = \{A'_t : t \in \tilde{F}\}$ be two (not necessarily distinct) τ^{-1}-normalized q-clans. Let $\theta = \theta(\sigma, \bar{A} \otimes \bar{B}, C, E)$ be an automorphism of $G^{\otimes}$ as in Eq. 10.26. We want to give **necessary** conditions on C and E for θ to map the members of $\mathcal{J}(\mathcal{C})$ to the members of $\mathcal{J}(\mathcal{C}')$. These conditions are worked out in S. E. Payne [67] to be the following:

$$C \equiv a_4^2 \bar{B} A'_{(a_2/a_4)^\tau} \bar{B}^T. \tag{10.34}$$

and

$$E \equiv a_3^2 \bar{B} A'_{(a_1/a_3)^\tau} \bar{B}^T. \tag{10.35}$$

From the above we obtain the following theorem (where we use $\bar{B}P\bar{B}^T = \Delta P$).

Theorem 10.8. *Let $\sigma \in AutF$, $\overline{A} = \begin{pmatrix} a_4 & a_2 \\ a_3 & a_1 \end{pmatrix} \in GL(2,q)$, $\overline{B} \in GL(2,q)$, $\mu = det(\overline{B})$. Suppose that $\theta = \theta(\sigma, \overline{A} \otimes \overline{B}, C, E)$ as in eq. 2.26 is an automorphism of $G^{\otimes}$ mapping the 4−gonal family $\mathcal{J}(\mathcal{C})$ to the 4−gonal family $\mathcal{J}(\mathcal{C}')$, where both $\mathcal{C}$ and $\mathcal{C}'$ are τ^{-1}−normalized. Then the following hold:*

1. $\begin{pmatrix} C & D \\ 0 & E \end{pmatrix} = (I \otimes \overline{B}) \begin{pmatrix} a_4^2 A'_{(a_2/a_4)^\tau} & a_2 a_3 P \\ 0 & a_3^2 A'_{(a_1/a_3)^\tau} \end{pmatrix} (I \otimes \overline{B})^T.$

1. 2. *$\theta(\sigma, \bar{A} \otimes \bar{B}) : ((\alpha,\beta), c) \mapsto ((\alpha^\sigma, \beta^\sigma)(\bar{A} \otimes \bar{B}),\ \mu\Delta c^\sigma + a_4^2 \alpha^\sigma (\bar{B} A'_{(a_2/a_4)^\tau} \bar{B}^T)(\alpha^\sigma)^T + \Delta a_2 a_3 (\alpha^\sigma \circ \beta^\sigma) + a_3^2 \beta^\sigma (\bar{B} A'_{(a_1/a_3)^\tau} \bar{B}^T)(\beta^\sigma)^T)$. And the permutation $A(t) \mapsto A'(t')$ is given by*

3. $t' = \dfrac{a_1^\tau t^\sigma + a_2^\tau}{a_3^\tau t^\sigma + a_4^\tau}.$

10.4. The Fundamental Theorem for q Even

Let $\mathcal{C} = \{A_t = \begin{pmatrix} x_t & y_t \\ 0 & z_t \end{pmatrix} : t \in F\}$ and $\mathcal{C}' = \{A'_t = \begin{pmatrix} x'_t & y'_t \\ 0 & z'_t \end{pmatrix} : t \in F\}$ be two q-clans with $A_o = A'_o = \begin{pmatrix} 0 & 0 \\ 0 & 0 \end{pmatrix}$. Let $GQ(\mathcal{C})$ and $GQ(\mathcal{C}')$ be their associated GQ of order (q^2, q) (see [2] and [61]). A full statement of the Fundamental Theorem includes the assertion that if $\theta : GQ(\mathcal{C}) \mapsto GQ(\mathcal{C}')$ is an isomorphism mapping $(\infty), (\underline{0}, 0, \underline{0}) \in GQ(\mathcal{C})$, to $(\infty), (\underline{0}, 0, \underline{0}) \in GQ(\mathcal{C}')$, respectively, then θ is induced by an automorphism (also called θ) of $G^{\otimes}$ mapping the 4−gonal family $\mathcal{J}(\mathcal{C})$ to $\mathcal{J}(\mathcal{C}')$. For out purposes here it suffices to give that part of the theorem that characterizes such automorphisms of $G^{\otimes}$ mapping $A(\infty)$ to $A'(\infty)$.

Theorem 10.9. *The Fundamental Theorem (Partial Version!) There is an automorphism $\theta : G^{\otimes} \to G^{\otimes}$ mapping $\mathcal{J}(\mathcal{C})$ to $\mathcal{J}(\mathcal{C}')$, and mapping $A(\infty)$ to $A'(\infty)$, for each 4-tuple $(\lambda, B, \sigma, \pi)$, $0 \neq \lambda \in F$, $B \in GL(2,q)$, $\sigma \in Aut\ (F)$, $\pi : F \to F : t \mapsto \bar{t}$ a permutation, for which*

(i) $A'_{\bar{t}} \equiv \lambda B^{-1} A^{\sigma}_t B^{-T} + A'_{\bar{0}}$ for all $t \in F$.

The corresponding automorphism $\theta : G^{\otimes} \to G^{\otimes}$ is given by (put $\Delta = \det(B)$):

(ii) $\theta = \theta(\sigma, \begin{pmatrix} 1 & y'_{\bar{0}} \\ 0 & \lambda/\Delta \end{pmatrix} \otimes B) : ((\alpha,\beta),c) \mapsto ((\alpha^{\sigma}, \beta^{\sigma})[\begin{pmatrix} 1 & y'_{\bar{0}} \\ 0 & \lambda/\Delta \end{pmatrix} \otimes B], \lambda c^{\sigma} + g'_{\bar{0}}(\alpha^{\sigma} B))$.

(iii) $\pi : t \mapsto \bar{t}$ satisfies $y'_{\bar{t}} = (\lambda/\Delta) y^{\sigma}_t + y'_{\bar{0}}$.

Conversely, all such automorphisms of G must be of this form.

For $0 \neq a \in F$, $\theta_a = \theta(id, I \otimes aI) : ((\alpha,\beta),c) \mapsto ((a\alpha, a\beta), a^2 c)$ is an automorphism of $G^{\otimes}$ that leaves invariant each $A(t)$ for each $t \in F$ and for each q-clan $\mathcal{C}$.

$\mathcal{N} = \{\theta_a : 0 \neq a \in F\}$ is the q-**clan kernel**, or just the **kernel** (cf. II.3 and II.4 of [70].

For any θ as in Theorem 10.9 , Eq. 10.26 has a particularly simple form.

$$\theta(\sigma_1, A_1 \otimes B_1) \cdot \theta(\sigma_2, A_2 \otimes B_2) = \theta(\sigma_1 \cdot \sigma_2, A_1^{\sigma_2} A_2 \otimes B_1^{\sigma_2} B_2). \tag{10.36}$$

So $\theta(\sigma, A \otimes B) \cdot \theta(id, I \otimes aI) = \theta(\sigma, A \otimes aB)$, and $\theta(id, I \otimes aI) \cdot \theta(\sigma, A \otimes B) = \theta(\sigma, A \otimes a^{\sigma} B)$. Hence $\theta(\sigma, A \otimes B)$ commutes with $\theta(id, I \otimes aI)$ for all nonzero $a \in F$ iff $\sigma = id$.

We say that $\theta = \theta(\sigma, A \otimes B)$ as in Theorem 10.9 is **linear** provided $\sigma = id$, and is **special** provided $\Delta = \det(B) = 1$. Clearly the product of two linear (resp. special) automorphisms θ is again linear (resp., special).

For any θ as in Theorem 10.9,

$$\theta(\sigma, A \otimes B) = \theta(\sigma, I \otimes I) \cdot \theta(id, A \otimes \Delta^{-\frac{1}{2}} B) \cdot \theta(id, I \otimes \Delta^{\frac{1}{2}} I), \tag{10.37}$$

where $\theta_{\sigma} = \theta(\sigma, I \otimes I) : ((\alpha,\beta),c) \mapsto ((\alpha^{\sigma}, \beta^{\sigma}), c^{\sigma})$, $\theta(id, A \otimes \Delta^{-\frac{1}{2}} B)$ is special linear, and $\theta(id, I \otimes \Delta^{\frac{1}{2}} I)$ is in the kernel. To study all automorphisms of $G^{\otimes}$ mapping $\mathcal{J}(\mathcal{C})$ to $\mathcal{J}(\mathcal{C}')$ it clearly suffices to study just those that are special. This suggests that in characteristic 2 we should use a variation on the notion of equivalence of q-clans. In the conditions of Theorem 4.1 replace λ with $\mu = \lambda/\Delta$ and B^{-1} with $\Delta^{\frac{1}{2}}B^{-1}$. So $\det(\Delta^{\frac{1}{2}}B^{-1}) = 1$. Then we obtain a variation of (part of) the Fundamental Theorem in the following simpler form ($q = 2^e$ only!).

Theorem 10.10. *The Fundamental Theorem (partial version for $q = 2^e$). For q-clans $\mathcal{C}$ and $\mathcal{C}'$ as in Theorem 4.1, there is an automorphism θ of $G^{\otimes}$ mapping $\mathcal{J}(\mathcal{C})$ to $\mathcal{J}(\mathcal{C}')$ and $A(\infty)$ to $A'(\infty)$ for each 4-tuple $(\mu, B, \sigma, \pi) \in F^{\bullet} \times SL(2,q) \times Aut(F) \times S_F$ for which*

(i) $A'_{\bar{t}} \equiv \mu B^{-1} A_t^{\sigma} B^{-T} + A'_{\bar{0}}$ for all $t \in F$.

The associated automorphism $\theta : G^{\otimes} \to G^{\otimes}$ is given by

(ii) $\theta = \theta(\sigma, \begin{pmatrix} 1 & y'_{\bar{0}} \\ 0 & \mu \end{pmatrix} \otimes B) : ((\alpha,\beta),c) \mapsto ((\alpha^{\sigma},\beta^{\sigma})[\begin{pmatrix} 1 & y'_{\bar{0}} \\ 0 & \mu \end{pmatrix} \otimes B], \mu c^{\sigma} + g'_{\bar{0}}(\alpha^{\sigma} B))$.

(iii) $\pi : t \mapsto \bar{t}$ satisfies $y'_{\bar{t}} = \mu y_t^{\sigma} + y'_{\bar{0}}$.

10.5. The Shift-Flip in $G^{\otimes}$

To begin with let $\mathcal{C} = \{A_t = \begin{pmatrix} x_t & y_t \\ 0 & z_t \end{pmatrix} : t \in F\}$ be a fixed q–clan with $A_o = \begin{pmatrix} 0 & 0 \\ 0 & 0 \end{pmatrix} = A_{\infty}$. Recall the shift-flip of a previous section, but for the group $G^{\otimes}$.

$$\textbf{Shift by } s : \tau_s = \theta(id, \begin{pmatrix} 1 & y_s \\ 0 & 1 \end{pmatrix} \otimes I) : ((\alpha,\beta),c) \mapsto ((\alpha, y_s\alpha + \beta), c + \alpha A_s \alpha^T). \tag{10.38}$$

$$\textbf{Flip} : \phi = \theta(id, P \otimes I) : ((\alpha,\beta),c) \mapsto ((\beta,\alpha), c + \alpha \circ \beta). \tag{10.39}$$

$$\textbf{Shift-Flip} : i_s = \tau_s \cdot \phi = \theta(id, \begin{pmatrix} y_s & 1 \\ 1 & 0 \end{pmatrix} \otimes I) : ((\alpha,\beta),c) \mapsto \tag{10.40}$$

$$((y_s\alpha + \beta, \alpha), c + \alpha A_s \alpha^T + \alpha \circ \beta).$$

Also define $i_{\infty} = id : G^{\otimes} \to G^{\otimes}$.

Each of the maps defined in Eqs. 10.38, 10.39 and 10.40 is an automophism of $G^{\otimes}$ that replaces the 4-gonal family $\mathcal{J}(\mathcal{C})$ with some 4-gonal family $\mathcal{J}'$. We want to compute a q-clan $\mathcal{C}^{i_s}$ so that i_s replaces $\mathcal{J}(\mathcal{C})$ with $\mathcal{J}(\mathcal{C}^{i_s})$.

Consider the image under i_s, $s \in F$, of the typical element of $A(t)$. Here

$$i_s : (\gamma_{y_t} \otimes \alpha, g_t(\alpha))) \mapsto ((\gamma_{y_t} \otimes \alpha[\begin{pmatrix} y_s & 1 \\ 1 & 0 \end{pmatrix} \otimes I], g_t(\alpha) + \alpha A_s \alpha^T)$$

$$\begin{aligned} &= (\gamma_{y_t} \begin{pmatrix} y_s & 1 \\ 1 & 0 \end{pmatrix} \otimes \alpha, g_t(\alpha) + g_s(\alpha)) = ((y_s + y_t, 1) \otimes \alpha, g_t(\alpha) + g_s(\alpha)) \\ &= (1, \{y_s + y_t\}^{-1}) \otimes (y_s + y_t)\alpha, g_t(\alpha) + g_s(\alpha)) \text{ (Put } \beta = (y_s + y_t)\alpha) \end{aligned}$$

$$= (\gamma_{(y_s+y_t)^{-1}} \otimes \beta, (y_s + y_t)^{-2}(g_t(\beta) + g_s(\beta)).$$

To go further we need some kind of additional normalization on $\mathcal{C}$. **So from now on suppose $\mathcal{C}$ is τ^{-1}-normalized,** $\tau^{-1} \in \mathrm{Aut}(F)$, i.e., $y_t = t^{\frac{1}{\tau}}$, $t \in F$. We want the image q-clan $\mathcal{C}^{i_s}$ to be τ^{-1}-normalized also. Then $(\gamma_{t^{\frac{-1}{\tau}}} \otimes \alpha, g_t(\alpha)) \in A(t)$ is mapped by i_s to $(\gamma_{(s+t)^{\frac{-1}{\tau}}} \otimes \beta, (s+t)^{\frac{-2}{\tau}}(g_t(\beta) + g_s(\beta)))$, for $\beta = (s+t)^{\frac{1}{\tau}}\alpha$. This image must be in $A^{i_s}((s+t)^{-1})$, since we want $\mathcal{C}^{i_s}$ also to be τ^{-1}-normalized. Hence $A^{i_s}_{(s+t)^{-1}} = (s+t)^{\frac{-2}{\tau}}(A_t + A_s)$. (Treat the case $s = t$ separately.) Put $x = (t+s)^{-1}$, so $t = s + x^{-1}$. Then

$$A^{i_s}_x = x^{2/\tau}(A_{s+x^{-1}} + A_s), \forall x, s \in F. \tag{10.41}$$

Here $i_s : 0 \mapsto \overline{0} = s^{-1}$, so $A^{i_s}_{\overline{0}} = A^{i_s}_{s^{-1}} = s^{-2/\tau}A_s$, for $0 \neq s \in F$. Then in $i_s = \theta(id, \begin{pmatrix} s^{1/\tau} & 1 \\ 1 & 0 \end{pmatrix} \otimes I)$, $a_4^2 A^{i_s}_{(a_2/a_4)^\tau} = s^{2/\tau}A^{i_s}_{s^{-1}} = A_s$, $a_2 a_3 = 1$, and $a_3^2 A^{i_s}_{(a_1/a_3)^\tau} = \begin{pmatrix} 0 & 0 \\ 0 & 0 \end{pmatrix}$. Hence, i_s does have the form prescribed by Theorem (10.9).

By the FT, the most general special automorphism of $G^{\otimes}$ mapping $\mathcal{J}(\mathcal{C})$ to $\mathcal{J}(\mathcal{C}^{i_s})$ and $A(\infty)$ to $A^{i_s}(\infty)$ is $\overline{\theta} = \theta(\sigma, \begin{pmatrix} 1 & \overline{0}^{1/\tau} \\ 0 & \mu \end{pmatrix} \otimes B)$, where $A^{i_s}_{\overline{t}} + A^{i_s}_{\overline{0}} \equiv \mu B^{-1} A^{\sigma}_t B^{-T}$, for $0 \neq \mu \in F, \sigma \in Aut(F), B \in SL(2,q)$. Using Eq. (2.41) (or (10.41)), we may compute

$$\begin{aligned} \bar{\theta} = \theta(\sigma, \begin{pmatrix} 1 & \bar{0}^{\frac{1}{\tau}} \\ 0 & \mu \end{pmatrix} \otimes B) : \mathcal{J}(\mathcal{C}) \to \mathcal{J}(\mathcal{C}^{i_s}) : A(\infty) \mapsto A^{i_s}(\infty) \\ \text{iff (use } \bar{t} = \mu^\tau t^\sigma + \bar{0}) \;\; \mu B^{-1} A^\sigma_t B^{-T} \equiv A^{i_s}_{\bar{t}} + A^{i_s}_{\bar{0}} \\ = (\mu^2 t^{\frac{2\sigma}{\tau}} + \bar{0}^{\frac{2}{\tau}}) A_{\left(\frac{s\mu^\tau t^\sigma + 1 + \bar{0}s}{\mu^\tau t^\sigma + \bar{0}}\right)} + \mu^2 t^{\frac{2\sigma}{\tau}} A_s + \bar{0}^{\frac{2}{\tau}} A_{s+\bar{0}^{-1}}, \end{aligned} \tag{10.42}$$

for all $t \in F$.

(Here $\bar{0}$ is determined by condition (i) of Theorem 4.2.)

In applications, it is usually easier to compute $A^{i_s}_t$ for all $t \in F$ before applying the FT. And then $\theta = \overline{\theta} \circ i_s^{-1} = \theta(\sigma, \begin{pmatrix} 1 & \overline{0}^{1/\tau} \\ 0 & \mu \end{pmatrix} \otimes B) \circ \theta(id, \begin{pmatrix} s^{1/\tau} & 1 \\ 1 & 0 \end{pmatrix} \otimes I) = \theta(\sigma, \begin{pmatrix} \overline{0}^{1/\tau} & (1+\overline{0}_s)^{1/\tau} \\ 0 & \mu s^{1/\tau} \end{pmatrix} \otimes B)$. We collect these results and slightly more in the following theorem. (Note: To achieve brevity in at least one aspect, we have concentrated on the group $G^{\otimes}$ and more or less ignored the associated $GQ(\mathcal{C})$, the flock $\mathcal{F}(\mathcal{C})$ of a quadratic cone, etc. But the reader should keep in mind that the FT also asserts something about isomorphisms of these q-clan geometries.)

Theorem 10.11. *The most general special automorphism of $G^{\otimes}$ leaving $\mathcal{F}(\mathcal{C})$ invariant and mapping $A(\infty)$ to $A(s)$, for $s \in F$, is given by*

(i) $\theta(\sigma, \begin{pmatrix} \overline{0} & (1+\overline{0}_s) \\ \mu^\tau & \mu^\tau s \end{pmatrix}^{1/\tau} \otimes B) : A(t) \to A(t')$, *where*

(ii) $t' = \dfrac{s\mu^\tau t^\sigma + \overline{0}s + 1}{\mu^\tau t^\sigma + \overline{0}}$,

subject to the condition that there exist $\sigma \in Aut(F)$, $B \in SL(2,q)$, $0 \neq \mu \in F$, and a permutation $\pi : t \mapsto \overline{t} = \mu^\tau t^\sigma + \overline{0}$ for which

(iii) $\mu B^{-1} A_t^\sigma B^{-T} \equiv A_{\overline{t}}^{i_s} + A_{\overline{0}}^{i_s}$

(as in Eq. 10.42), for all $t \in F$.

In this case,

(iv) $\theta(\sigma, \begin{pmatrix} \overline{0} & (1+\overline{0}_s) \\ \mu^\tau & \mu^\tau s \end{pmatrix}^{1/\tau} \otimes B) : ((\alpha,\beta),c) \mapsto$

$((\alpha^\sigma,\beta^\sigma)[\begin{pmatrix} \overline{0} & (1+\overline{0}_s) \\ \mu^\tau & \mu^\tau s \end{pmatrix}^{1/\tau} \otimes B], \mu c^\sigma +$

$(\alpha^\sigma,\beta^\sigma)(I \otimes B) \begin{pmatrix} \overline{0}^{2/\tau} A_{(s+\overline{0}^{-1})} & \mu(1+\overline{0}_s)^{1/\tau} P \\ 0 & \mu A_s \end{pmatrix} (I \otimes B^T)(\alpha^\sigma,\beta^\sigma)^T)$.

Moreover, if $\overline{\theta} = \theta(\sigma, \begin{pmatrix} 1 & \overline{0} \\ 0 & \mu^\tau \end{pmatrix}^{1/\tau} \otimes B)$ is a typical special automorphism of $G^{\otimes}$ mapping $\mathcal{J}(\mathcal{C}^{i_s})$ to $\mathcal{J}(\mathcal{C}^{i_s})$ and leaving $A^{i_s}(\infty)$ invariant, then

(v) $\theta = i_s \cdot \overline{\theta} \cdot i_s^{-1} = \theta(\sigma, \begin{pmatrix} \mu^\tau + \overline{0}s^\sigma & \mu^\tau s + s^\sigma + \overline{0}s^{\sigma+1}) \\ \overline{0} & 1+\overline{0}s \end{pmatrix}^{1/\tau} \otimes B)$ *is the typical special automoprhism of $G^{\otimes}$ leaving $\mathcal{J}(\mathcal{C})$ invariant and fixing $A(s)$.*

We now know that every automorphism of the generalized quadrangle $GQ(\mathcal{C})$ fixing (∞) and $(\underline{0},0,\underline{0})$ has the form $\theta(\sigma, A \otimes B)$, where by Theorem 10.9 this notation completly identifies the map $\theta(\sigma, A \otimes B)$ as an automorphism of $G^{\otimes}$.

10.6. Ovals in $\overline{R}_\alpha$

We continue to assume that all q-clans are τ^{-1}- normalized.

For α in $PG(1,q)$, $A(t) \cap R_\alpha = \{d(\gamma_{t^{1/\tau}} \otimes \alpha, g_t(\alpha)) = (\gamma_{t^{1/\tau}} \otimes d\alpha, g_t(d\alpha)) : d\epsilon F\}$. So, $A(t) \cap R_\alpha$ is the point $\overline{p}_\alpha(t) = (\gamma_{t^{1/\tau}} \otimes \alpha, g_t(d\alpha))$ in $\overline{R}_\alpha$. And, $O_\alpha = \{\overline{p}_\alpha(t) : t\epsilon F\}$ is an oval in $\overline{R}_\alpha$. The map (use $(a,b)^{(2)} = (a^2,b^2)$)

$$\pi_a : \overline{R}_\alpha \to PG(2,q) : (\gamma \otimes \alpha, c) \mapsto (\gamma^{(2)}, c)$$

is an isomorphism mapping $\overline{O}_\alpha$ to the oval

$$O_\alpha = \{p_a(t) = (\gamma_{t^{2/\tau}}, g_t(\alpha) : t \epsilon \widetilde{F}\} \text{ in } PG(2,q). \tag{10.43}$$

Suppose that

Theorem 10.12. $\theta = \theta(\sigma, A \otimes B) : ((\alpha,\beta),c) \mapsto$

$((\alpha^\sigma,\beta^\sigma)(A \otimes B), \mu c^\sigma +$

$(\alpha^\sigma, \beta^\sigma)(I \otimes B) \begin{pmatrix} a_4^2 A_{(a_2/a_4)^\tau} & a_2 a_3 P \\ 0 & a_3^2 A_{(a_1/a_3)^\tau} \end{pmatrix} (I \otimes B^T)(\alpha^\sigma, \beta^\sigma)^T)$ *is an automophism of* $G^\otimes$ *leaving* $\mathcal{J}(\mathcal{C})$ *invariant. Then* θ *induces an isomorphism from* $\overline{R}_\alpha$ *to* $\overline{R}_{\alpha^\sigma B}$ *mapping* O_α *to* $O_{\alpha^\sigma B}$. *Put* $\mu = \det(A)$, $A = \begin{pmatrix} a_4 & a_2 \\ a_3 & a_1 \end{pmatrix}$, $C = a_4^2 A_{(a_2/a_4)^\tau}$, $D = a_2 a_3 P, E = a_3^2 A_{(a_1/a_3)^\tau}$. *Then Fig. 1 gives the maps involved: Let*

$$F(A, B, \gamma, \mu) = ((\gamma^\sigma A \otimes \alpha^\sigma B), \mu c^\sigma + (\gamma^\sigma \otimes \alpha^\sigma B) \begin{pmatrix} C & D \\ 0 & E \end{pmatrix} (\gamma^\sigma \otimes \alpha^\sigma B)^T \text{ and}$$

$$G(A, B, \gamma, \mu) = ((\gamma^{(2\sigma)} A^{(2)}), \mu c^\sigma + (\gamma^\sigma \otimes \alpha^\sigma B) \begin{pmatrix} C & D \\ 0 & E \end{pmatrix} (\gamma^\sigma \otimes \alpha^\sigma B)^T.$$

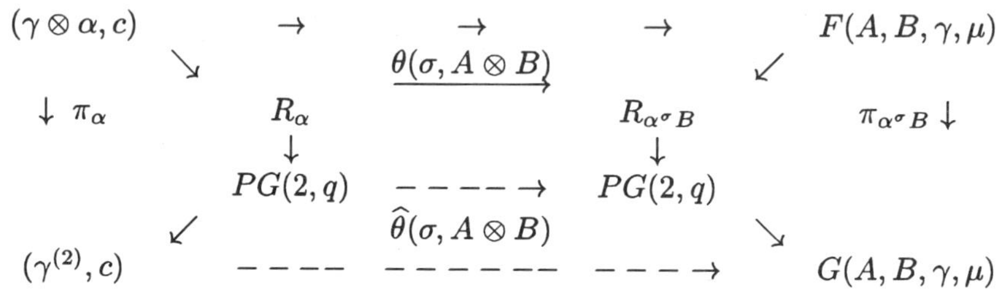

Fig. 1.

In this same context, suppose that $\alpha^\sigma B \equiv \lambda\alpha$, so $\theta(\sigma, A \otimes B)$ induces an automorphism of $\overline{R}_\alpha$ (resp. $(PG(2, q))$ stabilizing $\overline{O_\alpha}$ (resp. O_α). Making appropriate specializations in Fig. 1, we obtain Fig. 2.

Let $H(A, \gamma, \mu) = (\lambda(\gamma_{t^\sigma/\tau} A \otimes \alpha), \mu g(t)^\sigma + \lambda^2(\gamma_{t^\sigma/\tau} \otimes \alpha) \begin{pmatrix} C & D \\ 0 & E \end{pmatrix} (\gamma_{t^\sigma/\tau} \otimes \alpha)^T$ and

$$J(A, \gamma, \mu) = (\lambda^2(\gamma_{t^{2\sigma}/\tau} A^{(2)}), \mu g_t(\alpha)^\sigma + \lambda^2(\gamma_{t^\sigma/\tau} \otimes \alpha) \begin{pmatrix} C & D \\ 0 & E \end{pmatrix} (\gamma_{t^\sigma/\tau} \otimes \alpha)^T.$$

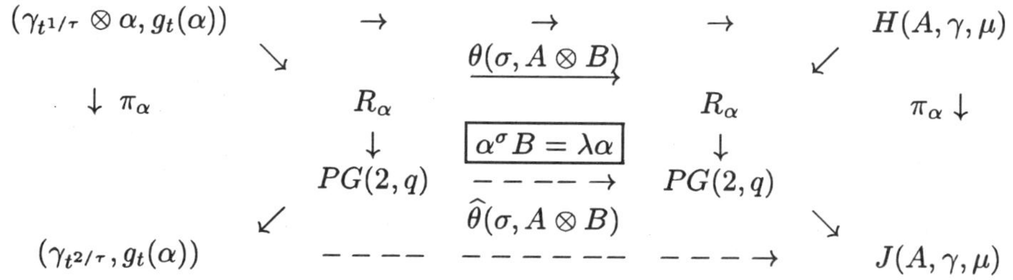

Fig. 2.

In the applications that follow for Subiaco q–clans, $\tau = 2$. In this case, the bottom row of Fig.2. appears as:

$$\theta(\sigma, A \otimes B) : (\gamma_t, g_t(\alpha)) \mapsto$$

$$
\begin{aligned}
&(\lambda^2(\gamma_{t^\sigma}A^{(2)}), \mu g_t(\alpha)^\sigma + \lambda^2(\gamma_{t^\sigma/2}\otimes\alpha)\begin{pmatrix} C & D \\ 0 & E \end{pmatrix}(\gamma_{\sigma/2}\otimes\alpha)^T \quad (10.44)\\
=\ &(\lambda^2\gamma_{t^\sigma}A^{(2)}, \mu g_t(\alpha)^\sigma + \lambda^2(\gamma_{t^\sigma})(\alpha C\alpha^T, \alpha E^{\alpha T})^T).
\end{aligned}
$$

From now on all q−clans are 1/2−normalized. Given a q−clan $\mathcal{C} = \{A_t = \begin{pmatrix} x_t & t^{1/2} \\ 0 & z_t \end{pmatrix} : t\epsilon F\}$, write $\mathcal{J}(\mathcal{C}) = \{A(t) : t\epsilon\widetilde{F}\}$, and $\widehat{\mathcal{J}}(\mathcal{C}) = \{\widehat{A}(t) : t\epsilon\widetilde{F}\}$ for the 4−gonal family in Gm. Note that $\psi : G^{\otimes} \to G : ((\alpha,\beta),c) \mapsto ((\alpha,\beta),\sqrt{c}) : (\gamma_{t^{1/2}}\otimes\alpha, \alpha A\alpha^T) \mapsto (\gamma_{t^{1/2}}\otimes\alpha, \sqrt{\alpha A_t\alpha^T})$. So, $\psi : A(t) \mapsto \widehat{A}(t)$. With a little care, we can use the same symbol for an automorphism $\theta(\alpha, A\otimes B)$ of $G^{\otimes}$ and the corresponding one $\psi^{-1}\cdot\theta(\alpha, A\otimes B)\cdot\psi$ of G.

In

$$
\begin{aligned}
G^{\otimes}\ :\ &\theta(\alpha, A\otimes B) : ((\alpha,\beta),c) \mapsto ((\alpha^\sigma,\beta^\sigma)(A\otimes B), \mu c^\sigma + \quad (10.45)\\
&(\alpha^\sigma,\beta^\sigma)(I\otimes B)(\begin{pmatrix} C & D \\ 0 & E \end{pmatrix}(I\otimes B)^T(\alpha^\sigma,\beta^\sigma)^T).
\end{aligned}
$$

In

$$
\begin{aligned}
G\ :\ &\theta(\alpha, A\otimes B) : ((\alpha,\beta),c) \mapsto ((\alpha^\sigma,\beta^\sigma)(A\otimes B), \mu^{1/2}c^\sigma + \quad (10.46)\\
&\sqrt{(\alpha^\sigma,\beta^\sigma)(I\otimes B)(\begin{pmatrix} C & D \\ 0 & E \end{pmatrix}(I\otimes B)^T(\alpha^\sigma,\beta^\sigma)^T)}.
\end{aligned}
$$

11. Known Examples of Finite Flocks and q-Clans

11.1. Derivation

We have discussed partial BLT sets in a previous section using the theory of Knarr. Here we re-introduce this concept in the finite case only.

Let q be odd and let $\mathcal{F} = \{C_1, \ldots, C_q\}$ be a flock of the quadratic cone K with vertex $V = V_0$. Embed the cone K in a nonsingular quadric Q of $PG(4,q)$ so that K is the intersection with Q of the hyperplane of $PG(4,q)$ tangent to Q at V_0. Then there are points $V_1, \ldots, V_q$ of Q, such that C_i is the set of points of Q collinear in Q with both V_0 and V_i, $1 \le i \le q$. The set $S = \{V_0, V_1, \ldots, V_q\}$ of points of Q has the property that no two points of S are collinear in Q and each point of $Q \setminus S$ is collinear in Q with either 0 or 2 points of S. This has the very interesting corollary that if K_i is the cone consisting of the points of Q collinear in Q with V_i, then the conics of the form $K_j \cap K_i$ for $j \neq i$ form a flock of K_i. The set S is a BLT-set as noted in a previous section. This process of obtaining a set of $q+1$ flocks from a given one was called "derivation" in [4], and it works only for q odd. Shortly after the appearance of [4], S. E. Payne and L. A. Rogers [71] showed that although derivation could yield new flocks (as first shown in [4]), it could never yield new GQ. In fact, the "new" flocks were just obtained by recoordinatizing the GQ so as to interchange roles of certain lines through (∞). Moreover, the algebraic process also worked in characteristic 2. The proof in [71] contains most the essential steps of the proof of the Fundamental Theorem associating a flock (or a q-clan) to each line through (∞). Unfortunately, the opportunity to discover the full Fundamental Theorem was missed. For characteristic 2 it first appeared in L. Bader, G. Lunardon and S. E. Payne [2], and in for all q in S. E. Payne [68].

In previously published surveys (for example see the Handbook [13]) of the known flocks and q-clans, the known examples were organized more or less by flocks and to some extent by the order in which they were discovered. Because of the Fundamental Theorem for q-clan geometry, we propose organizing them first into groups according to which generalized quadrangle (up to isomorphism) they belong to. (See S. E. Payne [65] for remarks concerning the automorphism groups of the various examples.) However, we follow the labeling of the flocks given in the Handbook. For additional results concerning flocks for q small, see [19, 20] and their references. Note: In this survey we have not attempted to distinguish carefully between the various overlaps in the construction of flocks for small q.

11.2. Classical $GQ(\mathcal{C})$

Let $x^2 + bx + c$ be a fixed irreducible polynomial over F.

Put $\mathcal{C} = \left\{ A_t = \begin{pmatrix} t & bt \\ 0 & ct \end{pmatrix} : t \in F \right\}$.

Then $GQ(\mathcal{C})$ is isomorphic to the classical GQ $H(3, q^2)$ arising from a nonsingular Hermitian surface in $PG(3, q^2)$ and is isomorphic to the point-line dual of the GQ $Q(5, q)$ arising from a nonsingular elliptic quadric in $PG(5, q)$. Se S. E. Payne and J. A. Thas [72] for a proof that $GQ(\mathcal{C})$ is classical. The automorphism group of $GQ(\mathcal{C})$ is transitive on the lines through (∞) and the unique flock is *linear*, i.e., the planes of the flock contain a common line. When q is even, the automorphism group of $GQ(\mathcal{C})$ is also transitive on the associated ovals and each of the ovals in the herd is a conic.

J. A. Thas [76] has also characterized these flocks as follows: If q is even and if the planes of the flock contain a common point, then the flock is linear. If q is odd and the planes of the flock contain a common interior point of the cone K, i.e., a point on no tangent plane of K, then the flock is linear.

11.3. GQ associated with $G_2(q)$

These GQ were given by W. M. Kantor [55] and probably deserve credit for starting the line of investigation that led to the notion of q-clan.

Flock: FTW (cf. [81, 76, 77]. Oval: translation oval associated with map $x \mapsto x^4$ (cf. [69]).

For $q \equiv 2 (\mathrm{mod}\ 3)$, put $\mathcal{C} = \left\{ A_t \equiv \begin{pmatrix} t & 3t^2 \\ 0 & 3t^3 \end{pmatrix} : t \in F \right\}$.

Here $Aut(GQ(\mathcal{C}))$ is triply transitive on the lines through (∞), implying that there is a unique flock, known as the Fisher-Thas-Walker flock (cf. [76]). Recently J. A. Thas [77] has shown that (except for a few small odd values of q for which the proof is incomplete) the FTW flock is characterized by the fact that no four planes of the flock have a point in common. When q is even, $\mathrm{Aut}(GQ(\mathcal{C}))$ is also transitive on the ovals, so that only one arises. It is the translation oval associated with the permuation $x \mapsto x^4$ (cf. S. E. Payne and C. Maneri [69]).

11.4. Dual TGQ of W. M. Kantor with arbitrary kernel

These GQ were constructed by W. M. Kantor [56]. For a discussion of the kernel of the point-line dual of the GQ see S. E. Payne [62].

Here q is an arbitrary odd prime power, m is a nonsquare of F, and σ is an automorphism of F. Then $\mathcal{C} = \left\{ A_t \equiv \begin{pmatrix} t & 0 \\ 0 & -mt^\sigma \end{pmatrix} : t \in F \right\}$.

$Aut(GQ(\mathcal{C}))$ is triply transitive on the lines through (∞). Moreover, the point-line dual of $GQ(\mathcal{C})$ is a translation GQ with respect to each line through the point ∞. (See [72] for definitions and results concerning transltation GQ.) The one flock that arises is denoted K_1. J. A. Thas [76] has characterized this flock as follows: If q is odd and the planes of the nonlinear flock contain a common exterior point of the cone K, i.e., a point on exactly two tangent planes of K, then the flock must be the flock K_1 of Kantor.

11.5. Two constructions with q a nonsquare modulo 5

The discovery of these examples by W. M. Kantor [56] for q odd inspired the discovery by S. E. Payne [61] of their analogues for q even. Flocks: $K2$ and JP for q odd; $P1$ and $P2$ for q even. Ovals for $q = 2^e$, e odd: Segre's oval associated with the map $x \mapsto x^6$ and the Payne oval associated with the map $x \mapsto x^{1/6} + x^{3/6} + x^{5/6}$.

Here q is a nonsquare mod 5, and $\mathcal{C} = \left\{ A_t \equiv \begin{pmatrix} t & 5t^3 \\ 0 & 5t^5 \end{pmatrix} : t \in F \right\}$.

In [65] it is shown that there is a group of automorphisms of $GQ(\mathcal{C})$ having two orbits on the lines through ∞, one of which is $\{A(\infty), A(0)\}$. This implies that at most two flocks arise. In [65] this group is alleged to be the entire automorphism group (when q is odd this is for $q \geq 7$), implying that exactly two flocks arise. For q even it is easy to see that this is the case, but for q odd a more complex proof is required. In N. L. Johnson [43] a proof is given for those q for which $(q+1)/2$ is not a prime power, using rather deep group theory. In G. Lunardon [58] there is a proof given of a more general result that relies heavily on the theory of groups acting on projective planes. In that article the result needed here does not appear explicitly, although it is rather easy to derive as a corollary. However, in S. E. Payne [68] there is a proof given that just relies on the intersection properties of the planes of the flock.

For q odd and $q \geq 13$ the nonlinear flock arising directly from the q-clan is denoted $K2$. And for q even with $q \geq 4$ this flock is denoted $P1$. For q odd, $q \geq 13$, the process of derivation gives the flock associated with the "other" orbit on the lines through (∞) and discovered independently by N. L. Johnson [43] and S. E. Payne [65, 68]. This flock is denoted JP in the Handbook, but because of the earlier reference [58] probably should be denoted JLP (rather than L, since it is not clear that the author was aware at the time that his result provided a new flock).

For q even (and nonsquare), the "other" flock, denoted $P2$, was first pointed out in S. E. Payne [65]. The automorphism group of $GQ(\mathcal{C})$ also has two orbits on the ovals, as indicated above.

11.6. Likeable GQ with $q = 5^e$

These GQ were first defined by H. Gevaert and N. L. Johnson [24] from some "likeable" planes of W. M. Kantor. Flocks: $K3$ and BLT.

Here $q = 5^e$ and k is a fixed nonsquare of F. Then

$$\mathcal{C} = \left\{ A_t \equiv \begin{pmatrix} t & 3t^2 \\ 3t^2 & k^{-1}t(1+kt^2)^2 \end{pmatrix} : t \in F \right\}.$$

The automorphisms of $GQ(\mathcal{C})$ have two orbits on the lines through (∞), one of which contains only $[A(\infty)]$. The original flock is denoted $K3$. The derived flock, denoted BLT, was first shown to be new at the time derivation was discovered in [4], but the proof used the classification of finite simple groups. However, in S. E. Payne and L. A. Rogers [71] the complete automorphism group of $GQ(\mathcal{C})$ was determined and a proof was given using more elementary methods.

11.7. Dual translation GQ with $q = 3^h$

These GQ, whose point-line duals are translation GQ, were constructed by H. Gevaert and N. L. Johnson [24] starting with some semifield planes of Ganley. Flocks: G and $PTJLW$.

Here $q = 3^h$ and n is a given nonsquare of F. Then

$$\mathcal{C} = \left\{ A_t \equiv \begin{pmatrix} t & t^3 \\ 0 & -nt - n^{-1}t^9 \end{pmatrix} : t \in F \right\}.$$

In S. E. Payne and J. A. Thas [73] the automorphism group of $GQ(\mathcal{C})$ was determined and shown to have two orbits on the lines through (∞), one of them being $[A(\infty)]$. The original flock is denoted G; the derived flock, denoted $PTJLW$, was discovered independently in [65] and in N. L. Johnson, G. Lunardon and F. W. Wilke [52].

11.8. The Thas-Fisher GQ

Flock: Fi.

Here q is odd. Let ζ be a primitive element of $GF(q^2)$, so $\omega = \zeta^{q+1}$ is a primitive element of F and hence a nonsquare of F. Put $i = \zeta^{(q+1)/2}$, so $i^2 = \omega$, $i^q = -i$. Put $z = \zeta^{q-1} = a + bi$, so z has order $q+1$ in the multiplicative group of $GF(q^2)$. Then the matrices A_t of the q-clan $\mathcal{C}$ are given by:

$A_t = \begin{pmatrix} t & 0 \\ 0 & -\omega t \end{pmatrix}$, for those $t \in F$ with $t^2 - 2(1+a)^{-1}$ a square in F, and the other matrices are those of the form

$\begin{pmatrix} -a_{2j} & b_j \\ b_j & -\omega a_{2j} \end{pmatrix}$, for $0 \le j \le (q-1)/2$ and with $a_k = (a^{k+1} + z^{-k})/(z+1)$ and $b_k = i(z^{k+1} - z^{-k})/(z+1)$.

The GQ were discovered by J. A. Thas [76] starting with some flocks Fi originally discovered by J. C. Fisher. See also S. E. Payne [63]. These GQ were studied further in S. E. Payne and J. A. Thas [73], where the automorphism group of $GQ(\mathcal{C})$ was shown to be transitive on the lines through (∞). This implies that only one flock arises by derivation, a result also found by L. Bader [1]. The Fisher flocks Fi were characterized in [73] as the only nonlinear flocks for which at least (and hence exactly) $(q-1)/2$ of the planes of the flock contain a common line.

11.9. The Subiaco geometries, q even

These examples as q-clans were first given by W. A. Cherowitzo, et al. [17], and exist for all even q. They were previously unknown for $q > 16$, except that certain of the smaller examples had been found by computer and the general construction was obtained in pieces. Since they were not discovered in time to be included in detail in the Handbook [13], we have decided to devote a short chapter to them, at least to the GQ and their ovals. At this point we mention only that the Subiaco construction provides the first known q-clans with q even, q a square larger than 16.

11.10. Sporadic examples

For $q = 11, 16, 17, 23$, F. De Clerck and C. Herssens [19] were able to find additional nonlinear flocks with the aid of a computer. Their example with $q = 16$ was rediscovered as a special case of the Subiaco construction [17]. To our knowledge, the sporadic constructions with q odd have received only minimal attention (cf. J. A. Thas, C. Herssens and F. De Clerck [78]).

For $q = 4^k$, with $3 \leq k \leq 7$, the computer has been used to find examples that seem very likely to belong to an infinite family. (For more on this topic see the article in this volume by S. E. Payne, T. Penttila and G. Royle).

12. Semifield skeletons

As mentioned, given a flock of a quadratic cone in $PG(3,q)$, there are $(q+1)-$associated flocks obtained via derivation of the flock and hence $(q+1)-$associated translation planes with spreads in $PG(3,q)$. Let π be a flock plane. Then the skeleton of π, $S(\pi)$, is the set of these $(q+1)-$planes. It follows that if τ is in $S(\pi)$ then $S(\tau) = S(\pi)$. It has been noted that in many cases, it can be determined if any of the planes of the skeleton of a given one are isomorphic to the original.

Suppose that π is a semifield flock plane with spread in $PG(3,q)$. That is, the associated generalized quadrangle is a translation generalized quadrangle. The following result is in Johnson [38].

Theorem 12.1. *A semifield flock plane of even order is Desarguesian; a translation generalized quadrangle corresponding to a $q-$clan for q even is classical.*

However, we have noted above that there are nonDesarguesian semifield flock planes of odd order. For example, if π is a Ganley semifield flock plane then $S(\pi)$

contains flock planes which are not isomorphic to π. On the other hand if ρ is the Knuth semifield flock plane, all the planes of $S(\rho)$ are isomorphic.

More generally, if π is a semifield flock plane and all of the planes of $S(\pi)$ are isomorphic then there exists an induced group acting doubly transitively on the associated BLT set. As noted, there is an algebraic procedure which produces the planes of $S(\pi)$ from a given π. However, it is computationally not always useful for the determination of isomorphism classes of translation planes. Never the less, it is theoretically possible to apply the classification of finite simple groups to determine the classes within the skeletons of semifield planes.

In particular, using this approach, Johnson, Lunardon and Wilke [52] prove

Theorem 12.2. *(i) Let π be a semifield plane of flock type and odd order q^2. Then the planes of the skeleton are all semifield planes if and only if π is the Knuth semifield plane of flock type.*

(ii) Let F be a flock of a quadratic cone in $PG(3,q)$, for q odd. If F admits an automorphism group which acts transitively on the conics of F and the flocks of the skeleton are all isomorphic then F either corresponds to the Walker plane or to the Knuth semifield plane of flock type.

(iii) Let π_1 and π_2 be nonisomorphic semifield planes of flock type and odd order q^2 which are not Knuth semifield planes. Then no plane in the skeleton of π_1 is isomorphic to any plane of the skeleton of π_2.

We have mentioned above that the dual translation generalized quadrangles of order 3^r produce various nonismorphic flocks. In Payne and Thas [65], there is a geometric argument that the planes of the skeleton of the Ganley semifield flock plane of order 3^{2r} are not all isomorphic. And, this may be seen using the above result.

Corollary 12.3. *If π is the Ganley semifield plane of order 3^{2r} for $r > 2$ then the planes of the skeleton are not isomorphic to a plane in the skeleton of a known flock plane (not in the skeleton of the Ganley plane).*

We have noted above that the group theoretic approach may also be used to determine the isomorphism classes of planes of the skeleton of planes that admit a collineation group that either acts transitively on the set of regulus nets or fixes one regulus net and acts transtively on the remaining regulus nets. The reader is referred to Johnson [43] and Bader, Lunardon and Thas [4] for further details.

13. The Subiaco Family of Geometries

13.1. Canonical Subiaco q-clans

For $q = 2^e$, $e \geq 4$, $F = GF(q)$, let $\delta \in F$ be chosen so that $x^2 + \delta x + 1$ is irreducible over F, i.e., $tr(\delta^{-2}) = tr(\delta^{-1}) = 1$. Then define the following functions on F.

$$\begin{array}{llcl} (i) & f(t) & = & \delta^2 t^4 + \delta^3 t^3 + (\delta^2 + \delta^4) t^2, \\ (ii) & g(t) & = & (\delta^2 + \delta^4) t^3 + \delta^3 t^2 + \delta^2 t, \\ (iii) & k(t) & = & (\nu(t))^2 = t^4 + \delta^2 t^2 + 1, \\ (iv) & F(t) & = & f(t)/k(t), \\ (v) & G(t) & = & g(t)/k(t). \end{array} \tag{13.1}$$

Then the canonical Subiaco $q-$clan ($cf.$ [70]) is the set $\mathcal{C}(\delta) = \mathcal{C} =$
$\{A_t = \begin{pmatrix} x_t & y_t \\ 0 & z_t \end{pmatrix} : t \in F\}$ where

Then the **canonical Subiaco** q **-clan** (cf. [70]) is the set $\mathcal{C}(\delta) = \mathcal{C} = \{A_t = \begin{pmatrix} x_t & y_t \\ 0 & z_t \end{pmatrix} : t \in F\}$ where

$$\begin{array}{llcl} (i) & x_t & = & F(t) + (t/\delta)^{\frac{1}{2}}, \\ (ii) & y_t & = & t^{\frac{1}{2}}, \\ (iii) & z_t & = & G(t) + (t/\delta)^{\frac{1}{2}}. \end{array} \tag{13.2}$$

13.2. Automorphisms of $GQ(\mathcal{C})$

Using the group G, the authors of [70] determined the full automorphism group of G leaving $\hat{\mathcal{J}}(\mathcal{C})$ invariant. We want to recall several of the results from [70] but using the efficient tensor product notation surveyed in section 10.3 – 10.6. For each member $\hat{A}(s)$ of $\hat{\mathcal{J}}(\mathcal{C})$, $s \in \tilde{F}$, there is an involution I_s of G leaving $\hat{\mathcal{J}}(\mathcal{C})$ invariant and fixing $\hat{A}(s)$.

$$I_\infty = \theta(id, \begin{pmatrix} 1 & \delta^{\frac{1}{2}} \\ 0 & 1 \end{pmatrix} \otimes \begin{pmatrix} 1+\delta+\delta^2 & \delta^{\frac{3}{2}} \\ \delta^{\frac{1}{2}} + \delta^{\frac{5}{2}} & 1+\delta+\delta^2 \end{pmatrix}). \tag{13.3}$$

Here $(a_2/a_4)2 = \delta$, and $A_\delta = \begin{pmatrix} 1+\delta^4+\delta^6 & \delta^{1/2} \\ 0 & 1+\delta^3+\delta^7 \end{pmatrix}$.
For $s \in F$, put $\bar{0} = \delta/\nu(s) = \delta/(s^2 + \delta s + 1)$. Then

$$\begin{array}{llcl} (i) & I_s & = & \theta(id, \begin{pmatrix} (1+\bar{0}s)^{\frac{1}{2}} & s\bar{0}^{\frac{1}{2}} \\ \bar{0}^{\frac{1}{2}} & (1+\bar{0}s)^{\frac{1}{2}} \end{pmatrix} \otimes B_s), \\ \text{where} & B_s & = & \begin{pmatrix} a(s) & b(s) \\ c(s) & a(s) \end{pmatrix} \text{ is defined by} \\ (ii) & a(s) & = & ((s^5+1)(1+\delta+\delta^2) + (s^4+s)(1+\delta))/\nu(s)^{\frac{5}{2}}, \\ (iii) & b(s) & = & (\delta^{\frac{3}{2}} s^5 + \delta^{\frac{1}{2}} s^4 + \delta^{3\ /2} s + \delta^{\frac{1}{2}} + \delta^{\frac{5}{2}})/\nu(s)^{\frac{5}{2}}, \\ (iv) & c(s) & = & ((\delta^{\frac{1}{2}} + \delta^{\frac{5}{2}}) s^5 + \delta^{\frac{3}{2}} s^4 + \delta^{\frac{1}{2}} s + \delta^{\frac{3}{2}})/\nu(s)^{\frac{5}{2}}. \end{array} \tag{13.4}$$

It is convenient to have a few specific involutions with which to compute.

$$
\begin{array}{llll}
(i) & I_0 & = & \theta(id, \begin{pmatrix} 1 & 0 \\ \delta^{\frac{1}{2}} & 1 \end{pmatrix} \otimes \begin{pmatrix} 1+\delta+\delta^2 & \delta^{\frac{1}{2}}+\delta^{\frac{5}{2}} \\ \delta^{\frac{3}{2}} & 1+\delta+\delta^2 \end{pmatrix}). \\
(ii) & I_1 & = & \theta(id, P\otimes P) : ((\alpha,\beta),c) \mapsto ((\beta P, \alpha P), c+\alpha\circ\beta). \\
(iii) & I_\delta & = & \\
 & \multicolumn{3}{l}{\theta(id, \begin{pmatrix} 1+\delta & \delta^{\frac{3}{2}} \\ \delta^{\frac{1}{2}} & 1+\delta \end{pmatrix} \otimes \begin{pmatrix} 1+\delta^4+\delta^6+\delta^7 & \delta^{\frac{1}{2}}+\delta^{\frac{9}{2}}+\delta^{\frac{13}{2}} \\ \delta^{\frac{15}{2}} & 1+\delta^4+\delta^6+\delta^7 \end{pmatrix}).}
\end{array}
\tag{13.5}
$$

In [70] it is shown that any two suitable δ give equivalent q-clans. So we lose no generality by choosing a particular one. Let ζ be a primitive element for $E = GF(q^2) \supseteq F$. Put $\lambda = \zeta^{q-1}$ and $\delta = \lambda + \lambda^{-1}$. Then $x^2 + \delta x + 1 = (x+\lambda)(x+\lambda^{-1})$ is irreducible over F. For each rational number a with denominator prime to $q+1$, put

$$
\begin{array}{ll}
(i) & [a] = \lambda^a + \lambda^{-a}. \text{ Then} \\
(ii) & [0] = 0, \text{ and } [1] = \delta. \\
(iii) & [a] = [b] \text{ iff } a \equiv \pm b (\bmod q+1). \text{ So } [a] = [-a]. \\
(iv) & [a][b] = [a+b] + [a-b]. \\
(v) & \text{For } \sigma = 2^i \in Aut(F), \ [a]^\sigma = [a\sigma].
\end{array}
\tag{13.6}
$$

13.3. Action on the Set of Ovals

In [70] it is observed that $I_0 \cdot I_\delta$ permutes the members of $\hat{\mathcal{J}}(\mathcal{C})$ in one cycle of length $q+1$. When $e \not\equiv 2 \pmod 4$, i.e., 5 does not divide $q+1$, $I_0 \cdot I_\delta$ permutes the $\mathcal{R}_\alpha$ and hence the ovals $\mathcal{O}_\alpha$, in one cycle of length $q+1$. But when $e \equiv 2 \pmod 4$, so $5|(q+1)$, it permutes the $\mathcal{R}_\alpha$ in five cycles of length $(q+1)/5$. Let Ω_1 be the orbit containing $\mathcal{R}_\alpha$ with $\alpha = (1,1)$. The group $\mathcal{G}_0$ of all automorphisms of G leaving $\hat{\mathcal{J}}(\mathcal{C})$ invariant has two orbits on the $\mathcal{R}_\alpha$: the orbit Ω_1 and a second orbit Ω_2 of length $4(q+1)/5$.

$$
\begin{array}{ll}
(i) & I_0 \cdot I_\delta = \theta(id, M\otimes N), \text{ where } M = \begin{pmatrix} 1+\delta & \delta^{3/2} \\ \delta^{3/2} & 1+\delta+\delta^2 \end{pmatrix}, \\
 & N = B_0 B_\delta. \\
(ii) & M^j = \delta^{-\frac{1}{2}} \begin{pmatrix} [2j-\frac{1}{2}] & [2j] \\ [2j] & [2j+\frac{1}{2}] \end{pmatrix}. \\
(iii) & N^j = \delta^{-\frac{1}{2}} \begin{pmatrix} [10j+\frac{1}{2}] & [10j] \\ [10j] & [10j-\frac{1}{2}] \end{pmatrix}.
\end{array}
\tag{13.7}
$$

Theorem 13.1. *For all $e \geq 4$, $N^{-1} = M^5$.*

Theorem 13.2. *When $e \equiv 2 \pmod 4$, $\mathcal{R}_\alpha \in \Omega_2$ for $\alpha = (1, \delta^{\frac{1}{2}})$.*

Proof. Since the orbit of $\mathcal{R}_{(1,1)}$ under $I_0 \cdot I_\delta$ is Ω_1, $\mathcal{R}_{(1,\delta^{\frac{1}{2}})} \in \Omega_1$ iff there is an integer j for which $(1,1)N^j \equiv (1,\delta^{\frac{1}{2}})$. This is iff
$([10j+\frac{1}{2}]+[10j], [10j]+[10j-\frac{1}{2}]) \equiv (1, [\frac{1}{2}])$ iff
$[10j]+[10j-\frac{1}{2}] = [\frac{1}{2}]([10j+\frac{1}{2}]+[10j])$ iff

$[10j] + [10j - \frac{1}{2}] = [10j] + [10j+1] + [10j - \frac{1}{2}] + [10j + \frac{1}{2}]$ i ff
$[10j+1] = [10j + \frac{1}{2}]$ iff $\frac{1}{2} \equiv 0 \pmod{q+1}$ (impossible!) or $20j + \frac{3}{2} \equiv 0 \pmod{q+1}$.
When $5|(q+1)$, $20j + \frac{3}{2} \equiv 0 \pmod{q+1}$ implies $0 + \frac{3}{2} \equiv 0 \pmod 5$, impossible!
■

13.4. Oval Stabilizer: $q = 2^e$, $e \equiv 2 \pmod 4$

Let $e \equiv 2 \pmod 4$, say $e = 2r$ with r odd.

In S. E .Payne et al. [70] it is shown that there are two ovals up to equivalence, that the associated hyperovals are absolutely irreducible algebraic curves of degree 10, and that the stabilizers of the hyperovals are precisely the collineations stabilizing the ovals that are induced by the automorphism of the associated GQ. To get one of the ovals we may put $\alpha = (1,1)$ and $\delta = \omega$, where $\omega \in F$ satisfies $\omega^2 + \omega + 1 = 0$. In this case

$$\mathcal{O}_\alpha = \left\{(1, t, \frac{\omega^2(t^4+t)}{t^4+\omega^2t^2+1} + t^{1/2}) : t \in F\right\} \cup \{(0,1,0)\}.$$

The subgroup of $P\Gamma L(3,q)$ stabilizing $\mathcal{O}_\alpha$ is a semidirect product C_5 semidirect C_{2e}.

Here $C_5 =< \phi >$, where $\phi : (x,y,z) \mapsto (y + \omega x, x, z + \omega^2 x)$, and $C_{2e} =< \psi >$, where $\psi : (x,y,z) \mapsto (x^2, x^2 + \omega^2 y^2, x^2 + \omega z^2)$.

The situation for the other oval is rather curious. When 5 does not divide e there is again a rather simple description of the oval and of its stabilizer. Here we may again put $\delta = \omega$ and then choose $\alpha = (0,1)$ to obtain

$$\mathcal{O}_\alpha = \left\{(1, t, \frac{t^3+t^2+\omega^2 t}{t^4+\omega^2t^2+1} + \omega t^{1/2}) : t \in F\right\} \cup \{(0,1,0)\}.$$

The stabilizer is C_5 semidirect $C_{e/2}$, where $C_5 =< \phi >$, where $\phi : (x,y,z) \mapsto (y + \sigma x, x, z + \omega x$, and $C_{e/2} =< \theta >$, where $\theta : (x,y,z) \mapsto (x^4, y^4, z^4)$.

Unfortunately, when 5 divides e, to get the oval with the stabilizer of order $5e/2$ we may not choose $\alpha = (0,1)$. If we choose $\delta = \lambda + \lambda^{-1}$ as above and put $\alpha = (1, \delta^{\frac{1}{2}})$, then the oval $\mathcal{O}_\alpha$ has the following appearance:

$$\begin{array}{ll} (i) & \mathcal{O}_\alpha = \{(1,t,h_\alpha(t)) : t \in F\} \cup \{(0,1,0)\}, \text{ where} \\ (ii) & h_\alpha(t) = \frac{\delta^2t^4+\delta^5t^3+\delta^2t^2+\delta^3t}{t^4+\delta^2t^2+1} + (t/\delta)^{\frac{1}{2}}. \end{array} \tag{13.8}$$

From [70] we know that the complete stabilizer of $\mathcal{O}_\alpha$ has order $5r$.

Clearly, $(I_o \cdot I_\delta)^{(q+1)/5} = \theta(id, \delta^{-1/2}(\begin{pmatrix} [(4q-1)/10] & [(2(q+1)/5 \\ [2(q+1)/5] & [(4q+9)/5 \end{pmatrix} \otimes I)$ induces a linear collineation φ of $PG(2,q)$ stabilizing $\mathcal{O}_\alpha$. Put $x = [4(q+1)/5]/[(4q-1)/5]$ and $y = [(49+9)/5]/[4(q+1)/5]$. Then from previous equation (10.44) we can work out the effect of φ to be

$$\begin{array}{ll} \varphi : (\gamma, c) \mapsto & (\delta^{-1}\gamma \begin{pmatrix} \left[\frac{4q-1}{5}\right] & \left[\frac{4(q+1)}{5}\right] \\ \left[\frac{4(q+1)}{5}\right] & \left[\frac{4q+9}{5}\right] \end{pmatrix}, \\ c + (\gamma^{(\frac{1}{2})} \otimes (1, \delta^{\frac{1}{2}})) & \begin{pmatrix} \left[\frac{4q-1}{5}\right] A_x & \left[\frac{4(q+1)}{5}\right] P \\ 0 & \left[\frac{4(q+1)}{5}\right] A_y \end{pmatrix} (\gamma^{(\frac{1}{2})} \otimes (1, \delta^{\frac{1}{2}}))^T). \end{array} \tag{13.9}$$

Put $\omega = [(q+1)/5]$, so $\omega^2 + \omega + 1 = 0$. Put $W_1 = [(4q-1)/5], W_2 = [(4q+9)/5]$. Note that $W_1 + W_2 = \omega\delta$ and $W_1 W_2 = \omega^2 + \delta^2$. So, W_1 and W_2 are the two roots in F of $x^2 + \omega\delta x + \omega^2 + \delta^2 = 0$. and then $x = \omega/W_1, y = \omega/W_2$. With the standard notation $A_t = \begin{pmatrix} x_t & y_t \\ 0 & z_t \end{pmatrix}$ for $t \epsilon F$, the description of φ in the above equation (13.9)

$$\begin{array}{l} \varphi : (a, b, c) \mapsto (aW_1/\delta + b\omega/\delta, a\omega/\delta + bW_2/\delta, \\ c + aW_1(x_{\omega/W_1} + \left(\frac{W_1+W_2}{W_1}\right)^{\frac{1}{2}} + \delta z_{\omega/W_1}) \\ +b\omega(x_{\omega/W_2} + (\frac{W_1+W_2}{W_2})^{\frac{1}{2}} + \delta z_{\omega/W_2})). \end{array} \tag{13.10}$$

This φ is a collineation of order 5 stabilizing $\mathcal{O}_\alpha$.

In principle we have an algorithm for computing that part of the stabilizer of $\mathcal{O}_\alpha$ contributed by Gal$(F/GF(4))$. If we put $\delta_1 = \omega$ and $\delta_2 = \delta$, Theorem IV.4 of [70] gives an isomorphism of G of the form $\theta(id, A \otimes B)$ mapping $\hat{\mathcal{J}}(\mathcal{C}(\delta_1))$ to $\hat{\mathcal{J}}(\mathcal{C}(\delta_2))$ and mapping $[A_{\delta_1}(\infty)]$ to $[A_{\delta_2}(\infty)]$. Eq.(38) of [70] is $\bar{0}^2 + \delta\bar{0} + \omega\delta^2 + 1 = 0$ and has solutions $\bar{0} = \omega^2 W$, for $W \in \{W_1, W_2\}$. For $\sigma \in$ Aut(F) with $\omega^\sigma = \omega$, $\theta_\sigma = \theta(\sigma, I \otimes I)$ is an automorphism of G stabilizing $\hat{\mathcal{J}}(\mathcal{C}(\delta_1))$ and mapping $\hat{A}(\infty)$ to itself. Therefore $\theta = \theta(id, A^{-1} \otimes B^{-1})\ cdot\theta(\sigma, I \otimes I) \cdot \theta(A \otimes B) = \theta(\sigma, A^{-\sigma} A \otimes B^{-\sigma} B)$ is the corresponding automorphism of G stabilizing $\hat{\mathcal{J}}(\mathcal{C}(\delta_2)) = \hat{\mathcal{J}}(\mathcal{C}(\delta)) = \hat{\mathcal{J}}(\mathcal{C})$ and mapping $\hat{A}(\infty)$ to itself. To compute θ is tedious but routine. But then one must compute the exponent i for which $\theta \cdot (I_0 \cdot I_\delta)^i$ leaves $\mathcal{R}_{(1,\delta^{\frac{1}{2}})}$ invariant. This seems more work than it is worth.

13.5. Oval Stabilizer: $q = 2^e$, $e \equiv 2$ (mod 4)

Since $e \not\equiv 2$ (mod 4), clearly 5 does not divide $q+1$. In this case it is shown in [70] that there is only one Subiaco oval $\mathcal{O}$ (up to projective equivalence) and that a group of order 2e stabilizing the oval is induced by the automorphism group of the GQ. In C. O'Keefe and J. A. Thas [59] it is shown that this group of order 2e is the complete stabilizer of $\mathcal{O}$, and indeed the complete stabilizer of the hyperoval containing $\mathcal{O}$. One step of the proof is to show that the hyperoval is an absolutely irreducible algebraic curve of degree 10. In S. E. Payne [67] the tensor product notation makes it possible to give this cyclic group C_{2e} of order 2e explicitly. First we give the oval.

$$\begin{array}{lll} (i) & \mathcal{O} = & \{(1, t, h(t)) : t \in F\} \cup \{(0, 1, 0)\}, \text{ where} \\ (ii) & h(t) = & \frac{(\delta^4+\delta^2)t^3+\delta^3 t^3+\delta^2 t}{t^4+\delta^2 t^2+1} + (t/\delta)^{1/2}. \end{array} \tag{13.11}$$

Let $\delta = \lambda + \lambda^{-1}$ as above. Put $\sigma = 2$ and $\bar{0} = \delta^{-1}$ in Theorem V.I of [70] to obtain

$$\bar{\theta} = \theta(\sigma = 2, \begin{pmatrix} 1 & \delta^{-\frac{1}{2}} \\ 0 & \delta^{-\frac{1}{2}} \end{pmatrix} \otimes \begin{pmatrix} \delta^{-\frac{1}{4}} & \delta^{\frac{1}{4}} \\ \delta^{-\frac{1}{4}} + \delta^{3/4} & \delta^{5/4} \end{pmatrix}). \tag{13.12}$$

Then $\bar{\theta}$ is an automorphism of G stabilizing $\hat{\mathcal{J}}(\mathcal{C})$ and mapping $\hat{A}(\infty)$ to itself. According to [70] for each $\alpha \in PG(1,q)$ there is a unique j such that $\bar{\theta} \cdot (I_0 \cdot I_\delta)^j$ stabilizes $\mathcal{R}_\alpha$. The involution $I_1 = \theta(id, P \otimes P)$ fixes $\mathcal{R}_{(1,1)}$ and $\mathcal{O}_{(1,1)}$ has a pleasant form. But the j for which $\bar{\theta} \cdot (I_0 \cdot I_\delta)^j$ stabilizes $\mathcal{R}_{(1,1)}$ leads to rather messy formulae. However, using Eqs. 13.5 and 13.6 it is easy to prove the following for $\alpha = (0,1)$.

(*i*) $\bar{\theta} \cdot (I_0 \cdot I_\delta)^{3/20}$ stabilizes $\mathcal{R}_{(0,1)}$ and has companion auto $\sigma = 2$.
(*ii*) $(I_0 \cdot I_\delta)^{-\frac{1}{40}}$ maps $\mathcal{R}_{(1,1)}$ to $\mathcal{R}_{(0,1)}$, so that
(*iii*) $\varphi = (I_0 \cdot I_\delta)^{\frac{1}{40}} \cdot I_1 \cdot (I_0 \cdot I_\delta)^{-\frac{1}{40}}$ is a (linear) involution fixing $\mathcal{R}_{(0,1)}$. (13.13)

In the following computations put $T = [\frac{1}{5}]$, so $T^5 + T^3 + T = \delta$. Note that since M and N are symmetric with determinant 1 (see Eq. 13.6), $PMP = M^{-1}$, $PNP = N^{-1}$. And now it is amusing to use the tensor algebra to compute

$$\begin{aligned}
\varphi &= \theta(id, M^{\frac{1}{40}} \otimes N^{\frac{1}{40}}) \cdot \theta(id, P \otimes P) \cdot \theta(id, M^{-\frac{1}{40}} \otimes N^{-\frac{1}{40}}) \\
&= \theta(id, M^{\frac{1}{40}}(PM^{-\frac{1}{40}}P)P \otimes N^{\frac{1}{40}}(PN^{-\frac{1}{40}}P)P) \\
&= \theta(id, M^{\frac{1}{20}}P \otimes N^{\frac{1}{20}}P) \\
&= \theta(id, \delta^{-\frac{1}{2}} \begin{pmatrix} [\frac{1}{10}] & [\frac{2}{5}] \\ [\frac{3}{5}] & [\frac{1}{10}] \end{pmatrix} \otimes \begin{pmatrix} 1 & \delta^{\frac{1}{2}} \\ 0 & 1 \end{pmatrix}) \\
&= \theta(id, \delta^{-\frac{1}{2}} \begin{pmatrix} T^{\frac{1}{2}} & T^2 \\ T^3 + T & T^{\frac{1}{2}} \end{pmatrix} \otimes \begin{pmatrix} 1 & \delta^{\frac{1}{2}} \\ 0 & 1 \end{pmatrix}).
\end{aligned}$$

In the context of Figs. 1 and 2, $C = \delta^{-1}TA_{T^3}, D = xP$ for some constant x, $E = \delta^{-1}(T^6 + T^2)A_{(\delta+T^3)^{-1}}, A^{(2)} = \delta^{-1}\begin{pmatrix} T & T^4 \\ T^6 + T^2 & T \end{pmatrix}$.
So applying Eq. (10.44) and multiplying through by δ we obtain:

$$\hat{\varphi} : (x,y,z) \mapsto (Tx + (T^6 + T^2)y, T^4x + Ty, \delta z + Tz_{T^3}x + (T^6 + T^2)z_{(\delta+T^3)^{-1}}y). \tag{13.14}$$

Actually, $\widehat{\varphi} = I_s$ where $s = [2/5]/[3/5] = T/(T^2 + I)$, because
$I_s = \theta(id, \begin{pmatrix} (1 + \bar{0}_s)^{1/2} & s\bar{0}^{1/2} \\ \bar{0}^{1/2} & (1 + \bar{0}s)^{1/2} \end{pmatrix} \otimes B_s)$ and s is easy to obtain as a ration of off-diagonal elements of A.

Also,

$$\bar{\theta} \cdot (I_0 \cdot I_\delta)^{3/20} = \theta(2, \frac{1}{\delta}\begin{pmatrix} [\frac{7}{10}] & [\frac{1}{5}] \\ [\frac{3}{10}] & [\frac{4}{5}] \end{pmatrix} \otimes \begin{pmatrix} \delta^{\frac{1}{4}} & \delta^{-\frac{1}{4}} \\ 0 & \delta^{-\frac{1}{4}} \end{pmatrix}) \tag{13.15}$$

yields a collineation $\hat{\theta}$ of $PG(2,q)$ stabilizing $\mathcal{O}_{(0,1)}$ and having companion automorphism $\sigma = 2$. In the context of Eq. 10.44,

$$A^{(2)} = \frac{1}{\delta^2}\begin{pmatrix} [\frac{7}{5}][\frac{2}{5}] \\ \\ [\frac{3}{5}][\frac{8}{5}] \end{pmatrix} = \frac{1}{\delta^2}\begin{pmatrix} T^7 + T^3 + \delta elta T^2 \\ T^3 + TT^8 \end{pmatrix} = \begin{pmatrix} a_4^2 a_2^2 \\ a_3^2 a_1^2 \end{pmatrix}.$$

Then $C = a_4^2 A_{(a_2/a_4)^2}$, $E = a_3^2 A_{(a_1/a_3)^2}$, and $\mu = \det(A) = \delta^{-\frac{1}{2}}$, so that

$$\hat{\theta} : (x,y,z) \mapsto ((x^2,y^2)A^{(2)},\ \ \delta^{-\frac{1}{2}} z^2 + x^2 a_4^2 z_{(a_2/a_4)^2} + y^2 a_3^2 z_{(a_1/a_3)^2}). \tag{13.16}$$

Then $\langle \hat{\theta}, \hat{\varphi} \rangle$ is the complete group of collineations of $PG(2,q)$ induced by $\mathcal{G}_0$ and stabilizing $\mathcal{O}_{(0,1)}$. In fact,
$(\hat{\theta})^e = \hat{\varphi}$ and $< \hat{\theta}, \hat{\varphi} > = < \hat{\theta} > \cong C_{2e}$.

14. Related semifield spreads.

14.1. Translation GQ; Translation Dual; Roman GQ

In this section we give an example of a GQ with parameters (q^2, q), where q is a power of 3, whose construction begins with a q-clan but requires somewhat more.

Let $\mathcal{C}$ be a q-clan for which the map $t \mapsto A_t$ is additive, i.e., $A_t + A_s = A_{t+s}$. Then for each $x \in F$,

$$\theta_x : (\alpha, c, \beta) \mapsto (\alpha, c + \alpha A_x \alpha^T, \beta + \alpha K_x)$$

induces a symmetry about $A^*(\infty)$ mapping the line $[A(t)]$ to $[A(t+x)]$. (A **symmetry** about a point p is a collineation fixing each point concurrent with p (points on lines incident with p), and necessarily acting regularly on the set of points not collinear with p.) The symmetries about the points on $[A(\infty)]$ generate an elementary abelian group $\mathcal{T}$ of order q^4. The q^2+1 groups of symmetries form a 4-gonal family for $\mathcal{T}$, so the point-line dual of $GQ(\mathcal{C})$ is a translation GQ (see [72]) with a kernel over which $\mathcal{T}$ and the members of the 4-gonal family are vector spaces. Taking the vector space dual yields a new translation GQ (TGQ).

Among the known examples of q-clans of this type there is only one case where the new TGQ is not isomorphic to the original. This example starts with the q-clan constructed by Gevaert and Johnson ([24]) from some planes found by Ganley. Here $q = 3^e$ and n is any nonsquare of F. For $t \in F$, $\gamma \in F^2$, put $\hat{g}_t(\gamma) =$

$$\gamma\begin{pmatrix} t & 0 \\ 0 & -nt \end{pmatrix}\gamma^T + \left[\gamma\begin{pmatrix} 0 & t \\ 0 & 0 \end{pmatrix}\gamma^T\right]^{1/3} + \left[\gamma\begin{pmatrix} 0 & 0 \\ 0 & -n^{-1}t \end{pmatrix}\gamma^T\right]^{1/9}.$$

Put $\mathcal{T} = F^4 = \{(r,c,b,d) : r,c,b,d \in F\}$ with coordinatewise addition. Define subgroups: $B(\infty) = \{(r,0,0,0) \in \mathcal{T} : r \in F\}$, $B^*(\infty) = \{(r,0,\gamma) \in \mathcal{T} : r \in F, \gamma \in F^2\}$. For $\gamma \in F^2$, $B(\gamma) = \{(-\hat{g}_c(\gamma), c, -c\gamma) \in \mathcal{T} : c \in F\}$. Write $\gamma = (g_1, g_2) \in f^2$.

Then $B^*(\gamma) = \{(r, c, b, d) \in \mathcal{T} : c, b, d \in F$ and $r = (cg_1^2 - cng_2^2 - bg_1 + dng_2) + (cg_1g_2 + bg_2 + dg_1)^{1/3} + (-cn^{-1}g_2^2 + dn^{-1}g_2)^{1/9}\}$.

Then $\mathcal{J} = \{B(\gamma) : \gamma \in F^2 \cup \{\infty\}\}$ is a 4-gonal family for $\mathcal{T}$ with tangent space $B^*(\gamma)$ at $B(\gamma)$. So in the usual way we have a TGQ $S = (S^{(\infty)}, \mathcal{T}) = S(\mathcal{T}, \mathcal{J})$ of order (q, q^2).

The new TGQ (called the Roman GQ by S. E. Payne [68]) have subquadrangles of order q isomorphic to $Q(4, q)$. This embedding of $Q(4, q)$ yields new ovoids of $Q(4, q)$, which correspond to symplectic spreads of $PG(3, q)$, giving new translation planes (cf. [79]) which are semifield planes due to the additivity of the associated functions. The planes constructed are not flock planes and do not appear to be previously known.

15. The Known finite examples of translation oval flocks, herds and the no-4 theorem.

The first known example of a flock of an oval cone which is not a quadratic cone is due to J.A. Thas and appears in Fisher and Thas [23]. The following explication may be found in Jha-Johnson [37].

Theorem 15.1. *(see [23] (3.11) and Jha-Johnson [37] (2.1)). Let q be even where $q \equiv -1 \bmod 3$ and let $\sigma \in Gal\ GF(q)$ such that $\{(1, t, t^\sigma), (0, 0, 1) : t \in GF(q)\} = O_1$ is a translation oval. Let homogenous coordinates (x_o, x_1, x_2, x_3) where $x_i \in GF(q)$,$i = 0, 1, 2, 3$ define the projective $3-$space $PG(3, q)$. Enbed the oval in the plane $x_3 = 0$ and form the translation oval cone by projective $(0, 0, 0, 1) = v_o$ to the oval O_1 in $x_3 = 0$.*

Then the following set of planes form a flock of the translation oval cone C_o :

$$\pi_s \ is \ s^{\sigma+1}x_o + s^\sigma x_1 + sx_2 + x_3 = 0 \ \forall s \in GF(q)\}.$$

Furthermore, this flock admits a doubly transitive group acting on the planes of the flock given as semi-direct product of S by T where

$$S = < \begin{bmatrix} 1 & s & s^\sigma & s^{\sigma+1} \\ 0 & 1 & 0 & s^\sigma \\ 0 & 0 & 1 & s \\ 0 & 0 & 0 & 1 \end{bmatrix} \forall s \in GF(q) >$$

and

$$T = < \begin{bmatrix} t^{\sigma+1} & 0 & 0 & 0 \\ 0 & t^{\sigma-1} & 0 & 0 \\ 0 & 0 & t^{\sigma-1} & 0 \\ 0 & 0 & 0 & t^{-1-\sigma} \end{bmatrix} \forall t \in GF(q) - \{0\} > .$$

15.1. Herds of translation oval cones.

Concerning flocks of arbitrary oval cones in $PG(3, 2^r)$, there may not be a corresponding herd of ovals. Following Cherowitzo, it is convenient to change the form of the planes of a translation oval flock. For such oval cones with corresponding automorphism α, the oval cone may be given the equation $y^\alpha = xz^{\alpha-1}$ while considering the points as homogeneous coordinates (x, y, z, w) in $PG(3, 2^r)$. When the order is even, the planes of an oval flock may be represented in the following form:

$\pi_t : xF(t) + t^{1/\alpha}y + \kappa zG(t) + w = 0$ for certain functions F and G and constant κ such that $Trace(\kappa) = 1$.

When $F(t) = t^{1\backslash\alpha^2}$ and $G(t) = t^{1\backslash\alpha+1\backslash\alpha^2}$ then Cherowitzo [15] shows that a herd of ovals is obtained exactly when $\{(1, t, t^{1\backslash\alpha+1\backslash\alpha^2})\ \forall t \text{ in } GF(q)\} \cup \{(0, 1, 0), (0, 0, 1)\}$ is a hyperoval in $PG(2, 2^r)$.

15.2. Flippable flocks.

Cherowitzo has considered an extension for translation oval flocks of what might be called algebraic derivation for even order flocks of quadratic cones. A translation oval flock with functions $(F(t), t^{1/\alpha}, G(t))$ is $\rho(t)-$ flippable exactly when the functions $(F(t)/\rho(t), t^{1/\alpha}/\rho(t), G(t)/\rho(t))$ provide a translation oval flock with the same element κ.

The following theorem connecting flippable flocks with herds is proved by Cherowitzo [15](we note that our terminology is slightly different from that of Cherowitzo).

Theorem 15.2. *If a translation oval flock provides a herd of ovals then the flock is $t-$flippable.*

Now consider the translation oval flock of Thas represented as above. Letting $G(t) = s^{\sigma+1}$, $F(t) = s$, $s^\sigma = t^{1/\alpha}$ where $\alpha = \sigma$, we may represent this flock using the Cherowitzo notation as:

$t^{1/\alpha^2}x + t^{1/\alpha}y + t^{1/\alpha+1/\alpha^2}z + w = 0$. We shall call these the $\alpha-$translation oval flocks of Thas.

Furthermore, Cherowitzo shows the equivalence of $t-$flippablity and the existence of oval-polynomials.

Theorem 15.3. *An $\alpha-$ translation oval flock of Thas is $t-$flippable if and only if $t^{1/\alpha+1/\alpha^2}$ is an oval-polynomial.*

In particular, if $q = 2^{2h-1}$ and $\alpha = \sigma = 2^h$ then the $\alpha-$translation oval flock of Thas is $t-$flippable. Furthermore, the $\sigma-$flocks are also $t^{2-1/\sigma}-$flippable as well as $t^{\sigma-3}-$flippable. Hence, three other classes of translation oval flocks are obtained by Cherowitzo all of which are somehow connected to the Thas translation oval flocks.

This theory is further extended by Cherowitzo and Storme [18] who prove general results of flipping, shifting, and herds of ovals. The following result is a special case and is phrased using our terminology.

Theorem 15.4. *(see [18] (2.2) and (2.3)) A translation oval flock of Thas is $t-$flippable if and only if there is a corresponding herd of ovals.*

Furthermore, Cherowitzo and Storme [18] completely determine the hyperovals of the form $\{(1,t,t^{2^i+2^j})\ \forall t \text{ in } GF(q)\} \cup \{(0,1,0),(0,0,1)\}$. In particular, the translation oval flocks of Thas which are $t-$flippable are completely determined. They also show that the only possible examples which are $t-$flippable are the previously known examples.

15.3. The no-4 theorem.

We have mentioned previously that with the exception of a few orders, Thas has completely determined the flocks of quadratic cones in $PG(3,q)$ such that no four of the planes containing the conics of the flock share a point (a no-4 flock). Considering the dual flocks, there is a set of q points no four of which are coplanar so there is an associated $q-$arc. By using the extension theory for $q-$arcs, Thas was able to determine the associated flocks. Furthermore, this can be done more generally and with a similar proof for flocks of oval cones in $PG(3,q)$.

Theorem 15.5. *(Thas [77], Jha-Johnson [37]).*
Let F be a no-4 flock of an oval cone in $PG(3,q)$. Assume that either q is even and ≥ 4 or odd with $q > 83$ or $q < 17$ or q in $\{27,81\}$.
Then either
(1) F is a Fisher-Thas-Walker flock of a quadratic cone or
(2) F is a translation oval flock of Thas and $q = 2^r$ where r is odd.

16. Examples of infinite flocks and maximal partial flocks.

In the infinite case, it is possible to construct flocks of oval cones over fields of even characteristic and some of these have herds of ovals as in the finite case (see Cherowitzo et al [16]).

16.1. Partial flocks of translation oval cones.

Let σ be an automorphism of an infinite field K of characteristic two.

Then $\{(t^\sigma,t,1),(1,0,0)\}$ is an oval with knot $(0,1,0)$ in $PG(2,K)$ represented by homogeneous coordinates (x,y,z) provided $\sigma-1$ is bijective. Embed the plane into $PG(3,K)$ and project from $(0,0,0,1)$ to form an oval cone. Representing $PG(3,K)$ by homogeneous coordinates (x,y,z,w) then it follows that $xz^{\sigma-1}=y^\sigma$ is the equation of the oval cone. Since the group

$$< (x,y,z) \to (x+a^\sigma z, y+az, z) \text{ for all } a \text{ in } K >$$

fixes $(1,0,0)$ (and $(0,1,0)$) and maps $(t^\sigma,t,1)$ onto $((t+a)^\sigma,(t+a),1)$ and the group is a translation group in the Pappian affine plane $z=1$, we call such an oval a translation oval and the corresponding cone a **translation oval cone**. Similarly, the corresponding Laguerre plane is called a **translation oval Laguerre plane.**

When σ is an automorphism of an infinite field K of characteristic two where $\sigma-1$ and $\sigma+1$ are bijective, it is possible to constuct infinite flocks analogous to the translation oval flocks of Thas as follows (see Cherowitzo et al [16]).

Theorem 16.1. *Let K be an infinite field of characteristic two. Let $\sigma \in Gal\ K$ such that $\sigma - 1$ is bijective so that $\{(1, t, t^{\sigma}), (0, 0, 1) : t \in GF(q)\} = O_1$ is a translation oval. Let homogenous coordinates (x_o, x_1, x_2, x_3) where $x_i \in GF(q)$,$i = 0, 1, 2, 3$ define the projective 3−space $PG(3, K)$. Enbed the oval in the plane $x_3 = 0$ and form the translation oval cone by projecting $(0, 0, 0, 1) = v_o$ to the oval O_1 in $x_3 = 0$.*

If $\sigma + 1$ and $x \mapsto x^2$ are both bijective mappings then the following set of planes form a flock of the translation oval cone C_o :

$$\pi_s \ is \ s^{\sigma+1}x_o + s^{\sigma}x_1 + sx_2 + x_3 = 0 \ \forall s \in GF(q)\}.$$

The condition that the mapping $x \mapsto x^2$ is bijective is only required to ensure that there is a doubly transitive group acting on the flock.

It is an open problem whether any of these flocks are t−flippable.

16.2. Quasifibrations and Partial flocks of oval cones

It is possible to construct partial flocks of such translation oval cones exactly as in the finite case. In the previous example, if $\sigma + 1$ is injective but not bijective then there is a maximal partial flock of a quadratic cone. More generally, if some line of the cone is covered by the planes of a partial flock then the partial flock is necessarily maximal. However, it is not always true that the partial flock is actually a flock. When the oval cone is a quadratic cone, one may translate everything over to the corresponding translation net to obtain a net with the property that some line of the net is covered by intersections with the set of lines incident with an affine point.

We mentioned previously class coverings with respect to partial flocks of Laguerre planes meaning that some parallel class is completely covered. Here we discuss a set of nets which are equivalent to certain maximal partial spreads of quadratic cones. The reader is referred to Jha-Johnson [31].

Definition 16.2. *Let N be an arbitrary net. Let P be an affine point and let L_P denote the set of lines which are incident with P. Let α be any parallel class and let M be a line of α which is not incident·with P. The net N shall be called class-covered with respect to M and P if and only if the points of M are contained in the intersections of M with L_P.*

Theorem 16.3. *A class-covered net with respect to line M and point P is either an affine plane or is a proper maximal net which contains no transversal containing P.*

The theorem of Abelian translation class-covered nets is similar to that of translation planes. In fact,

Theorem 16.4. *Let N be an Abelian translation class-covered net. Coordinates $(Q, +, *)$ may be chosen so that*

(0) there exists a sub-skewfield K such that $(Q, +)$ is a K−vector space,

(i) for each m in Q, there is an associated parallel class (m) so that the ternary functions is linear and defined for all m, b in $Q : T(x, m, b) = x * m + b$. Moreover, for $a \neq b$ and c, there is a unique solution to the equation $x * a = a * b + c$.

(ii) $(c+a)*m = c*m + a*m$ *for all* a, c *in* Q *and for all parallel class* (m).

Moreover, the lines through the zero vector are isomorphic $K-$*subspaces such that any two such subspaces direct sum to a vector space defining the net.*

Definition 16.5. *Let* V *be a vector space of the form* $W \oplus W$ *where* W *is a* $K-$*space for* K *a skewfield. A partial spread is a set of subspaces* $K-$*isomorphic to* W *such that the direct sum of any two distinct subspaces is* V.

Let B *be a basis for* W *and choose a fixed vector* e *from* $W \oplus 0$. *A quasifibration* Q *is a set of mutually disjoint subspaces* $K-$*isomorphic to* W *which contains* $W \oplus 0$ *and* $0 \oplus W$ *and which has the property that for each vector of the form* $e + w$ *for* w *in* $0 \oplus W$, *there is a subspace of* Q *which contains* $e + w$. *Furthermore,* K *is called the kernel of the quasifibration if and only if* K *is maximal among skewfields* L *such that the subspaces indicated are* $L-$*spaces.*

Coordinates systems just short of quasifields may be defined as follows:

Definition 16.6. *Let* $(Q, +, *)$ *be a triple so that*

(i) $(Q, +)$ *is an Abelian group,*

(ii) $(Q - \{0\}, *)$ *is a binary operation so that* $a * 0 = 0 * a = 0$ *and ther is an element* 1 *in* Q *so that* $1 * m = m$ *for all* a, m *in* Q,

(iii) $(a+b)*m = a*m + b*m$ *for all* a, b, m *on* Q,

(iv) for $a \neq b$, *in* Q *there is a unique solution to the equation* $x * a = x * b + c$.

Then we shall call $(Q, +, *)$ *a quasi-quasifield.*

Now consider a class-covered partial flock of a Laguerre plane. If the Laguerre plane is an oval plane then there is a line L which is completely covered by the planes of the partial flock. If the oval is a conic there is an associated partial spread π_L. The points of the line L become Baer subplanes of the corresponding partial spread π_L each of which intersects the common component in a $1-$dimensional space $< e >$. It turns out that the line $x = e$ and the zero vector provides a quasifibration.

Theorem 16.7. *The following are equivalent:*

(1) *Abelian translation class-covered nets,*

(2) *quasifibrations,*

(3) *quasi-quasifields.*

Corollary 16.8. *An Abelian translation class-covered net in* V_4 *over a field* K *which admits an elation group* E *such that any component orbit union the axis forms a regulus in* $PG(3, K)$ *is equivalent to a class-covered partial flock of a quadratic cone which is either a flock or a proper maximal partial flock.*

The following provides an example of this type of partial flock.

Theorem 16.9. *Let* K *be a field in which the function* $x \mapsto x^3$ *is injective but not surjective. Let* V_4 *denote a* $4-$*dimensional vector space over* K. *Then the following equations define a quasifibration which is a proper maximal partial spread. Further, there is a corresponding maximal partial flock of a quadratic cone.*

$$\{x = 0, y = x \begin{bmatrix} u - s^2 & -s^3/3 \\ s & u \end{bmatrix} \forall u, s \in K\}.$$

16.3. Examples of infinite flocks and derivation.

De Clerck and Van Maldeghem [21] have determined several examples of flocks of infinite quadratic cones. They have considered infinite flocks algebraically and have given conditions under which there is an associated generalized quadrangle. Furthermore, Jha-Johnson [33] continue this study and, in particular, construct some infinite nests of reguli which exist within a Pappian spread and which provide an infinite analog to the translation planes arising from the finite Fisher flocks mentioned above. (The reader interested in reading more about infinite nests of reguli and their construction is referred to Johnson [47].)

Using ideas on quasifibrations and class-covered nets, it is possible to show the following (see Jha-Johnson [33]). Recall that a full field of characteristic 0 or odd is a field such that the product of any two nonsquare elements is a square. If the characteristic is two let C_1 denote the elements k such that $x^2 + x + k$ is irreducible. A full field of characteristic two is a field such that the sum of two elements of C_1 is never in C_1. Note that finite fields are full fields.

Theorem 16.10. *Let F be a flock of a quadratic cone in $PG(3, K)$. If K is a full field then any derivation is either a flock or a maximal partial flock.*

16.4. The bilinear flocks.

In [22], Dembowski notes that the following provides a flock of an elliptic quadric in PG(3,R) where R denotes the field of real numbers: Let O denote an elliptic quadric and let π_o denote the equatorial plane. Let L, M be distinct affine lines of π_o which are exterior to O. Let N denote the set of planes which contain L and meet O in the positive (north) half and let S denote the set of planes which contain M and meet O in the negative (south) half. Then $N \cup S \cup \{\pi_o\}$ is a cover of $O - \{N_L, S_M\}$ by conics where N_L, S_M are the north and south points of tangency of the planes containing L, M respectively.

That is, this provides a "nonlinear" flock of the elliptic quadric in PG(3,R) in the sense that not all of the planes share the same line. (A linear flock is one for which the planes of the flock all share a common line.)

More generally, we shall call any flock of an elliptic quadric, hyperbolic quadric or quadratic cone in $PG(3, K)$,for K an arbitrary field, **a bilinear flock** if and only if the set of planes which contain the conics of the flock share one of two lines but do not share one line.

It is possible to use the little theme of Dembowski to explore some bilinear flocks of quadratic cones in $PG(3, K)$ for certain fields K. Basically, the idea is to consider two lines L,M on the hyperplane at infinity and construct a flock such that the positive affine cone is covered by planes containing L and the negative affine cone is covered by planes containing M. Biliotti and Johnson [9] show

Theorem 16.11. *Let K be an ordered field. Suppose two distinct polynomials $f(x)$ $= x^2 + a_2x - a_1$ and $g(x) = x^2 + b_2x - b_1$ are both always positive.*

Consider $PG(3, K)$ represented in homogeneous coordinates (x_o, x_1, x_2, x_3). Represent a quadratic cone C with equation $x_ox_1 = x_2^2$.

Define the following set $F(a_1,a_2,b_1,b_2)$ *of planes with the following equations*
$N_t : tx_o - a_1tx_1 + a_2tx_2 + x_3 = 0, t > 0,$
$S_t : tx_o - b_1tx_1 + b_2tx_2 + x_3 = 0, t < 0,$
$I_o : x_3 = 0.$
Then $F(a_1,a_2,b_1,b_2)$ *is a bilinear flock of the quadratic cone* C.

We denote the corresponding translation planes by $\pi(a_1,a_2,b_1,b_2)$. Recall any flock translation plane admits an elation group whose component orbits union the axis are the reguli corresponding to the flock. For the bilinear flock planes, there are two distinct elation groups which induce reguli. The following result is also in Biliotti and Johnson [9].

Theorem 16.12. *Any bilinear conical translation plane with spread in* $PG(3,K)$ *admits a collineation group in the translation complement isomorphic to* $SL(2,K)$ *generated by regulus inducing elation groups.*

16.5. Tangential packings

Since any bilinear conical translation plane admits $SL(2,K)$, there is not one corresponding quadratic cone but actually a set of quadratic cones which are permuted by $PSL(2,K)$. Moreover, there are some interesting ovoids associated with such spreads.

Definition 16.13. *Let* Q *be an ovoid of points of the Klein quadric in* $PG(5,K)$ *where* K *is a field. Assume that there is a conic* C *in* Q *such that, for each point* P *of the conic, there is a partition of* Q *into a set of conics containing* C *which mutually share the point* P. *In this case, we shall call the set of such partitions a tangential packing of the ovoid with respect to* C.

The following five results may be found in Biliotti and Johnson [9].

Theorem 16.14. *Let* K *be a field and let* π *denote a translation plane with spread in* $PG(3,K)$*which contains a regulus* R .

Suppose that there is a collineation group G *generated by elations in the net* N_R *corresponding to* R *which is isomorphic to* $SL(2,K)$ *and leaves* N_R *invariant.*

Then, for each component L *of the regulus net* N_R , *there is a set of regulus nets which share* L *and which are left invariant by an elation group with axis* L *which acts regularly on the components of each regulus net distinct from* L.

We now consider the associated flocks.

Theorem 16.15. *Under the assumptions of the previous theorem, for each of these components* L, *there is a flock of a quadratic cone* F_L *in some 3-dimensional projective space* Σ_L^3 *isomorphic to* $PG(3,K)$ *which has as its basic points the Baer subplanes of the regulus nets that share the component* L.

Hence, each of these flocks of quadratic cones share a conic.

Each of these projective spaces Σ_L^3 *lies in the same projective space* Σ^5 *isomorphic to* $PG(5,K)$*and there is a collineation group of* Σ^5 *which fixes the common conic and acts transitively on the cones and flocks.*

Furthermore, the ovoid in Σ^5 corresponding to the spread of π admits a tangential packing with respect to the conic corresponding to the regulus R.

The following theorem is the converse to the preceding result.

Theorem 16.16. *Assume that there is a flock of a quadratic cone F in a 3-dimensional projective space Σ^3. Embed Σ^3in the 5-dimensional projective space Σ^5 so that the cone is embedded within the Klein quadric. Take the intersection with the Klein quadric of the perpendicular planes of the planes which contains the conics of the flock. This forms an ovoid Q of points consisting of conics which share a point.*

For a fixed conic C , assume there is a tangential packing of the ovoid with respect to CThen there is a corresponding translation plane with spread in $PG(3,K)$ which admits $SL(2,K)$ generated by elations and which fixes a regulus net of the spread.

Finally, we note the connection with $SL(2,K)$ and coverings of the spread by sets of reguli that share a given component.

Theorem 16.17. *Let π be a translation plane with spread in $PG(3,K)$.*

If there is a regulus R in the spread of π such that for each component L of R, there is a set of reguli in the spread which share L and cover the component set, then π admits $SL(2,K)$ as a collineation group which is generated by elations leaving invariant the regulus R.

16.6. Double Covers

As we have seen, in the finite case, a flock of a quadratic cone corresponds to a translation plane with spread in $PG(3,K)$ which is covered by reguli that share a component L, and if the plane is not Desarguesian then there is a unique such cover. In the general case, we shall say there is a "double cover" if and only if there are two components L,M (possibly the same) such that the spread is covered by a set of reguli which share L and a distinct set of reguli which share M.

Theorem 16.18. *Let π be a translation plane with spread in $PG(3,K)$ whose spread admits a double cover of reguli.*

Then either π is Pappian or there is a group generated by elations isomorphic to $SL(2,K)$ which leaves a regulus invariant.

16.7. The derivations

There exist putative derivations for $F(a_1,a_2,b_1,b_2)$ (see (16.12)) which define flocks at least in the case when $b_2 = a_2 = 0$ and K is an ordered field such that the positive elements are always square. It follows from De Clerck and Van Maldeghem [21] that corresponding to these bilinear flocks in $PG(3,K)$ are classes of generalized quadrangles.

On the other hand, when K is a subfield of the reals such that not all positive elements are square, then conceivably none of the $s-$derivations produce flocks. In particular, none of these $s-$derivations can produce flocks over the field of rational numbers. **Hence, we have examples of flocks of quadratic cones in $PG(3,K)$ which have no corresponding generalized quadrangles.**

The reader is referred to Biliotti and Johnson for further details [9].

Furthermore, in Biliotti and Johnson [10], the construction of bilinear flocks is extended to $n-$linear flocks defined over subfields of the reals.

16.8. Infinite generalized quadrangles.

As mentioned earlier, given a flock of a quadratic cone, there is not necessarily a corresponding generalized quadrangle. However, we have also seen that when the associated field K is a full field, it is possible to show that there is a maximal partial generalized quadrangle (see Jha-Johnson [33]). The first attempt to construct flock quadrangles in the infinite case was by F. De Clerck and H. Van Maldeghem [21], who used the coordinatization of a generalized quadrangle by means of a quadratic quaternary ring (see [26]) to extend to the infinite case the theory of flocks of quadratic cones and their associated geometries, especially the generalized quadrangles. Starting with the notion of a full field K , they define a K-clan as a triple of functions that give a flock of a certain quadratic cone, imitating the situation for finite fields. Using coordinates they bypass the coset geometry construction of the generalized quadrangle (as given in our section 4.2) and proceed directly to give a description of the associated flock quadrangle, which in fact they show will exist if and only if all derivations of the original flock are well defined flocks. The approach in L. Bader and S. E. Payne is complementary to that in [21]. Bader and Payne first obtain an appropriate generalization to the infinite case of the notion of 4-gonal family (Kantor family) of subgroups of a given group and then interpret this in the context of the K-clan. This allows them to characterize infinite elation generalized quadrangles as group coset geometries and to give another proof of the result of De Clerck and Van Maldeghem [21] just mentioned. They also generalize the Fundamental Theorem of q-clan geometry to apply to the study of isomorphisms between infinite partial and complete generalized quadrangles constructed from flocks and partial flocks of quadratic cones. They conclude by surveying the known infinite families of finite flocks to see which ones easily generalize to give families of infinite generalized quadrangles.

16.9. Examples over the reals.

Let R be the field of real numbers. Then the union of two semi-ellipses can provide an oval which is not a translation oval and it is possible to construct partial flocks of such oval cones.

More generally, the following two results are in Johnson and Liu [49].

Proposition 16.19. *Let K be the field of real numbers. Consider the set* $O_a = \{(1,t^2,t) \text{ for } t \geq 0\} \cup \{(1,at^2,t) \text{ for } t<0\} \cup \{(0,1,0)\}$.

If a is positive then O_a is an oval in $PG(2,K)$.

Theorem 16.20. (1) *Let f be a continuous, strictly increasing function such that $lim_{t\mapsto\pm\infty} f(t) = \pm\infty$ over the field of real numbers R .*

Then the planes $\pi_t : tx_0 - x_1 f(t) + x_3 = 0$ for all t in R define a flock of a quadratic cone in $PG(3, R)$.

(2) *The planes of* (1) *define a flock of the semi-elliptic oval cone defined by O_a for any positive real number a.*

There is a tremendous variety of examples of infinite flocks. Instead of listing these here, we refer the reader instead to Johnson and Liu [48] of this volume where these examples are collected.

16.10. Flock type planes over skewfields.

Although not actually necessarily providing examples of infinite flocks, it is possible to consider generalizations of the translation planes associated with infinite flocks.

Given a regulus in $PG(3, F)$ for F a field, we have noticed that a standard form may be given so that over the corresponding vector space, the form is $x = 0, y = x \begin{bmatrix} u & 0 \\ 0 & u \end{bmatrix}$ for all u in F. A set of left 2–dimensional subspaces of the same form but over a skewfield K is called a pseudo-regulus. Hence, it may be possible to consider spreads which are unions of pseudo-reguli that share exactly one line (component) even though there may not be corresponding flocks of cones or generalized quadrangles. This is actually done in Jha-Johnson [36] where there is a general theory developed for such spreads. In particular, there is an easy construction of what might be called "skew-Hall" planes which are planes derived from planes which have the general appearance of a Pappian plane but which are defined more generally over quadratic extensions of skewfields. These planes not only have spreads in $PG(3, K)$ which are unions of pseudo-reguli that share a common line, the spreads also are unions of other sets of pseudo-reguli that mutually share two lines. It is interesting to note that the "skew-Hall" planes have kernels $Z(K)$, the center of K and hence may not all be defined by spreads in $PG(3, K)$.

Theorem 16.21. *Let K be a skewfield and let π denote a 4–dimensional left K–vector space.*

(1) Then the set of left 2–dimensional K–subspaces

$$\{x = 0, y = x \begin{bmatrix} u + \rho t & \gamma t \\ t & u \end{bmatrix} \textit{ for all } t, u \textit{ in } K\}$$

is a spread in $PG(3, K)$ if and only if $x^2 + x\rho - \gamma$ is irreducible over K where x and y are 2-vectors. Let $\pi_{\rho,\gamma}$ denote the corresponding translation plane.

(2) $\pi_{\rho,\gamma}$ is a Desarguesian plane if and only if both ρ and γ are in the center of K.

(3) $\pi_{\rho,\gamma}$ is a semifield plane which admits the elation group

$$E_u = < \begin{bmatrix} 1 & 0 & u & 0 \\ 0 & 1 & 0 & u \\ 0 & 0 & 1 & 0 \\ 0 & 0 & 0 & 1 \end{bmatrix} \textit{ for all } u \textit{ in } K > .$$

The component orbits the union the axis $x = 0$ of E defines a set C_u of pseudo-regulus nets that share exactly the component $x = 0$, and each of these nets may be derived to produce an associated translation plane.

(4) $\pi_{\rho,\gamma}$ also admits the elation group

$$T_t = < \begin{bmatrix} 1 & 0 & \rho t & \gamma t \\ 0 & 1 & t & 0 \\ 0 & 0 & 1 & 0 \\ 0 & 0 & 0 & 1 \end{bmatrix} \text{ for all } t \text{ in } K > .$$

The component orbits the union the axis $x = 0$ of E defines a set C_t of pseudo-regulus nets that share exactly the component $x = 0$, and each of these nets may be derived to produce an associated translation plane.

(5) $\pi_{\rho,\gamma}$ also admits the homology group

$$H = < \begin{bmatrix} 1 & 0 & 0 & 0 \\ 0 & 1 & 0 & 0 \\ 0 & 0 & u & 0 \\ 0 & 0 & 0 & u \end{bmatrix} \text{ for all } u \text{ in } K^* > .$$

The component orbits union the axis $(y = 0)$ and the coaxis $(x = 0)$ defines a set R of pseudo-regulus nets that share two components $x = 0, y = 0$ and each of these nets may be derived to produce an associated translation plane.

(6) For each derived translation plane, the kernel of the plane is $Z(K)$.

In particular, if K is not a field then the plane derived from a Desarguesian plane does not have its spread in $PG(3, K)$ and hence cannot be considered a Hall plane even in this case.

Also, note that if K is infinite dimensional over $Z(K)$ then the derived planes are infinite dimensional over their kernels.

17. Rigidity of flocks

Let Π be a partial flock of $t-$conics of a finite quadratic cone. If the planes of Π all share a common point exterior to the cone then it can be shown that there is an automorphism group of the cone which leaves the vertex invariant and which fixes each plane of Π. Conversely, a group acting as above forces the planes of the flock to share a point. We call such a partial flock a "star" partial flock and term the group "$t-$locally rigid". In Jha-Johnson [29], there is an analysis of such groups and star partial flocks. In particular,

Theorem 17.1. *(The Fundamental Theorem on Rigidity.) Let Π be a partial flock of $t-$conics of a finite quadratic cone in $PG(3, q)$. Let G be a $t-$locally rigid group on Π. If G contains a nontrivial linear subgroup then the t planes share a common point.*

Conversely, if there exists a partial flock Π of $t-$ conics such that the corresponding planes share a common point then there is a nontrivial linear $t-$locally rigid group on Π.

We have mentioned that Thas has classified all star flocks of quadratic cones. The theorem of Thas [76] is

Theorem 17.2. *Let $F = \{C_i$ for $i = 1, 2, ..., q\}$ be a flock of a quadratic cone Q in $PG(3, q)$ such that the planes π_i containing the conics C_i for $i = 1, 2, ..., q$ contain a common point; F is a star flock.*

(1) If q is even then F is linear.

(2) (a) If q is odd and the common point is interior then F is linear.

(b) If q is odd and the common point is exterior then either F is linear or a Kantor-Knuth flock (the generalized quadrangle is due to Kantor and the translation plane is due to Knuth).

When the order q is even there is an associated "axial" theorem for star flocks (see Jha-Johnson [29]). We say that an automorphism group of a flock is t−rigid if and only if it is t−locally rigid when acting on some subflock of t conics.

Theorem 17.3. *(The Axial Theorem.) Let F be a flock of a quadratic cone in $PG(3, q)$*

(1) If q is even and there exists a nontrivial t−rigid, linear automorphism group of F then the t planes of the associated partial flock share a line.

(2) If q is odd and there exists a nontrivial t−rigid linear automorphism group of F of order > 2 then the t planes of the associated partial flock share a line.

It is possible to skirt the linearity assumption when t is large enough.

Theorem 17.4. *Let F be a flock of a quadratic cone in $PG(3, q)$ that admits a t−rigid automorphism group with $t \geq (q - 1)/2 \geq 2$.*

(1) If q is even then F is linear.

(2) If q is odd and the order of the group is > 2 then F is linear or a Fisher flock.

Finally, it might be mentioned that the theorem of Thas for star partial flocks contained within flocks can be improved as follows (again see Jha-Johnson [29]).

Theorem 17.5. *Let F be a flock in $PG(3, q)$ which contains a star partial flock of t conics.*

(1) If q is even and $t \geq q/2$ then the flock is linear.

(2) If q is odd and $t \geq q - 1$ then the flock is linear or a Kantor-Knuth flock.

18. The Doubly Transitive Finite Flocks of oval cones

We have pointed that that the translation oval flocks of Thas admit a linear doubly transitive automorphism group acting on the planes of the flock. Furthermore, the Fisher-Thas-Walker flocks also admit such a group. Any semifield flock of a quadratic cone admits a transitive group on the conics and, in addition, the Kantor-Knuth flock admits a doubly transitive group. Furthermore, we have noted in the finite case certain situations which produce doubly transitive flocks. However, in all of the known examples, the cone is a translation oval cone. In fact, this is always the case when the flock is finite and there is evidence to suggest that this is the case when flock is infinite although here the situation is much more complex.

The following is due to Jha and Johnson and may be found in [30] and [32]. It might be noted that the proof uses the theory of rigidity and so does not necessarily extend to the infinite case.

Theorem 18.1. *Let F be a flock of an oval cone in $PG(3,q)$ that admits a linear group acting doubly transitively on the ovals of the flock. Then one of the following occurs:*

(1) *F is linear or*

(2) *the oval is a translation oval.*

(*a*) *If the translation oval is not a conic then the flock is one of the translation oval flocks of Thas.*

(*b*) *If the oval is a conic and the flock is not linear then q is odd and the flock is one of the following types:*

(*i*) *a semifield flock of Kantor-Knuth type.*

(*ii*) *a Fisher-Thas-Walker flock.*

18.1. Generalized Kantor-Knuth flocks

There is a natural way to generalize the Kantor-Knuth flocks to the infinite or arbitrary order situation. For example, some generalizations occur in De Clerck and Van Mandeghem [21]. Other generalizations occur in Johnson and Liu [48]. Also, in Johnson and Liu [49], there is a complete determination of the isomorphism classes analogous with the work of Kantor [55] in the finite case. It is noted that all of these flocks admit doubly transitive groups on the planes of the flock.

19. Transitive finite partial flocks of deficiency one and monomial flocks of even order.

Let Π be a partial flock of deficiency one of a finite quadratic cone. It has been mentioned previously that by Payne and Thas, Π has a unique extension to a flock. Furthermore, there is a Baer group of order q associated with any finite partial flock. If Π admits a transitive group acting on the $q-1$ planes of the partial flock then there is a description of such flocks by their associated functions due to Hiramine and Johnson [27]. For q odd, the results involve several cases, but for even order, it is possible to show that the associated functions are monomial.

Theorem 19.1. *Let Π be a partial flock of deficiency one in $PG(3,2^r)$. Represent the cone by $x_o x_1 = x_2^2$ where the projective space is represented by homogeneous coordinates (x_o, x_1, x_2, x_3).*

If Π admits a linear transitive group then the planes of the flock may be represented in the following form:

$\pi_t : tx_o + \alpha t^{2k+1}x_1 + \beta t^{k+1}x_2 + x_3 = 0$ *for all t in $GF(2^r)$ and α, β constants.*

The known examples are for $k = 1$ or 2 or when $k = 0$ and are the Fisher-Thas-Walker, Payne and linear flocks respectively.

Recently, Penttila and Storme [74] have shown that

Theorem 19.2. *Let F be any flock of a quadratic cone in $PG(3, 2^r)$ whose associated plane are given by monomial funtions. Then the flock is either one of the flocks of Fisher-Thas-Walker, Payne or is linear.*

Hence, we obtain the following

Theorem 19.3. *Let Π be a transitive partial flock of deficiency one of a quadratic cone in $PG(3, 2^r)$. Then Π may be uniquely extended to either a Fisher-Thas-Walker, Payne or linear flock.*

Proof: Combine the above results with the extension result of Payne and Thas [72].

20. MAXIMAL PARTIAL SPREADS

D. Jungnickel [54] has given some constructions of maximal partial spreads which have no transversals. They have cardinality $q^2 - q$, $q^2 - q + 1$, $q^2 - q + 2$. The idea for most of these partial spread constructions goes back to A. Bruen [11] who constructs maximal partial spreads as follows.

Let $\mathcal{S}$ be any non-regular spread of $PG(3, q)$. Then there exists a line m not in $\mathcal{S}$ such that the partial spread $\mathcal{P}_m = \{l \in \mathcal{S} \mid l \cap m \neq \emptyset\}$ is not a regulus, and there are exactly 1 or 2 transversal lines of $\mathcal{P}_m$. Let $\mathcal{S}_m$ denote the set of all lines of $\mathcal{S}$ not in $\mathcal{P}_m$ union m and another possible transversal line of $\mathcal{P}_m$. Then this produces a maximal partial spread of $PG(3, q)$ of cardinality $q^2 - q + 1$ or $q^2 - q + 2$. Of course, in general, it is difficult to determine the exact cardinality, and there is still the question of when the translation net defined by $\mathcal{S}_m$ can be embeddable in an affine plane $\pi_{\mathcal{S}_m}$. If so, then the plane $\pi_{\mathcal{S}_m}$ must be a translation plane. In this case, this is equivalent to determining if $\pi_{\mathcal{S}_m}$ is derivable and further, if so, this forces the cardinality of $\mathcal{S}_m$ to be $q^2 - q + 2$ (see [54]).

Now let $\mathcal{F}$ be a flock of a quadratic cone in $PG(3, q)$, $\mathcal{S} = S(\mathcal{F}) = \mathcal{R}_1 \cup \mathcal{R}_2 \cup ... \cup \mathcal{R}_q$ the associated spread which is the union of the q reguli $\mathcal{R}_i$ that have a common line l_∞. We will call $S(\mathcal{F})$ a flock spread and $\mathcal{R}_1$, $\mathcal{R}_2$,...$\mathcal{R}_q$ the base reguli.

Take any line m which is not a transversal line of one of the base reguli. If $\mathcal{F}$ is not linear, then there are exactly q reguli in the spread $\mathcal{S} = S(\mathcal{F})$ (see [25]). By the above remarks there is a maximal partial spread $\mathcal{S}_m$ of degree $q^2 - q + 1$ or $q^2 - q + 2$ containing m.

By the results on *rigidity* of V. Jha and N.L. Johnson [29] , Johnson and Lunardon [50] prove the following theorem.

Theorem 20.1. *Let $\mathcal{F}$ be a flock of a quadratic cone of $PG(3, q)$. Suppose there are two lines m and n of $PG(3, q)$ not transversal lines of a base regulus which share the same lines of $S(\mathcal{F})$. Then $\mathcal{F}$ is a linear flock or a Kantor-Knuth semifield flock.*

Therefore if $\mathcal{S}_m$ has degree $q^2 - q + 2$, then $\mathcal{F}$ is the Kantor-Knuth semifield flock.

It follows from [8] that in a derivable net of a translation plane with spread in $PG(3, K)$, where K is isomorphic to $GF(q)$, there are at least two $K-$Baer subspaces of the net. This means that there are two transversals m_1 and m_2 to the same set of $q+1$ components. If the elation axis L intersects m_i $(i = 1, 2)$, then N.L. Johnson [42] has proved that $\mathcal{F}$ is either linear or a Kantor-Knuth semifield flock. By the previous results, Johnson and Lunardon [51] prove the same result without the hypothesis that L and m_1 have a point in common.

It is also possible to characterize flock spreads by the existence of maximal partial spreads.

The following appears in Johnson and Lunardon [51].

Theorem 20.2. *Let $\mathcal{S}$ be a spread of $PG(3, q)$. There are exactly $q^4 + q^3$ lines m such that $\mathcal{S}_m$ has degree $q^2 - q + 1$ or $q^2 - q + 2$ if and only if there are q reguli in $\mathcal{S}$. Moreover if for some line l of $\mathcal{S}$ there are exactly $q^2(q + 1)$ maximal partial spreads produced from $\mathcal{S}$ which do not contain l, then $\mathcal{S}$ corresponds to a flock of a quadratic cone.*

There are various other consequences of the numbers of maximal partial spreads. We mention only the following result.

Theorem 20.3. *If the flock $\mathcal{F}$ of the quadratic cone is neither linear nor a Kantor-Knuth semifield flock, there are exactly $q^4 + q^3$ mutually distinct maximal partial spreads each of degree $q^2 - q + 1$ which may be obtained from the spread corresponding to $\mathcal{F}$.*

21. Partial flocks of quadratic cones in $PG(n, q)$

Finally, we might note that O'Keefe and Thas [60] recently study partial flocks in $PG(n, q)$ for n odd. There are generalizations of herds, partial flocks, partial ovoids and derivations. Furthermore, various interesting examples are introduced. The reader is referred directly to this article for complete details.

References

[1] L. Bader, Derivation of Fisher flocks. Preprint.

[2] L. Bader, G. Lunardon and S. E. Payne, On q-clan geometry, $q = 2^e$, *Bull. Belgian Math. Soc., Simon Stevin*, 1(1994), 301 - 328.

[3] L. Bader and S.E. Payne. On infinite K-clan geometry. Preprint.

[4] L. Bader, G. Lunardon and J. A. Thas, Derivation of flocks of quadratic cones, *Forum Math.*, 2(1990), 163 – 174.

[5] A. Barlotti. On the definition of Baer subplanes of infinite planes. J. Geom. 3 (1973), 87-92.

[6] A. Barlotti and R. C. Bose, Linear representation of a class of projective planes in four dimensional projective spaces, *Ann. Mat. Pura Appl.*, 88(1971), 9 – 31.

[7] M. Bernardi. Esistenza di fibrazioni in uno spazio proiettivo infinito. Ins. Lonbardo Accad. Sci. Lett. Rend. A 197 (1973), 528-542.

[8] M. Biliotti and G. Lunardon, Insiemi di derivazione e sottopiani di Baer in un piano di traslazione. Atti. Acc. Naz. Lencei. Cl. Sci. Fis. Mat. Nat. (8) 69 (1980).

[9] M. Biliotti and N.L. Johnson. Bilinear flocks of quadratic cones. J. Geometry (to appear).

[10] M. Biliotti and N.L. Johnson. Variations on a theme of Dembowski. **Mostly Finite Geometries**. Proceedings of a conference in honor of T.G. Ostrom, Marcel Dekker (to appear).

[11] A. Bruen. Inversive geometry and some translation planes, I. Geom. Ded. 7 (1978), 81-98.

[12] A. Bruen and J.C. Fisher. Spreads which are not dual spreads. Canad. Math. Bull. 12 (1969), 801-803.

[13] F. Buekenhout,editor, **HANDBOOK OF INCIDENCE GEOMETRY: Buildings and Foundations**, North Holland, 1995, xi + 1420 pp.

[14] P.J. Cameron. Projective and Polar Spaces. University of London, Queen Mary and Westfield College, QMW Notes 13.

[15] W. Cherowitzo. $\alpha-$flocks and hyperovals. Preprint.

[16] W. Cherowitzo, N.L. Johnson, X. Liu, L. Storme. Infinite oval flocks. Preprint.

[17] W. Cherowitzo, T. Penttila, I. Pinneri, G. Royle, Flocks and ovals, *Geom. Ded.*, to appear.

[18] W. Cherowitzo and L. Storme. $\alpha-$Flocks with oval herd and monomial hyperovals. Preprint.

[19] F. De Clerck and C. Herssens, Flocks of the quadratic cone in $PG(3,q)$, for q small, *Reports of the CAGe project*, 8(1993), 1 – 75.

[20] F. De Clerck, H. Gevaert and J. A. Thas, Flocks of a quadratic cone in $PG(3,q)$, $q \leq 8$, *Geom. Ded.*, 26(1988), 215 – 230.

[21] F. De Clerck and H. Van Maldeghem. Flocks of an infinite quadratic cone. Bulletin Belg. Mat. Soc/ Simon Stevin, (1994), 399-415.

[22] P.Dembowski. **Finite Geometries.** Springer Verlag, Berlin, Heidelberg, New York, n967.

[23] J. C. Fisher and J. A. Thas, Flocks in $PG(3,q)$, *Math. Zeit.*, 169(1979), 1 – 11.

[24] H. Gevaert and N. L. Johnson, Flocks of quadratic cones, generalized quadrangles and translation planes, *Geom. Ded.*, 27(1988), 301 – 317.

[25] H. Gevaert, N. L. Johnson and J. A. Thas, Spreads covered by reguli, *Simon Stevin*, 62(1988), 51 – 62.

[26] G. Hanssens and H. Van Maldeghem, Coordinatization of generalized quadrangles, Ann. Discrete Math., 37 (1988), 195 – 208.

[27] Y. Hiramine and N.L. Johnson, Regular partial conical flocks. Bull. Belg. Math. Soc. 2 (1995), 419-433.

[28] V. Jha and N.L. Johnson, Derivable nets defined by central collineations, Information and Systems Sci., vol. 11, nos.2-4 (l986),83-91.

[29] V. Jha and N.L. Johnson, Rigidity in Conical flocks. J. Comb. Theory (A), 73, (1996), 60-76.

[30] V. Jha and N.L. Johnson, Automorphism groups of oval cones . Geom. Ded. 61 (1996), 71-85.

[31] V. Jha and N.L. Johnson, Quasifibrations. Bull. Belg. Math. Soc. 3 (1996), 313-324.

[32] V. Jha and N.L. Johnson, The doubly transitive flocks of quadratic cones. Europ. J. Comb.(to appear).

[33] V. Jha and N.L. Johnson, Infinite flocks of a quadratic cone. J. Geometry (to appear).

[34] V. Jha and N.L. Johnson. Structure theory for point-Baer and line-Baer collineation groups in affine planes. **Mostly Finite Geometries**. Proceedings of a conference in honor of T.G. Ostrom, Marcel Dekker (to appear).

[35] V. Jha and N.L. Johnson. Infinite Baer nets. (Preprint).

[36] V. Jha and N.L. Johnson. Conical, ruled and deficiency one translation planes. (Submitted).

[37] V. Jha and N.L. Johnson, Flocks of oval cones and extensions of theorems of Thas. Note di Mat.(to appear).

[38] N.L. Johnson, Semifield flocks of quadratic cones. Simon Stevin, 61, no. 3-4(l987),313-326.

[39] N. L. Johnson, The derivation of dual translation planes, *J. Geom*, 36(1989), 63 – 90.

[40] N.L. Johnson, Flocks and partial flocks of quadric sets. COMTEMP.MATH. 111, (1990) 71-80.

[41] N.L. Johnson, Ovoids and translation planes revisited. Geom. Ded. 38 (1991), 13-57.

[42] N.L. Johnson, Conical flocks whose corresponding translation planes are derivable. Simon Stevin. Vol 65, 3/4(1991), 199-215.

[43] N. L. Johnson, Derivation of partial flocks of quadtatic cones, *Rend. Mat.*, 12(1992), 817 – 848.

[44] N.L. Johnson, Translation planes admitting Baer groups and partial flocks of quadric sets. Simon Stevin, vol. 63. no. 3(1989), 163-187.

[45] N.L. Johnson, Extending partial flocks containing linear subflocks. J. Geom. vol. 55 (1996), 99-106.

[46] N.L. Johnson, Derivable nets may be embedded in nonderivable planes. Groups and Geometries, Siena, '96 (submitted).

[47] N.L. Johnson, Infinite nests of reguli. Geom. Dedicata (to appear).

[48] N.L. Johnson and X. Liu. Flocks of quadratic and semi-elliptic cones. **Mostly Finite Geometries**. Proceedings of a conference in honor of T.G. Ostrom, Marcel Dekker (to appear).

[49] N.L. Johnson and X. Liu. The Generalized Kantor-Knuth Flocks. **Mostly Finite Geometries.** Proceedings of a conference in honor of T.G. Ostrom, Marcel Dekker (to appear).

[50] N. L. Johnson and G. Lunardon, On the Bose-Barlotti Δ-planes, *Geom. Ded.*, 49(1994), 173 – 182.

[51] N.L. Johnson and G. Lunardon, Maximal partial spreads and flocks. Designs, Codes and Cryptography (to appear).

[52] N. L. Johnson, G. Lunardon and F. W. Wilke, Semifield skeletons of conical flocks, *J. Geom.*, 40(1991), 105 – 112.

[53] N.L. Johnson and L. Storme. Spreads corresponding to flocks of quadrics. **Mostly Finite Geometries**. Proceedings of a conference in honor of T.G. Ostrom, Marcel Dekker (to appear).

[54] D. Jungnickel, Maximal partial spreads and translation nets of small deficiency. J. Alg. 90 (1984), 119-132.

[55] W. M. Kantor, Generalized quadrangles associated with $G_2(q)$, 29(1980), 212-219.

[56] W. M. Kantor, Some generalized quadrangles with parameters (q^2, q), *Math. Zeit.*, 192(1986), 45-50.

[57] N. Knarr, A geometric construction of generalized quadrangles from polar spaces of rank three. Resultate d. Math. 21 (1992), n. 3-4, 332-344.

[58] G. Lunardon, A remark on the derivation of flocks, *Advances in Finite Geometryies and Designs* (eds. J.W.P.Hirschfeld, et al.), Oxford University Press, 1991, 299 – 309.

[59] C. O'Keefe and J. A. Thas, Collineations of some hyperovals, to appear.

[60] C. O'Keefe and J.A. Thas. Partial flocks of quadratic cones with a point vertex in $PG(3, q)$, n odd. Preprint.

[61] S. E. Payne, A new infinite family of generalized quadrangles, *Congr. Numer.*, 49(1985), 115-128.

[62] S. E. Payne, A garden of generalized quadrangles, *Algebras, Groups and Geometries*, 3(1985), 323 – 354.

[63] S. E. Payne, The Thas-Fisher generalized quadrangles, *Ann. Discrete Math.*, 37(1988), 357 – 366.

[64] S. E. Payne, An essay on skew translation generalized quadrangles, **Geometriae Dedicata** 32 (1989), 93 – 118.

[65] S. E. Payne, Collineations of the generalized quadrangles associated with q-clans, *Ann. Discr. Math.*, 52(1992), 449-461.

[66] S. E. Payne, Collineations of the Subiaco generalized quadrangles, *Bull. Belgian Math. Soc., Simon Stevin*, 1(1994), 427 - 438.

[67] S. E. Payne, A tensor product action on q-clan generalized quadrangles with $q = 2^e$, *Lin. Alg. and its Appl.*, 226–228(1995), 115 – 137.

[68] S. E. Payne, The fundamental theorem of q-clan geometry, *Designs, Codes and Cryptography*, 8 (1996), 181 – 202.

[69] S. E. Payne and C. C. Maneri, A family of skew-translation generalized quadrangles of even order, *Congress. Numer.*, 36(1982), 127 – 135.

[70] S. E. Payne, T. Penttila, I. Pinneri, Isomorphisms between Subiaco q-clan geometries, *Bull. Belgian Math. Soc.*, 2 (1995), 197 – 222.

[71] S. E. Payne and L. A. Rogers, Local group actions on generalized quadrangles, *Simon Stevin*, 64(1990), 249-284.

[72] S. E. Payne and J. A. Thas, *Finite generalized quadrangles*, Pitman, 1984.

[73] S. E. Payne and J. A. Thas, Conical flocks, partial flocks, derivation and generalized quadrangles, *Geom. Ded.*, 38(1991), 229 – 243.

[74] T. Penttila and L. Storme, Monomial flocks and ovals, 1995, preprint.

[75] L. Storme and J. A. Thas, k-Arcs and partial flocks, to appear.

[76] J. A. Thas, Generalized quadrangles and flocks of cones, *European J. Combin.*, 8(1987), 441-452.

[77] J. A. Thas, A characterization of the Fisher-Thas-Walker flocks, Simon Stevin, 76 (1993), 219-226.

[78] J. A. Thas, C. Herssens and F. De Clerck, Flocks and partial flocks of the quadratic cone in $PG(3,q)$, in: *Finite Geometry and Combinatorics*, Vol. 77, Cambridge University Press, Cambridge, 379 – 393.

[79] J. A. Thas and S. E. Payne, Spreads and ovoids in finite generalized quadrangles, Geometriae Dedicata 52 (1994), 227 – 253.

[80] A.E. Schroth. *Topological circle planes and topological quadrangles.* Pitman Research Notes in Mathematics # 337, 1995.

[81] M. Walker, A class of translation planes, *Geom. Ded*, 5(1976), 135 – 146.

Ovoids and Translation Planes from Lattices

G. ERIC MOORHOUSE Dept. of Mathematics, University of Wyoming, Laramie WY 82071–3036, U.S.A. moorhous@UWyo.edu

Dedicated to Professor T. G. Ostrom

Abstract Translation planes admitting $2Alt_6 \cong SL(2,9)$, and those of 'extraspecial type' (admitting $2_-^{1+4}Alt_5$), have been studied by Ostrom, Mason, Shult and others. We show the existence of such planes of order p^2 for all odd primes p. We construct such planes using ovoids obtained from lattices by the constructions of Conway et al. [5] and this author [14], [15].

1 INTRODUCTION

The author is grateful to Ted Ostrom for motivating this research, in particular through his survey [21] which was the first paper the author read as a graduate student.

Throughout this paper, all translation planes considered are two-dimensional over $F = GF(q)$ where q is a power of an odd prime p. Translation planes of dimension two, the most extensively studied case, can be viewed in at least three equivalent ways (as described in [9], [8], [13]):

(i) π is an *affine translation plane* of order q^2 and kernel containing $F = GF(q)$. Thus π has point set $V = F^4$, and lines consisting of the cosets of certain mutually complementary two-dimensional subspaces $V_0, V_1, \ldots, V_{q^2} < V$ called the spread components of π. We may suppose $V_i = \{(x, xM_i) : x \in F^2\}$ for $0 \leq i < q^2$ and $V_{q^2} = \{(0,0)\} \oplus F^2$, where the *spread matrices* $M_0, M_1, \ldots, M_{q^2-1}$ are 2×2 matrices over F such that $M_i - M_j$ is nonsingular whenever $i \neq j$;

(ii) $\mathcal{S}$ is a *spread* of $PG(3,q)$, i.e. a collection of $q^2 + 1$ lines of $PG(3,q)$ which partition the point set; or

(iii) $\mathcal{O}$ is an *ovoid* of the Klein quadric, i.e. a set of $q^2 + 1$ points on a hyperbolic quadric in $PG(5,q)$, no two of which are on a line of the quadric.

Now consider a subgroup G of the linear translation complement of π; thus $G \leq GL(4,q)$ preserves the set of spread components. We are especially interested in the following two possibilities:

Type A $G \cong SL(2,9)$. We assume that $p > 3$, so that the choice of $G \cong SL(2,9)$ is unique up to conjugacy in $GL(4,q)$, and G acts irreducibly on V; see [17]. The corresponding subgroup of $PSL(4,q)$ given by $\overline{G} = G/\{\pm I\} \cong PSL(2,9) \cong Alt_6$ preserves the corresponding spread $\mathcal{S}$, and $\overline{G}$ acts irreducibly on $PG(5,q)$ preserving both the Klein quadric and the corresponding ovoid $\mathcal{O}$. (The restriction $p \neq 3$ eliminates Desarguesian planes of order 3^{2t}, among others, admitting $SL(2,9)$ in a less interesting representation.)

Type B ('Extraspecial type') G has a normal subgroup $Q \cong D_8 * Q_8$, a central product of a dihedral group of order 8 and a quaternion group of order 8, and $G/Q \cong Alt_5$. See [12], [13] for a more precise description of the isomorphism type of G and its representation on V. The corresponding group acting on $\mathcal{S}$ and on $\mathcal{O}$ is $\overline{G} = G/\{\pm I\} \cong 2^4 Alt_5$, a split extension of an elementary abelian group of order 2^4 by Alt_5. Although G acts irreducibly on V (preserving a symplectic form), $\overline{G}$ fixes a unique point of $PG(5,p)$ outside the Klein quadric.

Ostrom [19] shows that types A and B are the 'largest' two possibilities for G, assuming that $G' = G$ acts irreducibly on V, and $p \not| \, |G|$. Examples of such planes for small *prime* values of q have been given by Mason and Ostrom [12], Mason, Shult and Cabaniss [13], Mason [10], [11], Ostrom [20], [21], Biliotti and Korchmáros [1], and Nakagawa [18]. Remarkably, no examples are known except when $q = p$ is prime; more will be said about this mystery in Section 4. However, in Section 3 we construct examples over every odd prime field:

THEOREM 1 For every odd prime p there exist self-polar translation planes of order p^2 of types A and B. Valid examples are provided by Constructions 1.1 and 1.2 when $p \equiv 1 \mod 4$, and Construction 1.3 when $p \equiv 1 \mod 3$.

The self-polar property implies that the corresponding spread is invariant under a correlation of $PG(3,p)$, as we explain in the context of our examples.

1.1 Construction Let $p \equiv 1 \mod 4$ be prime. List all integer solutions of

$$x_1^2 + x_2^2 + \cdots + x_6^2 = 6p, \qquad x_i \equiv 1 \mod 4\,.$$

Here are the two smallest cases.

$p = 5$:			$p = 13$:		
	$(5, 1^5)$	6 vectors		$(5^3, 1^3)$	20 vectors
	$(-3^3, 1^3)$	20 "		$(-7, 5, 1^4)$	30 "
	total	$26 = 5^2 + 1$		$(5^2, -3^3, 1)$	60 "
				$(-7, -3^3, 1^2)$	60 "
				total	$170 = 13^2 + 1$

There are always $p^2 + 1$ solutions, and the resulting 6-tuples, taken modulo p, yield an ovoid $\mathcal{O}$ in $PG(5,p)$ with respect to the standard quadratic form. This ovoid is invariant under a group $\overline{G} \cong 2 \times Sym_6 < PO_6^+(p)$ generated by coordinate permutations and the reflection $x \mapsto x - \frac{1}{3}(\sum x_i)(1^6)$. By the Klein correspondence, we obtain a translation plane π of order p^2 of type A. We may identify $\overline{G} \cong 2 \times Sym_6$ with a subgroup of $Aut(PSL(4,p))$ preserving the corresponding spread $\mathcal{S}$, such that half of $\overline{G}$ (a subgroup $\cong Sym_6$) acts as collineations, and the remaining elements are correlations. The preimage of $\overline{G}$ given by $G < Aut(SL(4,p))$ has a subgroup $\cong \Sigma L(2,9)$ in the translation complement of π. Here $\Sigma L(2,9) = SL(2,9)\langle\sigma\rangle$ where σ is the Frobenius automorphism of $GF(9)$.

DEFINITION Let π, $\mathcal{S}$, $\mathcal{O}$ be a triple consisting of a plane, spread and ovoid as described above, and let π', $\mathcal{S}'$, $\mathcal{O}'$ be another such triple. We say that π' is the *polar* of π if any of the following three equivalent conditions is satisfied:

(i) π' is isomorphic to the translation plane with spread matrices $M_i^\top$, where M_i are the spread matrices of π;

(ii) $\mathcal{S}' = \mathcal{S}^\rho$ for some correlation ρ of $PG(3,q)$;

(iii) $\mathcal{O}' = \mathcal{O}^g$ for some orthogonal transformation g such that $\det(g) = -1$.

If in addition π' is isomorphic to π, we say π is self-polar.

If $p \equiv 3 \mod 4$ then the 'standard' quadratic form $\sum x_i^2$ is elliptic rather than hyperbolic. In this case, a modification (Section 3, case II) yields analogues of Construction 1.1.

1.2 Construction Let $p \equiv 1 \mod 4$ be prime. List all integer solutions of

$$x_1^2+x_2^2+\cdots+x_6^2 = p, \qquad x_1+1 \equiv x_2 \equiv x_3 \equiv \cdots \equiv x_6 \mod 2, \qquad \sum x_i \equiv 3 \mod 4.$$

Here are the two smallest cases.

$p=5$:		
	$(0\,\vert \pm 1^5)$	16 vectors
	$(1\,\vert \pm 2, 0^4)$	10 "
	total	$26 = 5^2+1$

$p=13$:		
	$(0\,\vert \pm 3, \pm 1^4)$	80 vectors
	$(1\,\vert \pm 2^3, 0^2)$	80 "
	$(-3\,\vert \pm 2, 0^4)$	10 "
	total	$170 = 13^2+1$

There are always p^2+1 solutions, and these vectors, taken modulo p, yield an ovoid $\mathcal{O}$ in $PG(5,p)$ with respect to the standard quadratic form. This ovoid is clearly invariant under a group of projective orthogonal transformations $\overline{G} \cong 2 \times 2^4 Sym_5$ generated by all permutations and sign changes of the last five coordinates. The corresponding translation plane π of order p^2 is of type B. We identify $\overline{G}$ with a subgroup of $Aut(SL(4,p))$ preserving the corresponding spread $\mathcal{S}$. Half of $\overline{G}$ (a subgroup $\cong 2^4 Sym_5$) consists of collineations, and the other half consists of correlations.

1.3 Construction Let $p \equiv 1 \mod 3$ be prime. The *root lattice of type* E_6 may be identified as

$$L = \{x = (x_1, x_2, \ldots, x_6) \;:\; x_i \in \mathbb{Z},\ \sum x_i \equiv 0 \mod 3\}$$

using the quadratic form $Q(x) = \sum x_i^2 - \frac{1}{9}\left(\sum x_i\right)^2$. Let $e = (1^6) \in L$. List all vectors $v \in e + 2L$ such that $Q(v) = 2p$, but omit the vector $-v$ if v has already been listed. Here are the two smallest cases.

$p=7:$		
	$(1^3,3^3)$	20 vectors
	$(3,1,-1^4)$	30 "
	total	$50=7^2+1$

$p=13:$		
	$(3^3,-1^3)$	20 vectors
	$(3^2,1^3,-3)$	60 "
	$(5,1^3,-1^2)$	60 "
	$(5,3^4,1)$	30 "
	total	$170=13^2+1$

In every case, p^2+1 vectors are obtained. Reducing modulo p gives an ovoid in $PG(5,p)$ with respect to Q (mod p) invariant under $\overline{G}\cong 2\times Sym_6$ generated by coordinate permutations and the reflection in $e^\perp$. The corresponding translation plane π of order p^2 is of type A.

If $p\equiv 1 \mod 12$ then Constructions 1.1 and 1.3 give two ovoids of type A in $O_6^+(p)$, and these are not necessarily equivalent under $PGO_6^+(p)$; in particular for $p=13$ our examples are inequivalent.

We remark that if $p\equiv 2 \mod 3$ then Q (mod p) gives instead a quadratic form of elliptic type. It is known (see [24]) that elliptic quadrics in $PG(5,q)$ (also known as generalized quadrangles of type $Q^-(5,q)$) do not admit ovoids. In this case (odd $p\equiv 2 \mod 3$) the above construction gives *caps* (sets of pairwise noncollinear points) of size $\frac{5}{4}(p^2-1)$ in the associated generalized quadrangles; however, we have found that larger caps than these are obtainable by other means.

2 THE BINARY OVOIDS

Let $V=F^{2n}$, $F=GF(q)$ where q is an odd prime power, and let $Q:V\to F$ be a nondegenerate quadratic form. Thus $Q(x)=\frac{1}{2}xAx^\top$ for some nonsingular symmetric $2n\times 2n$ matrix A over F. A point (one-dimensional subspace) $\langle v\rangle$ of $PG(V)$ is *singular* if $Q(v)=0$. A subspace $U\le V$ is *totally singular* if $Q(u)=0$ for all $u\in U$. We suppose that Q is of *hyperbolic type,* which is to say that V has totally singular subspaces of dimension n; equivalently, $(-1)^n\det(A)$ is a nonzero square in F. We denote by $O_{2n}^+(q)$ either the isometry type of the pair (V,Q), or the associated isometry group, depending on the context. An *ovoid* in $O_{2n}^+(q)$ is a collection $\mathcal{O}$ of singular points, such that every maximal totally singular subspace contains exactly one point of $\mathcal{O}$; equivalently, $\mathcal{O}$ consists of $q^{n-1}+1$ singular points, no two of which are perpendicular with respect to the bilinear form $(x,y):=xAy^\top$. Ovoids in $O_{2n}^+(q)$ are not known to exist for $n\ge 5$. Ovoids in $O_6^+(q)$ (projectively, the Klein quadric in $PG(5,q)$) were featured in Section 1. The known ovoids in $O_8^+(q)$ are listed in [8], [14]. Of these, the most important family we shall require are the *binary ovoids* of Conway et al. [5], which we proceed to construct.

Our terminology and basic facts regarding root systems, and their Weyl groups and lattices, are well-known; see [6], [7]. Consider the root lattice of type E_8 defined by

$$E=\left\{\tfrac{1}{2}(a_1,a_2,\ldots,a_8)\,:\,a_i\in\mathbb{Z},\ a_1\equiv a_2\equiv\cdots\equiv a_8 \mod 2,\ \textstyle\sum a_i\equiv 0 \mod 4\right\}.$$

It is well-known that the points of E determine the unique densest lattice packing of uniform spheres in $\mathbb{R}^8$. The *root vectors* of E are the 240 vectors $e \in E$ such that $\|e\|^2 = 2$. Among these we choose a system of *fundamental roots:*

$$\begin{array}{ll} & e_2 = (-1^2, 0^6) \\ & e_3 = (0, 1, -1, 0^5) \\ e_1 = \frac{1}{2}(1, -1^2, 1^5) & e_4 = (0^2, 1, -1, 0^4) \\ & e_5 = (0^3, 1, -1, 0^3) \\ & e_6 = (0^4, 1, -1, 0^2) \\ & e_7 = (0^5, 1, -1, 0) \\ & e_8 = (0^6, 1, -1) \end{array}$$

Edges of this Dynkin diagram represent pairs of roots at an angle $2\pi/3$; unjoined nodes represent perpendicular pairs of roots. Let r_i be the reflection $E \to E$, $x \mapsto x - (x \cdot e_i)e_i$. Then $r_1, \ldots, r_8$ generate the *Weyl group* $W(\mathsf{E}_8) \cong O_8^+(2)$.

For any prime p, $\overline{E} := E/pE$ is an 8-dimensional vector space over $F = GF(p)$, with quadratic form $Q(\overline{v})$ $(\overline{v} := v + pE,\ v \in E)$ of hyperbolic type, obtained by reducing $\frac{1}{2}\|v\|^2$ modulo p. Fix a root vector e and an odd prime p. There are exactly $2(p^3+1)$ vectors $v \in e + 2E$ such that $\|v\|^2 = 2p$, which come in pairs $\pm v$. These vectors determine exactly p^3+1 coset pairs $\pm(v+pE)$ in $\overline{E}$, forming a *binary ovoid* in $\overline{E} \simeq O_8^+(p)$. This ovoid, denoted $\mathcal{O}_{2,p}(e)$, is invariant under the stabilizer of $\mathbb{Z}e$ in $W(\mathsf{E}_8)$, namely $2 \times W(\mathsf{E}_7) \cong 2^2 \times Sp_6(2)$.

3 SLICING THE BINARY OVOIDS

Let $\mathcal{O}$ be an ovoid in $O_{2n}^+(q)$, and suppose $\langle w \rangle \notin \mathcal{O}$ is a singular point. Then Q induces a quadratic form of hyperbolic type on $w^\perp/\langle w \rangle$, and $\mathcal{O} \cap w^\perp$ yields an ovoid in $w^\perp/\langle w \rangle \simeq O_{2n-2}^+(q)$, called a *slice* of $\mathcal{O}$; see [8]. By appropriately slicing the binary ovoids $\mathcal{O}_{2,p}(e)$ of Section 3, we shall obtain ovoids in $O_6^+(p)$ of types A and B.

We require some facts about the representability of integers by integral quadratic forms.

LEMMA 1 Let p be a prime, and n a positive integer. Then

(i) $p = a^2 + b^2$ for some integers a, b, iff $p \equiv 1 \mod 4$;
(ii) $p = a^2 - ab + b^2$ for some integers a, b, iff $p \equiv 1 \mod 3$;
(iii) $n = a^2 + b^2 + c^2$ for some integers a, b, c, iff n is not of the form $4^k(8\ell - 1)$ where $k, \ell \in \mathbb{Z}$;
(iv) $n = a^2 - ab + b^2 + c^2$ for some integers a, b, c, iff n is not of the form $9^k(9\ell - 3)$ where $k, \ell \in \mathbb{Z}$.

Proof: Conclusions (i)–(iii) are well-known; see [4, pp.10,77], [23, p.45]. We prove (iv) using the theory of rational quadratic forms (see [23], especially Corollary 1 on p.43 therein, and the Corollary on p.37). The quadratic form $Q(a,b,c) = a^2 - ab + b^2 + c^2$ is rationally equivalent to the diagonal form $X^2 + 3Y^2 + Z^2$ where $a = X + Y$, $b = 2Y$, $c = Z$. Computing local invariants, we find that Q represents n in $\mathbb{Q}_p$ for every prime $p \neq 3$, and that Q represents n in $\mathbb{Q}_3$ if and only if $-3n$ is a nonsquare in the group of units of $\mathbb{Q}_3$. The latter condition on n is equivalent to

(*) $\quad n$ is not of the form $9^k(9\ell - 3)$ where $k, \ell \in \mathbb{Z}$.

Thus (*) is a necessary and sufficient condition for n to be represented as $Q(a,b,c)$ for some $(a,b,c) \in \mathbb{Q}^3$. To see that (a,b,c) may in fact be chosen in $\mathbb{Z}^3$, we apply the following:

LEMMA 2 (Davenport-Cassels) Consider a positive definite rational quadratic form $Q(x_1,\ldots,x_r) = \sum_{1\le i,j\le r} a_{ij}x_ix_j$ where $a_{ij} = a_{ji} \in \frac{1}{2}\mathbb{Z}$, $a_{ii} \in \mathbb{Z}$. Suppose furthermore that for all $x \in \mathbb{Q}^r$, there exists $x' \in \mathbb{Z}^r$ such that $Q(x' - x) < 1$. Then for any integer n, we may represent $n = Q(x)$ for some $x \in \mathbb{Q}^r$ if and only if $n = Q(x')$ for some $x' \in \mathbb{Z}^r$.

This is shown in [23, p.46] under the additional hypothesis that $a_{ij} \in \mathbb{Z}$, which is stronger than necessary, as the proof in [23] shows.

We show that our form $Q(a,b,c) = a^2 - ab + b^2 + c^2$ satisfies the hypothesis of Lemma 2. Fix $a,b,c \in \mathbb{Q}$. Choose $c' \in \mathbb{Z}$ such that $|c' - c| \le \frac{1}{2}$. Now observe that $Q(a,b,0)$ is the squared length of the Euclidean vector $ae + bf$ where $e = (1,0)$, $f = \frac{1}{2}(-1,\sqrt{3})$. Since every point of $\mathbb{R}^2$ lies at distance $\le 1/\sqrt{3}$ from some point of the root lattice $\mathbb{Z}e + \mathbb{Z}f$ of type A_2, we may choose $a', b' \in \mathbb{Z}$ such that $Q(a'-a, b'-b, c'-c) \le \frac{1}{3} + \frac{1}{4} < 1$. Now the statement of Lemma 1(iv) follows from Lemma 2. □

LEMMA 3 Let $\overline{w} = w + pE \in \overline{E}$, and suppose $\langle\overline{w}\rangle \in \mathcal{O}_{2,p}(e)$ where $e \in E$ is a root vector.

(i) If $p > 3$ and $f \in E \cap w^\perp$ is a root vector, then $e \cdot f$ is even.
(ii) If $p \equiv 3 \mod 4$, $\|w\|^2 = 2p$ and $w \cdot e$ is even, then $w \in e + 2E$.

Proof: By hypothesis, we have $v \in \lambda w + pE$ for some $v \in e + 2E$, $\|v\|^2 = 2p$, $p \nmid \lambda$.

If $f \in E \cap w^\perp$ is a root vector, then $v \cdot f \equiv \lambda w \cdot f \equiv 0 \mod p$. But $|v \cdot f| \le \|v\|\|f\| = 2\sqrt{p}$ and since $p > 3$, we must have $v \cdot f = 0$. Now $e \cdot f \equiv v \cdot f \equiv 0 \mod 2$.

Under the hypotheses of (ii), we have $|v \cdot w| \le \|v\|\|w\| = 2p$ and $v \cdot w \equiv 0 \mod p$, and so $v \cdot w \in \{0, \pm p, \pm 2p\}$. Also $v \cdot w \equiv e \cdot w \equiv 0 \mod 2$. If $v \cdot w = \pm 2p$ then $w = \pm v \equiv e \mod 2E$ and we are done. Hence we may assume that $v \cdot w = 0$. Then $v - \lambda w \in pE$ implies that $\|v - \lambda w\|^2 = \|v\|^2 + \lambda^2\|w\|^2 \equiv 0 \mod 2p^2$ and $1 + \lambda^2 \equiv 0 \mod p$, contradicting $p \equiv 3 \mod 4$. □

We proceed to prove Theorem 1 in five cases. In cases I and II we refer to the following decomposable root subsystem of E:

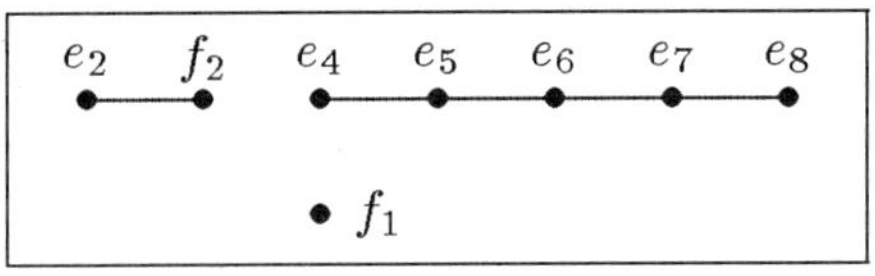

where $f_1 = (1, -1, 0^6)$,
$f_2 = \frac{1}{2}(1^8)$.

Case I Suppose $p \equiv 1 \mod 4$. By Lemma 1(i), we may write $p = a^2 + b^2$ for some integers a, b. We may suppose that a is odd, b is even, and $a + b \equiv 1 \mod 4$ (otherwise replace a by $-a$). Let $w = ae_2 + bf_1 = (-a + b, -a - b, 0^6) \in E$, so that $\|w\|^2 = 2p$. By Lemma 3(i), the singular point $\langle \overline{w} \rangle$ does not lie in the ovoid $\mathcal{O} := \mathcal{O}_{2,p}(f_2)$ (since $f_3 := (0^6 1^2) \in w^\perp$ but $f_3 \cdot f_2 = 1$). The slice $\mathcal{O} \cap \overline{w}^\perp$ is an ovoid invariant under $\langle r_4, r_5, \ldots, r_8 \rangle \cong W(\mathsf{A}_5) \cong Sym_6$ consisting of all permutations of the last six coordinates. To see that this is equivalent to Construction 1.1, it clearly suffices to establish:

$$\left.\begin{array}{l} v \in f_2 + 2E, \ \|v\|^2 = 2p, \\ v \cdot w \equiv 0 \mod p \end{array}\right\} \iff \left\{\begin{array}{l} v = \pm\frac{1}{2}(a - b, a + b, v_3, \ldots, v_8), \\ v_3 \equiv v_4 \equiv \cdots \equiv v_8 \equiv 1 \mod 4, \\ v_3^2 + \cdots + v_8^2 = 6p\,. \end{array}\right.$$

First suppose that $v = \frac{1}{2}(v_1, \ldots, v_8) \in f_2 + 2E$, $\|v\|^2 = 2p$, $v \cdot w \equiv 0 \mod p$. Then $|v \cdot w| \le \|v\|\|w\| = 2p$, so $v \cdot w \in \{0, \pm p, \pm 2p\}$. Also $v \cdot w \equiv f_2 \cdot w \equiv -a \equiv 1 \mod 2$, so $v \cdot w = \pm p$. We may assume that $v \cdot w = -p$. Since $2v \cdot w = (b-a)v_1 - (b+a)v_2 = -2p$ where $gcd(b - a, b + a) = 1$, we have

$$v_1 = a - b + (b + a)k, \qquad v_2 = a + b + (b - a)k$$

for some $k \in \mathbb{Z}$. Then $\|v\|^2 = \frac{1}{4}(2p + 2pk^2 + v_3^2 + \cdots + v_8^2) = 2p$ implies $|k| \le 1$. Now $v \in f_2 + 2E$ implies $v_1 \equiv v_2 \equiv \cdots \equiv v_8 \mod 4$. In particular, $v_1 \equiv 1 + k \equiv v_2 \equiv 1 - k \mod 4$, so $k = 0$ and $v_3^2 + \cdots + v_8^2 = 6p$ as required.

Conversely, we may suppose $v = \frac{1}{2}(a - b, a + b, v_3, \ldots, v_8)$ where $v_3 \equiv v_4 \equiv \cdots \equiv v_8 \equiv 1 \mod 4$ and $v_3^2 + \cdots + v_8^2 = 6p$. Clearly $v \cdot w = -p$, $\|v\|^2 = 2p$. We must show that

$$v - f_2 = \tfrac{1}{2}(a - b - 1, a + b - 1, v_3 - 1, \cdots, v_8 - 1) \in 2E.$$

Clearly $a - b - 1 \equiv a + b - 1 \equiv v_3 - 1 \equiv \cdots \equiv v_8 - 1 \equiv 0 \mod 4$. It remains to be shown that $2a + \sum v_i \equiv 0 \mod 8$ (sum over $i = 3, 4, \ldots, 8$). However, $\sum (v_i - 1)^2 \equiv 0 \mod 16$ implies that $2\sum v_i \equiv 6 + \sum v_i^2 \equiv 6(p + 1) \mod 16$. Therefore $2a + \sum v_i \equiv (2 - 2b) + 3(p + 1) \equiv 2 - 2b + 3a^2 + 3b^2 + 3 \equiv 3b(b + 2) \equiv 0 \mod 8$, as required.

Case II Suppose that $p \equiv 3 \mod 4$. By Lemma 1(iv), we may choose integers a, b, c such that $a^2 - ab + b^2 + c^2 = p$. Let $w = ae_2 + bf_2 + cf_1$, so that $\|w\|^2 = 2p$. Clearly a, b cannot both be even. We may suppose a is odd, so $(w - f_1) \cdot f_2 = -a + 2b \equiv 1 \mod 2$ and therefore $w - f_1 \notin 2E$. By Lemma 3(ii), the singular point $\langle \overline{w} \rangle$ is not in the ovoid $\mathcal{O} := \mathcal{O}_{2,p}(f_1)$. As in case I, the ovoid $\mathcal{O} \cap \overline{w}^\perp$ is invariant under $\langle r_4, r_5, \ldots, r_8 \rangle \cong Sym_6$.

In cases III and IV we refer to the following decomposable root subsystem of E:

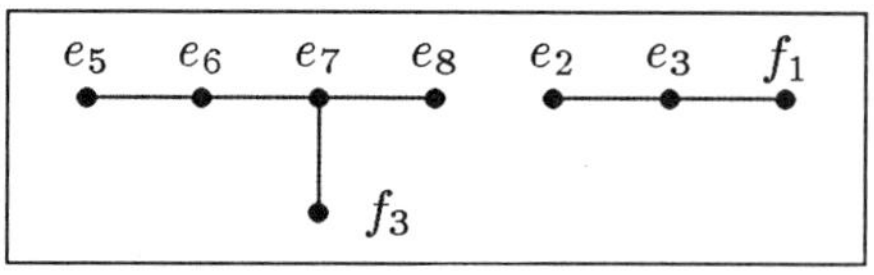

where $f_1 = (1, -1, 0^6)$,
$f_3 = (0^6 1^2)$.

Case III Suppose again that $p \equiv 1 \bmod 4$. As in case I, we have $p = a^2 + b^2$, a odd, b even, $a + b \equiv 1 \bmod 4$. Let $w = ae_2 + bf_1$, so that $\|w\|^2 = 2p$. By Lemma 3(i), the singular point $\langle \overline{w} \rangle$ does not lie in the ovoid $\mathcal{O} := \mathcal{O}_{2,p}(e_3)$ (since $e_4 \in w^\perp$ but $e_4 \cdot e_3 = -1$). The ovoid $\mathcal{O} \cap \overline{w}^\perp$ is invariant under the group $\cong 2^4 Sym_5$ consisting of all permutations, and an even number of sign changes, of the last five coordinates. Note that this group is the Weyl group of type D_5 generated by the reflections corresponding to the roots e_5, e_6, e_7, e_8, f_3. To see that this is equivalent to Construction 1.2, it clearly suffices to prove:

$$\left.\begin{array}{l} v \in e_3 + 2E,\ \|v\|^2 = 2p, \\ v \cdot w \equiv 0 \bmod p \end{array}\right\} \iff \left\{\begin{array}{l} v = \pm(a, b, v_3, \ldots, v_8) = (\text{OEEOOOOO}) \\ \text{or } \pm(b, a, v_3, \ldots, v_8) = (\text{EOOEEEEE}) \\ \qquad \text{where E} = \textit{even},\ \text{O} = \textit{odd}, \\ v_3 + \cdots + v_8 \equiv 3 \bmod 4, \quad v_3^2 + \cdots + v_8^2 = p\,. \end{array}\right.$$

First suppose that $v \in e_3 + 2E$, $\|v\|^2 = 2p$, $v \cdot w \equiv 0 \bmod p$. The first condition implies that $v = (v_1, v_2, \ldots, v_8) \in \mathbb{Z}^8$ where $v_1 \equiv v_2 + 1 \equiv v_3 + 1 \equiv v_4 \equiv \cdots \equiv v_8 \bmod 2$. As in case I, we may suppose $v \cdot w = -p$. Since $v \cdot w = (b-a)v_1 - (b+a)v_2 = -p$ where $gcd(b - a, b + a) = 1$, we have

$$v_1 = a + (b + a)k, \qquad v_2 = b + (b - a)k$$

for some $k \in \mathbb{Z}$. Then $\|v\|^2 = p + 2pk(k+1) + v_3^2 + \cdots + v_8^2 = 2p$ implies $k \in \{-1, 0\}$ and $v_3^2 + \cdots + v_8^2 = p$. Also, $a + b \equiv 1 \bmod 4$ implies that $v_3 + \cdots + v_8 \equiv 3 \bmod 4$. The converse is straightforward.

Case IV Assume that $p \equiv 3 \bmod 4$. Since $2p \equiv 6 \bmod 8$, by Lemma 3(iii) there exist odd integers a, b, c such that $2p = a^2 + b^2 + 4c^2$. We may assume that $a \equiv b \bmod 4$; otherwise replace b by $-b$. Then $\alpha = \frac{1}{2}(b-a) + c$ and $\beta = \frac{1}{2}(b-a) - c$ are odd integers. Let $w = \alpha e_2 + b e_3 + \beta f_1$, so that $\|w\|^2 = 2p$. Now $w \cdot e_3 = 2b - \alpha - \beta \equiv 0 \bmod 2$, and $(w - e_3) \cdot e_1 = \beta \equiv 1 \bmod 2$ so $w - e_3 \notin 2E$. By Lemma 3(ii), the singular point $\langle \overline{w} \rangle$ is not in the ovoid $\mathcal{O} := \mathcal{O}_{2,p}(e_3)$. As in case III, the ovoid $\mathcal{O} \cap \overline{w}^\perp$ is invariant under $2^4 Sym_5$.

It remains only to justify Construction 1.3, for which we consider the decomposable root subsystem

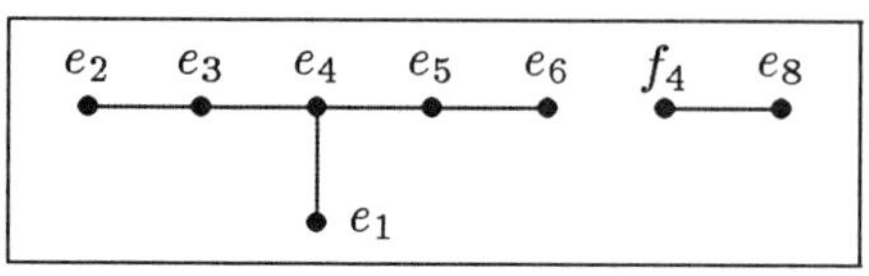

where $f_4 = \frac{1}{2}(1, -1^6, 1)$
$= -3e_1 - 2e_2 - 4e_3 - 6e_4$
$-5e_5 - 4e_6 - 3e_7 - 2e_8$.

Case V Suppose $p \equiv 1 \bmod 3$. Let $L = \mathbb{Z}e_1 + \mathbb{Z}e_2 + \cdots + \mathbb{Z}e_6$, a root sublattice of type E_6 isometric to the lattice of Construction 1.3. By Lemma 3.1(ii), there exist integers a, b such that $p = a^2 - ab + b^2$. Clearly a, b are not both even, so we may suppose that a is odd. Let $w = af_4 + be_8$, so that $\|w\|^2 = 2p$. By Lemma 3.3(i), $\langle\overline{w}\rangle \notin \mathcal{O} := \mathcal{O}_{2,p}(e_1)$ (since $e_4 \in w^\perp$ but $e_4 \cdot e_1 = -1$). We wish to show that the ovoid $\mathcal{O} \cap \overline{w}^\perp$ is equivalent to Construction 1.3. Since the Weyl group of type E_6 is transitive on its 72 roots, our root e_1 is equivalent to the root e chosen in Construction 1.3. Therefore it remains only to show that

$$\left.\begin{array}{l} v \in e_1 + 2E, \ \|v\|^2 = 2p, \\ v \cdot w \equiv 0 \mod p \end{array}\right\} \iff v \in e_1 + 2L, \quad \|v\|^2 = 2p.$$

We need only prove the '$\Rightarrow$' implication. Suppose $v \in e_1 + 2E$, $\|v\|^2 = 2p$, $v \cdot w \equiv 0 \mod p$. Then $|v \cdot w| \le \|v\|\|w\| = 2p$, so $v \cdot w \in \{0, \pm p, \pm 2p\}$. But $v \cdot w \neq \pm 2p$, since otherwise $w = \pm v$, contradicting $\langle\overline{w}\rangle \notin \mathcal{O}$. Also, $v \cdot w \equiv e_1 \cdot w \equiv 0 \mod 2$, so $v \cdot w = 0$. It is clear from the expression above for f_4 that $3E \subset L + \mathbb{Z}f_4 + \mathbb{Z}e_8$. Therefore $3v = \alpha f_4 + \beta e_8 + z$ for some $z \in L$. Now $0 = 3v \cdot w = (2a-b)\alpha + (2b-a)\beta$ where $gcd(2a-b, 2b-a) = 1$, and so

$$\alpha = (a - 2b)k, \qquad \beta = (2a - b)k$$

for some $k \in \mathbb{Z}$. Now $18p = \|3v\|^2 = 6pk^2 + \|z\|^2$ and so $|k| \le 1$. However, $3(v - e_1) \cdot e_8 = 2\beta - \alpha = 3ak$ is even since $v - e_1 \in 2E$. Thus $k = 0$ and $3v = z \in L$. This means that $v \in L$ and the desired conclusion follows.

4 REMARKS

Much more general examples of translation planes of types A and B are available than those presented in Section 3, using the more general ovoids in $O_8^+(p)$ constructed in [14], [15] using the E_8 root lattice. Choose a vector $x \in E$ and primes $r < p$ such that $-\frac{p}{2}\|x\|^2$ is a quadratic residue modulo r. List all vectors $v \in \mathbb{Z}x + rE$ such that $\|v\|^2 = 2i(r-i)p$ where $1 \le i \le \lfloor\frac{r}{2}\rfloor$. This gives an ovoid $\mathcal{O} = \mathcal{O}_{r,p}(x)$ in $\overline{E} = E/pE$. Let $\langle\overline{w}\rangle \notin \mathcal{O}$ be a singular point. Then the slice $\mathcal{O} \cap \overline{w}^\perp$ is invariant under the group of all $g \in W(\mathsf{E}_8)$ preserving both $\mathbb{Z}x + rE$ and $\mathbb{Z}w + pE$. We may arrange that this group is a suitably large Weyl group, for example of type A_5 or D_5 as in Section 3.

We find the computer implementation of all ovoid constructions described here to be very efficient using available techniques for finding short vectors in lattices; see [3, pp.102–104]. In general, the best available method for comparing isomorphism classes by computer is J. H. Conway's invariant known as the fingerprint; see [2], [16]. However, for ovoids of type A or B, isomorphism testing is greatly simplified; see [1, Step 4], [13, (2.8)].

In Table 4.1 we have listed the number of equivalence classes (under the general orthogonal group) of ovoids of type A or B in $O_6^+(p)$ for small primes p, as enumerated by computer. (Warning: There are slight disagreements between this table and similar lists found in at least three of the references.) The full stabilizers $\overline{G} < PO_6^+(p)$ of these ovoids, namely the five groups $2 \times Sym_6$, $2 \times Alt_6$, Sym_6, $2 \times 2^4 Sym_5$, $2^4 Sym_5$, all contain reflections. Each of the associated translation planes is self-polar, with full translation complement H and kernel K satisfying $H/K \cong Sym_6$, Alt_6, Alt_6, $2^4 Sym_5$ or $2^4 Alt_5$ respectively. Although for $p \le 37$ we

p	3	5	7	11	13	17	19	23	29	31	37
$2 \times Sym_6$		1	2	3	2	2	6	6	2	12	6
$2 \times Alt_6$		0	0	2	0	0	0	0	0	0	0
Sym_6		0	0	3	1	2	2	17	5	2	5
total type A		1	2	8	3	4	8	23	7	14	11
$2 \times 2^4 Sym_5$	1	1	1	3	2	2	5	7	8	6	7
$2^4 Sym_5$	0	0	0	1	0	1	0	7	3	4	2
total type B	1	1	1	4	2	3	5	14	11	10	9

Table 4.1 Number of ovoids of type A, B in $O_6^+(p)$ for small p

found that $\overline{G} \not\cong Alt_6$, $H/K \not\cong 2 \times 2^4 Alt_5$, and π is self-polar, we have no evidence that these must be true in general.

The lack of known examples for $q = p^t$, $t \geq 2$ is quite perplexing. By exhaustive computer search, we have concluded that there is no ovoid in $O_6^+(25)$ admitting Alt_6, and no ovoid in $O_6^+(q)$ admitting $2^4 Alt_5$ for $q = 9, 25, 27, 49$. Similarly, for $t \geq 2$, a lack of known ovoids in $O_8^+(p^t)$ invariant under $Sp_6(2)$ was observed in [15]. The best nonexistence result we have in this direction is the following:

LEMMA 4 There is no ovoid of type B in $O_6^+(3^{2t})$.

Proof: Suppose $\mathcal{O}$ is an ovoid in $O_6^+(q)$ invariant under $\overline{G} \cong 2^4 Alt_5$. We use the quadratic form $Q(x) = -x_0^2 + \sum_{j=1}^5 x_j^2$ of hyperbolic type, and $\overline{G}$ acts by even permutations, and an even number of sign changes, of the last five coordinates. If $q = 3^{2t}$ then $|\mathcal{O}| = q^2+1 \equiv 2 \bmod 5$, so $(12345) \in Alt_5$ fixes at least two points of $\mathcal{O}$. However, (12345) fixes exactly two singular points, namely $\langle(\theta, 1^5)\rangle$ and $\langle(-\theta, 1^5)\rangle$, where $\theta^2 = -1$. So both these points must belong to $\mathcal{O}$. But then $\mathcal{O}$ also contains $\langle(\theta, -1^2, 1^3)\rangle$, which is orthogonal to $\langle(-\theta, 1^5)\rangle$, a contradiction. □

Because of slicing, Lemma 4 implies the nonexistence of ovoids in $O_8^+(3^{2t})$ invariant under $Sp_6(2)$.

Does every known translation plane of type A or B arise from a lattice-type construction? We believe not. From the known ovoids in $O_8^+(q)$ for small q, as listed in [14], we have determined all possible $O_6^+(q)$-slices; and in general, these do not include all ovoids listed in Table 4.1. The smallest examples of this are for $q = 11$, where the five known ovoids in $O_8^+(11)$ yield 85 slices in $O_6^+(11)$, but these account for only $5 = 1+0+1+3+0$ of the $12 = 3+2+3+3+1$ ovoids of types A and B.

Finally, we emphasize the most striking feature of these constructions of ovoids and translation planes: their apparent dependence on some nonelementary number theory. In addition to Lemma 1, the construction of $\mathcal{O}_{r,p}(e)$ (see [5], [14], [15]) relies on the following theorem (see [23]). Here $N_\Lambda(n)$ denotes the number of vectors v in an integral lattice Λ such that $\|v\|^2 = n$. Also $\sigma_k(n) = \sum_{1 \leq d | n} d^k$.

THEOREM 2 For every positive integer n, we have $N_E(2n) = 240\sigma_3(n)$ and $N_{E\oplus E}(2n) = 480\sigma_7(n)$.

Sarnak's interesting book [22] describes similar 'applications of modular forms' to analysis, ergodic theory and graph theory, in particular making use of the following:

THEOREM 3 (Jacobi) $N_{\mathbb{Z}^4}(n) = 8\sum\{d : 1 \le d \mid n,\ 4 \nmid d\}$. In particular, $N_{\mathbb{Z}^4}(p) = 8(p+1)$ for every prime p.

Unlike Sarnak's examples, where the asymptotic behaviour of $N_{\mathbb{Z}^4}(p)$ is more important than the exact value $8(p+1)$, however, in the constructions of the ovoids $\mathcal{O}_{r,p}(x)$ the exact values $N_E(2p) = 240(p^3+1)$ and $N_{E\oplus E}(2p) = 480(p^7+1)$ are indispensable. A very interesting research problem is to find similar constructions in finite geometry which make use of the exact values $N_{\mathbb{Z}^4}(p) = 8(p+1)$.

REFERENCES

1. M. Biliotti and G. Korchmáros, Some finite translation planes arising from A_6-invariant ovoids of the Klein quadric, *J. Geom., 37:* 29–47 (1990).
2. C. Charnes, Quadratic matrices and the translation planes of order 5^2, in *Coding Theory, Design Theory, Group Theory,* ed. D. Jungnickel and S. A. Vanstone, pp.155–161, Wiley, New York (1993).
3. H. Cohen, *A Course in Computational Algebraic Number Theory,* Springer-Verlag, Berlin (1993).
4. D. A. Cox, *Primes of the Form $x^2 + ny^2$*, Wiley, New York (1989).
5. J. H. Conway, P. B. Kleidman and R. A. Wilson, New families of ovoids in O_8^+, *Geom. Dedicata, 26:* 157–170 (1988).
6. J. H. Conway and N. J. A. Sloane, *Sphere Packings, Lattices and Groups,* 2nd ed., Springer-Verlag, New York (1993).
7. J. E. Humphreys, *Introduction to Lie Algebras and Representation Theory,* Springer-Verlag, New York (1972).
8. W. M. Kantor, Ovoids and translation planes, *Canad. J. Math., 34:* 1195–1207 (1982).
9. H. Lüneburg, *Translation Planes,* Springer-Verlag, New York (1980).
10. G. Mason, Orthogonal geometries over $GF(2)$ and actions of extra-special 2-groups on translation planes, *Europ. J. Combinatorics, 4:* 347–357 (1983).
11. G. Mason, Some translation planes of order 7^2 which admit $SL_2(9)$, *Geom. Dedicata, 17:* 297–305 (1985).
12. G. Mason and T. G. Ostrom, Some translation planes of order p^2 and of extra-special type, *Geom. Dedicata, 17:* 307–322 (1985).
13. G. Mason and E. E. Shult, The Klein correspondence and the ubiquity of certain translation planes, *Geom. Dedicata, 21:* 29–50 (1986).
14. G. E. Moorhouse, Ovoids from the E_8 root lattice, *Geom. Dedicata, 46:* 287–297 (1993).
15. G. E. Moorhouse, Root lattice constructions of ovoids, in *Finite Geometry and Combinatorics,* ed. F. De Clerck et al., pp.269–275, Camb. Univ. Press, Cambridge (1993).
16. G. E. Moorhouse, Two-graphs and skew two-graphs in finite geometries, *Linear Algebra Appl., 226–228:* 529–551 (1995).

17. B. Mwene, On some subgroups of $PSL(4, q)$, q odd, *Geom. Dedicata, 12:* 189–199 (1982).
18. N. Nakagawa, Some translation planes of order 11^2 which admit $SL(2, 9)$, *Hokkaido Math. J., 20:* 91–107 (1991).
19. T. G. Ostrom, Collineation groups whose order is prime to the characteristic, *Math. Z., 156:* 59–71 (1977).
20. T. G. Ostrom, Elementary abelian 2-groups on the line at infinity of translation planes, *J. Geometry, 17:* 128–139 (1981).
21. T. G. Ostrom, *Lectures on Finite Translation Planes, Conf. Sem. Mat. Univ. Bari,* vol. 191, 31 pp. (1983).
22. P. Sarnak, *Some Applications of Modular Forms,* Camb. Univ. Press, Cambridge (1990).
23. J.-P. Serre, *A Course in Arithmetic,* Springer-Verlag, New York (1973).
24. J. A. Thas, Ovoids and spreads of finite classical polar spaces, *Geom. Dedicata, 10:* 135–144 (1981).

The Construction of Generalized m-gons from Blueprints

Michael A. Abramson Mathematics Department, Bowling Green State University, Bowling Green, Ohio 43402

The notion of a blueprint was introduced in a paper of Ronan and Tits [**5**]. Given a blueprint, one can construct a rank 2 geometry as a graph; in certain nice cases the geometry will be that of a generalized m-gon. This construction differs from more classical ones (see [**2**], [**3**] and [**4**] for instance) in that it emphasizes more the role of the lines. We describe here a natural blueprint for most of the classical generalized m-gons; these have the benefit of both being easy to describe and lending themselves to the construction of higher rank buildings.

Let $m \geq 2$ be an integer, and let L_1 and L_2 be sets (typically finite) with $|L_1| = |L_2|$ if m is odd. Consider the set Γ_m of alternating sequences of length at most m of 1's and 2's, and let ϕ represent the empty sequence; we have that $|\Gamma_m| = 2m+1$. If $f = iji\ldots$, let $L_f = L_i \times L_j \times L_i \times \ldots$, (where the number of terms in the product is the length of f as a word), and if $x = (a, b, c, \ldots) \in L_f$, let f^x be the labelled word $i^a j^b i^c \ldots$. Let $\Lambda_f = \{f^x | x \in L_f\}$. Finally, let $w_1 = 121\ldots$ and $w_2 = 212\ldots$ be the corresponding words of length m in Γ_m.

A *blueprint* for Γ_m is a bijection $\psi : \Lambda_{w_1} \bigcup \Lambda_{w_2} \to \Lambda_{w_1} \bigcup \Lambda_{w_2}$ such that

1. $\psi|_{\Lambda_{w_i}} : \Lambda_{w_i} \to \Lambda_{w_j}$ for $i \neq j$.

2. ψ^2 is the identity.

We identify x and $\psi(x)$ by an equivalence $\simeq$ and define

$$\Lambda_w = (\Lambda_{w_1} \bigcup \Lambda_{w_2})/\simeq .$$

We now define

$$\Lambda = \coprod_{|f|<m} \Lambda_f \bigcup \Lambda_w,$$

the disjoint union of the Λ_f's and Λ_w. From this, we can define a graph $\Delta = \Delta(L_1, L_2, \psi)$ as follows:

The edges of Δ will be the elements of Λ. Two edges are said to be *i-adjacent* if either

1. One is of the form f^x and the other is of the form $f^x i^a$ for some $x \in L_f$ and $a \in L_i$ (where$fi \in \Gamma_m$),

   or

2. They are respectively of the form $f^x i^a$ and $f^x i^b$ for some $x \in L_f$ and $a, b \in L_i$ (where $fi \in \Gamma_m$),

where if fi is either w_1 or w_2, then we mean the appropriate equivalence class.

Clearly i-adjacency is an equivalence relation. Define the vertices of Δ to be the equivalence classes under i-adjacency for both $i = 1$ and $i = 2$. An easy check shows that vertices arising from an i-equivalence class have valence $|L_i| + 1$.

It is easy to show that generalized m-gons are graphs of this form; in fact, each conforms to a large number of non-isomorphic blueprints. From the (B, N)-pair structure of the automorphism group of a classical m-gon, a *natural blueprint* arises.

The Constructions

The natural A_2 blueprint: Let K be any field. The blueprint is given by

$$1^a 2^b 1^c \simeq 2^c 1^{-b+ac} 2^a \qquad \text{for all } a, b, c \in K$$

and hence

$$2^a 1^b 2^c \simeq 1^c 2^{-b+ac} 1^a.$$

The natural B_2 blueprint: Let K be any field. The blueprint is given by

$$1^a 2^b 1^c 2^d \simeq 2^d 1^{-c+ad} 2^{b-2ac+a^2 d} 1^a \qquad \text{for all } a, b, c, d \in K$$

and hence

$$2^d 1^c 2^b 1^a \simeq 1^a 2^{b-2ac+a^2 d} 1^{-c+ad} 2^d$$

The natural G_2 blueprint: Let K be any field. The blueprint is given by

$$1^a 2^b 1^c 2^d 1^e 2^f \simeq \\ 2^f 1^{-e+af} 2^{d-3ae^2-a^3 f^2-bf+3a^2 ef+3ce} 1^{c-2ae+a^2 f} 2^{-b+3ac-3a^2 e+a^3 f} 1^a$$

and

$$2^f 1^e 2^d 1^c 2^b 1^a \simeq \\ 1^a 2^{-b+3ac-3a^2e+a^3f} 1^{c-2ae+a^2f} 2^{d-3ae^2-a^3f^2-bf+3a^2ef+3ce} 1^{-e+af} 2^f$$

for all $a, b, c, d, e, f \in K$.

The natural $^2A_3$ *blueprint:* Let K be a field, $K_0 \subset K$ with $dim(K/K_0) = 2$, and let $\tau : K \to K$ be an automorphism of order 2 with fixed field K_0. Given $a \in K$, define $\bar{a} = \tau(a)$. Suppose $a, b \in K$ and $\alpha, \beta \in K_0$. Then the blueprint is given by

$$1^a 2^\beta 1^b 2^\alpha \simeq 2^\alpha 1^{-\bar{b}+a\alpha} 2^{-\beta+ab+\bar{a}\bar{b}-a\bar{a}\alpha} 1^{\bar{a}}$$

and

$$2^\alpha 1^b 2^\beta 1^a \simeq 1^{\bar{a}} 2^{-\beta-ab-\bar{a}\bar{b}+a\bar{a}\alpha} 1^{-\bar{b}+a\alpha} 2^\alpha$$

Notice if $|K| = q^2$, then $|K_0| = q$ and this is a (q, q^2)-quadrangle.

The natural $^2A_4$ *blueprint:* Let K, K_0 and τ be as above, and let

$$K^+ = \{ (x, y) \in K \times K \mid x\bar{x} = y + \bar{y} \}.$$

Suppose $a, b \in K$ and $(e, f), (g, h) \in K^+$. Then the blueprint is given by

$$1^a 2^{(e,f)} 1^b 2^{(g,h)} \simeq 2^{(g,h)} 1^{\bar{b}+ah-e\bar{g}} 2^{(-e+ag, f+\bar{a}ah-\bar{a}e\bar{g}+\bar{a}\bar{b}-ab)} 1^{\bar{a}}$$

and

$$2^{(g,h)} 1^b 2^{(e,f)} 1^a \simeq 1^{\bar{a}} 2^{(-e+\bar{a}g, f+\bar{a}ah-\bar{a}\bar{e}g+\bar{a}\bar{b}-ab)} 1^{\bar{b}+ah-\bar{e}g} 2^{(g,h)}$$

Notice that this is a (q^2, q^3)-quadrangle.

References

[1] M. A. Abramson *Construction of Affine p-adic Buildings*, Ph.D. Thesis

[2] D. R. Hughes and F. C. Piper, *Projective Planes*, Springer-Verlag, Berlin, **1990**.

[3] W. M. Kantor, 'Generalized Polygons, SCABS and GABS' in *Lecture Notes in Mathematics (Buildings and the Geometry of Diagrams, Como 1984)*, Springer-Verlag, **1986**, pp. 79-158.

[4] S. E. Payne and J. A. Thas, 'Finite Generalized Quadrangles' in *Research Notes in Mathematics*, **1984**.

[5] M. A. Ronan and J. Tits, 'Building Buildings' *Math. Annalen*, **278**, 1987, pp. 291-306.

Variations on a Theme of Dembowski

Mauro Biliotti Dipartimento di Matematica, Università di Lecce, Via Arnesano, 73100 Lecce, Italy

Norman L. Johnson Mathematics Department, University of Iowa, Iowa City, Iowa 52242

This article is dedicated to Professor T.G. Ostrom on the occasion of his eightieth birthday.

1 Introduction.

In Dembowski [1] p.(255), it is observed in a footnote that it is possible to construct a flock of an elliptic quadric Q in $PG(3,R)$ where R is the field of real numbers such that the planes containing the conics of the flock do not all share a line. Recall that a flock is a covering of the quadric minus two points by a set of mutually disjoint conics. One sees the potential interest in such flocks if it is noted that in the finite case, Orr [8] for the odd order case and Thas [9] for the even order case showed that the only flocks of elliptic quadrics in $PG(3,q)$ must, in fact, share a line.

Actually, from the point of view of translation planes, the following theorem indicates why such objects are of interest.

Theorem 1 *Let Q be an elliptic quadric in $PG(3,K)$, for K a field, which admits a quadratic extension field K^+.*

(1) Let F be a flock of Q. Let Σ_{K^+} denote the Pappian plane coordinatized by K^+.

Then there is a set of mutually disjoint regulus nets R_F in Σ_{K^+} and a set of at most two components L, M such that the union of the components from R_F is $\Sigma_{K^+} - \{L,M\}$.

(2) Conversely, for a Pappian plane Σ_{K^+} and at most two components L,M, if a set of mutually disjoint regulus nets C forms a covering of $\Sigma_{K^+} - \{L,M\}$ then there is a corresponding flock F_C of an elliptic quadric Q in $PG(3,K)$.

(3) The translation plane obtained by multiply deriving the regulus nets is Pappian if and only if the planes containing the conics of the corresponding flock share a common line.

One may observe that using Plücker coordinates the spread of a Pappian plane in $PG(3,K)$ corresponds to the points of an elliptic quadric of a projective space isomorphic to $PG(3,K)$ such that the elliptic quadric sits in the Klein quadric Q^K of $PG(5,K)$.

In the finite case, it is only possible to have the two component situation. In the infinite case, it is conceivable that a Pappian plane could be the union of mutually disjoint regulus nets which either cover the spread or cover all but one component of the spread. In this article, we construct examples of all three types.

Furthermore, we show how to generalize the example of Dembowski to provide infinitely many flocks of elliptic quadrics in $PG(3,K)$ where K is any subfield of the field of real numbers such that the planes of the flock do not all share the same line. By the above observation, we obtain infinitely many non-Pappian translation planes.

All of the flocks constructed depend on various open line segments defined in $AG(3,K)$ for K an ordered field. We are able to construct flocks of elliptic quadrics such that the planes of the flock share one of n fixed lines where n is an arbitrary positive integer. We call such flocks n-interval flocks.

Recently in [2], we used a similar method as that of Dembowksi to construct what we call "bilinear flocks of quadratic cones". These are flocks whose designated planes contain one of two fixed lines. These flocks are quite interesting in that the associated translation planes admit an automorphism group isomorphic to $SL(2,K)$, when the spread is contained in $PG(3,K)$ for certain infinite fields K. Further, the group is generated by elation groups of the translation plane. Actually, there is a set of flocks of quadratic cones corresponding to a given translation plane which are in an orbit under a group isomorphic to $PSL(2,K)$. In the finite case, the translation planes of order q^2 which admit a collineation group isomorphic to $SL(2,q)$ are completely determined by Foulser and Johnson [3]. Moreover, when $SL(2,q)$ is generated by elations of the plane, it follows from Foulser,Johnson,Ostrom [4] that the plane must be Desarguesian.

In this article, we use the general construction idea for flocks of elliptic quadrics to construct what we call n-interval flocks of quadratic cones.[1]

Sections 2 thru 5 are devoted to flocks of elliptic quadrics. In section 6 and 7, we show that the general ideas developed for elliptic flocks may be adapted to generalize the construction of bilinear flocks given in [2] to construct a vast number of flocks of quadratic cones.

2 Translation Planes and Elliptic Flocks.

In this section, we provide some algebraic connections between flocks of elliptic quadrics and the corresponding translation planes connected via the preceeding theorem.

Let Σ denote the affine Pappian plane coordinatized by $K(\sqrt{\gamma})$ where K is a field admitting nonsquares and γ is a nonsquare in K.

We represent $K(\sqrt{\gamma})$ as a 2-dimensional vector space over K with a basis $\{e,1\}$ such that $e^2=\gamma$. Moreover, representing a vector $et+u$ as (t,u) the

[1]The ideas for this article originated when the second author was visiting the University of Lecce during May and June of 1995. The authors gratefully acknowledge the support of the University of Lecce and the $C.N.R.$.

corresponding matrix representation is $\begin{bmatrix} u & \gamma t \\ t & u \end{bmatrix}$.

Hence, letting x and y be 2-vectors, we obtain the following representation for Σ in terms of a spread in $PG(3, K)$:

$$\{x = 0, y = x \begin{bmatrix} u & \gamma t \\ t & u \end{bmatrix} \forall\ u,\ t \text{ in } K\}.$$

We recall some background which can be found in Johnson [6].

We assume that the Klein quadric is given by $x_o x_5 - x_1 x_4 + x_2 x_3$ in $PG(5, K)$ where the points of the projective space are given homogeneously by $(x_o, x_1, x_2, x_3, x_4, x_5)$.

Proposition 2 *Components of Σ correspond to points of the Klein quadric as follows:*

The component $x = 0$ correspond to the point $(0, 0, 0, 0, 0, 1)$ and the component $y = x \begin{bmatrix} u & \gamma t \\ t & u \end{bmatrix}$ corresponds to the point $(1, t, u, -u, -t, u^2 - \gamma t^2)$.

We also may obtain a representation of a regulus in the spread of Σ as follows:

Proposition 3 *A regulus of Σ is given by components $x = 0$ or $y = x \begin{bmatrix} u & \gamma t \\ t & u \end{bmatrix}$ if and only if the components satisfy an equation:*

$$(u^2 - \gamma t^2)\beta + 2a_1\gamma t - 2a_2 u - \alpha = 0 \text{ where } (a_2^2 - \gamma a_1^2) + \alpha\beta \neq 0.$$

$x = 0$ is contained within the regulus if and only if $\beta = 0$.

Proof: See Johnson [7] (4.2).

Now we consider an elliptic quadric Q in $PG(3, K)$ embedded in the projective space $PG(5, K)$ as follows. Consider a point (x, y, z, w) in $PG(3, K)$. Represent Q in the form $zw + y^2 = \gamma x^2$ where γ is the nonsquare in K as above.

We embed (x, y, z, w) of $PG(3, K)$ into $PG(5, K)$ by the injection g: $(x, y, z, w) \to (-z, x, y, -y, -\gamma x, w)$. Notice that $-zw + \gamma x^2 - y^2 = x_o x_5 - x_1 x_4 + x_2 x_3 = 0$ when considering $(-z, x, y, -y, -\gamma x, w)$ as $(x_o, x_1, x_2, x_3, x_4, x_5)$. Hence, we have an embedding of the quadric Q of $PG(3, K)$ into the Klein quadric as $PG(3, K)$ is embedded into $PG(5, K)$.

Hence, we obtain:

Proposition 4 *(1) The component $x = 0$ of Σ corresponds to the point $(0, 0, 0, 1)$ of Q within $PG(3, K)$.*

(2) The component $y = x \begin{bmatrix} u & \gamma t \\ t & u \end{bmatrix}$ of Σ corresponds to the point $(t, u, -1, u^2 - \gamma t^2)$ of Q within $PG(3, K)$.

(3) Consider the regulus of Σ *given by components* $x = 0$ *or* $y = x\begin{bmatrix} u & \gamma t \\ t & u \end{bmatrix}$ *which satisfies the equation*

$$(*)\text{: } (u^2 - \gamma t^2)\beta + 2a_1\gamma t - 2a_2 u - \alpha = 0 \text{ where } (a_2^2 - \gamma a_1^2) + \alpha\beta \neq 0.$$

Then the corresponding plane of $PG(3, K)$ *intersects* Q *in a conic given by the points* $(t, u, -1, u^2 - \gamma t^2)$ *or* $(0, 0, 0, 1)$ *satisfying the same equation.*

Hence, the plane satisfies the equation

$$(**): \ (2a_2\gamma)x - (2a_1)y + \alpha z + \beta w = 0.$$

(4) The tangent plane to Q *at the point* $(t, u, -1, u^2 - \gamma t^2)$ *has the following equation:*

$$2\gamma t x - 2uy - (u^2 - \gamma t^2)z + w = 0.$$

The tangent plane to Q *at the point* $(0, 0, 0, 1)$ *is* $z = 0$.

Proof: The component $x = 0$ correspond to $(0, 0, 0, 0, 0, 1)$ using the Klein correspondence which in turn corresponds to $(0, 0, 0, 1)$ under g^{-1}.

The component $y = x\begin{bmatrix} u & \gamma t \\ t & u \end{bmatrix}$ corresponds to $(1, t, u, -u, -\gamma t, u^2 - \gamma t^2)$ which, in turn corresponds to $(t, u, -1, u^2 - \gamma t^2)$ under g^{-1}.

A regulus satisifying the indicated equation $(*)$ then clearly corresponds to a plane which intersects Q in a conic defined by the same equation. The plane may easily be shown to satisfy the equation $(**)$ listed above.

To see that the tangent plane is as maintained, suppose that the plane with equation

$$ax + by + cy + dw = 0$$

contains the point $(t, u, -1, u^2 - \gamma t^2)$ so that $at + bu - c + d(u^2 - \gamma t^2) = 0$. We see from the above results that there is a corresponding regulus net and hence a conic of intersection unless we obtain:

$$(b/2)^2 - \gamma(a/2\gamma)^2 = -cd.$$

Since $(0, 0, 0, 1)$ is not on this plane, we may take $d = 1$. Note that

$$\begin{aligned} & at + bu - (b^2/4 + -a^2/4\gamma) + (u^2 - \gamma t^2) \\ = \ & -1(a - 2\gamma t)^2/4\gamma + (b + 2u)^2/4 \\ = \ & 0 \end{aligned}$$

if and only if $a = 2\gamma t$ and $b = -2u$. This proves the first part of (4).

The proof that the tangent plane to $(0, 0, 0, 1)$ is $z = 0$ follows similarly.

We may change bases again and choose a basis

$\{(1, 0, 0, 0), (0, 1, 0, 0), (0, 0, 1, 1), (0, 0, 1, -1)\}$. Note that $(x, y, z + w, z - w)$ in the old basis is (x, y, z, w) in the new basis. Hence, the equation of the elliptic quadric in the new basis is $(z + w)(z - w) + y^2 = \gamma x^2$ or rather $-\gamma x^2 + y^2 + z^2 = w^2$.

Thus,

Remark 1 *The points of the quadric* $(0,0,0,1)$, $(t,u,-1,u^2-\gamma t^2)$ *in the old basis have the form*

$$(0,0,1/2,-1/2) \equiv (0,0,-1,1),$$

$$(t,u,(-1+u^2-\gamma t^2)/2,-(1+u^2-\gamma t^2)/2)$$

in the new basis.

If $(1+u^2-\gamma t^2)$ *is not zero, an equivalent form is the following:*

$$(-2t/(1+u^2-\gamma t^2),-2u/(1+u^2-\gamma t^2),-(-1+u^2-\gamma t^2)/(1+u^2-\gamma t^2),1).$$

Let K be an ordered field such that all negative elements are nonsquares. Choose $\gamma = -1$. Starting from $PG(3,K)$ form the affine space by deleting $z=-w$.

If (x,y,x,w) represents a point of $PG(3,K)$, the affine points have the form $(x,y,-1,0)$ and $(x,y,-1,w)$ for $w \neq -1$. The associated vector space will be formed by the injection mapping g which takes $(x,y,-1,0)$ to $(x,y,-1)$ and $(x,y,-1,w)$ to (x,y,w) for $w \neq -1$.

Definition 5 *In a 3−dimensional vector space over* K*, define the open segment between the vectors* R,T*,* $oseg[R,T]$*, as* $\{R+\lambda(T-R)$ *such that* $0<\lambda<1\}$.

Proposition 6 *Let* $(x,y,-1,w)$ *and* $(x^*,y^*,-1,w^*)$ *be affine points represented as above.*

If w *and* w^* *are both* ≥ 0 *then the affine points on the open line segment are in*

$$\{(x+\lambda(x^*-x),y+\lambda(y^*-y^*),-1,\lambda(w^*-w)) \ni 0<\lambda<1\}.$$

Proof: $(x,y,-1,w)$ maps to (x,y,δ_w) where $\delta_w=-1$ if $w=0$ and $\delta_w=w$ if $w \neq 0$, -1. Similarly, $(x^*,y^*,-1,w^*)$ maps to (x,y,δ_{w^*}). The vectors of the open segment have the form

$$(x,y,\delta_w)+\lambda(x^*-x,y^*-y,\delta_{w^*}-\delta_w) \text{ for } 0<\lambda<1.$$

The pullback g^{-1} is as maintained provided $\delta_w+\lambda(\delta_{w^*}-\delta_w) \neq -1$.

First assume that $w^*w \neq 0$. Then since both elements are positive we have $\delta_w+\lambda(\delta_{w^*}-\delta_w)=w+\lambda(w^*-w)$ which is -1 if and only if $\lambda=(1+w)/(w-w^*)$ (note if $w=w^*$ then w would be forced to be -1 contrary to assumption). However, since $0<\lambda<1$ it must be that $w-w^*$ is positive so that $1+w<w-w^*$ a contradiction as $w^*>0$.

If $w^*=0$ then $w+\lambda(-w)=-1$ if and only if $\lambda=(1+w)/w$ contrary to the fact that $\lambda<1$.

If $w=0$ then λw^* can't be -1 since both elements are positive.

Corollary 7 *The open line segment of affine points between* $(0,0,0,1)$ *and* $(t,u,-1,u^2+t^2)$ *is*

$$\{(\lambda t, \lambda u, -1, 1+\lambda(u^2+t^2-1)) \ni 0<\lambda<1\}.$$

The open line segment of affine points between and $(t,u,-1,u^2+t^2)$ *and* $(t^*,u^*,-1,u^{*2}+t^{*2})$ *is*

$$\{(t+\lambda(t^*-t), u+\lambda(u^*-u), -1, u^2+t^2+\lambda(u^{*2}+t^{*2}-(u^2+t^2)) \ni 0<\lambda<1\}.$$

We choose a new basis as above so the quadric can be represented in the form $x^2+y^2+z^2=w^2$ with homogeneous points (x,y,z,w) and where the points of the quadric are represented in the form:

$$(-2t/(1+u^2+t^2), -2u/(1+u^2+t^2), -(-1+u^2+t^2)/(1+u^2+t^2), 1)$$
and $(0,0,-1,1)$
$\forall\ t,u$ *in* K.

Corollary 8 *In the above form, the tangent plane to* $(0,0,-1,1)$ *is* $z=-w$ *and the tangent plane to*

$$(-2t/(1+u^2+t^2), -2u/(1+u^2+t^2), -(-1+u^2+t^2)/(1+u^2+t^2), 1)$$

is

$$-2tx-2uy+(1-u^2-t^2)z-(1+u^2+t^2)w=0.$$

Theorem 9 *Let* K *be any ordered field where the negative elements are all nonsquare. Let* Q *be an elliptic quadric in* $PG(3,K)$ *whose points have the form*

$$(-2t/(1+u^2+t^2), -2u/(1+u^2+t^2), -(-1+u^2+t^2)/(1+u^2+t^2), 1)$$
and $(0,0,-1,1)$
$\forall\ t,u$ *in* K.

(1) Let $w=0$ *denote the hyperplane at infinity. Associate the* $1-$*space* $(x,y,z,1)$ *with the vector* (x,y,z) *of the affine space and recall that the open segment between two* $1-$*spaces* R *and* T *is* $R+\lambda(T-R)$ *for* $0<\lambda<1$.

Then this segment may be obtained simply by ignoring the 4-th. component 1 for any point $(x,y,z,1)$.

(2) Consider any two distinct points R,T of Q *and let* L *denote the line of intersection of the tangent plane to* R *and* T.

If W *is any point of* $Q-\{R,T\}$ *then the plane* $<W,L>$ *intersects the affine open line segment* $oseg[R,T]$ *in a unique point.*

Proof: We consider the affine space of coordinates (x, y, z) with quadric $x^2 + y^2 + z^2 = 1$. The tangent plane (affine) to (x_o, y_o, z_o) is $x_o x + y_o y + z_o z = 1$.

Since the elliptic quadric corresponds to an associated Pappian spread, we may choose coordinates in the spread in essentially any manner without disrupting the representation. Hence, we may choose coordinates so that any particular component of the spread may be regarded as $x = 0$. Thus, it follows that we may choose any particular point on the quadric as $(0, 0, 1)$ without disturbing our basic representation.

Hence, we let $R = (x_o, y_o, z_o)$ and $T = (0, 0, 1)$ in our affine form.

First assume that the tangent planes are parallel. Since the tangent plane to T is $z = 1$, it follows that the tangent plane to R is $z = z_o$ and so $R = (0, 0, -1)$. The affine line segment $oseg[R, T] = \{(0, 0, -1 + 2\lambda) \ni 0 < \lambda < 1\}$. If $W = (x_1, y_1, z_1)$ is in the segment then since we have an ordered field, it follows that $-1 < z_1 < 1$. Hence, the plane $z = z_1$ contains the line of intersection of the tangent planes (taken projectively) and this plane intersects $oseg[R, T]$ if and only if $-1 + 2\lambda = z_1$ which is valid if and only if $\lambda = (z_1 + 1)/2$. Since, $-1 < z_1 < 1$ if and only if $0 < (z_1 + 1)/2 < 1$ this proves the statement in this particular case.

Now assume that the tangent planes are not parallel, so they intersect in the set of points $(x, y, 1)$ given by the following equation: $x_o x + y_o y + z_o = 1$.

Assume that both x_o and y_o are zero. Then $R = T$. Hence, without loss of generality, assume that $y_o \neq 0$. Hence, the set of points on this line is $\{(x, (1 - x_o x - z_o)/y_o, 1)$ for all x in $K\}$.

Let the plane containing the line of intersection and W be denoted by $ax + by + cz = d$.

First assume that $d = 0$.

Then letting $x = 0, 1$ we obtain the following equations respectively.

$$b(1 - z_o)/y_o + c = 0,$$

$$a + b(1 - x_o - z_o)/y_o + c = 0.$$

Hence, we obtain the following plane: $x_o x + y_o y + (z_o - 1)z = 0$.

The line segment

$$oseg[(x_o, y_o, z_o), (0, 0, 1)] = \{(\lambda x_o, \lambda y_o, \lambda(z_o - 1) + 1) \ni 0 < \lambda < 1\}.$$

There is a point of intersection on this segment with the plane if and only if

$$\lambda x_o^2 + \lambda y_o^2 + (z_o - 1)(\lambda(z_o - 1) + 1) = 0.$$

Since $x_o^2 + y_o^2 + z_o^2 = 1$, it follows that the above equation is valid if and only if $\lambda(2 - 2z_o) = (1 - z_o)$ which is valid if and only if $\lambda = 1/2$.

Note that conversely if a plane thru $(0, 0, 0)$ contains the line of intersection of the two tangent planes then this plane is $x_o x + y_o y + (z_o - 1)z = 0$. Since this plane contains $(-x_o, -y_o, -z_o)$, it follows that for a point I on the open segment corresponding to $\lambda = 1/2$ then the plane containing I and the line of intersection must also intersect the quadric in a point W distinct from R and T.

Now assume that $d \neq 0$. Since (x_1, y_1, z_1) is on the plane, it follows that the plane has the following equation: $bx_ox + by_oy + (y_o - b(1 - z_o))z = y_o$ where $b = y_o(1 - z_1)/(x_ox_1 + y_oy_1 + (z_o - 1)z_1)$.

Note that y_o is not zero and $(1 - z_1)$ is not zero so that it must be that b is as maintained and also b is nonzero.

Now the plane has an intersection with the open segment between the two points if and only if

$$b\lambda x_o^2 + b\lambda y_o^2 + (y_o - b(1 - z_o))(\lambda(z_o - 1) + 1) = y_o.$$

This is equivalent to:

$$b\lambda((x_o^2 + y_o^2 + z_o^2) - 2z_o + 1) + (y_o\lambda + b)(z_o - 1) = 0.$$

Since $z_o \neq 1$, the above equation reduces to: $2b\lambda = y_o\lambda + b$.
Hence, $\lambda = b/(2b - y_o)$.
We need only to check that $2b - y_o$ is nonzero and that $0 < b/(2b - y_o) < 1$.
If $2b = y_o$ then $2(1 - z_1) = x_ox_1 + y_oy_1 + (z_o - 1)z_1$. Note that

$$1 - (x_ox_1 + y_oy_1 + z_oz_1) = ((x_o - x_1)^2 + (y_o - y_1)^2 + (z_o - z_1)^2)/2.$$

Hence, the previous equation implies that

$$((x_o - x_1)^2 + (y_o - y_1)^2 + (z_o - z_1)^2)/2 = z_1 - 1 < 0$$

which is a contradiction.

Now we shall show that $0 < b/(2b - y_o) = 1/(2 - y_o/b) =$
$1/(2 - (x_ox_1 + y_oy_1 + (z_o - 1)z_1)/(1 - z_1)) < 1$.
Clearly, if $(x_ox_1 + y_oy_1 + (z_o - 1)z_1)/(1 - z_1) \leq 0$, the inequality is valid.
Hence, assume that $(x_ox_1 + y_oy_1 + (z_o - 1)z_1/(1 - z_1) > 0$.

Note that $0 < 1/(2 - m)$ if and only if $m < 2$ and $1/(2 - m) < 1$ if and only if $m < 1$. Hence, it suffices to show that $(x_ox_1 + y_oy_1 + (z_o - 1)z_1)/(1 - z_1) < 1$ or equivalently
$x_ox_1 + y_oy_1 + (z_o - 1)z_1 < 1 - z_1$.
This equation is valid if and only if $1 - (x_ox_1 + y_oy_1 + z_oz_1) > 0$.
However,

$$1 - (x_ox_1 + y_oy_1 + z_oz_1) = ((x_o - x_1)^2 + (y_o - y_1)^2 + (z_o - z_1)^2)/2 > 0.$$

In the following section, we shall use the above algebraic connections with elliptic quadrics and corresponding Pappian planes to construct flocks of elliptic quadrics.

Theorem 10 *Let L be any field. Let Q denote an elliptic quadric of $PG(3, L)$ and let π denote a plane of $PG(3, L)$ which intersects Q in a nondegenerate conic. Let P be any point on $Q - \pi$ and let π_P denote the tangent plane to Q thru P. Then there exists a second tangent plane π_{P^*} to Q thru a point $P^* \neq P$ such that $\pi_P \cap \pi = \pi_{P^*} \cap \pi$.*

The previous result may be directly verified using coordinates. Note that the proof essentially depends on the fact that once a point of a geometry over a field satisfies a particular quadratic form then there are two distinct solutions .

Consider the geometry $PG(3, K)$ where K is a field of rational numbers and let K^+ denote the field of real numbers. Then there exist pairs of irrational points of $PG(3, K^+)$ on an elliptic quadric such that the line of intersection of the two tangent planes is a rational line. Obviously given this line, there are no rational tangent planes to the rational quadric which contain this line. On the other hand, once there is a rational tangent plane containing a given rational line then there are two rational tangent planes to the rational quadric which intersect in this line.

3 A Construction Method.

The example of Dembowski is as follows: Let Q be an elliptic quadric in $PG(3, R)$ where R is the field of real numbers. Let π denote the equatorial plane of Q. Let L and M denote two lines of π which are distinct and exterior to Q. Let N and S denote points of Q such that $< N, L >$ and $< S, M >$ are tangent planes to Q and N and S are in different hemispheres of Q defined by π. We refer to the part of Q containing N as the northern hemisphere and the part of Q containing S as the southern hemisphere. Let N_F denote the set of planes $< P, L >$ where P is in the northern hemisphere and not on π and S_F denote the set of planes $< T, M >$ where T is in the southern hemisphere and not on π. Then $N_F \cup S_F$ is a flock of Q which is not linear.

We consider a generalization of the example of Dembowski.

Remark 2 *Consider an elliptic quadric Q in $PG(3, K)$ where K is an ordered field such that the negative elements are always nonsquares. We have noted above that, given any plane π which intersects Q in a nondegenerate conic, it is possible to construct infinitely many lines M such that there exist two tangent planes to the conic which intersect in M.*

Definition 11 *Choose an affine subspace which contains Q. Let π_1 be any plane which intersects Q either in a nondegenerate quadric C_1 or is tangential to Q. Let L_1 be any line of π_1 which is external to Q and which is the intersection of two tangent planes to the quadric.*

Let N_{L_1} and S_{L_1} denote the two points of tangency on Q of planes which share L_1 lexicographically ordered (on the affine points) so that $S_{L_1} < N_{L_1}$. There is a flock of $Q - \{N_{L_1}, S_{L_1}\}$ consisting of conics in planes which contain L_1. We call this a linear flock of Q.

Definition 12 *Choose any point I. We call the set of points strictly between I and a point Z on the quadric Q the open segment and use the notation $oseg[I, Z]$. For any line L_1 and point Z external to L_1, we denote the plane containing Z and L_1 by $< Z, L_1 >$.*

Theorem 13 *Let K be any ordered field such that the negative elements are always nonsquares. Let L_1 be any exterior line which is the intersection of two tangent planes to the quadric.*

Then the linear flock F_{L_1} consists of planes sharing L_1 is $\cup\{< U, L_1 > \ni U \in oseg[S_{L_1}, N_{L_1}]$ and $< U, L_1 > \cap Q \neq \phi\}$.

Proof: Let W be a $K-$point on the quadric which is not one of two mentioned points of tangency. We have noted that $< W, L >$ must intersect the open segment. Hence, there is a cover of the quadric not equal to the two given points of tangency by this method.

Definition 14 *We define the north and south sides of Q relative to a plane π_1 and the line L_1 as follows:*

(1) The north side of Q : Let I_1 be the point on $oseg[S_{L_1}, N_{L_1}] \cup \{S_{L_1}, N_{L_1}\}$ which lies in π_1. (Note, π_1 could be tangential to the quadric.) Form $\cup\{< U, L_1 > \ni U \in oseg[S_{L_1}, N_{L_1}]$ where $I_1 < U < N_{L_1}$ and $< U, L_1 > \cap Q \neq \phi\}$. Note that this is a partial flock which we denote by $(\pi_1, L_1)^n$.

We shall refer to $(\pi_1, L_1)^n$ as the north partial flock of $Q - \{N_{L_1}\}$. The set of points of intersection with Q union $\{N_{L_1}\}$ shall be called the north side of Q relative to π_1 and L_1.

(2) Similarly, form $(\pi_1, L_1)^s = \cup\{< U, L_1 > \ni U \in oseg[S_{L_1}, N_{L_1}] \ni S_{L_1} < U < I_1$ and $< U, L_1 > \cap Q \neq \phi\}$.

We shall call this partial flock the south partial flock of $Q - \{S_{L_1}\}$. The set of points of intersection with Q union $\{S_{L_1}\}$ shall be called the south side of Q relative to the plane π_1 and the line L_1.

(3) If π_1 is not tangential then the linear flock of Q containing π_1 is the union of π_1 with the north and south partial partial flocks and is $\pi_1 \cup (\pi_1, L_1)^n \cup (\pi_1, L_1)^s$.

(4) Note if π_1 is tangential then $I_1 = S_{L_1}$ or N_{L_1} . If $I_1 = N_{L_1}$ then the north side of Q relative to π_1 and L_1 is N_{L_1} and the south side of Q is $Q - \{S_{L_1}\}$. Furthermore, the north partial flock consists only of π_1 and the south partial flock is the linear flock of Q containing L_1. Similarly, if $I_1 = S_{L_1}$ then the south side of Q relative to π_1 and L_1 is S_{L_1} and the north side of Q is $Q - \{S_{L_1}\}$.

(5) We denote the set of points of Q covered by the north side partial flock by $Q^n(\pi_1, L_1)$ and refer to these points as the north section. The set of those points of Q covered by the south side partial flock is called the south section and denoted by $Q^s(\pi_1, L_1)$. Note that the conic C_1 is not in either the south or north section of Q with respect to L_1 and π_1. Also, the north and south sections do not include the tangent points N_{L_1} and S_{L_1}.

We shall denote the north side of Q relative to π_1 and L_1 by $Q^{n+}(\pi_1, L_1)$ and the south side of Q relative to π_1 and L_1 by $Q^{s+}(\pi_1, L_1)$.

Theorem 15 *Let K be subfield of the field of real numbers. Assume that π_1 is not a tangential plane.*

Then the north section of Q relative to L_1 and π_1 is the set of all points Z of $Q - C_1$ such that $oseg[Z, N_{L_1}]$ does not intersect π_1.

Proof: This is a property of real 3-dimensional affine space which carries over to K 3-dimensional affine space. In particular, assume that $oseg[Z, N_{L_1}]$ does not intersect π_1 in a K -point but in an R-point when R is the field of real numbers. Represent the plane π_1 by $ax + by + cz + dw = 0$ for a, b, c, d in K. Representing the affine form by $w \neq 0$ then the points of $oseg[Z, L_1]$ are of the form

$$\lambda Z + (1 - \lambda)L_1 \text{ for } 0 < \lambda < 1$$

where we consider only the affine form of the vectors(points) and N_{L_1} . Note any intersection forces λ to be in K as all of the coordinates for Z, L_1, and π_1 are in K.

Definition 16 *Let K be any field. We shall say that an exterior line L to an elliptic quadric in $PG(3, K)$ is a $T-$line if and only if the line is the intersection of two distinct tangent planes to the quadric.*

Corollary 17 *Let K be a subfield of the field of real numbers. Let $\{L_1, L_2\}$ be any ordered set of two exterior $T-$lines to Q of π_1.*

Then $Q^{n+}(\pi_1, L_1) = Q^{n+}(\pi_1, L_2)$ or $Q^{s+}(\pi_1, L_2)$.

We further denote this partial flock by $F(\pi_1, L_1)^n$.

Definition 18 *The side of Q which is not covered by $Q^{n+}(\pi_1, L_1)$ is called the opposite side of $Q^{n+}(\pi_1, L_1)$ relative to L_2 and denoted by $oppQ^{n+}(\pi_1, L_1)$.*

The partial flock of $oppQ^{n+}(\pi_1, L_1)$ $-\{N_{L_2}\}$ or $-\{S_{L_2}\}$ which is either the north or south partial flock of Q relative to π_1 and L_2 is called the opposite partial flock of $F(\pi_1, L_1)^n$ relative to L_2. We denote this partial flock by $opp(F(\pi_1, L_1)^n, L_2)$.

We may now consider the example (extended to subfields of the reals) of Dembowski as follows:

Theorem 19 *Let K be a subfield of the field of real numbers. Let π_1 be a plane which intersects Q in more than a point. Let $\{L_1, L_2\}$ be any ordered set of two distinct $T-$lines of $PG(3, K)$ both of which are exterior to the elliptic quadric Q.*

Then $\pi_1 \cup F(\pi_1, L_1)^n \cup opp(F(\pi_1, L_1)^n, L_2)$ is a flock of $Q - \{N_{L_1}, S_{L_2}\}$.

We generalize this example as follows:

Let K be any subfield of the field of real numbers. Choose any fixed affine 3-space containing the elliptic quadric Q. Choose any two distinct points P,W and form the tangent planes to these points, $T(P)$ and $T(W)$. Let $L_1 = T(P) \cap T(W)$. Form S_{L_1},N_{L_1} and let I_1 be any point of $oseg[S_{L_1}, N_{L_1}]$ such that $\pi_1 =< I_1, L_1 >$ has a nontrivial intersection with Q. Form the partial flock $\pi_1 \cup F(\pi_1, L_1)^n$.

Now choose any $T-$line L_2 of π_1 distinct from L_1 and exterior to Q. Let $I_1^{L_2} = oseg[S_{L_2}, N_{L_2}] \cap \pi_1$. Exactly one segment $oseg[S_{L_2}, I_1^{L_2}]$ or $oseg[I_1^{L_2}, N_{L_2}]$ is in the previously considered partial flock and no point of the remaining segment is in this previous partial flock.

For notational purposes, we shall always consider that the uncovered segment is $oseg[S_{L_2}, I_1^{L_2}]$ although this notation is purely arbitrary.

We shall assume that whenever a point of a segment is chosen as above there is a nontrivial intersection with the corresponding plane and the quadric.

Take any point I_2 of $oseg[S_{L_2}, I_1^{L_2}]$. Form the partial flock of planes

$\cup\{< U, L_2 > \ni I_2 < U < I_1^{L_2}\}$ (recall our notation convention – the exact inequality could just as well be $I_1^{L_2} < U < I_2$). We shall call this partial flock the partial flock determined by the segment $oseg\ [I_1^{L_2}, I_2]$ and use the notation $F(oseg[I_1^{L_2}, I_2])$. Let $< I_2, L_2 >= \pi_2$.

Now form $\pi_1 \cup F(oseg[I_1, N_{L_1}]) \cup \pi_2 \cup F(oseg[I_1^{L_2}, I_2])$. Note that this partial flock union $\{N_{L_1}\}$ either covers the north or south side of π_2. For notational purposes, we shall assume that this partial flock always covers the north side of whatever subplane is being considered. We continue this process to cover the south side of the subplane under consideration always considering that the north side except for one point has been previously covered by partial flocks.

Choose a T–line $L_3 \neq L_2$ on π_2 and let $I_2^{L_3} = oseg[S_{L_3}, N_{L_3}] \cap \pi_2$. Using our notational device, choose any point I_3 on $oseg[S_{L_3}, I_2^{L_3}]$ and form the partial flock $F(oseg[I_2^{L_3}, I_3])$. Now let $\pi_3 =< I_3, L_3 >$ and form the partial flock

$$\pi_1 \cup F(oseg[I_1, N_{L_1}]) \cup \pi_2 \cup F(oseg[I_1^{L_2}, I_2]) \cup \pi_3 \cup F(oseg[I_2^{L_3}, I_3]).$$

Note that we have covered the north side (with our convention) of π_3 except for the point N_{L_1}.

We continue in this way to form the following partial flock: Let $N_{L_1}{=}I_0^{L_1}$, and $I_{i-1}^{L_i} = oseg[S_{L_i}, N_{L_i}] \cap \pi_{i-1}$ then the partial flock is

$\cup\{\pi_i \cup F(oseg[I_{i-1}^{L_i}, I_i]$ for $1 \leq n-1\}$ where $< I_j^{L_{j+1}}, L_{j+1} >= \pi_{j+1}$ for $1 \leq j \leq n-1$.

Note that this partial flock covers the north side of π_{n-1} except for N_{L_1}. To complete the cover, we choose a T–line L_n distinct from L_{n-1} on π_{n-1} and form the partial flock $F(oseg[S_{L_n}, I_{n-1}^{L_n}])$.where $I_{n-1}^{L_n} = oseg[S_{L_n}, N_{L_n}] \cap \pi_{n-1}$. Hence, we obtain:

Theorem 20 *For any subfield K of the reals, and for any finite integer n, there are card(K) flocks of an elliptic quadric Q in PG(3, K) such that planes of the flock share one of n lines.*

$\cup\{\pi_i \cup F(oseg[I_{i-1}^{L_i}, I_i]$ *for* $1 \leq i \leq n\}$ *where* $N_{L_1}{=}I_0^{L_1}$, $I_{i-1}^{L_i} = oseg[S_{L_i}, N_{L_i}\} \cap \pi_{i-1}$, *$I_i$ is any point on* $oseg[S_i, I_{i-1}^{L_i}]$, *for* $1 \leq i \leq n-1$ *and* $S_{L_n} = I_n$ *is a flock of* $Q - \{N_{L_1}, S'_{L_n}]$ *where* $S'_{L_n} \epsilon \{N_{L_n}, S_{L_n}\}$. *The planes share a line of the set* $\{L_1, L_2, ..., L_n\}$.

Theorem 21 *Let K be any subfield of the reals.For any finite integer n, there is a set R of mutually disjoint reguli of the Pappian plane* Σ *coordinatized by* $K[\sqrt{-1}]$ *which cover all but two components of* Σ *and a set of n Pappian spreads* Σ_i *for* $1 \leq i$ *such that the spread for the translation plane obtained by multiply derivation of the reguli of R has a spread which may be embedded into the union of the spreads for the* Σ_i .

Definition 22 *Any flock of an elliptic quadric in $PG(3,K)$, where K is a subfield of the reals constructed as in the above theorem, shall be called an $n-$interval flock.*

We may provide a characterization of sorts of this type of flock as follows:

Theorem 23 *Let K be any subfield of the reals. Let F be a flock of an elliptic quadric Q in $PG(3,K)$ which covers all but two points of Q.*

Assume that there is an ordered set of exterior $T-$lines $\{L_1, L_2, ...L_n\}$ with planes $\pi_i =< L_1, L_2 >$ for $i = 1, 2, ..., n$.

Assume that the planes of the flock contain the planes π_i and all planes of the flock contain one of the lines L_i .

Assume that the north side of π_j contains the north side of π_{j-1} for $j = 2, 3, ..., n-1$.

If the set of planes of F contains the following sets of planes:

(1) the set of planes containing L_1 defined by points on the north side of Q relative to π_1,

(2) the planes containing L_j defined by the set of points on the north side of Q relative to π_j which are not in the set of points on the north side of L_{j-1} for $j = 2, 3, ..., n-1$, and

(3) the set of planes containing L_n defined by the set of points on the south side of Q relative to π_{n-1}

then F is an $n-$interval flock.

Proof: If a plane on the north side of π_1 contains L_1 then it contains a point P_1 of the quadric which is not N_{L_1} on the north side of π_1. By the above results, for $I_1 = \pi_1 \cap oseg[N_{L_1}, S_{L_1}]$ then the indicated plane is $< P_1, L_1 >=< U, L_1 >$ for $I_1 < U < N_{L_1}$.

Consider planes π_j and π_{j-1}. These two planes intersect in the line L_j which is exterior to the quadric. It is clear that each plane partitions Q into sides such that one side relative to π_j is contained in a side relative to π_{j-1} and the other side Q relative to π_j contains the corresponding side relative to π_{j-1}.

For a plane containing L_j and a point P on the north side of Q relative to π_j which is not in the set of points of the north side of Q relative to π_{j-1} then $P \neq N_{L_j}$. If $P \neq S_{L_j}$ then the plane $< P, L_j >$ nontrivially intersects $oseg[S_{L_j}, N_{L_j}]$ say in T. If $< T, L_{j-1} >$ intersects $oseg[S_{L_{j-1}}, N_{L_{j-1}}]$ in a point T_1 between the intersection of this segment with π_{j-1} and $N_{L_{j-1}}$ then the plane is on the north side of π_{j-1}. This must occur unless T is between the intersection of $oseg[S_{L_j}, N_{L_j}]$ with π_{j-1} (notc that the point N_{L_j} must be in the north side of Q relative to π_{j-1} , so there is a nontrivial intersection) and the intersection of $oseg[S_{L_j}, N_{L_j}]$ with π_j. Hence, each of these planes is a plane of an $n-$interval flock. Now if $j = n-1$, the planes of the south side of Q with respect to π_{n-1} similarly have appropriate intersection so that every plane of F is a plane of an $n-$interval flock.

4 0- and 1-point flocks of elliptic quadrics.

In the previous section, a construction method was given which constructs flocks of elliptic quadrics for any subfield K of the field of real numbers. In this construction, the flocks are what might be called $2-$point flocks as the set of planes of the flock cover all but two points of the elliptic quadric.

Definition 24 *A flock of any elliptic quadric such that the plane covers all but i-points of the quadric shall be called an $i-$point flock.*

In the above construction, we consider the elliptic quadric over K to be embedded in the elliptic quadric over the field of real numbers. If we are trying to construct an $n-$linear flock, we choose the lines L_i in $PG(3, K)$ but to have the property that either the North or South point of tangency of the first and/or last lines of real elliptic quadric is not in $PG(3, K)$. In this case, the previous construction may used to produce both $0-$flocks and $1-$flocks.

To produce a $0-$flock, we choose the points of tangency so that the initial north and the final south points of tangency are not in $PG(3, K)$ for the proper subfield K of the reals but the line between the north and south points is a line of $PG(3, K)$. To produce a $1-$flock, we choose exactly one pair of these points of tangency so that neither of the points are in $PG(3, K)$. All other lines are chosen as in the previous method as $T-$lines. So, we choose n lines either one or two of which are not $T-$lines but have associated tangent planes that lie in $PG(3, R)$ but not in $PG(3, K)$. The method that we have been employing is to define planes via open line segments between points of tangency. If the points of tangency are in the supergeometry but not in the subgeometry then we need to formulate a slightly different method to generate the spread. For subfields of the reals, we may choose the first set of planes by generating the planes using the lines and a point of the north side of the given plane instead of the interval associated with the points of tangency. Alternatively, one can choose only the $K-$points of the interval and the external line to generate a set of planes which includes the set of planes generated using points of the quadric on the north side. We may then simply eliminate the planes of this second set which do not actually nontrivially intersect the quadric. We shall give some examples in the application section which uses the second method —determining the $K-$points on the open segment defined by the points of tangency which are not themselves $K-$points.

Theorem 25 *Let K be any proper subfield of the field of real numbers which admits positive nonsquares and let n be any positive integer. Then there exist both $0-$flocks and $1-$flocks of elliptic quadrics such that planes of the flock share one of n lines.*

Proof: It is sufficient to show that there exist lines of $PG(3, K)$ that have no tangent planes of intersection. Assume there exists an element t of R (reals)$-K$ such that t^2 is in K.

With the representation of the elliptic quadric as in (2.3), assume that $(\pm t, \pm t, -1, 2t^2)$ is a point of the real elliptic quadric. The tangent plane is given by $\pm 2tx - \pm 2ty - 2t^2 z + w = 0$ and the intersection of these two planes is the line of points $\{(x, -x, z, 2t^2)\}$. Since this may be consider a line of both projective spaces, this completes the proof of the theorem.

Corollary 26 *Let K be any subfield of the reals which admits positive non-squares. For any positive interger n, there exist partial partitions of a Pappian spread in $PG(3, K)$ as mutually disjoint reguli which cover the plane except for 0 or 1 lines and such that multiple derivation of the reguli produces a nonPappian translation plane whose components lie in the union of exactly n Pappian spreads.*

5 Applications.

First we provide an algebraic description of the flocks of Dembowski extended to include flocks over subfields K of the reals. We first construct

5.1 2−point flocks.

We represent the Pappian plane Σ coordinatized by $K[\sqrt{-1}]$ by the spread:

$$x = 0, y = x \begin{bmatrix} u & -t \\ t & u \end{bmatrix} \text{ for all } u, t \text{ in } K.$$

Represent $PG(3, K)$ by homogeneous coordinates (x, y, z, w) and the elliptic quadric Q by the equation $x^2 + y^2 + z^2 = w^2$. We may further represent the points of Q as follows:

$$(0, 0, -1, 0), (-2t/(1+u^2+t^2), -2u/(1+u^2+t^2), 1-(u^2+t^2)/(1+u^2+t^2), 1).$$

Note the obvious $1-1$ correspondence between the spread components of Σ and the points of the elliptic quadric.

Choose the affine space $w \neq 0$ and represent points here by (x, y, z) where the associated point in the projective space is $(x, y, z, 1)$. Hence, the quadric has the form $x^2 + y^2 + z^2 = 1$ taken in the affine space.

Choose the plane $\pi_1 =< L_1, L_2 >$ as $z = 0$ and let L_1 within $z = 0$ be denoted affinely by $x = -(1 + t_o^2)/2t_o$.

Note that the tangent planes to the points $(-2t_o/(1+t_o^2), 0, \pm(t_o^2-1)/(1+t_o^2))$ must intersect in the indicated line.

Furthermore, take L_2 to be the infinite line of intersection of the hyperplane at infinity with the plane $z = 0$.

Take $(1 - t_o^2) > 0$ to fix the north and south points. Hence,

$$N_{L_1} = (-2t_o/(1 + t_o^2), 0, (1 - t_o^2)/(1 + t_o^2))$$

and

$$S_{L_1} = (-2t_o/(1+t_o^2), 0, (t_o^2-1)/(1+t_o^2))\ (take\ t = 1/t_o).$$

The complete affine open line segment is:

$$(-2t_o/(1+t_o^2), 0, (1-2\lambda)(1-t_o^2)/(1+t_o^2))\ for\ 0<\lambda<1$$

and the points on the north part require that $(1-2\lambda) > 0$.

According to our construction process, we need to form the affine open line segment between the north and south tangent points of planes containing L_i and then form the north or south partial flock of planes containing L_i and a point of the affine open line segment respectively for $i = 1, 2$ for which there is nontrivial intersection with the corresponding plane and the quadric.

Note that $N_{L_2} = (0,0,1)$ and $S_{L_2} = (0,0,-1)$. Since the open line segment is simply $(0,0,1-2\lambda)\ for\ 0<\lambda<1$ then it follows that the planes of the south side of π_1 are the planes $z = (1-2\lambda)\ for\ -1 < 1-2\lambda < 0$.

Note that these planes intersect in conics with points:
$(-2t/(1+u^2+t^2), -2u/(1+u^2+t^2), 1-(u^2+t^2)/(1+u^2+t^2))$
$\ni u^2+t^2 = (1-\lambda)/\lambda\ for\ (1-2\lambda) < 0$.

Of course, if there is no solution, we do not include this plane within the flock.

To calculate the plane containing L_1 and a point of the affine open line segment between the north and south points of tangency, we note that the affine plane containing L_1 and $(-2t_o/(1+t_o^2), 0, (1-2\lambda)(1-t_o^2)/(1+t_o^2))$ for $(1-2\lambda) > 0$ is

$$\begin{aligned}(-2t_o/(1+t_o^2))x + cz &= 1\ where\\ c &= (1-4t_o^2/(1+t_o^2)^2)/((1-2\lambda)(1-t_o^2)/(1+t_o^2))\\ &= (1-t_o^2)/(1+t_o^2)(1-2\lambda).\end{aligned}$$

Hence, we have:

Theorem 27 *Let K be a subfield of the field of real numbers. Represent an elliptic quadric Q by the equation $x^2+y^2+z^2 = w^2$ in $PG(3,K)$ represented homogeneously by (x,y,z,w).*

Then the following planes provide a flock of
$Q - \{(0,0,-1,1), (-2t_o/(1+t_o^2), 0, (1-t_o^2)/(1+t_o^2), 1)\}$ where $t_o^2 - 1 < 0$,
(1) $z = (1-2\lambda)$ for all $-1 < 1-2\lambda \leq 0$ of K, which have a nontrivial solution with the quadric and
(2) $-2t_o/(1+t_o^2))x + (1-t_o^2)/((1+t_o^2)(1-2\lambda))z = w$ for all λ such that $1 > (1-2\lambda) > 0$ and such that there exists a nontrivial intersection with the quadric.

This flock is nonlinear as the planes of the flock share exactly two distinct lines.

We now translate the flock into a set of reguli in the Pappian plane Σ.

Theorem 28 *Let K be a subfield of the field of real numbers. Let Σ denote the Pappian plane coordinatized by $K[\sqrt{-1}]$.*

Represent the spread for Σ by the components:

$$x = 0, y = x\begin{bmatrix} u & -t \\ t & u \end{bmatrix} \text{ for all } u, t \text{ in } K.$$

Then the following set of reguli are mutually disjoint and partition the set of components not equal to $x = 0$ and $y = x\begin{bmatrix} 0 & -t_o \\ t_o & 0 \end{bmatrix}$:

$R_\rho = \{y = x\begin{bmatrix} u & -t \\ t & u \end{bmatrix} \ni u^2 + t^2 = \rho \geq 1\}$. *Note that R_ρ is an André net in Σ.*

$R_\lambda = \{y = x\begin{bmatrix} u & -t \\ t & u \end{bmatrix} \ni$
$(u^2+t^2)((1-t_o^2)/(1+t_o^2)(1-2\lambda)+1) - 4t_ot/(1+t_o^2)+$
$+(1-(1-t_o^2)/(1+t_o^2)(1-2\lambda)) \ = 0 \ \}$ *where* $1 > (1-2\lambda) > 0\}$.

Note that, of course, if the corresponding equation has no solution, there is no net defined.

Proof: We shall give two proofs one geometric and one algebraic.

Geometric proof:

As a corollary to the above theorem, we need only calculate the conics of intersection of the planes listed in the above theorem and use the algebraic connections to translate these conics back to reguli of Σ.

Considered affinely, the plane

$$(-2t_o/(1+t_o^2))x + ((1-t_o^2)((1+t_o^2)(1-2\lambda))z = 1$$

intersects the quadric $x^2 + y^2 + z^2 = 1$ in the points

$$(-2t/(1+u^2+t^2), -2u/(1+u^2+t^2), (1-(u^2+t^2))/(1+u^2+t^2))$$

if and only if $(u^2+t^2)((1-t_o^2)/((1+t_o^2)(1-2\lambda)+1)) - 4t_ot/(1+t_o^2)+$
$(1-(1-t_o^2)/(1+t_o^2)(1-2\lambda)) = 0$.

This completes the proof of the theorem.

Now that we have the form for the reguli in Σ, we may give a direct algebraic proof. We have to show that the given set of reguli union the two listed components give a spread of Σ and that they are mutually disjoint.

Algebraic proof:

Solve the above equation to obtain:

$$1 - 2\lambda = ((1-t_o^2)(u^2+t^2-1)/(1+t_o^2))/(-(u^2+t^2+1-4t_ot/(1+t_o^2))).$$

However,

$$(u^2+t^2+1-4t_ot/(1+t_o^2)) = (u^2+(t-2t_o/(1+t_o^2))^2+(1-t_o^2)^2/(1+t_o^2)^2) > 0.$$

Thus, we obtain:

$$1-2\lambda=((t_o^2-1)(u^2+t^2-1)/(1+t_o^2))/(u^2+(t-2t_o/(1+t_o^2))^2+(1-t_o^2)^2/(1+t_o^2)^2).$$

Note that to establish that we have a cover, we need only to show that for any $(u,t)\neq(0,t_o)$ such that $u^2+t^2<1$, there is a value λ with $0<\lambda<1/2$ such that the above equation is satisfied. Moreover, it easily follows that, in this case, all of the regulus nets are mutually disjoint.

Since, $0<1-2\lambda<1$ if and only if $0<\lambda<1/2$, we need to show

$$0<((t_o^2-1)(u^2+t^2-1)/(1+t_o^2))/(u^2+(t-2t_o/(1+t_o^2))^2+(1-t_o^2)^2/(1+t_o^2)^2)<1.$$

Note that the left hand inequality is clear since $t_o^2-1<0$ and $u^2+t^2-1<0$ by assumption.

Concentrating on the right hand side of the inequality and using the fact the denominator is positive, we have to show:

$$(u^2+t^2)((t_o^2-1)/(1+t_o^2)-1)<(t_o^2-1)/(1+t_o^2)+1-4t_ot/(1+t_o^2).$$

This inequality is equivalent to:

$$(u^2+t^2)(-2/(1+t_o^2))<2((t_o-t)^2-t^2).$$

Thus, we must show that

$$u^2>-(t_o-t)^2$$

which is always valid unless $u=0$ and $t_o=t$. This completes the algebraic proof.

5.2 1-point flocks.

Let R denote the field of real numbers and let K denote any subfield which does not contain a particular positive real number r but such that r^2 is in K.

We consider the points of the quadric $(\pm 2r/(1+r^2),0,(1-r^2)/(1+r^2),1)$ and note that the corresponding tangent plane in $PG(3,R)$ is $\pm 2rx+(1-r^2)z-(1+r^2)w=0$. If we take π_1 to be $x=0$ then the intersection is the line L_1 in $PG(3,K)$ within $x=0$ given by the equation $(1-r^2)z=(1+r^2)w$. We note that the real open line segment formed by the two tangent points consists of all real points $((2r/(1+r^2))(1-2\lambda),0,(1-r^2)/(1+r^2),1)$ for all real numbers λ such that $0<\lambda<1$. Call $(2r/(1+r^2))(1-2\lambda)=k$. We note that as λ varies between 0 and 1 then

$-2r/(1+r^2)<k<2r/(1+r^2)$. Hence, if we take all of the elements of K between these two points, we obtain the corresponding set of points of $PG(3,K)$ which lie on this open segment.

If we want to cover the positive or north side then we consider the interval points $(k,0,(1-r^2)/(1+r^2),1)$ such that $0<k<2r(1+r^2)$ and take the planes of intersection with the $K-$ quadric of the form $<L_1,(k,0,(1-r^2)/(1+r^2),1)>$.

We have noted that if W is any $K-$ point on the north side of π_1 then $< L_1, W >$ must intersect the $K-$line containing the interval in question in a $K-$point and hence this plane is defined by one of the planes above.

We may then cover the south side of π_1 say by taking $L_2 = (x = 0) \cap (w = 0)$ (the hyperplane at infinity) so that it follows that $N_{L_2}=(1, 0, 0)$ and $S_{L_2} = (-1, 0, 0)$ in the affine form. So, in order to cover the south side with the exception of S_{L_2} with lines containing L_2, we need only take the planes $x = \rho$ for $-1 > \rho > 0$ which intersect the $K-$quadric in a conic (i.e. a nontrivial intersection).

In order to obtain a flock, we merely need to adjoin the plane $x = 0$.

Theorem 29 *The planes listed above of intersection with the $K-$quadric form a flock of $Q - \{(-1, 0, 0)\}$ and hence define a $1-$point flock which is a $2-$line interval flock of the elliptic quadric in $PG(3, K)$.*

Although, we will not bother with a formal description of a $0-$point flock, it is clear how to proceed.

To give an algebraic description of an $n-$ interval flock or corresponding partition set of reguli when $n > 2$ is somewhat more complex but only technically so. To illustrate this, we shall give an algebraic description of a $3-$interval flock. We do this only for $2-$point$-n-$interval flocks, but again, the construction of $1-$point or $0-$point $-n-$interval flocks proceeds in a similar manner.

Theorem 30 *Let K be a subfield of the field of real numbers. Represent an elliptic quadric Q by the equation $x^2 + y^2 + z^2 = w^2$ in $PG(3, K)$ represented homogeneously by (x, y, z, w).*

Then the following planes provide a flock of

$Q - \{(-2t_1/(1+t_o^2), 0, (1-t_1^2)/(1+t_1^2)), (-2t_o/(1+t_o^2), 0, (1-t_o^2)/(1+t_o^2))\}$
where $t_o^2 - 1 < 0, 3 - t_1^2 < 0$.

(1) $z = \rho$ *for all* $-1/2 < \rho \leq 0$ *of* K,

(2) $\alpha x + \beta z = w$ *where* α *and* β *are uniquely determined by the system of simultaneous equations:*

$$(i)\ \alpha((3+t_1^2)/-4t_1) - \beta/1 = 1, \tag{1}$$

$$(ii)\ \alpha(-4t_1\lambda(t_1^2+3) - 2t_1(t_1^2+9)) + \beta(2\lambda(t_1^2-9) + (1-t_1^2)(t_1^2+9) =$$
$$(t_1^2+1)(t_1^2+9) \text{ where } 0 < \lambda < (t_1^2+9)/4(3+t_1^2).$$

and

(3) $-2t_o/(1+t_o^2))x + ((1-t_o^2)/(1+t_o^2)(1-2\lambda))z = w$ *for all* λ *such that* $1 > (1-2\lambda) > 0$.

This flock is nonlinear as the planes of the flock share exactly three distinct lines.

If there is no solution with a given plane and the quadric, the plane is not considered part of the flock.

Proof: We retain the north partial flock as in the previous theorem and take the subflock of the south partial flock as follows:

$z = \rho$ for $-1/2 \leq \rho \leq 0$. We now take a line L_3 in $z = -1/2$ distinct from the line L_2 which is the line at infinity on this plane. We form the north and south points of tangency relative to L_3 , form and affine open line segment and determine the λ for the point of intersection of this line segment and $z = -1/2$. We then take the planes that contain L_3 and a point of the line segment from the intersection point to (but not including) the south tangency point relative to L_3. This is a flock of Q.

We take the affine point $(-2t_1/(1+t_1^2), 0, (1-t_1^2)/(1+t_1^2))$ where $-1 < (1-t_1^2)/(1+t_1^2) < -1/2$ or equivalently $3 - t_1^2 < 0$ to ensure that this point is on the south side of the plane $z = -1/2$. We determine the affine tangent plane

$(-2t_1/(1+t_1^2))x + ((1-t_1^2)/(1+t_1^2))z = 1$

and intersect $z = -1/2$ to obtain the line $L_3 : x = (-(3+t_1^2)/4t_1)$ and $z = -1/2$.

A straightforward calculation shows that the north tangency point relative to L_3 is

$(-2(3/t_1)/(1+(3/t_1)^2), 0, (1-(3/t_1)^2)/(1+(3/t_1)^2)$.

After another calculation, we compute the afine open line segment between the north and south tangency points relative to L_3 consisting of the following points:

$((-4t_1\lambda(t_1^2-3) - 2t_1(t_1^2+9))/((t_1^2+1)), 0, (2\lambda(t_1^2-9) + (1-t_1^2)(t_1^2+9))/((t_1^2+9)(t_1^2+1), \forall\ \lambda \ni 0 < \lambda < 1$.

Furthermore, we determine the value for λ so that the point on the affine open line segment intersects $z = -1/2$ as $\lambda = (t_1^2+9)/4(3+t_1^2)$.

Note that when $\lambda = 0$, we obtain the south point of tangency relative to L_3.

Hence, we obtain the south side of $z = -1/2$ by forming intersection with the planes containing L_3 and a point on the affine open line segment determined by λ for $0 < \lambda < (t_1^2+9)/4(3+t_1^2)$.

Clearly, the affine plane containing L_3 and a point P_λ on the affine open line segment is of the general form $\alpha x + \beta z = 1$.

The condition given in the statement of the theorem is necessary and sufficient in order that the plane contains L_3 and the point P_λ.

From here, it is technically clear how to obtain an algebraic description of any $n-$interval flock.

6 n-interval flocks of quadratic cones.

Theorem 31 *Let K be an ordered field and elements a_1, a_2, b_1, b_2 such that $a_2^2 - 4a_1 < 0$, $b_2^2 - 4b_1 < 0$ and assume that $(a_1, a_2) \neq (b_1, b_2)$.*

Consider $PG(3, K)$ represented in homogeneous coordinates (x_0, x_1, x_2, x_3). Represent a quadratic cone C with equation $x_1x_2 = x_2^2$.

Define the following set $F(a_1, a_2, b_1, b_2)$ of planes with the following equations:

$N_t : tx_0 - a_1tx_1 + a_2tx_2 + x_3 = 0, t > 0,$

$S_t : tx_0 - b_1tx_1 + b_2tx_2 + x_3 = 0, t < 0,$
$I_o : x_3 = 0.$
Then $F(a_1, a_2, b_1, b_2)$ is a bilinear flock of the quadratic cone C.

Remark 3 *We shall refer to the partial flocks N_t and S_t as the North and South partial flocks respectively.*

Lemma 32 *Let K be a subfield of the reals and C the cone $x_ox_1 = x_2^2$ in $PG(3,K)$. We call $x_3 = 0$ the hyperplane at infinity.*

Let $L_\infty =< (1,0,0,1),(0,0,0,1)$ for all t of $K - \{0\} >$. Let L be any line of $PG(3,K)$ which is exterior to C and form the planes

$$\pi_t^L :< (-1/t, 0, 0, 1), L > .$$

(1) There exist constants $\alpha_1, \beta_1, \alpha_2, \beta_2$ of K such that

π_t^L has equation: $tx_o + (t\alpha_1 + \beta_1)x_1 + (t\alpha_2 + \beta_2)x_2 + x_3 = 0 \; \forall \; t \in K - \{0\}$.

(2) There exist non-negative real numbers a, b, possibly equal, such that the subset $\{t\}$ of $K - \{0\}$ such that

$$\pi_t^L \cap (x_3 = 0) \cap C \neq \phi$$

is $((-\infty, -a) \cup (b, \infty)) \cap (K - \{0\})$.

Proof: We assert that the set of planes π_t^L: have the general form:

$$tx_o + (t\alpha_1 + \beta_1)x_1 + (t\alpha_2 + \beta_2)x_2 + x_3 = 0 \text{ for all nonzero } t \text{ in } K.$$

If L is in $x_3 = 0$, it is easy to see that the subset is $((-\infty, 0) \cup (0, \infty)) \cap (K - \{0\})$. Hence, Let $L =< (b_o, b_1, b_2, 1), (c_o, c_1, c_2, 0) >$ such that
$(b_o + zc_o)(b_1 + zc_1) \neq (b_2 + zc_2)^2$ for all z in K.

To see this, note that π_t^L must have the form $tx_o + mx_1 + nx_2 + x_3 = 0$. First assume that $c_1 = 0$. Then $c_0c_2 \neq 0$ as the line L is exterior to the cone. Hence, it follows that $n = t\alpha + \beta$ for certain contants α and β. Then either $b_1 = 0$ or we may obtain $m = t\delta + \gamma$ for some constants δ and γ. However, if $c_1 = b_1 = 0$ then L cannot be exterior to the cone.

Hence, we may assume that $c_1b_1 \neq 0$ and similarly $c_2b_2 \neq 0$.

Consider the equations $c_1(tb_0 + mb_1 + nb_2 + 1) = 0$ and $b_1(tc_o + mc_1 + nc_2) = 0$.

Now if $c_1b_2 - b_1c_2 \neq 0$ then we may solve for n in the form required and then subsequently solve for m in the required form.

Hence, we are finished unless $c_1b_2 - b_1c_2 = 0 = c_1b_1 - b_2c_1$.

However, in this latter case, it is easily verified that the line L contains a point of L_∞.

Thus, we have the proof to (1).

There is a nonempty intersection $(x_o, x_1, x_2, 0)$ on $\pi_t^L \cap (x_3 = 0) \cap C$ if and only if we have the following cases:

(i) $x_1 = 0$ and

(ii) $x_1 \neq 0$.

If $x_1 = 0$ then $x_2 = 0$, so π_t^L contains the point $(1,0,0,0)$ and hence $t(-1/t,0,0,1)+(1,0,0,0)) = (0,0,0,t) \equiv (0,0,0,1)$. Hence, the plane contains L_∞ so that L cannot be exterior to C. Thus, we may assume case (ii) $x_1 \neq 0$.

Note that $\cap\pi_t^L = L$ is given by $\{(-(\alpha_1x_1+\alpha_2x_2), x_1, x_2, -(\beta_1x_1+\beta_2x_2))$ for all x_1, x_2 of $K\}$.

Consider when $-(\alpha_1x_1+\alpha_2x_2)x_1 = x_2^2$. Since we may assume that $x_1 \neq 0$, letting $z = (x_2/x_1)$ we obtain the quadratic $z^2+\alpha_2z+\alpha_1 = 0$ if and only if $\alpha_2^2 - 4\alpha_1 \geq 0$. Hence, we must have $\alpha_2^2 - 4\alpha_1 < 0$.

Now ant intersection point satisfies

$$tx_ox_1 + (t\alpha_1+\beta_1)x_1^2 + (t\alpha_2+\beta_2)x_2x_1 = 0$$

Letting $x_ox_1 = x_2^2$ and $y = (x_2/x_1)$ we obtain the following quadratic in y :

$$ty^2 + (t\alpha_2+\beta_2)y + (t\alpha_1+\beta_1) = 0.$$

There is a solution only if

$$(t\alpha_2+\beta_2)^2 - 4t(t\alpha_1+\beta_1) \geq 0.$$

Since $\alpha_2^2 - 4\alpha_1 < 0$, this inequality is equivalent to

$$t^2 + t(2\alpha_2\beta_2 - 4\beta_1)/(\alpha_2^2-4\alpha_1) + \beta_2^2/(\alpha_2^2-4\alpha_1) \leq 0.$$

Hence, $(t+(\alpha_2\beta_2-2\beta_1)/(\alpha_2^2-4\alpha_1))^2 \leq ((\alpha_2\beta_2-2\beta_1)/(\alpha_2^2-4\alpha_1))^2 - \beta_1^2/(\alpha_2^2-4\alpha_1)$.

Let $\delta = (\alpha_2\beta_2-2\beta_1)/(\alpha_2^2-4\alpha_1)$ and $\rho = -\beta_1^2/(\alpha_2^2-4\alpha_1)$ and note that ρ must be nonnegative. Hence, Letting $-a = -\delta - \sqrt{\delta^2+\rho}$ and $b = \sqrt{\delta^2+\rho}$ $-\delta$ (note that b and a are real numbers) and taking the complement, we have $((-\infty,-a) \cup (b,\infty)) \cap (K - \{0\})$.

Definition 33 *The set $((-\infty,-a)\cup(b,\infty))\cap(K-\{0\})$ for a given exterior line L shall be called the affine intersection of L.*

Definition 34 *Let K be a subfield of the reals. Choose a line L_1 on $x_3 = 0$ exterior to the cone C. Choose any line L_2 not on $x_3 = 0$ and exterior to C and let the $K-$ interval $[d_2,\infty)$ be contained in the affine intersection of L_2 for $d_2 > 0$ ans d_2 in K. Choose any exterior line L_3 not on $x_3 = 0$ and not equal to L_2 in $<(-1/d_3,0,0,1), L_2>$ whose affine intersection contains $[d_3,\infty)$ where $d_2 < d_3$. Continuing this process, let L_{k+1} be any exterior line, distinct from L_k , of $<(-1/d_{k+1},0,0,1), L_k>$ for $k = 1,2,3,...n-1$ which is not on $x_3 = 0$ and whose affine intersection contains $[d_{k+1},\infty)$ where $d_k < d_{k+1}$.*

Any such ordered set of lines $\{L_1, L_2, ..., L_n\}$ with the above property shall be called a south $n-$interval set of lines.

Similarly, when $(-\infty,d_2)$ is contained in the affine intersection of L_2 for $d_2 < 0$ and $d_{k+1} < d_k < ... < 0$, we obtain a north $n-$interval set of lines.

Definition 35 *Let K be any subfield of the reals. Consider the quadratic cone C given by $x_o x_1 = x_2^2$ in $PG(3,K)$ where the points are given by homogeneous coordinates (x_o, x_1, x_2, x_3). Let $x_3 = 0$ denoe the hyperplane at infinity of the affine space 3−spaces $AG(3,K)$.*

We define the north cone $N(C)$ as $\{(x_o, x_1, x_2, 1)$ such that either $x_o > 0$ or $x_o = 0$ and $x_1 > 0\} \cap C$. The south cone $S(C)$ is similarly defined as $\{(x_o, x_1, x_2, 1)$ such that either $x_o < 0$ or $x_o = 0$ and $x_1 < 0\} \cap C$.

Let $C_A = N(C) \cup S(C)$. Let π be any plane which intersects $N(C)$ (resp. $S(C)$) in a conic. The south side $S(C,\pi)$ (resp. north side $N(C,\pi)$) of π is set of cone points Q not in π such that $oseg[Q, <(0,0,0,1)>]$ does not intersect (resp. does intersect) π and Q. The north side $N(C,\pi)$ (resp. south side $S(C,\pi)$) is the set of cone points of $C_A - S(C,\pi)$ (resp. $C_A - N(C,\pi)$).

Let $\{(1/t, 0, 0, 1)$ for t in $K - \{0\}\} = L_\infty^-$ where the line $<(1,0,0,0),(0,0,0,1)> = L_\infty$.

Remark 4 *Under the previous assumptions, choose any plane π_o that intersects $S(C)$ in a conic. Assume that $\pi_o = L_\infty^- = (-1/d, 0, 0, 1)$. Choose any affine line L on π_o which is exterior to C.*

(1) Then π_o is a plane of π_t^L and $[d,\infty)$ is contained in the affine intersection of L.

(2) The subflock $\{\pi_t^L : d \le t < \infty\}$ covers the points of the north side of π_o in the sourth cone. That is, the subflock covers $N(C,\pi_o) \cap S(C)$.

Proof: (1). Any line L of π_o has an affine intersection $(-\infty, -a) \cup (b, \infty)$. Hence, if $d > 0$ is in (b,∞) then so is $[d,\infty)$. This proves (1).

Assume the conditions of (2).

Note that $\{\pi_t^L : d \le t < \infty\}$ covers the points $(-1/e, 0, 0, 1)$ such that $-1/d < -1/e$. Hence,

$\{\pi_t^L : d \le t < \infty\}$ overs $L_\infty^- \cap S(C)$.

The points $N(C,\pi) \cap S(C) \cup \{(0,0,0,1)\}$ form a bounded cone B with base $\pi_o \cap S(C)$ and vertex $(0,0,0,1)$.

Let M_∞ and N_∞ be any two K−lines of the cone. Then, considering everything over the field R of real numbers, then $\{<L,P>: P \epsilon M_\infty \cap B\} = \{<L,Q>: Q \epsilon N_\infty \cap B\}$ by properties of $AG(3,R)$.

If L is a line of $AG(3,K)$ and P is a point of $AG(3,K)$ of $M_\infty \cap B$ and if $<L,P> = <L,Q>$ then we assert that Q is a point of $AG(3,K)$. To see this, note that we may choose an R−basis so that $M_\infty = L_\infty$ and $N_\infty = <(0,1,0,0),(0,0,0,1)>$. Hence, $P = (d,0,0,1)$ for d in K and $Q = (0,e,0,1)$ for e in R. Since L is a line of K then $<L,P>$ has the form $ax_o + bx_1 + cd_2 + x_3 = 0$. Since Q is incident with this plane, it follows that e must be in K so that Q is a K−point.

Definition 36 *Let K be a subfield of the reals. Choose a line L_1 on $x_3 = 0$ exterior to the cone C and a point $(-1/d_2, 0, 0, 1)$ for $d_2 > 0$ on L_∞. Form $\pi_2 = <L_1, (-1/d_2, 0, 0, 1)>$. Choose any affine line L_2 on π_2 which is exterior to the cone and note that $[d_2, \infty)$ is in the affine intersection of L_2 relative* *to* $\pi_t^{L_2}$. *Choose any point* $(-1/d_3, 0, 0, 1)$ *on* L_∞ *such that* $d_2 < d_3$ *and form* $\pi_3 =< L_2, (-1/d_3, 0, 0, 1) >$. *Choose any lines* L_3 *on* π_3 *and note that* $[d_3, \infty)$ *is contained in the affine intersection of* L_3 *relative to* $\pi_t^{L_3}$. *Continuing this process* n *times for* $n < \infty$, *we have a set of* $n+2$ *elements* $\{0 = d_1, d_2, d_3, ..., d_{n+1} = \infty\}$ *such that* $d_i < d_{i+1}$ *for* $i = 0, 1, ..., n$ *and a set of* n *lines* $\{L_1, L_2, ..., L_n\}$ *such that* $[d_i, \infty)$ *is contained in the affine intersection of* L_i *for* $i = 1, 2, ..., n$.

The set $\{L_1, L_2, ..., L_n\}$ *is called a south* $n-$*interval set.*

Similarly, a north $m-$*interval set of lines* $\{M_1, M_2, ..., M_m\}$ *may be defined with compansion set* $\{-\infty = e_{m+1}, e_m, ..., e_2, e_1 = 0\}$ *such that* $e_{i+1} < e_i$.

Theorem 37 *Let* K *be a subfield of the reals. Let* $\{L_1, L_2, ..., L_n\}$ *be a south* $n-$*interval set of lines with companion set* $\{0 = d_1, d_2, d_3, ..., d_{n+1} = \infty\}$ *such that* $d_i < d_{i+1}$ *for* $i = 0, 1, 2, ..., n$.

Then $\cup_{i=1}^n \{\pi_t^{L_i} : d_i \leq t < d_{i+1}\}$ *is a partial flock of the quadratic cone* C *which covers the south cone and the oval at infinity* $S(C) \cup \{(x_3 = 0) \cap C\}$.

Proof: Note that $\{\pi_t^{L_1} \ : 0 \leq t < d_2\}$ will cover the subset of point of L_∞ which is $\{(-1/t, 0, 0, 1) : -\infty < -1/t < -1/d_1\}$. Note that the affine intersection of L_1 is $(-\infty, 0) \cup (0, \infty)$ since L_1 is exterior to the cone and on the plane $x_3 = 0$. It follows from properties of $AG(3, R)$, for R the field of real numbers that $\{\pi_t^{L_1} : 0 \leq t < d_2\}$ covers all points of $S(C)$ of $AG(3, K)$ which lie on the south side of $\pi_{d_2}^{L_1}$. Hence, it reamins to cover the bounded cone B on the north side of $\pi_{d_2}^{L1}$.

Let P be any point of B. If P is not in any $\pi_{i+1}^{L_i}$ and on the north side of $\pi_{d_n}^{L_{n-1}}$ then P is clearly covered by $\{\pi_t^{L_n} : d_n \leq t < \infty\}$. Hence, there is a unique integer h such that P is on the north side of $\pi_{d_h}^{L_k}$ and on the south side of $\pi_{d_{h+1}}^{L_{k+1}}$. Hence, by properties of the real affine space, P is in $\{\pi_t^{L_k} : d_k \leq t < \infty\}$.

We may also select a set of lines and planes in a similar way to cover the nouth partial cone. The union of these two sets define our flock of a quadratic cone.

Hence, we obtain:

Theorem 38 *Let* K *be a subfield of the reals. Assume there exist a south* $n-$*interval set of lines* $\{L_1, L_2, ..., L_n\}$ *with companion set* $\{0 = d_1, d_2, d_3, ..., d_{n+1} = \infty\}$ *such that* $d_i < d_{i+1}$ *for* $i = 0, 1, ..., n$ *and a north* $m-$*interval set of lines* $\{M_1, M_2, ..., M_m\}$ *with compansion set compansion set* $\{-\infty = e_{m+1}, e_m, ..., e_2, e_1 = 0\}$ *such that* $e_{i+1} < e_i$.

Then $\cup_{i=1}^n \{\pi_t^{L_i} \ni d_i \leq t < d_{i+1}\} \cup_{j=2}^m \{\pi_t^{M_j} \ni e_j \geq t > e_{j+1}\}$ *is a flock of the quadratic cone.*

Furthermore, the planes of the flock share exactly $n + m - 1$ *lines.*

Proof: The south $n-$interval set of lines covers the sorth partial cone and $(x_3 = 0) \cap C$ and the nouth $m-$interval set covers the nouth partial cone and $(x_3 = 0) \cap C$. The indexing does not use the line M_1. Since the planes intersecting the cone in points covered by the north partial flock cannot intersect the points of the cone covered by the south partial flock, we have a flock of a quadratic cone.

Definition 39 *A flock of a quadratic cone constructed as above, where the planes share one of n -lines, shall be call an n−interval flock or n−linear flock.*

In the next section, we show that it is always possible to apply this construction process to produce 3− and 4−linear flocks for any ordered field K which is standard for one or two polynomials.

Note that the planes of a flock over an ordered field previously constructed using the above method completely cover the line

$L_\infty =< (-1/t, 0, 0, 1), (0, 0, 0, 1) >$ of the cone. A natural question to ask is whether any partial flock of a quadratic cone whose planes completely cover L_∞ must be actually be a flock. Although the answer to this question seems to be trivially yes, it is actually false.

In [5], Jha and Johnson give examples of maximal and proper partial flocks whose planes completely cover L_∞. For example, consider the set of planes $tx_o + t_1^3 x_1/3 - t^2 x_2 + x_3 = 0$ for all t in K. If the mapping $x \rightarrow x^3$ is $1-1$ but not onto in K then this set of planes determines a maximal and proper partial flock which obviously covers the points on L_∞. It turns out that no three planes of this partial spread contain a common line. Hence, this is a maximal partial spread where there does not exist a finite set of lines such that planes share one of the lines of the set.

Thus, a cover of L_∞ by a partial flock does not necessarily determine a flock. However, in the construction give above, this is the case when K is a subfield of the reals.

7 3-linear and 4-linear conical flocks.

In this section, we consider converings of the North partial flock by a North n−interval set. We shall consider only ordered fields K which are standard for a finite set of polynomials.

Proposition 40 *Let K be an ordered field K which is standard for a polynomial such that $\delta^2 - 4\alpha < 0$.*

Theorem 41 *Then the set of planes*

$$tx_o - (t\alpha + \beta)x_1 + (t\delta + \gamma)x_2 + x_3 = 0 \text{ for all } t \text{ in } K$$

defines a linear flock for any fixed elements β and γ of K.

Proof: We consider the corresponding set of subspaces

$$x = 0, y = x \begin{bmatrix} u + (t\delta + \gamma) & (t\alpha + \beta) \\ t & u \end{bmatrix}.$$

Change bases by $\sigma : (x,y) \to (x, x\begin{bmatrix} -\gamma & -\beta \\ 0 & 0 \end{bmatrix} + y)$. σ maps the set of subspaces onto the set

$$x = 0, y = x\begin{bmatrix} u + t\delta & t\alpha \\ t & u \end{bmatrix}.$$

Since the set of the matrices involved forms a field, it follows that the set of planes defines a linear flock.

Proposition 42 *Let K be an ordered field such that $a_{2i}^2 - 4a_{1i} < 0$ for $i = 1, 2, ..., n$.*

Theorem 43 *Then any $n-$interval flock of a quadratic cone has the following representation:*

The planes are $tx_o - (ta_{1i} + b_{1i})x_1 + (ta_{2i} + b_{2i})x_2 + x_3 = 0$ all t in K.

Proof: The basic construction begins with an infinite line and uses a linear flock indexed by t on some interval. The line contained in the subsequent linear flock has a suitable affine intersection assuming such exists for the ordered fields in question. In any case, the flock consists of n pieces of linear flocks indexed by intervals. It follows that any $n-$interval flock has the above form.

Theorem 44 *Let K be an ordered field which is standard for two polynomials and $a_2^2 - 4a_1 < 0$ and $b_2^2 - 4b_1 < 0$.*

Consider PG(3,K) represented in homogeneous coordinates (x_o, x_1, x_2, x_3). Represent a quadratic cone C with equation $x_o x_1 = x_2^2$.

Define the following set of planes with the following equations
$N_t^d : tx_o - a_1 tx_1 + a_2 tx_2 + x_3 = 0, 0 < t < d$, *for some d in K,*
$R_t : tx_2 - a_1 tx_1 + a_2 dx_2 + x_3 = 0$ *for* $d \leq t < \infty$
$S_t : tx_o - b_1 tx_1 + b_2 tx_2 + x_3 = 0, t < 0,$
$I_o : x_3 = 0.$

Then this set of planes defines a $3-$linear flock of a quadratic cone such that the planes contain one of the lines $\cap N_t^d$, $\cap R_t$ or $\cap S_t$.

Proof: First we show that two planes π and ρ of N_t^d, R_t, or S_t cannot intersect in a point of the cone. Assume that $\pi = S_k$ for some $k < 0$ and that $\rho = R_s$ for some s where $d \leq s < \infty$ (where the notation indicates all such elements $s \geq d$ of K).

Then $\pi \cap \rho$ is the set of points (x_o, x_1, x_2, x_3) such that

$$(s - k)x_o - (sa_1 - kb_1)x_1^2 + (da_2 - kb_2)x_2 = 0.$$

And, $(0, 1, 0, x_3)$ is such a point if and only if $sa_1 = kb_1$. However, since both a_1 and $b_1 < 0$ and $s \geq d > 0$ and $k < 0$ this leads to a contradiction. Similarly, $(1, 0, 0, x_3)$ cannot be a point of $\pi \cap \rho$.

Hence, any point (x_o, x_1, x_2, x_3) of $\pi \cap \rho \cap C$ must have x_o, x_1, x_2 all nonzero. Multiplying the above equation by x_1, dividing by x_2^2 and letting $y = (x_1/x_2)$ we obtain the following polynomial in y:

$$y^2(sa_1 - kb_1) - (da_2 - kb_2)y - (s - k) = 0.$$

By the above assumed properties of K this polynomial cannot have a root if

$$(da_2 - kb_2)^2 + 4(s - k)(sa_1 - kb_1) < 0.$$

Recall that

$$d^2a_2^2 + 4s^2a_1 < s^2(a_2 + 4a_1) < 0,$$

$$k^2(b_2^2 + 4b_1) < 0,$$

and in [2], it was noted that

$$a_2b_2 + 2(a_1 + b_1) < 0.$$

Hence,

$$da_2b_2 + s(2(a_1 + b_1) \leq d(a_2b_2 + 2(a_1 + b_1)) < 0 \textit{ as } 2(a_1 + b_1) < 0 \textit{ and } d \leq s.$$

So,

$$(da_2 - kb_2)^2 + 4(s - k)(sa_1 - kb_1) = (d^2a_2^2 + 4s_2^2a_1) +$$

$$k^2(b_2^2 + 4b_1) - 2k(da_2b_2 + s(2(a_1 + b_1)) < 0 \text{ as } -2k > 0.$$

Hence, a plane of type S_s and a plane of type R_k cannot have an intersection on C.

Now assume that $\pi = N_t$ for some t such that $0 < t < d$ and $\rho = R_s$ for $d \leq s$.

By a similar argument to above, there is a nontrivial intersection of $\pi \cap \rho \cap C$ only if

$$a_1(t - s)y^2 - (t - d)a_2y - (t - s) = 0$$

has a solution. However, we assert that

$$(t - d)^2a_2^2 + 4(t - s)^2a_1 < 0.$$

Note that $(t - d)^2a_2^2 + 4(t - s)^2a_1 \leq (t - s)^2(a_2^2 + 4a_1) < 0$ since $2t \leq s + d$.

Thus, planes of the north and south partial flocks cannot intersect in a point of C.

Hence, we have a flock of a quadratic cone provided we have a cover of $C - \{(0, 0, 0, 1)\}$ by planes of thc indicated type.

It suffices to show that $\cup\{N_t \cap C \text{ for } d \leq t < \infty\} = \cup\{R_t \cap C \text{ for } d \leq t < \infty\}$.

Note that $(x_o, 0, 0, 1)$ is on N_t if and only if the point is on R_t. Hence, assume that a point $(x_o, x_1, x_2, 1)$ of C is incident with N_t. This is valid if and only if

$$t(x_o - a_1x_1 + a_2x_2) = -1$$

which is valid if and only if

$$t = -x_1/(x_2^2 - a_1x_1^2 + a_2x_1x_2).$$

Note that

$$(x_2^2 - a_1x_1^2 + a_2x_1x_2) = (x_2 + a_2x_1/2) - x_1^2(a_2^2 + 4a_1)/4 > 0.$$

The same point of C is on R_s if and only if

$$s(x_o - a_1x_1) + a_2dx_2 = -1.$$

Hence, we must show that

$$d \leq -x_1(1 + da_2x_2)/(x_2^2 - a_1x_1^2).$$

We are assuming that

$$d \leq -x_1/(x_2^2 - a_1x_1^2 + a_2x_1x_2)$$

and we have noticed above that the denominator is always positive. Multiply the inequality by $(x_2^2 - a_1x_1^2 + a_2x_1x_2)$, subtract $d(a_2x_1a_2)$ and divide by $(x_2^2 - a_1x_1^2)$ (which is always positive) to obtain

$$d \leq -x_1(1 + da_2x_2)/(x_2^2 - a_1x_1^2).$$

This shows that we have a cover by the partial flock listed above so we have a flock of the quadratic cone.

This argument also proves the following two theorems, providing various 2–linear and 4–linear flocks.

Theorem 45 *Let K be an ordered field and assume that $a_2^2 - 4a_1 < 0$ and $b_2^2 - 4b_1 < 0$.*

Consider PG(3,K) represented in homogeneous coordinates (x_o, x_1, x_2, x_3). Represent a quadratic cone C with equation $x_ox_1 = x_2^2$.

Define the following set of planes with the following equations
$N_t^d : tx_o - a_1tx_1 + a_2tx_2 + x_3 = 0, 0 < t <\neq d,$ *for some d in K,*
$R_t : tx_o - a_1tx_1 + a_2dx_2 + x_3 = 0, d \leq t < \infty,$ *for $a_2 \neq 0$,*
$S_t : tx_o - b_1tx_1 + b_2tx_2 + x_3 = 0, e < t < 0,$ *for $(a_1, a_2) \neq (b_1, b_2)$.*
$T_t^e : tx_o - b_1tx_1 + b_2ex_2 + x_3 = 0,$ *for $-\infty < t \leq e$,*
$I_o : x_3 = 0.$

Then this set of planes defines a 4–linear flock of a quadratic cone such that the planes contain one of the lines $\cap N_t^d$, $\cap R_t$, $\cap T_t^e$ or $\cap S_t$.

Theorem 46 *Let K be an ordered field such that $a_2^2 - 4a_1 < 0$.*

Consider PG(3,K) represented in homogeneous coordinates (x_o, x_1, x_2, x_3). Represent a quadratic cone C with equation $x_o x_1 = x_2^2$.

Define the following set of planes with the following equations

$N_t^{d,-} : tx_o - a_1 tx_1 + a_2 tx_2 + x_3 = 0, 0 < t < d,$ *or* $t < 0$ *for some* d *in* K,

$R_t : tx_o - a_1 tx_1 + a_2 dx_2 + x_3 = 0, d \leq t < \infty,$ *for* $a_2 \neq 0$,

$I_o : x_3 = 0$.

Then this set of planes defines a 2−linear flock of a quadratic cone such that the planes contain one of the lines $\cap N_t^{d,-}$, $\cap R_t$.

Actually, if one changes bases, it is possible to show that the flock defined directly above is one of the "bilinear" flocks constructed in [2]. Hence, such flocks correspond to translation planes which admit $SL(2, K)$ as a collineation group.

Note that potentially, one might define an $n-$linear flock in a similar way for arbitrary ordered fields K which are standard for one or two polynomials.

8 Generalized Quadrangles and Derivations.

We have given a construction of infinitely many flocks of quadratic cones over suitable ordered fields. In particular, if K is a subfield of the reals then there are infinitely many flocks of quadratic cones whose planes contain exactly n-lines for a fixed positive integer n. Furthermore, we have shown some examples of 3−linear and 4−linear flocks of quadratic cones over certain ordered fields. All of the constructed flocks may be potentially derived to produce corresponding flocks. In [2], it has been shown that at least some of the "bilinear" flocks produce derivations which are flocks over certain fields K. Furthermore, there are subfields of the reals over which there is no associated derivations. Hence, in this case, there are examples when there exists a corresponding generalized quadrangle and examples where associated generalized quadrangles do not exist.

However, for the more general construction process considered above, we have not determined whether generalized quadrangles exist or not. We shall leave this as an open problem.

References

[1] P. Dembowski. **Finite Geometries**. Springer-Verlag, New York, Inc. 1968.

[2] M. Biliotti and N.L. Johnson. Bilinear flocks of quadratic cones. J. Geom. (to appear).

[3] D. A. Foulser and N. L. Johnson, The translation planes of order q^2 that admit $SL(2, q)$ as a collineation group. I. Even order, J. Algebra **86**, (1984), 385-406. The translation planes of order q^2 that admit $SL(2, q)$ as a collineation group. II. Odd order, J. Geometry **18**, (1982), 122-139.

[4] D.A. Foulser, N.L. Johnson, T.G. Ostrom. Characterization of the Desarguesian planes of order q^2 by $SL(2,q)$. Internat. J. Math. and Math. Sci. 6 (1983), 605-608.

[5] V. Jha and N.L. Johnson. Quasifibrations. Bull. Belgian Math. Soc./Simon Stevin, vol. 3, no. 3 (1996), 313-324.

[6] N.L. Johnson. Flocks of hyperbolic quadrics and translation planes admitting affine homologies. J. Geom. vol. 34 (1989), 50-73.

[7] N.L. Johnson. Infinite nests of reguli. Geom. Ded. (to appear).

[8] W.F. Orr. The Miquellian inversive plane $IP(q)$ and the associate projective planes. Ph.D. Thesis (Madison, Wisconsin., May 1973).

[9] J.A. Thas. Flocks of finite egglike inversive planes. In: Finite geometric structures and theore applications. (CIME, Bressanone 1972), pp. 189-191. Rome: Cremonese 1973.)

On a Theorem of Hering and Two-Transitive Ovals with a Fixed External Line

A. BONISOLI Dipartimento di Matematica, Università della Basilicata, Potenza, Italy

0 INTRODUCTION

Let π be a finite projective plane of order n. An oval Ω in π is said to be 2–transitive if π admits a collineation group G fixing Ω and acting 2–transitively on its points.

In case n is even, one of the three possibilities for Ω and G determined in [3] is that G also fixes a line ℓ which is external to Ω. In this situation some extra information can be given. The group G contains no Baer involutions, the set Δ of all involutory elations in G has size $n+1$ and is sharply transitive on ℓ, the subgroup $J = \langle\Delta\rangle$ is the semidirect product of $O(J)$ by a group of order 2, see Theorem C in [3]. We further know from [4] that either $n \in \{2,4\}$ or $n \equiv 0 \mod 8$ and the Sylow 2–subgroups of G are generalized quaternion groups.

Each one of the planes of order 2 and 4 does contain a 2–transitive oval with a fixed external line. They are the unique such planes of even order that the author is aware of. Although the possibility for additional examples is not excluded here, we do restrict the possibilities for the group G somewhat further, we show namely that G has to be a subgroup of the affine semilinear group $A\Gamma L(1,q)$ in its natural permutation representation. That is achieved by applying a classification theorem for 2–transitive groups with a regular normal subgroup and such that no involution fixes more than one point: this result is due to Hering [6] and is presented in Section 3. Some properties of the action of the collineation group G on the plane π are illustrated in the next two Sections; they are required in the concluding four Sections in order to rule out the other cases in Hering's theorem.

Research performed within the activity of G.N.S.A.G.A. of C.N.R. with the support of the Italian Ministry for Research and Technology

1 PRELIMINARY PROPERTIES

PROPOSITION 1.1 The stabilizer $G_{P,Q}$ of two distinct points P, Q on Ω has further fixed points on Ω.
Proof: Let P' and Q' be the intersections of the lines PK and QK with ℓ respectively. Since the lines PK and QK are fixed by $G_{P,Q}$, so are the points P' and Q'. The lines $P'Q$ and $Q'P$ are secant to Ω and are fixed by $G_{P,Q}$. Setting $\Omega \cap P'Q = \{A, Q\}$ and $\Omega \cap Q'P = \{B, Q\}$ we have that A and B are two further fixed points of $G_{P,Q}$ on Ω. $\square$

Assume G is 2–transitive on Ω but not sharply. In other words, if P and Q are distinct points on Ω then $G_{P,Q}$ does not reduce to the identity. If Γ is the set of fixed points of $G_{P,Q}$ on Ω then $3 \leq |\Gamma| < |\Omega|$ by the previous Proposition. Clearly $\mathrm{Fix}(G_{P,Q})$ is then a proper subplane π_0 of π and $\Gamma = \Omega \cap \pi_0$ is an oval in π_0.

Denote by $G_{(\Gamma)}$ and $G_{\{\Gamma\}}$ the pointwise and setwise stabilizer of Γ in G respectively. We have $G_{(\Gamma)} \trianglelefteq G_{\{\Gamma\}}$ and we define $G^{\Gamma} = G_{\{\Gamma\}}/G_{(\Gamma)}$. We sometimes say that G induces the group G^{Γ} on Γ. Clearly G^{Γ} also acts on π_0 as a collineation group.

REDUCTION LEMMA The group G^{Γ} is sharply 2–transitive on Γ.
Proof: See Theorem 9.4 in [9] or Section 2 in [8]. $\square$

2 POINT– AND LINE–ORBITS

The set of all lines of π will be denoted by $\mathcal{L}$, while $\mathcal{S}$ and $\mathcal{E}$ will denote the subsets of $\mathcal{L}$ consisting of all secant and external lines to Ω respectively. For each point P of the plane π we denote by $\mathcal{L}_P$ the set of all lines through P. If, in addition, $P \notin \Omega$ we define $\mathcal{E}_P = \mathcal{L}_P \cap \mathcal{E}$ and $\mathcal{S}_P = \mathcal{L}_P \cap \mathcal{S}$. By Proposition 4.1 in [4] the group $N = O(J)$ is an elementary abelian minimal normal subgroup of G acting regularly on Ω. If K is the knot of the oval Ω then K is fixed by G and the action of J on $\pi \setminus (\ell \cup \{K\})$ yields $n-1$ orbits $\Omega_1, \ldots, \Omega_{n-1}$ forming a G–invariant partition $\mathcal{F}$. Each orbit is an oval with knot K and $\Omega \in \mathcal{F}$, see also [1, Prop.3].

Each involution in G is an elation whose axis is some tangent line to Ω and whose center is the point of intersection of ℓ with this tangent line.

Select a point $O \in \Omega$ and let C be the point of intersection of the line OK with ℓ. Denote by η the involutory elation with center C and axis OK. The subgroup $\langle \eta \rangle$ is the kernel of the action of G_O on the points of the line OK and is also the kernel of the action of G_O on the lines in $\mathcal{L}_C$.

PROPOSITION 2.1 Given any two distinct external lines to Ω other than ℓ, there are at least two ovals in $\mathcal{F} \setminus \{\Omega\}$ meeting the two lines simultaneously.
Proof: Let $\mathcal{F}_1$ and $\mathcal{F}_2$ denote the subsets of $\mathcal{F}$ consisting of the $n/2$ ovals in $\mathcal{F}$ meeting the first and second line respectively. We have $\mathcal{F}_1 \cup \mathcal{F}_2 \subseteq \mathcal{F} \setminus \{\Omega\}$, whence $|\mathcal{F}_1 \cup \mathcal{F}_2| \leq n-2$. Consequently $|\mathcal{F}_1| + |\mathcal{F}_2| - |\mathcal{F}_1 \cap \mathcal{F}_2| \leq n-2$ and so $|\mathcal{F}_1 \cap \mathcal{F}_2| \geq (n/2) + (n/2) - (n-2) = 2$. $\square$

We recall that a line–oval in π is simply an oval in the dual plane π^*, that is a set Σ of $n+1$ lines of π no three of which are concurrent. The dual knot of Σ is the unique line of π such that through any one of its points there passes precisely one line of Σ.

PROPOSITION 2.2 The set of lines of π different from ℓ and from the $n+1$ tangents to Ω admits a partition $\mathcal{F}^*$ into $n-1$ line–ovals sharing the line ℓ as common dual knot.

Proof: The assertion is contained in [1, Prop.4]. For an explicit proof let m be one of the lines in question. Set $m \cap \ell = \{P\}$ and denote by Σ the J–orbit of m. Since Δ is sharply transitive on ℓ we see that Σ consists of at least $n+1$ lines. Since J has order $2(n+1)$ and m is fixed by the involutory elation with center P, we obtain $|\Sigma| = n+1$ and $\Sigma = \{m^\delta;\ \delta \in \Delta\} = \{m^\nu;\ \nu \in N\}$.

Assume m, m_1, m_2 are three lines in Σ with the point Q in common. We may set $m_1 = m^{\delta_1}$, $m_2 = m^{\delta_2}$ with $\delta_1, \delta_2 \in \Delta$. If a_1, a_2 are the axes of δ_1, δ_2 respectively, then a_1 is the line joining the knot K to the point of intersection of m with m_1 and a_2 is the line joining K to the intersection of m with m_2, in other words $a_1 = KQ = a_2$ and consequently also $\delta_1 = \delta_2$, a contradiction.

The $n+1$ lines in Σ form thus a line–oval in π. We have seen that Σ contains a line through an arbitrary point of ℓ and therefore ℓ is the dual knot of Σ. □

For each point $P \in \pi \setminus (\ell \cup \{K\})$ we denote by Ω_P the unique oval in $\mathcal{F}$ containing P; for each line m of π which is not a tangent to Ω and is different from ℓ we denote by Σ_m the unique line–oval in $\mathcal{F}^*$ containing m.

PROPOSITION 2.3 If E is a Sylow 2–subgroup of G contained in G_O then E (and consequently the group G_O) acts fixed–point–free on $OK \setminus \{O, C, K\}$. In particular E fixes no oval in $\mathcal{F} \setminus \{\Omega\}$. Finally, the group E fixes precisely one line $\ell' \in \mathcal{E}_C \setminus \{\ell\}$.

Proof: Assume E fixes $P \in OK \setminus \{O, C, K\}$. Let $\mathcal{U}$ be the set of common secants to Ω and Ω_P; we have $|\mathcal{U}| = n(n+1)/4$, see [4, Prop.2.3]. Let $\mathcal{U}_C$ be the subset of all lines in $\mathcal{U}$ through C. For each $u \in \mathcal{U}_C$ the stabilizer F of u in E has order 2, namely $F = \langle \eta \rangle$. Let 2^a be the highest power of 2 dividing n. If $\gamma \in E$ fixes a line $v \in \mathcal{U} \setminus \mathcal{U}_C$, then it also fixes the point of intersection of v with ℓ, yielding $\gamma = \mathrm{id}$. Consequently E acts semiregularly on $\mathcal{U} \setminus \mathcal{U}_C$ and all its orbits there have length $|E| = 2^a$, whence $|\mathcal{U} \setminus \mathcal{U}_C| = b \cdot 2^a$. From $|\mathcal{U}| = n(n+1)/4 = 2^{a-2} \cdot d$ with d odd, we obtain $|\mathcal{U}_C| = 2^{a-2} \cdot c$ with $c = d - 4b$ odd. Each orbit of E on $\mathcal{U}_C$ has length $|E|/2 = 2^{a-1}$, showing that $|\mathcal{U}_C|$ must be divisible by 2^{a-1}, a contradiction.

The 2–group E acts on the set $\mathcal{E}_C \setminus \{\ell\}$, which has cardinality $(n/2) - 1$, an odd number. Consequently E fixes a line $\ell' \in \mathcal{E}_C \setminus \{\ell\}$ and therefore also fixes the line–oval Σ' in $\mathcal{F}^*$ to which ℓ' belongs. Assume E fixes $\ell'' \in \mathcal{E}_C \setminus \{\ell, \ell'\}$ and denote by Σ'' the line–oval in $\mathcal{F}^*$ to which ℓ'' belongs. Denote by $\mathcal{U}^*$ the set of common dual–secants to Σ' and Σ'' (a dual–secant to a line–oval is a point such that precisely two lines through it belong to the line–oval). For $P \in \mathcal{U}^*_C = \mathcal{U}^* \cap CK$ the stabilizer F^* of P in E has order 2, namely $F^* = \langle \eta \rangle$. Arguing as above, we obtain that E acts semiregularly on $\mathcal{U}^* \setminus \mathcal{U}_C$ and $|\mathcal{U}^*_C| = 2^{a-2} \cdot c^*$ with c^* odd. Each orbit of E on $\mathcal{U}^*_C$ has length $|E|/2 = 2^{a-1}$, showing that $|\mathcal{U}^*_C|$ must be divisible by 2^{a-1}, a contradiction. □

PROPOSITION 2.4 Each point–orbit of G on $\pi \setminus (\ell \cup \{K\})$ is the union of ovals in $\mathcal{F}$. More precisely, taken arbitrary points $P_1, \ldots, P_s$ of $OK \setminus \{C, K\}$ we have that $\Omega_{P_1} \cup \cdots \cup \Omega_{P_s}$ is a G–orbit if and only if $\{P_1, \ldots, P_s\}$ is a G_O–orbit (equivalently: if and only if $\{\Omega_{P_1}, \ldots, \Omega_{P_s}\}$ is a G_O–orbit on $\mathcal{F}$).
Proof: A G–orbit of points Θ containing a point Q must also contain the oval Ω_Q, as this is a J–orbit. The assertion follows by observing that a collineation $\gamma \in G$ mapping a point P of $OK \setminus \{C, K\}$ to another such point necessarily fixes O. □

In a similar vein we have

PROPOSITION 2.5 Each line–orbit of G on $\mathcal{L} \setminus (\mathcal{L}_K \cup \{\ell\})$ is the union of line–ovals in $\mathcal{F}^*$. More precisely, if $m_1, \ldots, m_s$ are lines in $\mathcal{L}_C \setminus \{OK, \ell\}$ then $\Sigma_{m_1} \cup \cdots \cup \Sigma_{m_s}$ is a G–orbit if and only if $\{m_1, \ldots, m_s\}$ is a G_O–orbit (equivalently: $\{\Sigma_{m_1}, \ldots \Sigma_{m_s}\}$ is a G_O–orbit on $\mathcal{F}^*$). □

Write $\mathcal{F} = \{\Omega\} \cup \mathcal{F}_1 \cup \cdots \cup \mathcal{F}_h$, where $\mathcal{F}_1, \ldots, \mathcal{F}_h$ are G_O–orbits of ovals in $\mathcal{F}$, and $\mathcal{E}_C = \{\ell\} \cup \mathcal{E}_1 \cup \cdots \cup \mathcal{E}_k$, where $\mathcal{E}_1, \ldots, \mathcal{E}_k$ are G_O–orbits on $\mathcal{E}_C$.

PROPOSITION 2.6 We have $h = k$.
Proof: The G–orbits of points are $\{K\}$, ℓ, Ω and $\mathcal{P}_i = \cup_{\Omega_P \in \mathcal{F}_i} \Omega_P$ for $i = 1, 2, \ldots, h$. The G–orbits of lines are $\{\ell\}$, $\mathcal{L}_K$, $\mathcal{S}$ and $\mathcal{L}_j = \cup_{m \in \mathcal{E}_j} \Sigma_m$ for $j = 1, 2, \ldots, k$. The collineation group G of π has $h + 3$ point–orbits and $k + 3$ line–orbits. Since any collineation group of a finite projective plane has as many point–orbits as line–orbits, we have $h + 3 = k + 3$. □

PROPOSITION 2.7 Assume Θ to be a point–orbit of G on $\pi \setminus (\Omega \cup \ell \cup \{K\})$. There exists an even positive integer c with $|m \cap \Theta| = c$ for each secant m to Ω.
Proof: The 2–transitivity of G on Ω forces G to be transitive on the secants to Ω. Therefore $|m \cap \Theta|$ is a constant as m runs over all secants to Ω. If $Q \in \Theta$ then at least one line through Q is secant to Ω: all we need to do is join Q to a point $P \in \Omega \setminus QK$. We have thus $c > 0$. The fact that c is even follows from the fact that each secant to Ω is either a secant or an external line to any other oval in the family $\mathcal{F}$. □

PROPOSITION 2.8 If G_O fixes a line $\ell' \in \mathcal{E}_C \setminus \{\ell\}$ then G_O has a point–orbit of length $n/2$ on $OK \setminus \{O, C, K\}$ and consequently an orbit of length $n/2$ on $\mathcal{F}$.
Proof: For P on ℓ' the unique oval Ω_P of $\mathcal{F}$ through P meets ℓ' in a further point; the line ℓ' is thus secant to precisely $n/2$ ovals of $\mathcal{F}$. Since G_O is transitive on $\Omega \setminus \{O\}$, G_O is also transitive on $\ell' \setminus \{C\}$. We conclude that G_O is transitive on the $n/2$ ovals. Since ℓ' is fixed by G_O, we see that these $n/2$ ovals form a G_O–invariant set and therefore a full orbit under G_O. The points at which these ovals touch OK form thus a point–orbit under G_O. □

PROPOSITION 2.9 If E is a Sylow 2–subgroup of G contained in G_O then the group E acts semiregularly on the points of π off OK and on the lines not through C.
Proof: Assume $\varphi \in E \setminus \{\mathrm{id}\}$ fixes a point P off OK; the point of intersection of the tangent PK with Ω is different from O and is fixed from φ, contradicting the

fact that E acts semiregularly on $\Omega \setminus \{O\}$, see [4, Prop. 3.1].

A line m not through C meets ℓ in a point D which is different from C. If m is fixed by φ then so is D and that cannot be the case by the previous part. □

PROPOSITION 2.10 Assume $\mathcal{F}_i$ to be a G_O-orbit on $\mathcal{F} \setminus \{\Omega\}$. Then $|\mathcal{F}_i|$ is even and each secant to Ω meets precisely $|\mathcal{F}_i|/2$ ovals in $\mathcal{F}_i$.
Proof: Setting $\mathcal{F}_i = \{\Omega_{P_1}, \Omega_{P_2}, \ldots, \Omega_{P_s}\}$, we know $\Theta = \Omega_{P_1} \cup \Omega_{P_2} \cup \cdots \cup \Omega_{P_s}$ to be a G-orbit of points. Let us determine the number of flags (Q, m) where $Q \in \Theta$ and $m \in \mathcal{S}$.

On the one hand we know from Proposition 2.7 that each secant to Ω meets Θ in a constant even number c of points. There are thus $\binom{n+1}{2} \cdot c$ flags of the required kind.

On the other hand, each secant to Ω through a point $Q \in \Theta$ must also be a secant of the unique oval $\Omega_Q \in \mathcal{F}$ containing Q. Since $n(n+1)/4$ is the number of common secants to any pair of ovals in $\mathcal{F}$, see [4, Prop. 2.3], we have that the required number is $\big(n(n+1)/4\big) \cdot 2s$.

The equality $\binom{n+1}{2} \cdot c = \big(n(n+1)/4\big) \cdot 2s$ yields $c = s$. In particular s is even and s is also the number of points from Θ on any given line m from $\mathcal{S}$; since the points from Θ on m occur in pairs from the ovals of $\mathcal{F}_i$ meeting m, we conclude that m meets precisely $s/2$ ovals from $\mathcal{F}_i$. □

As G is 2-transitive on Ω, G_O is transitive on $\Omega \setminus \{O\}$. Since G_O fixes C we see that $\mathcal{S}_C$ is a G_O-orbit of lines. Each line in $\mathcal{S}_C$ is secant to precisely $|\mathcal{F}_i|/2$ ovals in $\mathcal{F}_i$. By [4, Prop.2.3] precisely $n(n+1)/4$ lines in $\mathcal{S}$ are secant to any given oval in $\mathcal{F} \setminus \{\Omega\}$; as G is transitive on ℓ there are thus precisely $n/4$ lines in $\mathcal{S}_C$ which are secant to any given oval in $\mathcal{F} \setminus \{\Omega\}$.

For $m \in \mathcal{E}_j$ the number of ovals in $\mathcal{F}_i$ to which m is secant is independent of the choice of m in $\mathcal{E}_j$: we denote this number by c_{ij}. Since each line in $\mathcal{E}_C \setminus \{\ell\}$ is secant to $n/2$ ovals in $\mathcal{F}$, we have

$$\sum_{i=1}^{h} c_{ij} = \frac{n}{2}$$

for each $j = 1, 2, \ldots, h$.

For $\overline{\Omega} \in \mathcal{F}_i$ the number of lines in $\mathcal{E}_j$ which are secant to $\overline{\Omega}$ is independent of the choice of $\overline{\Omega}$ in $\mathcal{F}_i$: we denote this number by d_{ij}. There are precisely $n/2$ lines in $\mathcal{L}_C$ which are secant to a given oval in $\mathcal{F}$; we have thus

$$\frac{n}{4} + \sum_{j=1}^{h} d_{ij} = \frac{n}{2}$$

that is

$$\sum_{j=1}^{h} d_{ij} = \frac{n}{4}$$

for each $j = 1, 2, \ldots, h$. Define

$$\mathcal{U}_{ij} = \{(m, \overline{\Omega});\ m \in \mathcal{E}_j, \overline{\Omega} \in \mathcal{F}_i, |m \cap \overline{\Omega}| = 2\}.$$

We have

$$|\mathcal{F}_i| \cdot d_{ij} = |\mathcal{U}_{ij}| = |\mathcal{E}_j| \cdot c_{ij}. \tag{1}$$

Summing up both sides with respect to different indices we obtain

$$|\mathcal{F}_i| \cdot \frac{n}{4} = \sum_{j=1}^{h} |\mathcal{F}_i| \cdot d_{ij} = \sum_{j=1}^{h} |\mathcal{E}_j| \cdot c_{ij}, \tag{2}$$

$$\sum_{i=1}^{h} |\mathcal{F}_i| \cdot d_{ij} = \sum_{i=1}^{h} |\mathcal{E}_j| \cdot c_{ij} = |\mathcal{E}_j| \cdot \frac{n}{2}. \tag{3}$$

3 HERING'S THEOREM

The group G satisfies all the assumptions of Theorem 6.7 in [6]. We can therefore define on Ω the structure of a vector space V over the finite field $L = GF(q)$ (where q is of course a power of p) in such a way that G is a group of L–semilinear transformations on V, i.e. $G \leq A\Gamma L(V) = A\Gamma L(k,q)$ where k is the dimension of V over L and one of the following five cases occur:

(a) $k = 1$;
(b) $k = 2$ and G contains $ASL(2,q)$.
(c) $k = 2$, $q \in \{9, 11, 19, 29, 59\}$ and if $\mathbf{0} \in V$ then the stabilizer $G_\mathbf{0}$ contains $SL(2,5)$;
(d) $k = 2$, $q \in \{5, 7, 11, 23\}$ and $G_\mathbf{0}$ contains a normal quaternion group of order 8;
(e) $k = 6$, $q = 3$ and $G_\mathbf{0} \cong SL(2,13)$.

It is the purpose of the present paper to prove that cases (b), (c), (d), (e) cannot occur. Throughout the rest of the paper we choose O to be the point of Ω corresponding to the zero vector $\mathbf{0}$ in V.

4 CASE (a)

Although we are not able to handle this case completely, some useful information is available in a special circumstance.

PROPOSITION 4.1 If G contains $AGL(1,q)$ then $n \in \{2,4\}$.
Proof: The subgroup $H \cong AGL(1,q)$ is still 2–transitive on V and so by the result in [4] we either have $n \in \{2,4\}$ or $n \equiv 0 \mod 8$ and a Sylow 2–subgroup of H is a generalized quaternion group. Since H is sharply 2–transitive on V we have $|H| = (n+1)n$. Now $n+1$ is odd and a Sylow 2-subgroup of H is thus contained in $H_\mathbf{0}$, a cyclic group in this case, and the assertion follows. □

5 CASE (b)

Set $H = ASL(2,q)$. Then $H \leq G$ and H is still 2–transitive on V. Since $n+1 = q^2$ we see that q is odd in our case. We have $|ASL(2,q)| = q^2(q^2-1)q$, $|SL(2,q)| =$

$(q^2-1)q$ and so, since q is odd, we see that a Sylow 2–subgroup of H can already be found in $H_{\mathbf{0}}$.

Setting $\mathbf{w} = (1,0)$ and denoting by W the 1–dimensional vector subspace of V spanned by $\mathbf{w}$, we have that since $H_{\mathbf{0}}$ consists of linear transformations the stabilizer $H_{\mathbf{0},\mathbf{w}}$ fixes W elementwise. By the Reduction Lemma the setwise stabilizer $H_{\{W\}}$ of W in H induces a sharply 2–transitive group $\overline{H}$ on W. The transformations $(x,y) \mapsto (ax, a^{-1}y)$ as a varies in $GF(q)^*$ form a cyclic subgroup of $\overline{H}_{\mathbf{0}}$ acting regularly on $W \setminus \{\mathbf{0}\}$, hence $\overline{H}_{\mathbf{0}}$ consists precisely of these transformations and a Sylow 2–subgroup of $\overline{H}$ is therefore cyclic. As remarked in the Introduction we have then $q-1 \in \{2,4\}$ by [4], i.e. $q \in \{3,5\}$.

PROPOSITION 5.1 We cannot have $q=3$.
Proof: If $q=3$ then $n = q^2-1 = 8$. An oval Ω in a projective plane of order 8 is either a non–degenerate conic $\mathcal{C}$ or it is obtained from such a conic $\mathcal{C}$ by exchanging one of its points, say K, with the knot K' of $\mathcal{C}$, see [7, Thm.9.2.5].

Assume the latter case occurs. Since K is the knot of Ω, we have that K is fixed by G. For each $g \in G$ there exist distinct points $P_1, \ldots, P_5$ on $\mathcal{C} \setminus \{K\}$ such that $P_1^g, \ldots, P_5^g$ are again points of $\mathcal{C} \setminus \{K\}$. A non–degenerate conic is uniquely determined by any five of its points; furthermore, each collineation of $PG(2,n)$ maps a non–degenerate conic to a non–degenerate conic. Consequently g fixes $\mathcal{C}$ and therefore K'. We conclude that G cannot act 2–transitively on Ω.

In order to exclude the former case, write $n = p^h$, p prime, and let $\mathcal{E}$ denote the set of all external lines to the conic $\mathcal{C}$ in $PG(2,n)$. If we denote by G the setwise stabilizer of $\mathcal{C}$ in the full collineation group of $PG(2,n)$, then G acts transitively on $\mathcal{E}$ and we have $G \cong P\Gamma L(2,p^h)$. The stabilizer G_ℓ of a line $\ell \in \mathcal{E}$ in G has thus order $p^h(p^h-1)(p^h+1)h/(p^h(p^h-1)/2) = 2(p^h+1)h$. If G_ℓ is 2–transitive on $\mathcal{C}$ then the order $2(p^h+1)h$ must be divisible by $(p^h+1)p^h$ and consequently p^h must divide $2h$, yielding $p=2$ and $h \in \{1,2\}$. The assertion now follows. □

PROPOSITION 5.2 We cannot have $q=5$.
Proof: Let D be the subgroup of order 5 of $H_{\mathbf{0}} \cong SL(2,5)$ consisting of the transformations $(x,y) \mapsto (x+dy, y)$ as d varies in $L = GF(5)$. Setting $\mathbf{w} = (1,0)$ we have $W = \{(x,0);\, x \in L\}$ and W is pointwise fixed by D. The collineation group D fixes a subplane π_0 of order 4 pointwise and we have $\pi_0 \cap \Omega = W$. The tangent line to Ω at $\mathbf{0}$ lies in π_0 and the points of π_0 on this tangent line are precisely $\mathbf{0}$, the knot K of the oval Ω, the point of intersection with the external line ℓ and two further points A, B.

The secant line joining A to one of the points in $\{(1,0),(2,0),(3,0),(4,0)\}$ meets Ω in a further point of this set. The remaining two points lie on another secant through A, yielding thus one of the following three possibilities:

$$\begin{aligned}
\text{i)}\ \ & \mathcal{B} = \{\{(1,0),(2,0)\},\{(3,0),(4,0)\}\},\\
\text{ii)}\ \ & \mathcal{B} = \{\{(1,0),(3,0)\},\{(2,0),(4,0)\}\},\\
\text{iii)}\ \ & \mathcal{B} = \{\{(1,0),(4,0)\},\{(2,0),(3,0)\}\}.
\end{aligned}$$

The transformation $(x,y) \mapsto (3x,y)$ in $GL(2,5)$ maps the pair $\{(1,0),(2,0)\}$ to the pair $\{(3,0),(1,0)\}$ and the pair $\{(3,0),(4,0)\}$ to the pair $\{(4,0),(2,0)\}$. Since $SL(2,5)$ is a normal subgroup of $GL(2,5)$, we see that possibilities i) and ii) are equivalent.

The transformation $\eta : (x, y) \mapsto (-x, -y)$ fixes the subplane π_0 and so it generates together with D a subgroup D^* of order $(1, 0)$ fixing A. Since η is in the center of H_0 we have that η is induced by the involutory elation of π having ℓ as axis and the point of intersection of ℓ with the line AB as center. In particular, if E is a Sylow 2–subgroup of H_0 then η is the unique involution in E, see [4].

Let $(\bar{x}, \bar{y})$ be the further point of Ω on the chord joining A with the point $P = (0, 1)$. Clearly $\bar{y} \neq 0$. Furthermore, $\bar{y} \neq 1$, otherwise, setting $b = -\bar{x}$, the transformation $(x, y) \mapsto (x + by, y)$ maps $(\bar{x}, 1)$ to $(0, 1)$, whence $(0, 1)$ to $(b, 1)$ yielding $\bar{x} = b$, i.e. $b = -b$, a contradiction.

Consider the secants obtained as images of the line AP under the transformations in D^*. They intersect Ω in the pairs $\{(d, 1), (\bar{x} + \bar{y}d, \bar{y})\}$, $\{(-d, -1), (-\bar{x} - \bar{y}d, -\bar{y})\}$, as d varies in L: we denote by $\mathcal{D}$ the set of all such pairs. If $\bar{y} = 2$ or $\bar{y} = 3$ then these pairs are all distinct, hence so are the corresponding secants through A and we have thus obtained ten secants through A.

We have determined a set $\mathcal{P} = \mathcal{B} \cup \mathcal{D}$ of 12 pairs of points of Ω on which of course the group G acts. For each transformation $g \in G$ such that $\mathcal{P} \neq \mathcal{P}^g$ we necessarily have $|\mathcal{P} \cap \mathcal{P}^g| \leq 1$, since the corresponding set $\mathcal{P}^g$ represents the 12 secants through $A^g \neq A$. The following list shows that for each one of the above candidates there exists a transformation $g \in G$ violating this condition.

$\mathcal{B} = \{\{(1, 0), (2, 0)\}, \{(3, 0), (4, 0)\}\}$,

$\bar{x} = 0$,

$\bar{y} = 2$, $(x, y) \mapsto (4x + 4y, x)$;

$\bar{y} = 3$, $(x, y) \mapsto (4x + 4y, x)$;

$\bar{x} = 1$,

$\bar{y} = 2$, $(x, y) \mapsto (4x + 4y + 2, x + 1)$;

$\bar{y} = 3$, $(x, y) \mapsto (3x + 3y + 3, 3x + 1)$;

$\bar{x} = 2$,

$\bar{y} = 2$, $(x, y) \mapsto (4x + 4y, x + 2)$;

$\bar{y} = 3$, $(x, y) \mapsto (3x + 3y + 2, 3x + 1)$.

$\bar{x} = 3$,

$\bar{y} = 2$, $(x, y) \mapsto (4x + 4y, x + 2)$;

$\bar{y} = 3$, $(x, y) \mapsto (3x + 3y, 3x + 1)$;

$\bar{x} = 4$,

$\bar{y} = 2$, $(x, y) \mapsto (4x + 4y + 1, x + 1)$;

$\bar{y} = 3$, $(x, y) \mapsto (3x + 3y + 3, 3x + 1)$;

$\mathcal{B} = \{\{(1, 0), (4, 0)\}, \{(3, 0), (2, 0)\}\}$,

$\bar{x} = 0$,

$\bar{y} = 2$, $(x, y) \mapsto (4x + 4y, x)$;

$\bar{y} = 3$, $(x, y) \mapsto (4x + 4y, x)$;

$\bar{x} = 1$,

$\bar{y} = 2$, $(x, y) \mapsto (4x + 4y, x + 1)$;

$\bar{y} = 3$, $(x, y) \mapsto (4x + 4y, x + 2)$;

$\bar{x} = 2$,

$\bar{y} = 2$, $(x, y) \mapsto (3x + 3y + 3, 3x + 1)$;

$\bar{y} = 3$, $(x, y) \mapsto (3x + 3y + 1, 3x + 2)$;

$\bar{x} = 3$,

$\bar{y} = 2$, $(x, y) \mapsto (3x + 3y + 3, 3x + 1)$;

$\bar{y} = 3$, $(x, y) \mapsto (3x + 3y + 1, 3x + 2)$;

$\bar{x} = 4$,
$\bar{y} = 2$, $(x, y) \mapsto (4x + 4y, x + 1)$;
$\bar{y} = 3$, $(x, y) \mapsto (4x + 4y, x + 2)$.

If $\bar{y} = 4$ then the D^*–images of AP account for only five secants through A. We must have a further secant through A meeting Ω in the pair $\{(0, 2), (x', y')\}$, where $y' \in \{2, 3\}$. We cannot have $y' = 2$ otherwise, taking d such that $x' + 2d = 0$, the transformation $(x, y) \mapsto (x + dy, y)$ maps $(x', 2)$ to $(0, 2)$ hence also $(0, 2)$ to $(x', 2)$, yielding $2d = x'$, i.e. $2x' = 0$, a contradiction. We conclude that $y' = 3$ and the D^*–images of the given secant yield the five more pairs $\{(2d, 2), (x' + 3d, 3)\}$ as d varies in L. Again for each candidate set of 12 pairs we list here a transformation $g \in G$ violating the condition.

$\mathcal{B} = \{\{(1, 0), (2, 0)\}, \{(3, 0), (4, 0)\}\}$,
$\bar{x} = 0$,
$x' = 0$, $(x, y) \mapsto (4x + 4y, x)$;
$x' = 1$, $(x, y) \mapsto (4x + 4y, x)$;
$x' = 2$, $(x, y) \mapsto (4x + 4y, x)$;
$x' = 3$, $(x, y) \mapsto (4x + 4y, x)$;
$x' = 4$, $(x, y) \mapsto (4x + 4y, x)$;
$\bar{x} = 1$,
$x' = 0$, $(x, y) \mapsto (4x + 4y, x)$;
$x' = 1$, $(x, y) \mapsto (4x + 4y + 1, x + 2)$;
$x' = 2$, $(x, y) \mapsto (4x + 4y + 1, x + 2)$;
$x' = 3$, $(x, y) \mapsto (4x + 4y + 2, x + 1)$;
$x' = 4$, $(x, y) \mapsto (4x + 4y, x + 2)$;
$\bar{x} = 2$,
$x' = 0$, $(x, y) \mapsto (4x + 4y, x)$;
$x' = 1$, $(x, y) \mapsto (4x + 4y + 1, x + 2)$;
$x' = 2$, $(x, y) \mapsto (4x + 4y + 2, x + 4)$;
$x' = 3$, $(x, y) \mapsto (4x + 4y, x + 1)$;
$x' = 4$, $(x, y) \mapsto (4x + 4y + 3, x + 3)$;
$\bar{x} = 3$,
$x' = 0$, $(x, y) \mapsto (4x + 4y, x)$;
$x' = 1$, $(x, y) \mapsto (4x + 4y + 2, x + 1)$;
$x' = 2$, $(x, y) \mapsto (4x + 4y, x + 1)$;
$x' = 3$, $(x, y) \mapsto (4x + 4y + 3, x + 1)$;
$x' = 4$, $(x, y) \mapsto (4x + 4y + 1, x + 1)$;
$\bar{x} = 4$,
$x' = 0$, $(x, y) \mapsto (4x + 4y, x)$;
$x' = 1$, $(x, y) \mapsto (4x + 4y, x + 2)$;
$x' = 2$, $(x, y) \mapsto (4x + 4y + 3, x + 3)$;
$x' = 3$, $(x, y) \mapsto (4x + 4y + 1, x + 1)$;
$x' = 4$, $(x, y) \mapsto (4x + 4y + 4, x + 3)$;
$\mathcal{B} = \{\{(1, 0), (4, 0)\}, \{(3, 0), (2, 0)\}\}$,
$\bar{x} = 0$,
$x' = 1$, $(x, y) \mapsto (4x + 4y, x)$;
$x' = 2$, $(x, y) \mapsto (4x + 4y, x)$;
$x' = 3$, $(x, y) \mapsto (4x + 4y, x)$;
$x' = 4$, $(x, y) \mapsto (4x + 4y, x)$;

$\bar{x} = 1$,
$x' = 0$, $(x, y) \mapsto (4x + 4y, x)$;
$x' = 1$, $(x, y) \mapsto (4x + 4y + 3, x)$;
$x' = 2$, $(x, y) \mapsto (4x + 4y + 1, x + 2)$;
$x' = 3$, $(x, y) \mapsto (4x + 4y + 2, x + 1)$;
$x' = 4$, $(x, y) \mapsto (4x + 4y, x + 2)$;
$\bar{x} = 2$,
$x' = 0$, $(x, y) \mapsto (4x + 4y, x)$;
$x' = 1$, $(x, y) \mapsto (4x + 4y + 1, x + 2)$;
$x' = 2$, $(x, y) \mapsto (4x + 4y + 1, x)$;
$x' = 3$, $(x, y) \mapsto (4x + 4y, x + 1)$;
$x' = 4$, $(x, y) \mapsto (4x + 4y + 3, x + 3)$;
$\bar{x} = 3$,
$x' = 0$, $(x, y) \mapsto (4x + 4y, x)$;
$x' = 1$, $(x, y) \mapsto (4x + 4y + 2, x + 1)$;
$x' = 2$, $(x, y) \mapsto (4x + 4y, x + 1)$;
$x' = 3$, $(x, y) \mapsto (4x + 4y + 4, x)$;
$x' = 4$, $(x, y) \mapsto (4x + 4y + 1, x + 1)$;
$\bar{x} = 4$,
$x' = 0$, $(x, y) \mapsto (4x + 4y, x)$;
$x' = 1$, $(x, y) \mapsto (4x + 4y, x + 2)$;
$x' = 2$, $(x, y) \mapsto (4x + 4y + 3, x + 3)$;
$x' = 3$, $(x, y) \mapsto (4x + 4y + 1, x + 1)$;
$x' = 4$, $(x, y) \mapsto (4x + 4y + 2, x)$;

The unique candidate left is given by $\mathcal{B} = \{\{(1,0),(4,0)\},\{(3,0),(2,0)\}\}$, $\bar{x} = 0$, $x' = 0$ and consists of all pairs $\{(a,b),(-a,-b)\}$ as (a,b) varies in $V \setminus \{(0,0)\}$. But this is precisely the set of pairs obtained from the secants to Ω through the point of intersection of the lines ℓ and AB. The proof of the assertion is now complete. □

6 CASES (c) and (d)

Apart from the case $q = 9$, the order of the field L is a prime and so the identity is the unique field–automorphism of L. Consequently G is a group of affine linear transformations on V. If G is not sharply 2–transitive, then for each non-zero vector $\mathbf{w} \in V$ the stabilizer $G_{\mathbf{0},\mathbf{w}}$ fixes the one–dimensional vector subspace $W = \langle \mathbf{w} \rangle$ pointwise and no further vector in V. Clearly the setwise stabilizer $G_{\{W\}}$ induces on W a group $\overline{G}$ of affine L–linear transformations and we have already observed several times that $\overline{G}$ is sharply 2–transitive on W, i.e. $\overline{G} \cong AGL(1,q)$. We have thus reduced our problem to an oval Ω_0 in a subplane π_0 on which $\overline{G}$ acts as $AGL(1,q)$ in its natural permutation representation: we have seen that for $q \in \{7, 11, 19, 23, 29, 59\}$ this cannot be the case.

The following result will be useful in order to rule out the case $q = 5$.

PROPOSITION 6.1 Let q be an odd prime power and let $\zeta \in GF(q^2) = V$ be a primitive element. Consider the group $\widetilde{N}$ consisting of the $GF(q)$–linear transformations $V \to V$, $v \mapsto \zeta^k v^{q^i}$ with $k = 0, 1, \ldots, q^2 - 2$, $i = 0, 1$ ($\widetilde{N}$ is the normalizer of a Singer cycle in $GL(2,q) = GL(V)$).

The group $\widetilde{N}$ contains precisely three subgroups which are transitive on the non-zero vectors of V, namely $\widetilde{N}$ itself, the cyclic group C consisting of the transformations $V \to V$, $v \mapsto \zeta^k v$ with $k = 0, 1, \ldots, q^2 - 2$ and the subgroup D consisting of the transformations $V \to V$, $v \mapsto \zeta^{2k+i} v^{q^i}$ with $k = 0, 1, \ldots, (q^2-1)/2 - 1$, $i = 0, 1$.

A Sylow 2–subgroup of $\widetilde{N}$ contains at least two involutions; a Sylow 2–subgroup of C is obviously cyclic and for $q \equiv 1 \mod 4$ a Sylow 2–subgroup of D is cyclic.

Proof: We only prove the final statement. The group $\widetilde{N}$ contains two commuting involutions, namely $v \mapsto -v$ and $v \mapsto v^q$. If 2^m is the highest power of 2 dividing $(q^2-1)/2$ and ξ is an element of order 2^m in $GF(q^2)$ then ξ lies in the subgroup $\langle \zeta^2 \rangle$ of all squares in $GF(q^2)^*$. Furthermore, if $q \equiv 1 \mod 4$ then ξ is a $(q+1)$–th power in $GF(q^2)$. Setting $\xi = \alpha^{q+1}$ we have $\alpha \in GF(q^2)^* \setminus \langle \zeta^2 \rangle$. Hence $v \mapsto \alpha v^q$ is a transformation in D, the order of which is 2^{m+1}. Since 2^{m+1} is also the highest power of 2 dividing $|D| = q^2 - 1$ the assertion follows. □

PROPOSITION 6.2 We cannot have $q = 5$.

Proof: Assume now $q = 5$. We have that G_0 is a subgroup of $GL(2,5)$ acting transitively on the non-zero vectors in V, hence of order at least 24. In other words, if $\widehat{\ }: GL(2,5) \to PGL(2,5)$ is the canonical epimorphism, then $\widehat{G_0}$ is a transitive subgroup of $PGL(2,5)$.

Since a Sylow 2–subgroup of $GL(2,5)$ has at least two involutions, we cannot have $G_0 = GL(2,5)$. More generally, Proposition 5.2 shows that we cannot have $SL(2,5) \leq G_0$. Consequently $\widehat{G_0}$ is either cyclic, dihedral, A_4 or S_4.

If $\widehat{G_0} \cong A_4$ or S_4 then $\widehat{G_0}$ is the normalizer in $PSL(2,5)$ or in $PGL(2,5)$ of a Klein 4–group $\widehat{K}$. We may assume up to conjugation that $\widehat{K}$ is generated by the linear fractional transformations $x \mapsto -x$, $x \mapsto 1/x$.

We may assume that a Sylow 2–subgroup of $\widehat{G_0}$ containing $\widehat{K}$ is $\widehat{K}$ itself if $\widehat{G_0} \cong A_4$ or is generated by $x \mapsto 2x$ and $x \mapsto 1/x$ if $\widehat{G_0} \cong S_4$. Two possibilities arise: either a Sylow 2–subgrop of G_0 contains the group

$$\left\{ \begin{pmatrix} 1 & 0 \\ 0 & 1 \end{pmatrix}, \begin{pmatrix} -1 & 0 \\ 0 & 1 \end{pmatrix}, \begin{pmatrix} 0 & 1 \\ 1 & 0 \end{pmatrix}, \begin{pmatrix} 0 & -1 \\ 1 & 0 \end{pmatrix} \right\},$$

or G_0 has order 24 and G is sharply 2–transitive on V. The former possibility contradicts the fact that a Sylow 2–subgroup of G_0 contains a unique involution. In the latter case G is the group of affine linear transformations over a nearfield of order 25. There are three nearfields of order 25, namely the field, the Dickson nearfield and the irregular nearfield: the first two cases are excluded by the fact that a Sylow 2–subgroup of G_0 would be cyclic in these cases, the third case is ruled out in Section 8 below.

If $\widehat{G_0}$ is cyclic or dihedral then $\widehat{G_0}$ lies in the normalizer of a Singer cycle. In other words, G_0 lies in the normalizer of a Singer cycle in $GL(2,5)$. Since $5 \equiv 1 \mod 4$, Proposition 6.1 shows that a Sylow 2–subgroup of G_0 is either cyclic or contains at least two involutions: we have seen that this is not the case [4]. □

PROPOSITION 6.3 We cannot have $q = 9$.

Proof: Assume $q = 9$. We prove first of all that G is sharply 2–transitive on Ω. Since G_0 contains $SL(2,5)$ we have, in particular, that G_0 contains transformations of order 3.

Assume G is not sharply 2–transitive on $V = \Omega$. If $\mathbf{w}$ is a non–zero vector in V then from $|G_\mathbf{0}| = n|G_{\mathbf{0},\mathbf{w}}| = (9^2 - 1)|G_{\mathbf{0},\mathbf{w}}|$ we see that a Sylow 3–subgroup of $G_{\mathbf{0},\mathbf{w}}$ is also a Sylow 3–subgroup of $G_\mathbf{0}$. Now $G_\mathbf{0}$ is a subgroup of $\Gamma L(2,9)$ and the relation $|\Gamma L(2,9)| = 2|GL(2,9)|$ shows that a Sylow 3–subgroup of $GL(2,9)$ is also a Sylow 3–subgroup of $\Gamma L(2,9)$. A Sylow 3–subgroup S of $GL(2,9)$ is the elementary abelian group of order 9 consisting of the transformations $(x,y) \mapsto (x+cy, y)$ as c varies in $GF(9)$. For an arbitrary subgroup S^* of order 3 of S we have $\mathrm{Fix}(S) = \mathrm{Fix}(S^*) = \{(x,0);\ x \in GF(9)\}$. We can therefore conclude that for a Sylow 3–subgroup X of $G_{\mathbf{0},\mathbf{w}}$ we have $|\mathrm{Fix}(X)| = 9$. If Y is the normalizer of X in G, then Y is 2–transitive on $\Omega_0 = \mathrm{Fix}(X)$, see Theorem 9.4 in [9]. The usual argument yields that Ω_0 is a 2–transitive oval in a subplane π_0 with the fixed external line ℓ. From $|\Omega_0| = 9$ we conclude that π_0 has order 8 and we have a contradiction as in case (b).

There are two nearfields of order 81 containing $GF(9)$, namely the field and the Dickson nearfield; both of them have a cyclic Sylow 2–subgroup and so the assertion is proved. □

7 CASE (e)

By Proposition 2.8 in [2] the group $GL(6,3)$ contains a unique conjugacy class of subgroups isomorphic to $SL(2,13)$. We may therefore assume that $G_\mathbf{0}$ is generated by the matrices H, C, W given in [2]. If $\mathbf{w}$ is a non–zero vector in V then from $|G_\mathbf{0}| = n|G_{\mathbf{0},\mathbf{w}}| = (3^6 - 1)|G_{\mathbf{0},\mathbf{w}}|$ we see that a Sylow 3–subgroup of $G_{\mathbf{0},\mathbf{w}}$ is also a Sylow 3–subgroup of $G_\mathbf{0}$. A Sylow 3–subgroup of $G_\mathbf{0}$ has order 3. In the above representation for $SL(2,13)$ a transformation of order 3 is

$$C^4 = \begin{pmatrix} 0 & 1 & -1 & 0 & 0 & 0 \\ 1 & 0 & 1 & 0 & 0 & 0 \\ 1 & 0 & 0 & 0 & 0 & 0 \\ 0 & 0 & 0 & 0 & 1 & -1 \\ 0 & 0 & 0 & 1 & 0 & 1 \\ 0 & 0 & 0 & 1 & 0 & 0 \end{pmatrix}$$

and it has 9 fixed vectors as it is easily checked. Let $\mathbf{w}$ be any one of the non–zero fixed vectors of C^4. The subgroup $X = \langle C^4 \rangle$ is a Sylow 3–subgroup of $G_{\mathbf{0},\mathbf{w}}$. Again Theorem 9.4 in [9] shows that the normalizer Y of X in G is 2–transitive on $\Omega_0 = \mathrm{Fix}(X)$. The usual argument yields that Ω_0 is a 2–transitive oval with a fixed external line in a subplane π_0 of order 8, a contradiction as in case (b).

8 THE IRREGULAR NEARFIELD OF ORDER 25

Throughout this Section we assume that $n+1 = 25$ and G acts on Ω as the sharply 2–transitive group of affine linear transformations over the irregular nearfield of order 25 [5, 5.2.6]: it is our purpose here to show that this cannot be the case and thus complete the proof of Proposition 6.2.

We have $G_O \cong SL(2,3)$ and the kernel of the action of G_O on the points of the line OK is $\langle \eta \rangle$ and for the induced group $\overline{G}_O = G_O/\langle \eta \rangle$ we have $\overline{G}_O \cong A_4$; the subgroup $\langle\eta\rangle$ is also the kernel of the action of G_O on $\mathcal{L}_C$ and A_4 is the induced group there as well. We want to determine the orbits of G_O (equivalently: of $\overline{G}_O$) on OK and on $\mathcal{L}_C$ (alternatively: the orbits of G_O on $\mathcal{F}$ and on $\mathcal{F}^*$).

PROPOSITION 8.1 G_O fixes a line $\ell' \in \mathcal{E}_C \setminus \{\ell\}$.
Proof: Let ℓ' be the unique fixed line of E among the lines in $\mathcal{E}_C \setminus \{\ell\}$. We know E to be normal in G_O, because $\overline{G}_O$ has a normal Sylow 2–subgroup. Consequently ℓ' is fixed by the whole group G_O. □

PROPOSITION 8.2 G_O has three orbits on $OK \setminus \{O, C, K\}$ of lengths 12, 6 and 4 respectively.
Proof: Since G_O fixes a line $\ell' \in \mathcal{E}_C \setminus \{\ell\}$ we know from Proposition 2.8 that G_O has an orbit $\mathcal{P}_1$ of length 12 on $OK \setminus \{O, C, K\}$. Set $\mathcal{P} = OK \setminus (\{O, C, K\} \cup \mathcal{P}_1)$. The proof is achieved through the following three Lemmas.

LEMMA 8.2.1 Each Sylow 3–subgroup of $\overline{G}_O$ fixes precisely one point in $\mathcal{P}$.
Proof: Denote by M_1, M_2, M_3, M_4 the Sylow 3–subgroups of $\overline{G}_O$. For $i = 1, 2, 3, 4$ let $\mathcal{R}_i$ be the subset of $\mathcal{P}$ consisting of the fixed points of M_i. Since M_1, M_2, M_3, M_4 are conjugate in $\overline{G}$ we have $|\mathcal{R}_1| = |\mathcal{R}_2| = |\mathcal{R}_3| = |\mathcal{R}_4|$ and this common number can be 1, 4 or 7 because M_i is cyclic of order 3. If $|\mathcal{R}_i| = 4$ or 7 then we should have $\mathcal{R}_j \cap \mathcal{R}_k \neq \emptyset$ for a pair of distinct indices j, k. Since M_j and M_k generate $\overline{G}_O$, we see that $\overline{G}_O$ should fix each point in $\mathcal{R}_j \cap \mathcal{R}_k$, a contradiction. □

With the previous notation we have $\mathcal{R}_i = \{X_i\}$ for $i = 1, 2, 3, 4$ and X_1, X_2, X_3, X_4 are all distinct (otherwise $\overline{G}_O$ should again fix one of them).

LEMMA 8.2.2 The set $\mathcal{P}_2 = \{X_1, X_2, X_3, X_4\}$ is an orbit of $\overline{G}_O$ and $\overline{G}_O$ acts on $\mathcal{P}_2$ as A_4 in its natural action.
Proof: If $\gamma \in \overline{G}_O$ then to each $i \in \{1, 2, 3, 4\}$ there corresponds $j \in \{1, 2, 3, 4\}$ with $\gamma^{-1} M_i \gamma = M_j$, consequently $X_i^\gamma = X_j$ and $\mathcal{P}_2$ is $\overline{G}_O$–invariant. Since M_1, M_2, M_3, M_4 are all conjugate in $\overline{G}_O$ we have that $\overline{G}_O$ is transitive on $\mathcal{P}_2$. □

LEMMA 8.2.3 Setting $\mathcal{P}_1 = \mathcal{P} \setminus \mathcal{P}_2$ we have that $\overline{G}_O$ is transitive on $\mathcal{P}_1$.
Proof: Denote by δ_1, δ_2, δ_3 the involutions in $\overline{G}_O$. They are all conjugate in $\overline{G}_O$ and so their cycle structure is of the same kind. Let $V = \{\mathrm{id}, \delta_1, \delta_2, \delta_3\}$ be the Klein 4–group in $\overline{G}_O$. No point of $\mathcal{P}_1$ can be fixed by V (otherwise E should have fixed points on $\mathcal{P}_1$ and we know this is not the case). Hence the orbits of V on $\mathcal{P}_1$ have length 2 or 4. In any case at least one orbit of length 2 exists and a subgroup of index 2 of V must fix this orbit pointwise, i.e. at least one involution in V has at least two fixed points on $\mathcal{P}_1$. But then all of δ_1, δ_2, δ_3 have at least two fixed points and since no two of δ_1, δ_2, δ_3 can fix one and the same point (otherwise V fixes a point of $\mathcal{P}_1$), we conclude that each involution fixes precisely two points of $\mathcal{P}_1$ and these two points are exchanged by the other two involutions. The transitivity of $\overline{G}_O$ on $\mathcal{P}_1$ follows again from the fact that δ_1, δ_2 and δ_3 are conjugate in $\overline{G}_O$. □

Set $\Omega^{(i)} = \Omega_{X_i}$ for $i = 1, 2, 3, 4$. Then $\mathcal{F}_r = \{\Omega^{(1)}, \Omega^{(2)}, \Omega^{(3)}, \Omega^{(4)}\}$ is a G_O–orbit on $\mathcal{F}$. By Proposition 2.6 the group G_O also has three orbits on $\mathcal{E}_C \setminus \{\ell\}$. One of them is $\mathcal{E}_1 = \{\ell'\}$. There are thus two further orbits $\mathcal{E}_2$, $\mathcal{E}_3$ whose lengths must have sum 10. Since these lengths must also be equal to indices of subgroups of $\overline{G}_O \cong A_4$, we only have one possibility up to relabelling, namely $|\mathcal{E}_2| = 4$, $|\mathcal{E}_3| = 6$. We also know from Proposition 2.10 that each secant line to Ω meets precisely two of the ovals $\Omega^{(1)}, \Omega^{(2)}, \Omega^{(3)}, \Omega^{(4)}$.

PROPOSITION 8.3 We have $c_{r1} = d_{r1} = 0$.
Proof: The stabilizer of each oval $\Omega^{(i)}$ in G_O is a (cyclic) group of order 6. Each oval $\Omega^{(i)}$ is thus fixed by some collineation $\gamma \in G_O$ of order 3. If ℓ' were secant to $\Omega^{(i)}$, then γ should fix $\ell' \cap \Omega^{(i)}$ pointwise, a contradiction. Hence ℓ' misses each one of the ovals $\Omega^{(1)}, \Omega^{(2)}, \Omega^{(3)}, \Omega^{(4)}$, i.e. $c_{r1} = 0$, whence also $d_{r1} = 0$. □

PROPOSITION 8.4 Each pair of ovals in $\mathcal{F}_r$ has four common secants in $\mathcal{E}_1 \cup \mathcal{E}_2$.
Proof: Each line in $\mathcal{S}_C$ is secant to precisely two ovals in $\mathcal{F}_r$. Since G_O is 2–transitive on $\mathcal{F}_r$, there are thus $(n/2)/\binom{4}{2} = 2$ lines in $\mathcal{S}_C$ which are secant to any given pair of ovals in $\mathcal{F}_r$. Since both ℓ and ℓ' miss all four ovals in $\mathcal{F}_r$, we must have $(n/4) - 2 = 4$ common secants in $\mathcal{E}_1 \cup \mathcal{E}_3$. □

PROPOSITION 8.5 We have $c_{r2} \in \{0, 3\}$. For each pair of ovals in $\mathcal{F}_r$ there are 0 or 2 common secants in $\mathcal{E}_2$ according as $c_{r2} = 0$ or $c_{r2} = 3$ respectively.
Proof: The stabilizer in G_O of a line $m \in \mathcal{E}_2$ has order 6, consequently m is fixed by a collineation of order 3 in G_O. This collineation fixes an oval $\Omega^{(i)} \in \mathcal{F}_r$ and permutes the remaining three ovals in a 3–cycle. As before m cannot be secant to $\Omega^{(i)}$ and if m is secant to one of the remaining three ovals, then it is secant to all three of them.

If $c_{r2} = 0$ then no line in $\mathcal{E}_2$ can be a common secant to a pair of ovals in $\mathcal{F}_r$. If $c_{r2} = 3$ then for each oval in $\mathcal{F}_r$ there is precisely one line in $\mathcal{E}_2$ missing it. There are thus 2 lines in $\mathcal{E}_2$ which are secant to a given pair. □

PROPOSITION 8.6 We have $c_{r3} = 2$. For each pair of ovals in $\mathcal{F}_r$ there is precisely one common secant in $\mathcal{E}_3$.
Proof: The stabilizer in G_O of a line $m \in \mathcal{E}_3$ is a cyclic group of order 4. Since $\overline{G}_O$ acts on $\mathcal{F}_r$ as the alternating group A_4 in its natural action on four objects, we see that a collineation of order 4 in G_O fixing m yields two cycles of length 2 on $\mathcal{F}_r$, showing that $c_{r3} \in \{0, 2, 4\}$.

If $c_{r3} = 4$ then all six lines in $\mathcal{E}_3$ meet all four ovals in $\mathcal{F}_r$. These six lines are common secants of any given pair of ovals in $\mathcal{F}_r$, in contradiction to Proposition 8.4.

If $c_{r3} = 0$ then each line in $\mathcal{E}_2$ must be a common secant to any given pair of ovals in $\mathcal{F}_r$, which means that each line in $\mathcal{E}_2$ meets all four ovals in $\mathcal{F}_r$, contradicting the fact established in Proposition 8.5 that each line in $\mathcal{E}_2$ misses at least one oval in $\mathcal{F}_r$.

Since G_O is 2–transitive on $\mathcal{F}_r$ we have that the number of lines in $\mathcal{E}_3$ meeting a given pair of ovals in $\mathcal{F}_r$ is $6/\binom{4}{2} = 6/6 = 1$. □

Propositions 8.5 and 8.6 show that the number of common secants in $\mathcal{E}_2 \cup \mathcal{E}_3$ to a given pair of ovals in $\mathcal{F}_r$ is either $0 + 1$ or $2 + 1$. Both possibilities are excluded by Proposition 8.4, the final contradiction.

REFERENCES

1. L.M. Abatangelo, Piani affini di ordine pari e disegni di Hadamard, *Boll. Un. Mat. Ital. A* (7) *8*: 377–382 (1994).
2. O.E. Barriga, R. Pomareda, The uniqueness of Hering's translation plane of order 27, *Geom. Dedicata 19*: 237–245 (1985).
3. M. Biliotti, G. Korchmáros, Collineation groups strongly irreducible on an oval, *Ann. Discrete Math. 30*: 85–98 (1986).
4. A. Bonisoli, G. Korchmáros, On two–transitive ovals in projective planes of even order, *Arch. Math. 65*: 89–93 (1995).
5. P. Dembowski, *Finite Geometries*, Springer, Berlin (1968).
6. C. Hering, Transitive linear groups and linear groups which contain irreducible subgroups of prime order, *Geom. Dedicata 2*: 425–460 (1974).
7. J.W.P. Hirschfeld, *Projective Geometries over Finite Fields*, Oxford Univ. Press, Oxford (1979).
8. W.M. Kantor, On 2–transitive groups in which the stabilizer of two points fixes additional points, *J. London Math. Soc. 5*: 114–122 (1972).
9. H. Wielandt, *Finite Permutation Groups*, Academic Press, New York (1964).

Inductively Minimal Flag-Transitive Geometries

FRANCIS BUEKENHOUT C.P. 216, Université Libre de Bruxelles, Bd. du Triomphe, B-1050 Bruxelles, Belgium, e-mail: fbueken@ulb.ac.be

MICHEL DEHON C.P. 216, Université Libre de Bruxelles, Bd. du Triomphe, B-1050 Bruxelles, Belgium

PHILIPPE CARA Departement Wiskunde, Vrije Universiteit Brussel, Pleinlaan 2, B-1050 Brussel, Belgium, e-mail: pcara@vub.ac.be

ABSTRACT

We classify all pairs (Γ, G) where Γ is a firm, finite geometry with a connected digon diagram and G is a group of automorphisms of Γ acting flag-transitively, assuming that the pair is inductively minimal.

1. INTRODUCTION

We recall some definitions from [1] and fix notation. Let I be a finite set of $n \geq 1$ elements called **types**. A **geometry over** I is a triple $\Gamma = (X, *, t)$ where X is a set whose members are called **elements** of Γ, $*$ is a symmetric, reflexive binary relation on X called the **incidence relation**, and t is a mapping of X onto I called the **type function**, such that $a * b$ and $t(a) = t(b)$ implies $a = b$. The **rank** of Γ is $|I| = n$. If $A \subseteq X$, the **type** of A is the set $t(A)$. A **flag** F of Γ is a complete subgraph of $(X, *)$. We assume that Γ is **firm**, that is, every flag F with $t(F) \neq I$ is contained in at least two flags of type I.

The **residue** Γ_F of a flag F is the firm geometry over $I \setminus t(F)$ whose elements are all $a \in X \setminus F$ such that $a \cup F$ is also a flag, together with the restrictions of $*$ and t to these elements.

The **digon diagram** of Γ is the graph $(I, \sim)$ defined by $i \sim j$, for $i \neq j$ in I, if and only if there is a flag F of type $I \setminus \{i, j\}$ such that Γ_F is not a complete bipartite graph, that is, Γ_F has elements a, b with $t(a) = i$, $t(b) = j$, and $(a, b) \notin *$. We assume that the digon diagram I is **connected**. A subgraph of $(I, \sim)$ is a subset J of I provided with the restriction of $\sim$ to J. From here on, **adjacency** in a graph is meant to be $\sim$. We call Γ **thin** if every flag F with $|t(F)| = n - 1$ has a residue containing exactly two elements. The geometry Γ is **residually connected** if for any residue Γ_F with $|t(\Gamma_F)| \geq 2$, the graph induced by $*$ on Γ_F is connected. An **automorphism** of Γ is an automorphism α of the graph $(X, *)$ mapping every element x of X onto an element $\alpha(x)$ such that $t(x) = t(\alpha(x))$.

Let G be a group of automorphisms of Γ acting **flag-transitively**, that is, G acts transitively on the flags F with $t(F) = I$.

Let (Γ, G) be called **minimal** if $|G| \leq (n+1)!$ We call (Γ, G) **inductively minimal** if for any connected subgraph J of I and any flag F of Γ, with $t(F) = I \backslash J$, the pair (Γ_F, G_F) where G_F is the group induced on Γ_F by the stabilizer of F, is minimal.

We shall show that a full control can be achieved on inductively minimal pairs although their number grows with n in a fairly wild way.

We obtain the following results.

THEOREM 1. *Let I be a finite set of $n \geq 1$ elements. Let Γ be a finite, firm geometry over I with a connected digon diagram and let G be a flag-transitive automorphism group of Γ. Assume that (Γ, G) is an inductively minimal pair. Then:*

(1) *Γ is thin and residually connected;*
(2) *G is isomorphic to $Sym(n+1)$ and for each $i \in I$ such that $I \backslash i$ is connected, Γ has $n+1$ elements of type i on which G acts faithfully;*
(3) *I has no minimal circuits of length $\ell > 3$;*
(4) *every edge of I is on a unique maximal clique;*
(5) *each vertex p of I is either on one or two maximal cliques of the diagram.*

THEOREM 2. *For any connected graph I as in (3), (4), (5) of Theorem 1, there is, up to isomorphism, one and only one inductively minimal pair (Γ, G) admitting I for diagram.*

How do we get control over the graphs as in (3), (4), (5)? Assume that maximal cliques are called lines. Then we obtain a connected incidence structure of points and lines, with no circuits, any two points are incident with at most one line, every line is incident with at least two points and any point is incident with at most two lines. This structure is trivially under control. The results of the present paper have been announced in [2].

Our results were partially inspired by experimental work with $n = 3$ and $n = 4$, carried out in [3].

2. ENDS OF GRAPHS

Let $(X, *)$ be a finite connected graph. A vertex e of X is called an **end** (or noncutting point according to Harary [3]) if the subgraph induced by $*$ on $X \setminus e$ is connected.

LEMMA 1. *There exist two ends a, b of X such that $X \setminus \{a, b\}$ is connected.*

Proof. Let a and a' be vertices of X at maximal distance. Then a and a' are ends (Harary [4], Theorem 3.4) since otherwise, provided a is not an end, there exists a vertex p such that p and a' are in different components of $X \setminus \{a\}$. Then $d(p, a') > d(a, a')$, a contradiction. Now consider $X \setminus \{a\}$ and let b, b' be at maximal distance in it. Hence they are ends of $X \setminus \{a\}$ by the preceding argument. Assume b' is not an end of X. Then $\{a, b'\}$ is an edge and it is the unique edge on a. Therefore b is an end of X. By the construction $X \setminus \{a, b\}$ is connected.

3. PROOF OF THEOREM 1

3.1. We proceed by induction on n. For $n - 1$ everything is obvious. Hence, let $n > 1$. Since rank 1 residues satisfy the theorem, Γ is thin as required in (1).

3.2. Let e, f be ends of I such that $I \setminus \{e, f\}$ is connected, whose existence results from Lemma 1. Let p (respectively q) be an element of Γ of type e (respectively f). In view of the induction hypothesis, G_p is isomorphic to $Sym\,(n)$ and Γ_p has n elements of type f. The same holds true for G_q and Γ_q. Let N_e (resp. N_f) be the number of elements of type e (resp. f). Then $|G| = (n!)N_e = (n!)N_f \leq (n+1)!$ and so $N_e \leq n+1$. Thus $N_e = N_f \in \{n, n+1\}$.

3.3. Assume first that $N_e = N_f = n+1$. Then $|G| = (n+1)!$, G_q induces $Sym\,(n)$ on n of the $n+1$ elements of type e, hence G induces $Sym\,(n+1)$ on these elements and the action of G is faithful. For any end x of I, $|G| = (n!)N_x$, thus $N_x = n+1$ and (2) holds.

3.4. Assume that $N_e = N_f = n$. Then $G_p = Sym\,(n)$ acts on the $n-1$ elements of type e other than p, hence this action has a nontrivial kernel K. Now K is also the kernel of the action of G on the elements of type e, hence $K \trianglelefteq G$. By the induction hypothesis, G_p, hence K, acts faithfully on the elements of type f and so K acts transitively on those. Now $K \cap K' = 1$, $K \times K' \trianglelefteq G$. If K and K' contain *Alt* (n) we get $(\frac{n!}{2})^2 \leq (n+1)!$ hence $n! \leq 4(n+1)$ and $n \leq 3$. For $n = 2$, $N_e = N_f = 2$ and the firmness of Γ forces a nonconnected digon diagram, a contradiction. For $n = 3$, all elements of type e must be incident to all elements of type f and the diagram must be

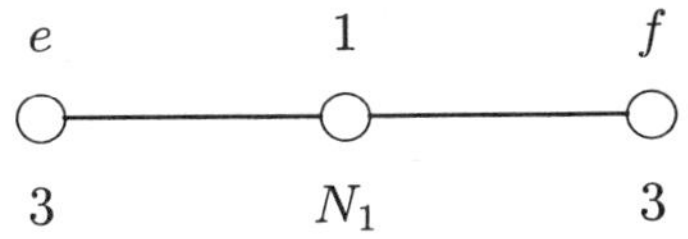

In view of the induction hypothesis, the number N_1 of elements of type 1 is $N_1 = 3^2/2$, a contradiction. If one of K, K' does not contain *Alt* (n), then $n = 4$, $|K| = 4 = |K'|$. Then all elements of distinct end types must be incident because $N_e = 4$. Hence there is a nonend vertex. The diagram is one of:

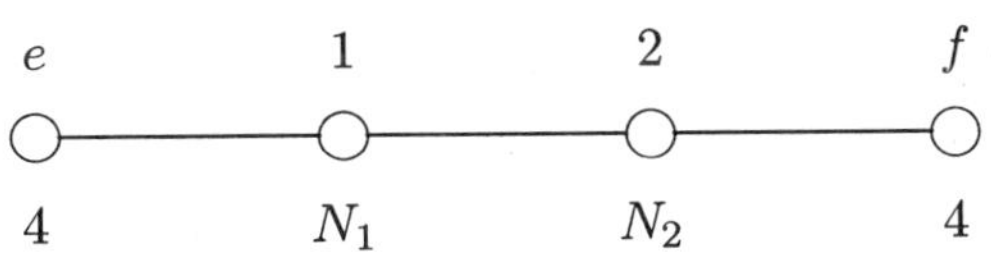

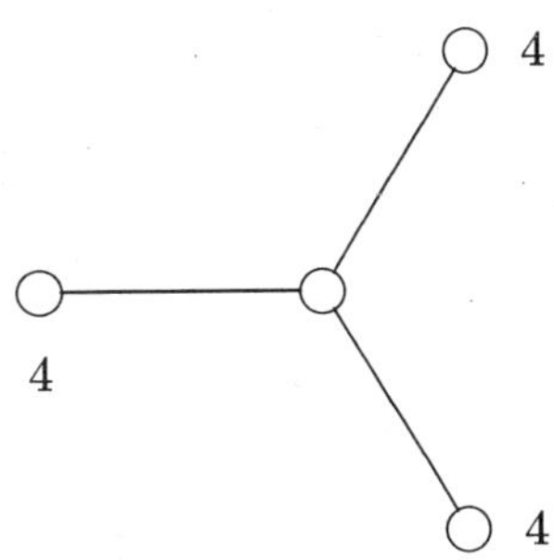

In the first, $N_1 = N_2 = 8$ but $Sym\,(4)$ acts on the 4 elements of type e and so N_1 must be a multiple of 6, the number of pairs in a 4-set, a contradiction. In the third

diagram there is a residue of shape

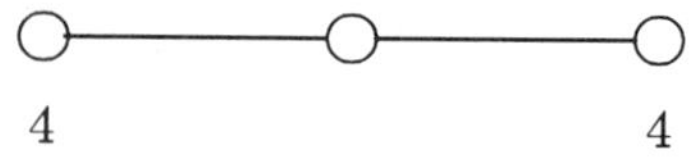

and all end elements are incident to each other, a contradiction.

3.5. We show (1), namely that Γ is residually connected.

First, the incidence graph $(X, *)$ of Γ is connected. Indeed, if e and f are as above, every element in X is incident to some element of type e and if p, q are two elements of type e, then Γ_p, Γ_q have $n - 1 \geq 1$ common elements of type f.

Next, let F be a nonempty flag, let Γ_F be its residue and let $J = t(\Gamma_F)$. If J is connected, (Γ_F, G_F) is inductively minimal and so the induction hypothesis forces it to be residually connected. If e is an end of I in J, there is an element p of type e completing F to a flag of type $I \setminus J \cup e$. Then the property follows by induction in Γ_p. If e is an end of I not in J, F has an element p of type e, the induction hypothesis applies to Γ_p, and since Γ_F is a flag residue in it, Γ_F must be connected. Therefore (1) holds.

3.6. We prove (3). Assume that C is a minimal circuit of I of length ℓ. In view of the definition of inductive minimality (replacing J by C in the definition) we may assume without loss of generality that $I = C$ and so all vertices of C are ends. Let e, e', e'' be consecutive vertices of C. In Γ over C, there are $\ell + 1$ elements of each type by (2). If p is an element of type e, Γ_p has a string diagram of rank $\ell - 1$ and therefore, there are respectively ℓ and $\ell(\ell - 1)/2$ elements of types e', e'' in Γ_p. Therefore $\ell(\ell - 1) \leq 2\ell + 2$, forcing $\ell \leq 3$. Therefore (3) holds.

3.7. We prove (4). Suppose the contrary. Then I has a subgraph as in Figure 1, with a, b, c, d in I. By the definition of inductive minimality, there exists an inductively minimal pair (Γ, G) over $I = \{a, b, c, d\}$ as in Figure 1.

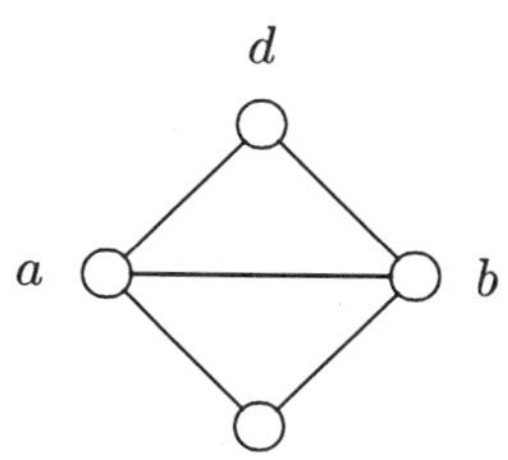

Figure 1.

Since each vertex is an end of I there are 5 elements of each type in Γ, by (2). Moreover $G \cong Sym\,(5)$. If p is some element, $G_p \cong Sym\,(4)$ fixes a unique element of each type. Therefore, the incidence graph of Γ is the complement of a (5×4)-grid. This geometry does however admit the complete graph on $\{a, b, c, d\}$ as a digon diagram: a contradiction. Therefore (4) holds.

3.8. We prove (5). Assume by way of contradiction that p is on at least three maximal cliques of the diagram. By (4), this means that there is a subgraph in I as in Figure 2 and an inductively minimal pair (Γ, G) over this rank 4 diagram.

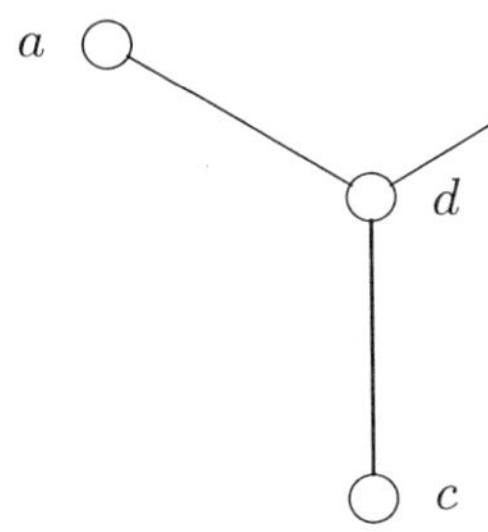

Figure 2.

By (2), there are 5 elements of each type and $G \cong Sym\,(5)$. If p is some element, G_p fixes a unique element of each type and so the incidence graph of Γ over $\{a, b, c\}$ is the complement of a (4×3)-grid. Then an element of type d is incident to 2 elements of type a (resp. b, c) and those are incident to all those of the other types, a contradiction. Therefore (5) holds and Theorem 1 is proved.

4. THE UNIQUENESS OF Γ FOR GIVEN I

4.1. Consider I of size n with p end vertices and Γ inductively minimal over I. We shall show that the control over the structure of Γ results from an $(n+1) \times p$-grid. For each end e there is a set of $n+1$ elements of type e in Γ that we see as a column of the future grid.

If e, e' are distinct ends and $C_e, C_{e'}$ are the corresponding columns, each $x \in C_e$ is incident to n elements of $C_{e'}$ and the unique one to whom it is incident is fixed by the stabilizer $Sym\,(n)$ of x in the automorphism group $Sym\,(n+1)$ of Γ. Thus, the stabilier of x in C_e fixes a unique element in each column and these constitute, by definition, the row of x in the grid.

In the $(n+1) \times p$-grid $\mathcal{G}$ two elements are incident if and only if they are neither in the same column nor in the same row of $\mathcal{G}$.

4.2. Assume that f_1 is a non-end in I. Then f_1 decomposes $I \setminus f_i$ in two connected components I_1 and I_2. Since Γ is residually connected, the direct sum theorem (see [1]) applies to the residue of an element x of type f_1: all elements of type in I_1 are incident to all elements of type in I_2. The elements of type in I_1 (resp. I_2) constitute an inductively minimal geometry. If I_1 (resp. I_2) has size n_1 (resp. n_2) and I_1 (resp. I_2) contains p_1 (resp. p_2) ends of I then $p_i \geq 1$, $p_1 + p_2 = p$, $n_1 + n_2 = n - 1$, and so the residue of x on the elements of end type are two grids $(n_1 + 1) \times p_1$ over I_1 and $(n_2 + 1) \times p_2$ over I_2 constituting orthogonal blocks in $\mathcal{G}$. By the action of $Sym\,(n+1)$ on $\mathcal{G}$, all pairs of orthogonal blocks of the same size over I_1 and I_2 occur as residues of some element of type f over the end types.

4.3. Distinct elements of type f are thus represented by distinct pairs of orthogonal blocks, a feature we can establish by induction on n.

4.4. As to the incidence of an element x of type f represented by subgrids $\mathcal{G}_1$ and $\mathcal{G}_2$ with an element x' of type f' represented by subgrids $\mathcal{G}'_1$ and $\mathcal{G}'_2$, it reduces to the inclusion of $\mathcal{G}_1$ in $\mathcal{G}'_1$ and of $\mathcal{G}'_2$ in $\mathcal{G}_2$ (or vice versa) provided I'_1 contains f.

4.5. The uniqueness part of Theorem 2 is thus established.

5. PROOF OF THEOREM 2

5.1. The uniqueness of Γ is established in Section 4. There we see also a unique way to construct it in order to prove its existence. Does the construction of Γ provide the required properties? The construction provides the automorphism group $G = Sym\ (n+1)$ leaving each column of the grid invariant.

5.2. The existence requires to check the following parts:

(i) Γ is a geometry which reduces to check that every flag can be completed to a flag of type I;
(ii) Γ is firm;
(iii) Γ admits I as digon diagram;
(iv) G is flag-transitive on Γ;
(v) (Γ, G) is an inductively minimal pair.

5.3. We proceed by induction on $|I| = n$ and may assume $n \geq 2$.

5.4. For every end vertex e of I and every element x of type e of Γ, the residue Γ_x conforms to the construction of Γ and so the induction hypothesis applies to it. Moreover, its group is $Sym\ (n)$.

5.5. Therefore, all of (i)–(v) follow at once.

REFERENCES

[1] F. Buekenhout, editor, *Handbook of Incidence Geometry,* North-Holland, Amsterdam, 1995.
[2] F. Buekenhout, Minimal flagtransitive geometries, *Combinatorics '90* (Gaeta, 1990) (A. Barlotti, A. Bichara, P.V. Ceccherini, and G. Tallini, eds.), Ann. Discrete Math., vol. 52, North-Holland, Amsterdam, 1992, pp. 69–70.
[3] F. Buekenhout, The geometry of the finite simple groups, *Buildings and the Geometry of Diagrams* (L.A. Rosati, ed.), Lecture Notes in Mathematics, vol. 1181, Springer-Verlag, Berlin–New York, 1986, pp. 1–78.
[4] F. Harary, *Graph Theory,* Addison-Wesley Publishing Co., Reading, Mass., 1969.

Transitive Autotopism Groups and the Generalized Twisted Field Planes

MINERVA CORDERO Department of Mathematics, Texas Tech University, Lubbock, Texas 79409.

RAÚL F. FIGUEROA Department of Mathematics, University of Puerto Rico, PO Box 23355, Río Piedras, PR 00931.

1 Introduction

In this article we present a survey of the results obtained in our study of semifield planes with an autotopism group that acts transitively on the non-vertex points on a side of the autotopism triangle. In doing this study we found a characterization of the generalized twisted field planes that is also presented.

Our motivation for this research came from two different sources. Kallaher and Ostrom [9] proved that any finite odd dimensional translation plane of rank 3 is a semifield plane. A finite affine plane $\mathcal{U}$ is a *rank 3 affine plane* if it admits a collineation group acting as a rank 3 permutation group on the affine points of $\mathcal{U}$. The other source was a result of Ganley and Jha [6]. They showed that if a finite affine non-Desarguesian translation plane π admits a collineation group acting transitively on $\ell_\infty - \{L\}$, where L is a point on the line at infinity, ℓ_∞, then π is a semifield plane.

Notice that if π is a semifield plane of rank 3, then the autotopism group of π, relative to an autotopism triangle $\triangle$, acts transitively on the non-vertex points on each affine side of $\triangle$. However, if π admits a collineation group as in the result of Ganley and Jha, then the autotopism group acts transitively on the non-vertex points on the side at infinity of the autotopism triangle. It is not difficult to see that, in any of the above cases, $p^n - 1$ divides the order of the autotopism group of π, where the order of π is p^n, p a prime. When $n \geq 3$ and $p^n \neq 2^6$, then by [10] there is a p-primitive prime divisor h of $p^n - 1$, i.e., $h \mid p^n - 1$ and $h \nmid p^i - 1$, for $1 \leq i < n - 1$. Hence π admits an autotopism of order h.

In the following section we first state two results on semifield planes with an autotopism of order h as in last sentence in the preceding paragraph. Then we apply these results to the study of semifields of order 3^4 and 3^6 (Results 3 and 4). Section 3 contains the main results of this article.

2 Preliminaries

Let V denote the n-dimensional vector space over the Galois field GF(p) consisting of all the n-tuples of the form $(x^{(0)}, x^{(1)}, \ldots, x^{(n-1)})$, where $x \in K = \mathrm{GF}(p^n)$, $x^{(i)} = x^{p^i}$ for $i = 0, 1, \ldots, n-1$, and p is a prime number. The group Aut(V) of all the automorphisms of V over GF(p) consists of the non-singular matrices of the form

$$M = \begin{pmatrix} a_0 & a_{n-1}^{(1)} & \cdots & a_1^{(n-1)} \\ a_1 & a_0^{(1)} & & a_2^{(n-1)} \\ \vdots & \vdots & & \vdots \\ a_{n-1} & a_{n-2}^{(1)} & \cdots & a_0^{(n-1)} \end{pmatrix} \tag{1}$$

where $a_0, a_1, \ldots, a_{n-1} \in K$; we denote this matrix M by $\mathrm{col}(a_0, a_1, \ldots, a_{n-1})$. Also, for $0 \leq k \leq n-1$ and $c \in K$, let $L_k(c) = \mathrm{col}(0, \ldots, c, \ldots 0)$, where c is in the k–th position and the rest of the entries are zero. The matrix $\begin{pmatrix} A & 0 \\ 0 & B \end{pmatrix}$ will be denoted by $\mathrm{diag}(A, B)$. Let $V(\infty) = \{(0, v) | v \in K\}$ and $V(M) = \{(v, vM) | v \in V\}$ for M in Aut(V) or $M = 0$ (the zero matrix).

In [3] we proved the following.

RESULT 1 Let π be a non-Desarguesian semifield plane of order p^n. If π admits a collineation g_0 of order a h, a p-primitive prime divisor of $p^n - 1$, then

(i) There exists a spread set $\{V(0), V(\infty)\} \cup \{V(M(y)) | y \in K - \{0\}\}$ for π in $V \oplus V$ such that $V(\infty)$ is the shear axis and for $y \in K$, $y \neq 0$,

$$M(y) = \mathrm{col}(a_0 y^{(e_0)}, a_1 y^{(e_1)}, \ldots, a_{n-1} y^{(e_{n-1})}) \in \mathrm{Aut}(V).$$

The coefficients $a_0 = 1, a_1, a_2, \ldots, a_{n-1}$ belong to K and the integers $e_0 = 0, e_1, \ldots, e_{n-1}$ are such that $0 \leq e_i \leq n-1$, $e_i \neq i$ if $a_i \neq 0$, and $e_i \neq e_j$ when $i \neq j$ and $a_i \neq 0 \neq a_j$.

(ii) $g_0 = \mathrm{diag}(L_0(\gamma), L_0(\delta))$, where γ and δ are two different elements in K, both of order h.

(iii) $(\delta/\gamma)^{(e_i)} = \delta/\gamma^{(i)}$, for each i such that $a_i \neq 0$.

A generalized twisted field [1] is a pre-semifield $(K, +, \circ')$, where "$+$" is the addition on the field K and $x\circ' y = xy + cx^{p^d}y^{p^e}$. A semifield plane coordinatized by such a semifield is called a *generalized twisted field plane*. In Result 1 the plane π is coordinatized by the pre-semifield $(K, +, \circ)$, where

$$x \circ y = \sum_{i=0}^{n-1} a_i x^{(i)} y^{e_i}, \tag{2}$$

for $x, y \in K$. Hence either π is a generalized twisted field plane or there exist at least two non-zero indices u, v such that $a_u \neq 0 \neq a_v$; so by Result 1(i) we have that $e_u \neq u \neq v$ and $e_u \neq e_v \neq v$. In the latter case, let c be the smallest among u, v, e_u, and e_v and denote the remaining ones by a, b and d in such a way that $a \neq c \neq b$ and $a \neq d \neq b$. In [5] we proved the following result.

RESULT 2 Let π be a non-Desarguesian semifield plane of order p^n, p a prime number, and let $\mathcal{G}$ be the autotopism group relative to an autotopism triangle Δ. If the group induced by $\mathcal{G}$ on a side of Δ has an element $\bar{g}$ whose order is divisible by a p - primitive prime divisor of $p^n - 1$, then either π is a generalized twisted field plane, or there are integers a, b, c and d such that

$$1 \leq a, b, c, d \leq n-1, \quad a \neq c \neq b, \ a \neq d \neq b \tag{3}$$

and

$$p^d + p^{a-c} + p^{b-c} \equiv p^w + 1 + p^{d-c} \bmod o(\bar{g}), \tag{4}$$

where $0 \leq w \leq n-1$ and $w \equiv a + b - c \bmod n$.

In same cases the mere existence of an element of order a p-primitive prime divisor of $p^n - 1$ forces the plane to be Desarguesian or a generalized twisted field plane. Result 3 below illustrates this situation and Result 4 shows that in general that condition is not sufficient.

RESULT 3 Let π be a non-Desarguesian semifield plane of order 3^4 with an autotopism of order 5, a 3-primitive prime divisor of $3^4 - 1$. Then π is a generalized twisted field plane.
Proof: in [2]

RESULT 4 Let π be a non-Desarguesian nor a generalized twisted field plane of order 3^6 with an autotopism g_0 of order 7, a 3-primitive prime divisor of $3^6 - 1$. Then π can be coordinatized by the pre-semifield $\mathrm{GF}(3^6, +, \circ)$, where

$$x \circ y = xy + \gamma x^{(1)} y^{(3)} + \gamma^{13} x^{(3)} y^{(1)}, \tag{5}$$

and γ is a primitive element in $\mathrm{GF}(3^6)$ that satisfies $\gamma^6 + \gamma + 2 = 0$. Moreover, π has autotopism of order 3 that normalizes g_0.

Proof: By Result 1 we may assume that $g_0 = \text{diag}(L_0(\alpha), L_0(\beta))$, for some $\alpha \neq \beta$ in GF(3^6), $|\alpha| = |\beta| = 7$. Since $\alpha = \beta^r$, for some integer r, $1 \leq r \leq 6$, from Result 1(iii) we get that $3^{e_i} - 1 = r(3^{e_i} - 3^i) \bmod 7$. A direct computation shows that if $r = 4$, then $(i, e_i) = (1,3), (3,1), (4,5)$ or $(5,4)$. In this case, with the aid of a computer we found that, up to isotopism, the multiplication (2) yields a pre-semifield when $a_0 = 1$, $a_1 = \gamma$, $a_3 = \gamma^{13}$ and $a_2 = a_4 = a_5 = 0$. We also found that $g = \text{diag}(L_1(\gamma^1), L_1(\gamma^1))$ is an autotopism of the plane associated to this pre-semifield. This element has order $|g| = 6$ and normalizes the subgroup generated by g_0 (with $\alpha = \gamma^{104}$ and $\beta = \gamma^{208}$).

The cases when $r \neq 4$ can be reduced to the previous one. Thus, if $r = 6$, then the multiplication (2) takes the form $xy + a_1 x^{(1)} y^{(2)} + a_2 x^{(2)} y^{(5)} + a_4 x^{(4)} y^{(3)} + a_5 x^{(5)} y^{(1)}$. By replacing x by $x^{(5)}$ and y by $y^{(4)}$ we obtain the multiplications associate to the case $r = 4$.

3 Main results

The proof of our main result on semifield planes with a transitive autotopism group depends on the following result of Ganley and Jha [7].

RESULT 5 Let π be a non-Desarguesian semifield plane of order p^n, where p is a prime number. Let $\mathcal{G}$ denote the autotopism group relative to an autotopism triangle Δ. Suppose that $\mathcal{G}$ induces a group $\bar{\mathcal{G}}$ transitive on the non-vertex points of a side of Δ. Then we have that either

(i) $\bar{\mathcal{G}} \leq \Gamma L(1, p^n))$, or

(ii) $p^n = 3^4$ (and $\bar{\mathcal{G}}$ is a certain group), or

(iii) $p^n = 3^6$ and $\bar{\mathcal{G}} \cong$SL$(2, 13)$.

In [2] (case p odd) and [5] we first showed that if case (i) occurs, then $\bar{\mathcal{G}}$ has an element of order $(p^n - 1)/\mu$, where μ is a divisor of n, and then we proved the following

RESULT 6 Let π be a non-Desarguesian semifield plane of order $p^n \neq 2^n$, with p a prime number and $n \geq 3$. Let Δ be an autotopism triangle of π and let $\bar{\mathcal{G}}$ be the group induced by the autotopism group of π on a side of Δ. If $\bar{\mathcal{G}}$ has an element of order $(p^n - 1)/\mu$, where μ is an integer dividing n, then π is a generalized twisted field plane.

In SL$(2, 13)$ the order of the normalizer of a subgroup of order 7 is 28 (see [8, Salz 8.4]). This fact together with Result 4 imply that the case (iii) in Result 6 does not occurs. If case (ii) in Result 6 holds, then we apply Result 3. Thus we have

THEOREM 1 Let π be a non-Desarguesian semifield plane of order $p^n \neq 2^6$, where p is a prime number and $n \geq 3$. Let $\mathcal{G}$ be the autotopism group relative to an autotopism triangle Δ of π. If $\mathcal{G}$ induces a group $\bar{\mathcal{G}}$ transitive on the non-vertex points of a side of Δ, then π is a generalized twisted field plane.

From this theorem, and Theorem A in [6] in the second corollary, we infer the following two corollaries.

COROLLARY 1 Let π be a non-Desarguesian semifield plane of order $p^n \neq 2^6$. If π is of rank 3, then π is a generalized twisted field plane.

COROLLARY 2 Let π be a translation plane of order $p^n \neq 2^6$. If π admits a collineation group that fixes a point L on the line at infinity, ℓ_{infty}, and acts 2-transitively on $\ell_\infty - L$, then π is Desarguesian or a generalized twisted field plane.

Our proof of Result 6 consisted on showing that there are not integers a, b, c and d that satisfy conditions (3) and (4) in Result 2. The technique used in solving congruence (4) was applied by Figueroa [4] and by Cordero and Figueroa [5] to prove:

THEOREM 2 Let π be a non-Desarguesian semifield plane of order $p^n \neq 2^6$, with p a prime number and $n \geq 3$. Let $\mathcal{G}$ be the autotopism group of π relative to an autotopism triangle Δ and let $\bar{\mathcal{G}}$ denote the group induced by $\mathcal{G}$ on the line at infinity. Then π is a generalized twisted field plane if, and only if, $\bar{\mathcal{G}}$ has an element of order $(p^k-1)(p^n-1)/(p^m-1)$ for some integers k and m, where $k|m$, $m|n$, and $k \geq 2$ when $p=2$, except possibly when $n = 2m$ and $p = 2$ or $p = 3$

The exceptional case does occurs when $p = 3$. For the semifield plane exhibited in Result 4 we found that $g_1 = (L_0(\gamma^{117}), L_0(\gamma^{52}))$ is an autotopism that induces an element of order 56.

References

[1] Albert, A. A., Generalized twisted fields, *Pacific. J. Math. 11*: 1-8 (1961).

[2] Cordero, M. and Figueroa, R.F., Semifield planes with a transitive autotopism group, *Arch. Math. 63*: 17-22 (1994).

[3] Cordero, M. and Figueroa, R., Towards a characterization of generalized twisted field planes, *J. Geom. 52*: 54-63 (1995).

[4] Figueroa, R., A characterization of the generalized twisted field planes of characteristic ≥ 5, *Geom. Dedicata 50*: 205-216 (1994).

[5] Figueroa, R.F. and Cordero, M., On the generalized twisted field planes of characteristic 2, *Geom. Dedicata 56*: 197-208 (1995).

[6] Ganley, M.J. and Jha, V., On translation planes with a 2-transitive orbit on the line at infinity, *Arch. Math. 47*: 379-384 (1986).

[7] Ganley, M.J. and Jha, V., On a conjecture of Kallaher and Liebler, *Geom. Dedicata 21*: 277-289 (1986).

[8] Huppert, B., *Endliche Gruppen I*, Spinger-Verlag, Berlin, Heidelberg, New York, 1967.

[9] Kallaher, M.J. and Ostrom, T.G., Collineation groups irreducible on the components of a translation plane, *Geom. Dedicata 9*: 153-194 (1980).

[10] Zsigmondy, K., Zur Theorie der Potenzreste, *Monatsh. Math. Phys. 3*: 265-284 (1892).

A Family of Non-Buekenhout Unitals in the Hall Planes

JEREMY M. DOVER, University of Delaware, Newark, DE

1 INTRODUCTION

Consideration of unitals embedded in non-Desarguesian translation planes first emerged with the seminal paper of Buekenhout [8] in which it is shown that every translation plane with dimension at most two over its kernel admits an embedded unital. Indeed, Buekenhout gives two explicit constructions, which we will discuss later. Much recent work has been devoted to studying the various types of unitals in the Hall planes which arise from Buekenhout's constructions; see the papers of Barwick [4], [3] and Rinaldi [14], [15].

Our motivation comes from Brouwer [7]. In this paper, Brouwer discovered a large number of unitals of order 3. Exactly three of these are embeddable in the Hall plane of order 9. Of these three, one is the unital investigated by Grüning [11], which is also embeddable in the dual Hall plane of order 9. A second is almost undoubtedly a parabolic Buekenhout unital. The third (Brouwer's E.2) seems to defy classification. It cannot be a parabolic Buekenhout unital since it is not resolvable as a design (see [9]), and one can easily verify that the only hyperbolic Buekenhout unital in the Hall plane of order 9 is isomorphic to Grüning's. It is this unital that we seek to build upon to obtain an infinite family of non-Buekenhout unitals in the Hall planes.

2 THE HALL PLANES

In this section, we would like to develop two models of the Hall plane of order q^2, which we denote $H(q^2)$. In particular, the notation used in this section will be used throughout the paper.

First, recall the Bose/André construction for two-dimensional translation planes. Let Σ be a projective four-space $\mathcal{PG}(4,q)$, and let Σ^* be any hyperplane of Σ. If $\mathcal{S}$ is a spread of Σ^*, we can define an incidence structure as follows. The points will be the points of $\Sigma \setminus \Sigma^*$, together with the lines of $\mathcal{S}$, and the lines are the planes of Σ which meet Σ^* in a line of $\mathcal{S}$, together with $\mathcal{S}$ itself. Incidence is given by containment. The resulting incidence structure is a translation plane of order q^2, and will be denoted $\pi(\Sigma, \Sigma^*, \mathcal{S})$.

We now wish to construct a spread $\mathcal{S}'$ of Σ^* such that the translation plane $\pi(\Sigma, \Sigma^*, \mathcal{S}')$ is a Hall plane. Start with a regular spread $\mathcal{S}$ of Σ^*, and let $\mathcal{R}$ be a regulus contained in $\mathcal{S}$. Finally, let $\mathcal{R}'$ denote the opposite regulus to $\mathcal{R}$. Define $\mathcal{S}' = (\mathcal{S} \setminus \mathcal{R}) \cup \mathcal{R}'$. With this definition, the translation plane $\pi(\Sigma, \Sigma^*, \mathcal{S}')$ will be isomorphic to the Hall plane $H(q^2)$ for any prime power q. It should be noted that if $q = 2$, the Hall plane obtained in this manner is isomorphic to $\mathcal{PG}(2,4)$.

We now wish to talk about how the Hall planes can be obtained from the Desarguesian plane by derivation. Let $\Pi = \mathcal{PG}(2, q^2)$. Let l be any line of Π, and let $\mathcal{L}$ be a Baer subline of l. We define a new incidence structure as follows: The points of the structure are the points of $\Pi \setminus l$. The lines of the structure come in two types. The first type of lines are lines of Π which meet l in a point of $l \setminus \mathcal{L}$. The second type of lines are the Baer subplanes of Π which contain the Baer subline $\mathcal{L}$. Incidence is given by containment. This gives us a second model of the Hall plane, but in affine form. We can then complete this plane to a projective Hall plane.

We would like to conclude this section with some general comments about the automorphism group of the Hall plane $H(q^2)$. At least for $q > 3$, the following theorem tells us how to construct the full automorphism group of $H(q^2)$, using the derivation model. The first part is a standard result, while the second part is due to Hughes [12].

THEOREM 2.1 Let l be a line of $\mathcal{PG}(2, q^2)$, and let $\mathcal{L}$ be a Baer subline of l. Model the Hall plane $H(q^2)$ by performing derivation on the Baer subline $\mathcal{L}$. If $q > 3$, the following hold:

1. Every collineation of $\mathcal{PG}(2, q^2)$ which leaves the Baer subline $\mathcal{L}$ invariant induces a collineation of $H(q^2)$.

2. The automorphism group of $H(q^2)$ is inherited, i.e. every collineation of $H(q^2)$ is induced by some collineation of $\mathcal{PG}(2, q^2)$ leaving $\mathcal{L}$ invariant.

3 BUEKENHOUT'S CONSTRUCTION TECHNIQUES

In this section, we wish to recall Buekenhout's two methods for constructing unitals in translation planes.

If $\mathcal{U}$ is a unital embedded in a projective plane, then every line of that projective plane meets $\mathcal{U}$ in either one point or $q + 1$ points. Lines with these intersection sizes are called *tangents* and *secants*, respectively. A unital in a non-Desarguesian translation plane is called *parabolic* or *hyperbolic* as the translation line of the plane is a tangent or secant, respectively.

Buekenhout [8] presented two methods for constructing unitals in translation planes of dimension two or less over their kernel. Let Π be a translation plane of dimension less than two with Bose/André representation $\pi(\Sigma, \Sigma^*, \mathcal{S})$, where $\Sigma = \mathcal{PG}(4, q)$.

A parabolic Buekenhout unital can be obtained by considering a cone $\mathcal{O}$ over an ovoid in Σ, such that the cone meets Σ^* in a line of the spread $\mathcal{S}$. The points of

Π which form the unital are the q^3 points represented by $\mathcal{O} \setminus \Sigma^*$, together with the point represented by the line of $\mathcal{S}$ which lies in $\mathcal{O}$.

We would like to point out that every parabolic Buekenhout unital contains Baer sublines. Recall that in the Bose/André presentation of Π, every line l which meets Σ^* in a point represents a Baer subline of Π whose points consist of the points represented by $l \setminus \Sigma^*$ together with the point represented by the unique line of the spread $\mathcal{S}$ which contains the point $l \cap \Sigma^*$. Since $\mathcal{O}$ represents our unital, every line in $\mathcal{O}$ except the spread line contained therein represents a Baer subline contained in the corresponding unital. In fact, one can easily see that any parabolic Buekenhout unital is the union of q^2 Baer sublines, all of which meet in one point. Quinn and Casse [13] have proven that in the Desarguesian plane, this property characterizes parabolic Buekenhout unitals.

A hyperbolic Buekenhout unital can be obtained by considering a nondegenerate quadric surface $\mathcal{P}$ in Σ, of which there is only one up to projective equivalence. If $\mathcal{P}$ meets Σ^* in a set of $(q+1)^2$ points which are covered by a regulus contained in $\mathcal{S}$, then $\mathcal{P}$ will represent a unital in Π. Its points in Π will be represented by the $q^3 - q$ points of $\mathcal{P} \setminus \Sigma^*$ together with the $q+1$ points represented by the lines of the regulus in $\mathcal{S}$ which cover $\mathcal{P} \cap \Sigma^*$.

4 THE CONSTRUCTION METHOD

Before we give our construction, we need to know what all Baer subplanes look like in the Bose/André model of $\mathcal{PG}(2,q^2)$. The following result is given in Bose, Freeman, and Glynn [6].

THEOREM 4.1 Let $\pi(\Sigma, \Sigma^*, \mathcal{S})$ be the Bose/André representation of the Desarguesian plane $\mathcal{PG}(2,q^2)$, and l be the line of $\mathcal{PG}(2,q^2)$ which is represented by the spread $\mathcal{S}$. Then, there are two types of Baer subplanes in $\mathcal{PG}(2,q^2)$:

1. The Baer subplanes of $\mathcal{PG}(2,q^2)$ which meet l in a Baer subline are represented in the Bose/André model by planes of Σ which meet Σ^* in a line which is not a spread line. The points on l which are contained in the Baer subplane are represented by the $q+1$ lines of $\mathcal{S}$ which meet the representative plane.

2. The Baer subplanes of $\mathcal{PG}(2,q^2)$ which meet l in a single point are represented in the Bose/André model by certain cubic surfaces. These cubic surfaces are called <u>twisted ladders</u> and have the following structure. Call such a twisted ladder $\mathcal{V}$. Then, there exists a line m of the spread $\mathcal{S}$ which is contained in $\mathcal{V}$ and represents the point of intersection of the Baer subplane and l. There are $q+1$ other lines contained in $\mathcal{V}$, one through each point of m, such that no two of these lines are coplanar and none of these lines lie in Σ^* other than m.

We should point out that not every object which has the structure given in part 2 of this theorem represents a Baer subplane.

THEOREM 4.2 Let $\mathcal{U}$ be a parabolic Buekenhout unital embedded in $\mathcal{PG}(2,q^2)$. Then there exists a unique point P of $\mathcal{U}$ such that $\mathcal{U}$ is the union of q^2 Baer sublines through P, unless $\mathcal{U}$ is Hermitian, in which case P may be any point of $\mathcal{U}$. Let l

be any line through P which meets $\mathcal{U}$ in a Baer subline, and let $\mathcal{L} = l \cap \mathcal{U}$ be that Baer subline. Let $H(q^2)$ be the Hall plane obtained by performing derivation on the subline $\mathcal{L}$, and completing the resulting affine plane. Let $\mathcal{U}'$ be the set $\mathcal{U} \setminus \mathcal{L}$, together with the $q + 1$ points which are slope points of the lines represented by the Baer subplanes of $\mathcal{PG}(2, q^2)$ which contain $\mathcal{L}$. Then $\mathcal{U}'$ is a hyperbolic unital in $H(q^2)$.

Proof: In any projective plane, any set of $q^3 + 1$ points with the property that any line meets the set in 1 or $q + 1$ points is a unital. We adopt this approach now.

First, we note that $\mathcal{U}'$ has q^3+1 points. Since $|\mathcal{L}| = q+1$, we have $|\mathcal{U}\setminus\mathcal{L}| = q^3 - q$. But, there are $q + 1$ points of $H(q^2)$ which are slope points of lines represented by Baer subplanes of $\mathcal{PG}(2, q^2)$. This adds back the number lost, and we have $|\mathcal{U}'| = q^3 + 1$.

To look at lines of $H(q^2)$, we will consider the situation in the Bose/André model of $\mathcal{PG}(2, q^2)$. Since $\mathcal{U}$ is a parabolic Buekenhout unital, there exists a Bose/André representation $\pi(\Sigma, \Sigma^*, \mathcal{S})$ of $\mathcal{PG}(2, q^2)$ such that $\mathcal{U}$ is represented by an ovoidal cone $\mathcal{O}$ in Σ which meets Σ^* in a line m of the spread $\mathcal{S}$.

Recall that to do derivation, we need a particular line l and a Baer subline $\mathcal{L}$ on l. In the Bose/André model, it is usual to take the line represented by the spread to be the line on which we do derivation. However, we would like to pick another line instead.

The line l of $\mathcal{PG}(2, q^2)$ which we will use is represented by a plane Π_l which meets Σ^* in the line m, and the Baer subline $\mathcal{L}$ is represented by a line L of Σ which meets Σ^* in a point of m, which we call V. Note that V is the vertex of $\mathcal{O}$, and that every line of $\mathcal{O}$ must pass through V.

We now wish to consider how the lines of $H(q^2)$ are represented in this model. In $\mathcal{PG}(2, q^2)$, these lines come in two kinds: the lines of $\mathcal{PG}(2, q^2)$ which meet l in a point not contained in $\mathcal{L}$; and the Baer subplanes which contain $\mathcal{L}$. For now, we ignore the ideal line of $H(q^2)$.

First, consider a line of $H(q^2)$ which is represented by a line of $\mathcal{PG}(2, q^2)$ which meets l in a point off $\mathcal{L}$. Since $\mathcal{U}$ is a unital in $\mathcal{PG}(2, q^2)$, every such line meets $\mathcal{U}$ in either 1 or $q + 1$ points. Since none of these lines contains a point of $\mathcal{L}$, they must meet $\mathcal{U} \setminus \mathcal{L}$ in 1 or $q+1$ points. Finally, none of these lines can contain a slope point of a line represented by a Baer subplane of $\mathcal{PG}(2, q^2)$ by their very definition. So, each line of $H(q^2)$ which is obtained in this manner meets $\mathcal{U}'$ in either 1 or $q + 1$ points.

Now, we wish to consider the lines of $H(q^2)$ which are represented by Baer subplanes of $\mathcal{PG}(2, q^2)$ which contain the Baer subline $\mathcal{L}$. When considered as structures in Σ, these Baer subplanes take on three forms:

1. Planes of Σ which meet $\mathcal{O}$ in only the line L.
2. Planes of Σ which meet $\mathcal{O}$ in a pair of intersecting lines, one of which is L. Note that the other line cannot be m, since this would force the plane under consideration to represent the line l.
3. Twisted ladders which contain the line L, and meet Σ^* in the line m.

We consider each of these cases separately.

Let Π be a plane of Σ which meets $\mathcal{O}$ in only the line L. Then Π contains no points of $\mathcal{O} \setminus L$. In $\mathcal{PG}(2, q^2)$, this implies that the Baer subplane represented by

Π contains no points of $\mathcal{U} \setminus \mathcal{L}$. So, when considered as a line of $H(q^2)$, this Baer subplane contains only one point of $\mathcal{U}'$, namely a slope point.

Now, let Π be a plane of Σ which meets $\mathcal{O}$ in a pair of intersecting lines. Then, Π contains q points of $\mathcal{O} \setminus L$. Further, since m is not one of these lines, all q of these points are points of $\Sigma \setminus \Sigma^*$. Therefore, the Baer subplane of $\mathcal{PG}(2, q^2)$ represented by Π contains q points of $\mathcal{U} \setminus \mathcal{L}$. So, when considered as a line of $H(q^2)$, this Baer subplane contains $q + 1$ points of $\mathcal{U}'$, namely the q points of $\mathcal{U} \setminus \mathcal{L}$ together with a slope point.

Finally, let $\mathcal{V}$ be a twisted ladder which contains L and meets Σ^* in the line m. Then, $\mathcal{V}$ contains $q + 1$ lines other than m, one of which is L. Call the other q lines $n_1, n_2, \ldots, n_q$. These lines are pairwise skew, and each meets Σ^* in a point of m.

For each $i \in \{1 \ldots q\}$, the plane $\Pi_i =< m, n_i >$ represents a line of $\mathcal{PG}(2, q^2)$. Since the n_i's are mutually skew, these planes are distinct. Further, Π_i must meet $\mathcal{O}$ in a pair of intersecting lines for all $i \in \{1 \ldots q\}$, one of which must be m. For each i, let l_i be the line of $\mathcal{O}$ distinct from m which is contained in Π_i.

Since n_i is a line of $\mathcal{V}$, n_i cannot contain the vertex V of $\mathcal{O}$; indeed, the line L is a line of $\mathcal{V}$ which contains V, and L and n_i must be disjoint. Further, n_i is a line of Π_i. Both m and l_i are lines of Π_i which are distinct from n_i, as both contain V. In particular, n_i meets l_i in a point of $\Sigma \setminus \Sigma^*$. While n_i also meets m in a point, this point is in Σ^*, and will not represent a point of $\mathcal{PG}(2, q^2)$.

So, each line n_i meets $\mathcal{O}$ in two points, but only one of these points represents a point of $\mathcal{PG}(2, q^2)$. As a Baer subplane of $\mathcal{PG}(2, q^2)$, $\mathcal{V}$ thus contains q points of $\mathcal{U} \setminus \mathcal{L}$, one for each line n_i. As a line of $H(q^2)$, this Baer subplane meets $\mathcal{U}'$ in $q + 1$ points: the q points of $\mathcal{U} \setminus \mathcal{L}$ together with a slope point.

Finally, we need to consider the ideal line. As mentioned before, there are $q + 1$ ideal points which are slope points of the lines represented by Baer subplanes of $\mathcal{PG}(2, q^2)$. Thus, the ideal line also meets $\mathcal{U}'$ in $q + 1$ points.

Therefore, every line of $H(q^2)$ meets $\mathcal{U}'$ in either one or $q + 1$ points, and we have that $\mathcal{U}'$ is a unital in $H(q^2)$. Since the ideal line meets $\mathcal{U}'$ in $q + 1$ points, we can conclude that it is hyperbolic. □

We now would like to show that a unital obtained via this construction is a hyperbolic Buekenhout unital if and only if the unital with which you begin the construction is Hermitian. Before doing this, we need the following result. The implication that every Hermitian unital is a hyperbolic Buekenhout unital is due to Buekenhout [8]. The converse is due to Barwick [5].

THEOREM 4.3 Let $\mathcal{U}$ be a unital embedded in the Desarguesian plane $\mathcal{PG}(2, q^2)$. Then, $\mathcal{U}$ is Hermitian if and only if $\mathcal{U}$ is a hyperbolic Buekenhout unital.

THEOREM 4.4 Let $\mathcal{U}$ be a parabolic Buekenhout unital in $\mathcal{PG}(2, q^2)$, and let $\mathcal{U}'$ be the unital in $H(q^2)$ derived from $\mathcal{U}$ via the procedure in Theorem 4.2. Then, $\mathcal{U}'$ is a hyperbolic Buekenhout unital if and only if $\mathcal{U}$ is Hermitian.

Proof: Let us set up a Bose/André model for this situation. Let $\Sigma = \mathcal{PG}(4, q)$ and let Σ^* be a special hyperplane in Σ. Let $\mathcal{S}$ be a regular spread of Σ^*, so our model for $\mathcal{PG}(2, q^2)$ will be $\pi(\Sigma, \Sigma^*, \mathcal{S})$. Let l be the line of $\mathcal{PG}(2, q^2)$ represented by the spread $\mathcal{S}$. Without loss of generality, we can assume the Baer subline $\mathcal{L}$ of our construction lies on l, since $Aut(\mathcal{PG}(2, q^2))$ is transitive on Baer sublines.

Now, let $\mathcal{R}$ be the regulus in $\mathcal{S}$ which represents the Baer subline $\mathcal{L}$, and let $\mathcal{R}'$ be its opposite regulus. Letting $\mathcal{S}' = (\mathcal{S} \setminus \mathcal{R}) \cup \mathcal{R}'$, we have that $\pi(\Sigma, \Sigma^*, \mathcal{S}')$ is a model of the Hall plane.

Suppose first that $\mathcal{U}$ is Hermitian. Since $\mathcal{U}$ contains the Baer subline $\mathcal{L}$, $\mathcal{U}$ is hyperbolic with respect to the line l. By Theorem 4.3, $\mathcal{U}$ is represented in Σ by a four-dimensional nondegenerate quadric $\mathcal{Q}$ which meets Σ^* in the points covered by the regulus $\mathcal{R}$.

In $H(q^2)$, the unital $\mathcal{U}'$ has the same affine points as $\mathcal{U}$, but now contains the slope points of the Baer sublines through $\mathcal{L}$. Since the slope points of the Baer subplanes are represented in Σ by the lines of $\mathcal{R}'$, the unital $\mathcal{U}'$ is also represented in Σ by $\mathcal{Q}$. Therefore, $\mathcal{U}'$ is a hyperbolic Buekenhout unital in $H(q^2)$, as it is obtainable from a four-dimensional quadric which meets Σ^* in the points covered by a regulus of $\mathcal{S}'$.

Now, suppose $\mathcal{U}'$ is a hyperbolic Buekenhout unital in $H(q^2)$. Then, $\mathcal{U}'$ is represented by a four-dimensional quadric $\mathcal{Q}$ in Σ which meets Σ^* in the points covered by the regulus $\mathcal{R}'$. Then, reversing the above argument, $\mathcal{U}$ is represented by the same quadric $\mathcal{Q}$. Therefore, $\mathcal{U}$ is a hyperbolic Buekenhout unital in $\mathcal{PG}(2, q^2)$. Again using Theorem 4.3, $\mathcal{U}$ must be Hermitian. □

In light of this theorem, we can claim that the construction method presented here is really a generalization of Buekenhout's hyperbolic construction, at least for the Hall plane. If one starts with a Hermitian unital, one obtains the unitals discussed by Grüning; however if one starts with a non-classical parabolic Buekenhout unital, we get a seemingly new family of unitals.

We would also like to point out that this construction is a special case of one given by Barlotti and Lunardon [2]. However, their presentation uses a nonstandard model of the Hall plane, and the properties we would like to discuss here will be much easier to obtain using our new viewpoint.

5 SOME PROPERTIES OF $\mathcal{U}'$

Let $\mathcal{U}$ be any Buekenhout-Metz unital embedded in $\mathcal{PG}(2, q^2)$. In particular, we mean that $\mathcal{U}$ is a parabolic Buekenhout unital in $\mathcal{PG}(2, q^2)$ which can be obtained from the cone over an elliptic quadric. These unitals have been studied extensively in Baker and Ebert [1] and Ebert [10]. We will assume $\mathcal{U}$ is not Hermitian, as this case has been thoroughly studied by Grüning [11].

5.1 Collineation Group

Let $\mathcal{U}$ be a Buekenhout-Metz unital in $\mathcal{PG}(2, q^2)$, and let $\mathcal{U}'$ be a unital in $H(q^2)$ obtained from $\mathcal{U}$ using the procedure in Theorem 4.2. We begin with the following results about the collineation group of $\mathcal{U}$, whose proofs can be obtained in Baker and Ebert [1] and Ebert [10].

PROPOSITION 5.1.1 Let $\mathcal{U}$ be a Buekenhout-Metz unital embedded in $\mathcal{PG}(2, q^2)$ such that $\mathcal{U}$ is not Hermitian. Let G be the linear collineation group of $\mathcal{PG}(2, q^2)$ which leaves $\mathcal{U}$ invariant. Then, the following results hold.

1. There exists a point P of $\mathcal{U}$ which is left invariant by G. Further, every secant to $\mathcal{U}$ through P meets $\mathcal{U}$ in a Baer subline.

2. $|G| = q^3(q-1)$, unless q is odd and $\mathcal{U}$ is a union of conics unital, in which case $|G| = 2q^3(q-1)$.

3. G acts transitively on the points of $\mathcal{U} \setminus \{P\}$.

COROLLARY 5.1.2 Let $\mathcal{U}$ be a Buekenhout-Metz unital in $\mathcal{PG}(2, q^2)$ with linear collineation group G, and let P be the point of $\mathcal{U}$ fixed by G. Then, G acts transitively on the Baer sublines contained in $\mathcal{U}$ through P.

Proof: This statement follows immediately from Part 3 of Proposition 5.1.1, since P and any other point Q of $\mathcal{U}$ determine a unique subline through P contained in $\mathcal{U}$. □

We can now compute the collineation group of our unital $\mathcal{U}'$.

THEOREM 5.1.3 Let $\mathcal{U}$ be a non-Hermitian Buekenhout-Metz unital embedded in $\mathcal{PG}(2, q^2)$, with $q > 3$. Let P be the point of $\mathcal{U}$ fixed by its collineation group, and let $\mathcal{U}'$ be the unital in $H(q^2)$ obtained by doing derivation on a Baer subline $\mathcal{L}$ through P. Then, the linear collineation group of $H(q^2)$ which leaves $\mathcal{U}'$ invariant has order $2q(q-1)$ or $q(q-1)$, as $\mathcal{U}$ is a union of conics or not.

Proof: By Theorem 2.1, every automorphism of the Hall plane $H(q^2)$ is inherited from $\mathcal{PG}(2, q^2)$. In particular, this implies that the collineation group of our unital $\mathcal{U}'$ in $H(q^2)$ is exactly the group of collineations of $\mathcal{PG}(2, q^2)$ which leave $\mathcal{U}$ invariant and fix the Baer subline $\mathcal{L}$.

If G is the collineation group of $\mathcal{PG}(2, q^2)$ leaving $\mathcal{U}$ invariant, then $|G| = q^3(q-1)$ unless $\mathcal{U}$ is a union of conics, in which case $|G| = 2q^3(q-1)$. In either case G fixes the point P, which implies G maps the set of q^2 Baer sublines of $\mathcal{U}$ passing through P onto itself. But by Corollary 5.1.2, this action is transitive. Therefore, by the Orbit-Stabilizer theorem, the subgroup H of G which leaves the Baer subline $\mathcal{L}$ invariant has order $2q(q-1)$ or $q(q-1)$ as $\mathcal{U}$ is or is not a union of conics unital. However, this subgroup H is exactly the group of collineations which induce collineations of $H(q^2)$ which leave $\mathcal{U}'$ invariant, which yields the result. □

We note that if $\mathcal{U}$ is a classical unital, then the collineation group fixing $\mathcal{U}'$ is isomorphic to $\mathrm{P\Gamma L}(2, q^2)$, if $q > 3$. This was determined by Grüning [11].

5.2 Projective Equivalence

We would now like to sort out projective equivalence amongst the unitals in $H(q^2)$ obtained from our construction. The question is answered quite easily by the following theorem.

THEOREM 5.2.1 Let $\mathcal{U}$ and $\mathcal{V}$ be two Buekenhout-Metz unitals in $\mathcal{PG}(2, q^2)$, $q > 3$, which share a common Baer subline $\mathcal{L}$ which satisfies the conditions of Theorem 4.2 for both unitals. Let $\mathcal{U}'$ and $\mathcal{V}'$ be the unitals in $H(q^2)$ obtained from the construction in Theorem 4.2 by doing derivation on the Baer subline $\mathcal{L}$. Then, the unitals $\mathcal{U}'$ and $\mathcal{V}'$ are projectively equivalent in $H(q^2)$ if and only if the unitals $\mathcal{U}$ and $\mathcal{V}$ are projectively equivalent in $\mathcal{PG}(2,q^2)$.

Proof: Suppose first that $\mathcal{U}'$ and $\mathcal{V}'$ are projectively equivalent. Then, there exists a collineation ϕ of $H(q^2)$ which maps $\mathcal{U}'$ onto $\mathcal{V}'$. By Theorem 2.1, since $q > 3$, ϕ is induced by a collineation ψ of $\mathcal{PG}(2,q^2)$ which leaves the Baer subline $\mathcal{L}$ invariant. Further, in $\mathcal{PG}(2,q^2)$, ψ maps the set $\mathcal{U} \setminus \mathcal{L}$ onto the set $\mathcal{V} \setminus \mathcal{L}$. Thus, ψ is a collineation of $\mathcal{PG}(2,q^2)$ which maps $\mathcal{U}$ onto $\mathcal{V}$, and $\mathcal{U}$ and $\mathcal{V}$ are projectively inequivalent.

Suppose now there exists a collineation ψ of $\mathcal{PG}(2,q^2)$ which maps $\mathcal{U}$ onto $\mathcal{V}$. If $\mathcal{U}$ and $\mathcal{V}$ are both Hermitian, then $\mathcal{U}'$ and $\mathcal{V}'$ are both Grüning unitals, and are projectively equivalent. If $\mathcal{U}$ and $\mathcal{V}$ are not Hermitian, then there exists a point $P \in \mathcal{U}$ and a point $Q \in \mathcal{V}$ such that P is fixed by the collineation group of $\mathcal{U}$ and Q is fixed by the collineation group of $\mathcal{V}$. This implies $P\psi = Q$.

By the hypotheses of Theorem 4.2, we must have that $P \in \mathcal{L}$ and $Q \in \mathcal{L}$. Now, $\mathcal{L}\psi = \mathcal{M}$ is some Baer subline contained in $\mathcal{V}$ passing through Q. By Corollary 5.1.2, there exists a collineation γ which leaves $\mathcal{V}$ invariant, and maps $\mathcal{M}$ onto $\mathcal{L}$. Define $\phi = \psi\gamma$. Then, $\mathcal{U}\phi = \mathcal{V}$ and $\mathcal{L}\phi = \mathcal{L}$. Again using Theorem 2.1, ϕ must induce a collineation χ of $H(q^2)$. Since $q > 3$, χ must fix the $q+1$ slope points of the lines of $H(q^2)$ defined by Baer subplanes of $\mathcal{PG}(2,q^2)$ through $\mathcal{L}$, and further χ maps the affine part of $\mathcal{U}'$ onto the affine part of $\mathcal{V}'$. Therefore, χ is a collineation of $H(q^2)$ which maps $\mathcal{U}'$ onto $\mathcal{V}'$, and $\mathcal{U}'$ and $\mathcal{V}'$ are projectively equivalent. □

COROLLARY 5.2.2 The number of projectively inequivalent unitals in $H(q^2)$ obtained from our construction beginning with Buekenhout-Metz unitals is precisely the number of projectively inequivalent Buekenhout-Metz unitals in $\mathcal{PG}(2,q^2)$.

A formula for the number of such inequivalent unitals can be found in Baker and Ebert [1] and Ebert [10].

6 CONCLUSION

There are many interesting questions yet to be answered about these unitals. As was mentioned, the construction given here is a special case of a construction given by Barlotti and Lunardon. However, their construction does not seem to generalize to other translation planes. One interesting question is whether the pointset of the unitals given here can form a unital in a plane derived from the Hall plane. Recall that Grüning's unitals are embeddable in both the Hall plane and the dual Hall plane. One might ask if the unitals given here can be embedded in other planes as well.

An interesting design theoretic question is whether the unitals given here are resolvable. One can easily show that they admit spreads. Indeed, let $\mathcal{U}$ be any parabolic Buekenhout unital in $\mathcal{PG}(2,q^2)$, and let $\mathcal{U}'$ be the unital in $H(q^2)$ derived from $\mathcal{U}$ by doing derivation on some Baer subline $\mathcal{L}$ contained in $\mathcal{U}$. If $\mathcal{S}$ is any spread of $\mathcal{U}$ which contains the block $\mathcal{L}$, then it is easy to see that this spread inherits to a spread of $\mathcal{U}'$ by taking $\mathcal{S}'$ to be the set of blocks $\mathcal{S} \setminus \{\mathcal{L}\}$ together with the block which is the intersection of the ideal line of $H(q^2)$ with $\mathcal{U}'$. That every such unital $\mathcal{U}$ admits a spread containing the Baer subline $\mathcal{L}$ follows from the fact that $\mathcal{U}$ must be resolvable (see Dover [9]). Another interesting question is whether these unitals admit spreads which are not inherited in this manner.

Finally, there are still some geometric question to be answered. For example, the existence of O'Nan configurations in these unitals is unknown. An important first step would be constructing O'Nan configurations in the Buekenhout-Metz unitals in the Desarguesian plane, which has also not been done.

References

[1] R.D. Baker and G.L. Ebert. On Buekenhout-Metz unitals of odd order. *J. Combin. Theory, Ser. A*, 60:67–84, 1992.

[2] A. Barlotti and G. Lunardon. Una classe di unitals nei Δ-piani. *Riv. Mat. Univ. Parma*, 4:781–785, 1979.

[3] S.G. Barwick. A class of Buekenhout unitals in the Hall plane. *to appear, Bull. Belg. Math. Soc., Simon Stevin.*

[4] S.G. Barwick. Unitals in the Hall plane. *to appear, J. Geom.*

[5] S.G. Barwick. A characterization of the classical unital. *Geom. Ded.*, 52:175–180, 1994.

[6] R.C. Bose, J.W. Freeman, and D.G. Glynn. On the intersection of two Baer subplanes in a finite projective plane. *Utilitas Math*, 17:65–77, 1980.

[7] A.E. Brouwer. Some unitals on 28 points and their embeddings in projective planes of order 9. In *Geometries and Groups*, volume 893 of *Lecture Notes in Mathematics*, pages 183–188. Springer-Verlag, New York/Berlin, 1981.

[8] F. Buekenhout. Existence of unitals in finite translation planes of order q^2 with kernel of order q. *Geom. Ded.*, 5:189–194, 1976.

[9] J. Dover. Some design theoretic properties of Buekenhout unitals. *preprint.*

[10] G.L. Ebert. On Buekenhout-Metz unitals of even order. *Europ. J. Combinatorics*, 13:109–117, 1992.

[11] K. Grüning. A class of unitals of order q which can be embedded in two different planes of order q^2. *J. Geom.*, 29:61–77, 1987.

[12] D.R. Hughes. Collineation groups of non-Desarguesian planes, I. the Hall Veblen-Wedderburn systems. *Am. J. Math*, 81:921–938, 1959.

[13] C.T. Quinn and R. Casse. Concerning a characterization of Buekenhout-Metz unitals. *J. Geom.*, 52:159–167, 1995.

[14] G. Rinaldi. Construction of unitals in the Hall planes. *Geom. Ded.*, 56:249–255, 1995.

[15] G. Rinaldi. Hyperbolic unitals in the Hall planes. *J. Geom.*, 54:148–154, 1995.

4-Caps in Orthogonal Spaces

JOHN FUELBERTH Division of Math and Sciences, Wayne State College, Wayne NE 68787, U.S.A. jfuelber@wscgate.wsc.edu

ATHULA GUNAWARDENA[1] Division of Math and Sciences, Wayne State College, Wayne NE 68787, U.S.A. agunawar@wscgate.wsc.edu

Dedicated to Professor T. G. Ostrom

Abstract We study 4-caps in nondegenerate finite orthogonal spaces over $GF(q)$. It is known that 3-caps form at most 2 orbits under the group of isometries [4]. We classify 4-caps up to the equivalence of their matrices corresponding to the bilinear form. We also find the number of orbits of nondegenerate 4-caps which span 4-dimensional subspaces under the group of isometries.

1. NOTATION

More details of the following definitions can be found in Artin [1], Aschbacher [2] and Kantor [3]. An *orthogonal space* is a pair (V, Q) such that V is a finite dimensional vector space of dimension n over $GF(q)$ and $Q : V \longrightarrow GF(q)$ is a quadratic form, i.e.

$$Q(\lambda x) \;=\; \lambda^2 Q(x)$$

and

$$Q(x+y) \;=\; Q(x) + Q(y) + f(x,y),$$

for all $\lambda \in F; x, y \in V$, where f is a bilinear form. Q is *nondegenerate* if $f(x,y) = 0$ for $Q(x) = 0$ and all $y \in V$ implies $x = 0$. The *points* of an orthogonal space (V, Q) are one-dimensional subspaces and $\langle v \rangle$ is called a *singular* point if $Q(v) = 0$. Two points $\langle v_1 \rangle$ and $\langle v_2 \rangle$ are *perpendicular* if $f(v_1, v_2) = 0$. A subspace S is *totally singular* if $Q(v) = 0$ for all $v \in S$. When $n = 2m$ and Q is nondegenerate, there are two types of quadratic forms up to equivalence. One form is called *hyperbolic* which produces maximal totally singular subspaces with dimension m. The other

[1]The authors are grateful to Eric Moorhouse for valuable discussions.

form is called *elliptic* and it produces maximal singular subspaces with dimension $m-1$. When Q is nondegenerate, (V,Q) is called an O_{2m-1} space if $n=2m-1$, or an $O^{\pm}_{2m}(q)$ space if $n=2m$, using superscript $+$ or $-$ according as Q is hyperbolic or elliptic.

An *r-cap* in an orthogonal space is a set of pairwise nonperpendicular singular points with cardinality r. *The group of isometries* is the subgroup of $GL(V)$ which fixes the quadratic form.

2. CAPS

Let (V,Q) be a nondegenerate orthogonal space over $GF(q)$ with f as its bilinear form. Define the following equivalence relation on $r\times r$ matrices over $GF(q)$. Two matrices A and B are *equivalent* if there is a diagonal matrix (with nonzero entries in the diagonal) D such that $A=DBD$. Let $C=(\langle x_1\rangle,\langle x_2\rangle,\ldots,\langle x_r\rangle)$ be an ordered r-cap in (V,Q). Select representative vectors $x_1,x_2,\ldots,x_r$ for C. Construct an $r\times r$ matrix $A=(a_{ij})$ over $GF(q)$ such that $a_{ij}=f(x_i,x_j)$. It is clear that A is not unique and different representative vectors give different matrices. One can verify that the set of all possible matrices from a given ordered cap forms an equivalence class under the above defined equivalence relation. With the following Lemma, we can select a *standard matrix* for the equivalence class of a given ordered cap.

LEMMA 1 Let A be an $r\times r$ matrix over $GF(q)$ such that $a_{ii}=0$ and $a_{ij}=a_{ji}\neq 0$ if $i\neq j$. If q is odd, fix a nonsquare ϵ in $GF(q)$. Then A is equivalent to a matrix B of one of the following types.

1. $b_{1i}=b_{i1}=1$ for $i\neq 1$ and $b_{23}=1$. When q is even, this is the only type.
2. $b_{1i}=b_{i1}=1$ for $i\neq 1$ and $b_{23}=\epsilon$.

Proof: We use the notation $diag(d_{11},d_{22},\ldots,d_{rr})$ to represent an $r\times r$ diagonal matrix D. Let $D_1=diag(1,a_{12}^{-1},a_{13}^{-1},\ldots,a_{1r}^{-1})$. So $C=D_1AD_1$ has $c_{1i}=c_{i1}=1$ for $i\neq 1$. If q is even, c_{23} is λ^2 for some nonzero scalar λ. If q is odd, c_{23} is λ^2 or $\lambda^2\epsilon$ for some nonzero scalar λ. Let $D_2=diag(\lambda,\lambda^{-1},\lambda^{-1},\ldots,\lambda^{-1})$. Thus $B=D_2CD_2$ is one of the above two types.∎

Hereafter we consider 4-caps. Any matrix of the equivalence class of a given ordered 4-cap has the form of matrix A mentioned in Lemma 1. It can be verified that there is only one matrix in the equivalence class having the form of matrix B mentioned in Lemma 1. We select this standard matrix to represent the equivalence class of a given ordered cap. Let A be a standard matrix. Since this symmetric standard matrix is determined by the entries a_{23},a_{24} and a_{34}, we use an ordered triple to represent A as shown below.

$$\begin{pmatrix} 0 & 1 & 1 & 1\\ 1 & 0 & a & b\\ 1 & a & 0 & c\\ 1 & b & c & 0 \end{pmatrix} \Leftrightarrow (a,b,c)$$

Here a is 1 or ϵ and $b,c\neq 0$. So the set of standard matrices has $2(q-1)^2$ distinct matrices when q is odd and $(q-1)^2$ distinct matrices when q is even.

Let A be the standard matrix of an ordered 4-cap. With an element of the symmetric group S_4 we can permute the points of the cap. Let P be the corresponding permutation matrix. One can verify that the resulting standard matrix of the new ordered cap is $DPAPD$ for some diagonal matrix D.

THEOREM 2 Let C be an ordered 4-cap in an orthogonal space over $GF(q)$. If q is odd, let ϵ be a nonsquare in $GF(q)$. Assume that the symmetric group is permuting the points of C and hence the corresponding equivalence classes of matrices. Then the orbit of the standard matrix of C under S_4 on the set of standard matrices is one of the following where neither a, b, c or d is zero.

1. When $q = 2^{2e-1}$, there are $q(q+1)/6$ orbits.
   (a) 1 orbit, $|\text{orbit}| = 1$. The orbit is $\{(1,1,1)\}$.
   (b) $(q-2)$ orbits, $|\text{orbit}| = 3$. The orbits are $\{(1,1,c),(1,c,1),(1,1/c,1/c)\}$, where $c \neq 1$.
   (c) $(q-2)(q-3)/6$ orbits, $|\text{orbit}| = 6$. $\{(1,c,d),(1,d,c),(1,c/d,1/d),(1,1/d,c/d),(1,1/c,d/c),(1,d/c,1/c)\}$, where $c,d \neq 1$ and $c \neq d$.
2. When $q = 2^{2e}$, there are $(q^2+q+4)/6$ orbits.
   (a) -(b) Same as in 1(a)–1(b).
   (c) $((q-2)(q-3)-2)/6$ orbits, $|\text{orbit}| = 6$. $\{(1,c,d),(1,d,c),(1,c/d,1/d),(1,1/d,c/d),(1,1/c,d/c),(1,d/c,1/c)\}$, where $(c^2 \neq d$ or $cd \neq 1)$, $c,d \neq 1$ and $c \neq d$.
   (d) 1 orbit, $|\text{orbit}| = 2$. $\{(1,c,d),(1,d,c)\}$, where $c^2 = d$ and $cd = 1$ and $c,d \neq 1$. (Only one pair of c,d satisfies this condition.)
3. When q is odd and $q \not\equiv 1$ *mod* 3, there are $(5q+3)(q+1)/24$ orbits.
   (a) 2 orbits, $|\text{orbit}| = 1$. $\{(\lambda,\lambda,\lambda)\}$, where $\lambda \in \{1,\epsilon\}$.
   (b) $(q-3)$ orbits, $|\text{orbit}| = 3$. $\{(\lambda,\lambda,\lambda a^2),(\lambda,\lambda a^2,\lambda),(\lambda,\lambda/a^2,\lambda/a^2)\}$, where $a^2 \neq 1$ and $\lambda \in \{1,\epsilon\}$.
   (c) $(q-1)/2$ orbits, $|\text{orbit}| = 6$. $\{(1,\epsilon a^2,1),(1,1/(\epsilon a^2),1/(\epsilon a^2)),(1,1,\epsilon a^2),(\epsilon,1/a^2,1/a^2),(\epsilon,\epsilon^2 a^2,\epsilon),(\epsilon,\epsilon,\epsilon^2 a^2)\}$.
   (d) $(q-1)(q-3)/8$ orbits, $|\text{orbit}| = 12$. $\{(1,\epsilon a^2,b^2),(1,(\epsilon a^2)/b^2,1/b^2),(1,1/(\epsilon a^2),b^2/(\epsilon a^2)),(1,b^2/(\epsilon a^2),1/(\epsilon a^2)),(1,b^2,\epsilon a^2),(1,1/b^2,\epsilon a^2/b^2),(\epsilon,1/a^2,b^2/a^2),(\epsilon,b^2/a^2,1/a^2),(\epsilon,\epsilon^2 a^2,\epsilon b^2),(\epsilon,\epsilon^2 a^2/b^2,\epsilon/b^2),(\epsilon,\epsilon b^2,\epsilon^2 a^2),(\epsilon,\epsilon/b^2,\epsilon^2 a^2/b^2)\}$, where $b^2 \neq 1$.
   (e) $(q-3)(q-5)/12$ orbits, $|\text{orbit}| = 6$. $\{(\lambda,\lambda a^2,\lambda b^2),(\lambda,\lambda b^2,\lambda a^2),(\lambda,\lambda a^2/b^2,\lambda/b^2),(\lambda,\lambda/b^2,\lambda a^2/b^2),(\lambda,\lambda/a^2,\lambda b^2/a^2),(\lambda,\lambda b^2/a^2,\lambda/a^2)\}$, where $a^2,b^2 \neq 1$, $a^2 \neq b^2$ and $\lambda \in \{1,\epsilon\}$.
4. When q is odd and $q \equiv 1$ *mod* 3, $(5q^2+8q+35)/24$ orbits.
   (a) -(d) Same as in 3(a)–3(d).

(e) $((q-3)(q-5)-8)/12$ orbits. Orbits are same as in 3(e) with $\lambda \in \{1, \epsilon\}$, $a^2, b^2 \neq 1$, $a^2 \neq b^2$ and ($a^2 \neq b^4$ or $a^2b^2 \neq 1$).

(f) 2 orbits, $|\text{orbit}| = 2$. $\{\lambda, \lambda a^2, \lambda b^2), (\lambda, \lambda b^2, \lambda a^2)\}$, where $\lambda \in \{1, \epsilon\}$, $a^2 = b^4$, $a^2b^2 = 1$, and $a^2 \neq b^2$. (Only one pair of a, b satisfies this condition.)

Proof: Let C, A be as mentioned above. Then A can be represented by a triple $(\alpha, \beta a^2, \gamma b^2)$, for some $\alpha, \beta, \gamma \in \{1, \epsilon\}$ and $a, b \in GF(q)$. The images of A under S_4 are as follows. When q is even, omit the cases that involve ϵ and substitute $c = a^2$ and $d = b^2$.

Group Element	Image	
(1)	$(\alpha, \beta a^2, \gamma b^2)$	
(12)	$(\alpha, \alpha^2/(\beta a^2), \alpha\gamma b^2/(\beta a^2))$	
(13)	$(\alpha, \alpha\beta a^2/(\gamma b^2), \alpha^2/(\gamma b^2))$	
(14)	$(\epsilon, \epsilon\gamma b^2/\alpha, \epsilon\beta a^2/\alpha)$	if $\epsilon\alpha = \beta\gamma$
	$(1, \epsilon b^2, \epsilon a^2)$	if $\epsilon^2\alpha = \beta\gamma$
	$(\alpha/(\beta\gamma), b^2/\beta, a^2/\gamma)$	otherwise
(23)	$(\alpha, \gamma b^2, \beta a^2)$	
(24)	$(\gamma, \beta a^2/b^2, \alpha/b^2)$	
(34)	$(\beta, \alpha/a^2, \gamma b^2/a^2)$	
(123)	$(\alpha, \alpha^2/(\gamma b^2), \alpha\beta a^2/(\gamma b^2))$	
(124)	$(\gamma, \gamma^2 b^2/(\beta a^2), \alpha\gamma/(\beta a^2))$	
(134)	$(\beta, \alpha\beta/(\gamma b^2), \beta^2 a^2/(\gamma b^2))$	
(132)	$(\alpha, \alpha\gamma b^2/(\beta a^2), \alpha^2/(\beta a^2))$	
(142)	$(\epsilon, \epsilon\alpha/(\gamma b^2), \epsilon\beta a^2/(\gamma b^2))$	if $\epsilon\gamma = \alpha\beta$
	$(1, \epsilon/b^2, \epsilon a^2/b^2)$	if $\epsilon^2\gamma = \alpha\beta$
	$(\gamma/(\alpha\beta), 1/(\beta b^2), a^2/(\alpha\beta^2))$	otherwise
(143)	$(\epsilon, \gamma\alpha b^2/(\beta a^2), \epsilon\alpha/(\beta a^2))$	if $\epsilon\beta = \alpha\gamma$
	$(1, \epsilon b^2/a^2, \epsilon/a^2)$	if $\epsilon^2\beta = \alpha\gamma$
	$(\beta/(\alpha\gamma), b^2/(\alpha a^2), 1/(\gamma a^2))$	otherwise
(234)	$(\beta, \gamma b^2/a^2, \alpha/a^2)$	
(243)	$(\gamma, \alpha/b^2, \beta a^2/b^2)$	
(12)(34)	$(\beta, \beta^2 a^2/\alpha, \gamma\beta b^2/\alpha)$	
(13)(24)	$(\gamma, \beta\gamma a^2/\alpha, \gamma^2 b^2/\alpha)$	
(14)(23)	$(\epsilon, \epsilon\beta a^2/\alpha, \epsilon\gamma b^2/\alpha)$	if $\epsilon\alpha = \beta\gamma$
	$(1, \epsilon a^2, \epsilon b^2)$	if $\epsilon^2\alpha = \beta\gamma$
	$(\alpha/(\beta\gamma), a^2/\gamma, b^2/\beta)$	otherwise
(1234)	$(\beta, \beta^2 a^2/(\gamma b^2), \alpha\beta/(\gamma b^2))$	
(1432)	$(\epsilon, \epsilon\beta a^2/(\gamma b^2), \epsilon\alpha/(\gamma b^2))$	if $\epsilon\gamma = \alpha\beta$
	$(1, \epsilon a^2/b^2, \epsilon/b^2)$	if $\epsilon^2\gamma = \alpha\beta$
	$(\gamma/(\alpha\beta), a^2/(\alpha b^2), 1/(\beta b^2))$	otherwise
(1324)	$(\gamma, \alpha\gamma/(\beta a^2), \gamma^2 b^2/(\beta a^2))$	
(1423)	$(\epsilon, \epsilon\alpha/(\beta a^2), \epsilon\gamma b^2/(\beta a^2))$	if $\epsilon\beta = \alpha\gamma$
	$(1, \epsilon/a^2, \epsilon b^2/a^2)$	if $\epsilon^2\beta = \alpha\gamma$
	$(\beta/(\alpha\gamma), \gamma/a^2/, b^2/(\alpha a^2))$	otherwise
(1243)	$(\gamma, \gamma^2 b^2/\alpha, \gamma\beta a^2/\alpha)$	
(1342)	$(\beta, \beta\gamma b^2/\alpha, \beta^2 a^2/\alpha)$	

With the above images, it is a direct calculation to verify the orbits mentioned in Theorem 2. When $a^2 = b^4$, $a^2b^2 = 1$, and $a^2, b^2 \neq 1$, the above images collapse to an orbit with 2 elements (the restrictions imply that $a^4 = b^2$). Let us consider this situation. Suppose ρ is a primitive element in the field. Since a^2 is the inverse of b^2, we can write $a^2 = \rho^k$ and $b^2 = \rho^{q-1-k}$ for some integer k such that $1 < k < q-1$. Since we look for a pair $a^2 = \rho^k$ and $b^2 = \rho^{q-1-k}$, without loss of generality we assume that $k < q-1-k$. $a^4 = b^2$ implies that $\rho^{2k} = \rho^{q-1-k}$ and hence $q = 3k+1$. Thus $q \equiv 1 \mod 3$, and one pair of a, b satisfies the given restrictions. ∎

We say *a set of points is nondegenerate* if the subspace spannned by the given set is nondegenerate. We need the following Lemma to classify nondegenerate 4-caps which span 4-dimensional subspaces.

LEMMA 3 Let (Q, V) be an orthogonal space over $GF(q)$ with the bilinear form f. Let dimension of V be equal to n, $(n \geq 4)$. Let A be a 4×4, symmetric matrix with $a_{ii} = 0$, $a_{ij} \neq 0$ for $i \neq j$. Suppose that the determinant of A is not equal to zero. Then there is an ordered nondegenerate 4-cap, $(\langle x_1\rangle, \langle x_2\rangle, \langle x_3\rangle, \langle x_4\rangle)$, which spans a 4-dimensional subspace such that $a_{ij} = f(x_i, x_j)$.

Proof: Without loss of generality we can represent a vector x in V by an n-tuple $(x_1, x_2, x_3, x_4, x_5, \ldots, x_n)$. Select x_L and x_R such that $x_L = (x_1, x_2, x_3, x_4)$ and $x_R = (x_5, x_6, \ldots, x_n)$. Assume q is odd. Let L and R be the subspaces such that $L = \{x \in V | x_R = 0\}$ and $R = \{x \in V | x_L = 0\}$. When q is odd, we can define a quadratic form Q_L on L such that $Q_L = x_L A x_L^t$. When q is even, we can define a quadratic form $Q_L = x_L B x_L^t$, where B is a 4×4 matrix such that $b_{ij} = a_{ij}$ if $i \leq j$ and $b_{ij} = 0$ otherwise. In both cases of q, the resulting bilinear form has the matrix A. Since determinant of A is nonzero, Q_L is nondegenerate and (L, Q_L) is an $O_4^+(q)$ or $O_4^-(q)$. Now we can find a nondegenerate quadratic form Q_R on R satisfying the following. If n is odd any nondegenerate quadratic form for Q_R works. Suppose n is even. If Q_L is hyperbolic and Q is elliptic (or hyperbolic), select Q_R to be elliptic (or hyperbolic). If Q_L is elliptic and Q is elliptic (or hyperbolic), select Q_R to be hyperbolic (or elliptic). Then $\overline{Q}(x) = Q_L(x_L) + Q_R(x_R)$ is a nondegenerate quadratic form on V and (V, Q) and $(V, \overline{Q})$ are of same type. Since the standard basis is an ordered 4-cap in $(V, \overline{Q})$ which produces A, there must be an ordered 4-cap in (V, Q) that produces A. ∎

THEOREM 4 Consider a nondegenerate orthogonal space over $GF(q)$. Let u be the number of orbits of nondegenerate 4-caps which span 4-dimensional subspaces under the group of isometries. Then u is given below for different cases of q.

1. If q is even then $u = (q^2 + 2)/6$.
2. If $q = 3^e$ then $u = (5q^2 - 21)/24$.
3. If $q \neq 3^e$ and q is odd then $u = (5q^2 - 5)/24$.

Proof: Let C be a nondegenerate 4-cap which spans a 4-dimensional subspace. Fix an order for C and select the corresponding standard matrix A. In dimension 4, Q is nondegenerate implies that f is nondegenerate (i.e. If q is odd this is generally true. If q is even, A is skew symmetric and the rank of A cannot go down by 2.). Hence the determinant of A ($det(A)$) is nonzero. Now consider two nondegenerate 4-caps C_1 and C_2. For fixed orders of C_1 and C_2, select the corresponding standard matrices A_1 and A_2. If A_1 and A_2 are in the same orbit then there is an isometry between the 4-dimensional subspaces spanned by C_1 and C_2. By Witt's Lemma [1, page 121], this isometry can be extended to an isometry in the space and hence C_1 and C_2 are in the same orbit. By Lemma 3, for any standard matrix with nonzero determinant, there exists a 4-cap that produces the matrix. Thus we need to select the orbits which contain matrices with nonzero determinants in Theorem 2.

We check the cases in Theorem 2. Consider case 1 where $q = 2^{2e-1}$. Assume $A \leftrightarrow (1, c, d)$. So $det(A) = (1 + c + d)^2$. Only 1(c) has degenerate orbits when $c+d+1 = 0$. Thus $u = 1+(q-2)+(q-2)(q-4)/6$. Consider case 2 where $q = 2^{2e}$. 2(a), 2(b), 2(c) are same as case 1. We need to check 2(d). Here $A \leftrightarrow (1, c, d)$, $c^2 = d$ and $cd = 1$. Since $c(c+d+1) = (c+d+1)$ and $c \neq 0, 1$, $c+d+1 = 0$. Thus we exclude 2(d). So u has the same value as in case 1.

When q is odd, assume $A \leftrightarrow \lambda(1, a^2, b^2)$. So $det(A) = \lambda^2[(a+b)^2 - 1][(a-b)^2 - 1]$. Assume $q = 3^e$. We check the orbits in case 3. It is not hard to check that 3(a) is degenerate and 3(b), 3(c), 3(d) are nondegenerate. Consider 3(e). $det(A) = 0$ implies that $a^2 = (1 + b)^2$ or $a^2 = (1 - b)^2$. Since $a^2, b^2 \neq 0, 1$ and $a^2 \neq b^2$, for each b^2 there exist two distinct values of a^2 with $det(A) = 0$. Thus $u = (q - 3) + (q - 1)/2 + (q - 1)(q - 3)/8 + (q - 3)(q - 5)/12 - (q - 3)/3$. Assume $q = 3k + 2$. We check the orbits in case 3. It is not hard to check that 3(a), 3(c), 3(d) are nondegenerate. 3(b) is nondegenerate when $q \neq 4$. 3(e) is similar to the case which we have considered above for $q = 3^e$ with the exception that $b^2 = 4$ and $b^2 = 4^{-1}$ produce one degenerate case each. Thus $u = 2 + (q - 5) + (q - 1)/2 + (q - 1)(q - 3)/8 + (q - 3)(q - 5)/12 - (q - 5)/3$. Assume $q = 3k + 1$. We check the orbits in case 4. Here 4(a), 4(b), 4(c), 4(d) and 4(e) are similar to 3(a), 3(b), 3(c), 3(d) and 3(e) when $q = 3k + 2$. 3(f) is similar to 2(d) and hence degenerate. Thus $u = 2 + (q - 5) + (q - 1)/2 + (q - 1)(q - 3)/8 + ((q - 3)(q - 5) - 8)/12 - (q - 7)/3$. ∎

REFERENCES

1. E. Artin, *Geometric Algebra*, J. Wiley & Sons, New York, 1957.
2. M. Aschbacher, *Finite Group Theory*, Cambridge University Press, 1986.
3. W. M. Kantor, "Ovoids and translation planes," *Can. J. Math.* **34** (1982), 1195–1207.
4. G.E. Moorhouse, "Two-graphs and skew two-graphs in finite geometries," *Linear Algebra and Its Applications*, **226–228** (1995), 529–551.

A Remark on a Theorem of T. G. Ostrom

Christoph Hering Universitaet Tuebingen, Mathematisches Institut, Auf der Morgenstelle 10, D-72076 Tuebingen, Germany

Hans-Jörg Schaeffer Universitaet Tuebingen, Mathematisches Institut, Auf der Morgenstelle 10, D-72076 Tuebingen, Germany

Let E be a Klein 4-group contained in the linear translation complement G of a translation plane $\mathfrak{A}$ of finite odd order. Assume that all involutions in E are Baer involutions. By a theorem of Ostrom [1], the dimension of $\mathfrak{A}$ over its kern is divisible by 4, and also, if a and b are two different involutions in E, then b induces a Baer involution on the fixed point subplane of a.

If G does not contain Klein 4-groups of the type described above, then G has cyclic or quaternion Sylow 2-subgroups, unless G contains involutory homologies, in which case the usually well accessible subgroup generated by perspectivities in G will provide much information about $\mathfrak{A}$ and G. Therefore it seems important to know if in the linear translation complement G Klein 4-groups of the type described above can exist at all. To decide this question, we construct the following example.

Let K be a field of order 3 and $F = \{\alpha I + \beta i \mid \alpha, \beta \in K\}$, where

$$I = \begin{pmatrix} 1 & 0 \\ 0 & 1 \end{pmatrix} \quad \text{and} \quad i = \begin{pmatrix} 0 & 1 \\ -1 & 0 \end{pmatrix}.$$

Then F is a subfield of the ring of 2×2-matrices over K. Furthermore, let D be the set consisting of all 8×8-matrices over K of the form

$$\begin{pmatrix} a & 0 & 0 & 0 \\ 0 & a^5 & 0 & 0 \\ 0 & 0 & b & 0 \\ 0 & 0 & 0 & a^4 b \end{pmatrix}$$

where $a, b \in F \backslash \{0\}$ and $b^4 = I$, let t be the matrix

$$\begin{pmatrix} 0 & c & 0 & 0 \\ -c & 0 & 0 & 0 \\ 0 & 0 & 0 & c \\ 0 & 0 & -c & 0 \end{pmatrix}$$

where $c = \left(\begin{smallmatrix} 1 & 1 \\ -1 & 1 \end{smallmatrix}\right)$, and Φ the 8×8-diagonal matrix $\mathrm{diag}(1, -1, 1, -1, 1, -1, 1, -1)$. Finally, denote by G the group generated by D, t and Φ. It turns out that $D \trianglelefteq G$,

$t^2 \in D$, $\langle D, t\rangle \trianglelefteq G$ and $|G| = 128$. Let $V = K^8$ with standard basis $e_1, \ldots, e_8$. We introduce 7 subspaces of V. Let $U_1, \ldots, U_4$ be the row spaces of the matrices

$$\begin{pmatrix} 1 & 0 & 0 & 0 & 0 & 0 & 0 & 0 \\ 0 & 1 & 0 & 0 & 0 & 0 & 0 & 0 \\ 0 & 0 & 0 & 0 & 1 & 0 & 0 & 0 \\ 0 & 0 & 0 & 0 & 0 & 1 & 0 & 0 \end{pmatrix} \quad \begin{pmatrix} 1 & 0 & 1 & 0 & 0 & 0 & 0 & 0 \\ 0 & 1 & 0 & 1 & 0 & 0 & 0 & 0 \\ 0 & 0 & 0 & 0 & 1 & 0 & 1 & 0 \\ 0 & 0 & 0 & 0 & 0 & 1 & 0 & 1 \end{pmatrix}$$

$$\begin{pmatrix} 0 & 1 & 1 & 0 & 0 & 0 & 0 & 0 \\ -1 & 0 & 0 & 1 & 0 & 0 & 0 & 0 \\ 0 & 0 & 0 & 0 & 0 & -1 & 1 & 0 \\ 0 & 0 & 0 & 0 & 1 & 0 & 0 & 1 \end{pmatrix} \quad \begin{pmatrix} 1 & 1 & 1 & 0 & 0 & 0 & 0 & 0 \\ -1 & 1 & 0 & 1 & 0 & 0 & 0 & 0 \\ 0 & 0 & 0 & 0 & -1 & -1 & 1 & 0 \\ 0 & 0 & 0 & 0 & 1 & -1 & 0 & 1 \end{pmatrix}$$

respectively and W_i be the row spaces of the matrices $(X_i I_4)$ for $i = 1, 2, 3$, where I_4 is the 4×4-identity matrix and

$$X_1 = \begin{pmatrix} 1 & 0 & -1 & 0 \\ 0 & -1 & 0 & 1 \\ -1 & 0 & -1 & 0 \\ 0 & 1 & 0 & 1 \end{pmatrix}, X_2 = \begin{pmatrix} -1 & 0 & -1 & 0 \\ 0 & 0 & 0 & -1 \\ -1 & 0 & 1 & 0 \\ 0 & 1 & 0 & 0 \end{pmatrix}, X_3 = \begin{pmatrix} -1 & 1 & -1 & 0 \\ -1 & 0 & 0 & 1 \\ -1 & 0 & 1 & -1 \\ 0 & 1 & 1 & 0 \end{pmatrix}$$

The set $\mathfrak{F}$ consisting of all images of these 7 subspaces $U_1, \ldots, W_3$ under the group G forms a fibration of V and hence determines a translation plane $\mathfrak{T}$ of order 81. (This can be verified by a tiny computer calculation.) By the way, the lengths of the orbits $U_1^G, \ldots, W_3^G$ of G in $\mathfrak{F}$ are 2, 2, 2, 4, 8, 32, and 32 respectively. The collineation group G contains the two elements Φ and

$$u = \begin{pmatrix} I & 0 & 0 & 0 \\ 0 & I & 0 & 0 \\ 0 & 0 & i & 0 \\ 0 & 0 & 0 & i \end{pmatrix},$$

which generate a dihedral group D_8 of order 8. Clearly, this group leaves invariant pointwise the subspace S generated by e_1 and e_3. As both $S \cap U_1$ and $S \cap U_2$ are non-trivial, this subspace is a subplane of order 3. Thus all involutions in D_8 are Baer involutions. Furthermore, it is easy to see that the kern of $\mathfrak{T}$ has order 3. Therefore, both Klein 4-groups in D_8 satisfy the assumption of Ostrom's theorem.

We mention that there also are planar subgroups of G isomorphic to the quaternion group of order 8, e.g., the stabilizer in G on the vector $e_2 + e_3$.

REFERENCE

[1] T.G. Ostrom, Elementary abelian 2-groups in finite translation planes, *Arch. Math. (Basel)* **36** (1981), 21–22.

A Note on the Aschbacher Biplanes of Order 11

YUTAKA HIRAMINE Department of Mathematics, Faculty of Education, Kumamoto University, Kurokami, Kumamoto, Japan

CHIHIRO SUETAKE Amagasaki-minami High School, Amagasaki, Hyogo 660, Japan

Dedicated to Prof. T.G. Ostrom on his 80th birthday

ABSTRACT

Biplanes of order 11 admitting automorphisms of order 5, 11 or 13 are completely classified. The tactical decompositions of biplanes give a crucial computational approach of this problem. Any biplane of order 11 admitting an automorphism of order 5 or 11 is isomorphic to one of the two biplanes constructed by Aschbacher [1] and there are no biplanes of order 11 admitting automorphisms of order 13.

1. INTRODUCTION

A symmetric design $\mathfrak{D}$ with parameters $(v, k, 2)$ is called a biplane of order $k-2$. It is conjectured that there exist (up to isomorphism) only finitely many biplanes (see [3, Conjecture 2.4]). Only for $k = 3, 4, 5, 6, 9, 11$ and 13 are biplanes known. The more the order of biplanes increases, the more the number of biplanes increases for $k \leq 11$ (see [2]). But for $k = 13$ only two biplanes constructed by Aschbacher [1] are known and we want to know whether other biplanes with $k = 13$ exist or not. In [1] he also determined possible prime divisors p of $|\operatorname{Aut} \mathfrak{D}|$. One of his results says that $p \in \{2, 3, 5, 11, 13\}$.

In this note we consider the case that $p \in \{5, 11, 13\}$ and show the following.

THEOREM.

(i). *The only biplanes of order* 11 *that admit automorphisms of order* 5 *or* 11 *are the two biplanes constructed by Aschbacher.*

(ii). *There are no biplanes of order* 11 *that admit automorphisms of order* 13.

2. PRELIMINARIES

For a positive integer n let I_n be the identity matrix of size n and J_n the $n \times n$ all 1 matrix.

PROPOSITION 1. *Let $\mathfrak{D} = (\mathfrak{P}, \mathfrak{B})$ be a symmetric (v,k,λ)-design and G a subgroup of* Aut $\mathfrak{D}$ *of order s. Let $\mathfrak{P}_1, \ldots, \mathfrak{P}_r$ be some G-orbits on $\mathfrak{P}$ and let $\mathfrak{B}_1, \ldots, \mathfrak{B}_r$ be some G-orbits on $\mathfrak{B}$, where G acts regularly on each $\mathfrak{P}_i$ and $\mathfrak{B}_i, 1 \le i \le r$. Choose a block $B_i \in \mathfrak{B}_i$ for each $i, 1 \le i \le r$. Set $m_{ij} = |(B_i) \cap \mathfrak{P}_j|$ for i,j, $1 \le i,j \le r$ and $N = (m_{ij})$. Let $A = \begin{pmatrix} B & B_{12} \\ B_{21} & B_{22} \end{pmatrix}$ be an incidence matrix of $\mathfrak{D}$, where the labelling of points or blocks of B corresponds to the r point-orbits $\mathfrak{P}_i$'s or r block-orbits $\mathfrak{B}_j$'s, respectively. Suppose $B_{12} = {}^tB_{21}$. Then $B\,{}^tB = {}^tBB = \lambda J_{rs} + (k-\lambda) I_{rs} - B_{12}B_{21}$ and the absolute value of any nonzero eigenvalue of N is equal to $\sqrt{|y|}$ for some eigenvalue y of $B\,{}^tB$.*

Proof. Since $A\,{}^tA = {}^tAA = \lambda J_v + (k-\lambda)I_v$ and $B_{12} = {}^tB_{21}$, it follows that $B\,{}^tB = {}^tBB = \lambda J_{rs} + (k-\lambda)I_{rs} - B_{12}B_{21}$. Hence B is a normal matrix. Moreover $BJ_0 = J_0B = N \otimes J_s$ as $\mathfrak{P}_i$'s and $\mathfrak{B}_j$'s are regular orbits on $\mathfrak{P} \cup \mathfrak{B}$. Here $J_0 = I_r \otimes J_s = \begin{pmatrix} J_s & & 0 \\ & J_s & \\ 0 & & \ddots \quad J_s \end{pmatrix}$ and $\otimes$ is the Kronecker product. Thus the eigenvalues of J_0B are $0, \ldots, 0, sx_1, \ldots, sx_r$, where $x_1, \ldots, x_r$ are some eigenvalues of B.

Let $0, \ldots, 0, \lambda_1, \ldots, \lambda_v$ ($\lambda_i \ne 0$, $1 \le i \le v$) be the eigenvalues of N. Then one can easily check that $0, \ldots, 0, s\lambda_1, \ldots,$ and $s\lambda_v$ are the eigenvalues of $N \otimes J_s$. From this, $s\lambda_j \in \{sx_1, \ldots, sx_r\}$ for each j and so $|\lambda_j| \in \{|x_1|, \ldots, |x_r|\}$. Therefore the proposition holds. ∎

In the rest of the article we assume that $\mathfrak{D} = (\mathfrak{P}, \mathfrak{B})$ is a biplane of order 11 that admits an automorphism φ of order $p \in \{5, 11, 13\}$. Set $F(\varphi) = \{X \mid X \in \mathfrak{P} \cup \mathfrak{B}, X\varphi = X\}$ and $G = \langle \varphi \rangle$ $(\simeq Z_p)$.

3. THE CASE $p = 11$

In this section we consider the case that $p = 11$.

STEP 1. $|F(\varphi) \cap \mathfrak{P}| = |F(\varphi) \cap \mathfrak{B}| = 2$.

Proof. Since $\mathfrak{D}$ is a symmetric design, $|F(\varphi) \cap \mathfrak{P}| = |F(\varphi) \cap \mathfrak{B}|$. By Lemmas 2.5 and 2.6 of [1], $F(\varphi)$ is a biplane of order t for some integer $t \le 1$. Therefore $|F(\varphi) \cap \mathfrak{P}| \le 4$. But $|\mathfrak{P}| = 79 \equiv 2 \pmod{11}$. Thus $|F(\varphi) \cap \mathfrak{P}| = 2$. ∎

Let $\mathfrak{P}_1, \ldots, \mathfrak{P}_7$ be the G-orbits on $\mathfrak{P}$ of length 11 and choose an element $P_i \in \mathfrak{P}_i$ for each $i, 1 \le i \le 7$. Similarly we define the G-orbits $\{\mathfrak{B}_i\}_{1 \le i \le 7}$ of length 11 on $\mathfrak{B}$ and blocks $B_i \in \mathfrak{B}_i$; ($|\mathfrak{B}_i| = 11$, $1 \le i \le 7$). Set $F(\varphi) \cap \mathfrak{P} = \{P_8, P_9\}$ and $F(\varphi) \cap \mathfrak{B} = \{B_8, B_9\}$. Here we may assume that $\mathfrak{P}_6 = (B_8) - \{P_8, P_9\}, \mathfrak{P}_7 = (B_9) - \{P_8, P_9\}$ and $\mathfrak{B}_6 = (P_8) - \{B_8, B_9\}, \mathfrak{B}_7 = (P_9) - \{B_8, B_9\}$. Moreover, set $\mathfrak{P}_8 = \{P_8\}, \mathfrak{P}_9 = \{P_9\}, \mathfrak{B}_8 = \{B_8\}, \mathfrak{B}_9 = \{B_9\}$. Then $\mathfrak{P}_i$'s and $\mathfrak{B}_i$'s , $1 \le i \le 9$, are the G-orbits on $\mathfrak{P} \cup \mathfrak{B}$.

Set $m_{ij} = |(B_i) \cap \mathfrak{P}_j|$ for i,j with $1 \le i,j \le 9$ and $M = (m_{ij})$. Clearly m_{ij} depends only on $\mathfrak{B}_i$ and $\mathfrak{P}_i$. We may assume that

$$m_{11} \ge m_{12} \ge m_{13} \ge m_{14} \ge m_{15} \tag{1}$$

Then M has the following form.

STEP 2.

$$M = \begin{pmatrix} * & * & * & * & * & 2 & 2 & 0 & 0 \\ * & * & * & * & * & 2 & 2 & 0 & 0 \\ * & * & * & * & * & 2 & 2 & 0 & 0 \\ * & * & * & * & * & 2 & 2 & 0 & 0 \\ * & * & * & * & * & 2 & 2 & 0 & 0 \\ 2 & 2 & 2 & 2 & 2 & 1 & 1 & 1 & 0 \\ 2 & 2 & 2 & 2 & 2 & 1 & 1 & 0 & 1 \\ 0 & 0 & 0 & 0 & 0 & 11 & 0 & 1 & 1 \\ 0 & 0 & 0 & 0 & 0 & 0 & 11 & 1 & 1 \end{pmatrix}.$$

Proof. It suffices to show that the sixth and seventh rows have the form given in the assertion. The others are clear. We consider the sixth row. By definition $m_{66} = m_{67} = m_{68} = 1$ and $m_{69} = 0$.

Let $j \in \Omega = \{1,2,3,4,5\}$ and set $m = m_{6j}$, $B = B_6$. We define an incidence structure $(\mathfrak{P}^*, \mathfrak{B}^*)$, where $\mathfrak{P}^* = \mathfrak{P}_j, \mathfrak{B}^* = \{B\sigma \cap \mathfrak{P}^* \mid \sigma \in G\}$. Suppose $m \geq 3$. Since $(\mathfrak{P}^*, \mathfrak{B}^*)$ is a tactical configuration and $|\mathfrak{P}^*| = |\mathfrak{B}^*| = 11$, it follows that $m \leq \lambda = 2$. Hence $m_{6j} \leq 2$ for any $j \in \Omega$. On the other hand, $\sum_{1\leq j\leq 5} m_{6j} = 13 - \sum_{6\leq j\leq 9} m_{6j} = 10$. Thus $m_{6j} = 2$ for any $j \in \Omega$. Similarly, $m_{7j} = 2$ for any $j \in \Omega$. ∎

We now define subsets D_{ij}'s $(1 \leq i,j \leq 7)$ by $D_{ij} = \{r \mid P_j r \in (B_i), r \in G\}$. Clearly $|D_{ij}| = m_{ij}$. For a subset H of G we set $\widehat{H} = \sum_{r\in H} r$ $(\in \mathbf{Z}[G])$ and $H^{-1} = \{r^{-1} \mid r \in H\}$.

STEP 3.

$$\sum_{1\leq j\leq 7} \widehat{D_{ij}}\widehat{D_{i'j}^{-1}} = \begin{cases} 13 + 2(\widehat{G} - \{1\}) & \textit{if } 1 \leq i = i' \leq 5, \\ 12 + (\widehat{G} - \{1\}) & \textit{if } 6 \leq i = i' \leq 7, \\ 2\widehat{G} & \textit{if } i \neq i'. \end{cases}$$

Proof. We only show the case $i \neq i'$ (one can obtain the others by a similar argument).

Set $B = B_i$ and $C = B_{i'}$. Let t be any element of G. Then there exist exactly two pairs $(j,s),(k,u) \in \{1,\ldots,7\} \times G$ such that $(Bt^{-1}) \cap (C) = \{P_j s, P_k u\}$. This is equivalent to $s \in D_{i'j}$, $u \in D_{i'k}, st \in D_{ij}$, $ut \in D_{ik}$. Since $t = (st)s^{-1} = (ut)u^{-1}$, we have the assertion when $i \neq i'$. ∎

One can easily check the following fact.

PROPOSITION 2. *Let D_{ij}'s $(1 \leq i,j \leq 7)$ be subsets of G satisfying Step 3 and let P and Q be distinct symbols not contained in G. We define a set of points $\mathfrak{P} = \{(j,r) \mid 1 \leq j \leq 7, r \in G\} \cup \{P,Q\}$, and a set of blocks $\mathfrak{B} = \{B_{i,s} \mid 1 \leq i \leq 7,\ s \in G\} \cup \{B_8, B_9\}$ in the following manner:*

$$B_{i,s} = \begin{cases} \{(j,rs) \mid 1 \leq j \leq 7, r \in D_{ij}\} & \textit{if } 1 \leq i \leq 5, \\ \{(j,rs) \mid 1 \leq j \leq 7, r \in D_{ij}\} \cup \{P\} & \textit{if } i = 6, \\ \{(j,rs) \mid 1 \leq j \leq 7, r \in D_{ij}\} \cup \{Q\} & \textit{if } i = 7, \end{cases}$$

$$B_8 = \{(6,r) \mid r \in G\} \cup \{P,Q\},$$

$$B_9 = \{(7,r) \mid r \in G\} \cup \{P,Q\}.$$

Then $\mathfrak{D} = (\mathfrak{P}, \mathfrak{B})$ *is a biplane of order* 11 *and* G *is an automorphism group of* $\mathfrak{D}$ *of order* 11 *in such a way that* $(j,r)t = (j,rt)$ *and* $Pt = P$, $Qt = Q$ *for* $t \in G$.

STEP 4.

(i). $\sum_{1\le j\le 9} m_{ij} = 13$ *for each* i, $1 \le i \le 7$, *and*

$$\sum_{1\le j\le 7} m_{ij}^2 = \begin{cases} 33 & \text{if } 1 \le i \le 5, \\ 22 & \text{if } 6 \le i \le 7. \end{cases}$$

(ii). $\sum_{1\le j\le 7} m_{ij}m_{i'j} = 22$ *for distinct* $i, i' \in \{1, 2, \dots 7\}$.

Proof. (i) follows immediately from the definition of m_{ij}'s. We get the other equations by considering the action of the trivial character of G on the equations of Step 3. ∎

STEP 5. *Set* $N = (m_{ij})_{1\le i,j\le 7}$. *Then*

$$N = \begin{pmatrix} 4 & 2 & 2 & 1 & 0 & 2 & 2 \\ 0 & 4 & 2 & 2 & 1 & 2 & 2 \\ 1 & 0 & 4 & 2 & 2 & 2 & 2 \\ 2 & 1 & 0 & 4 & 2 & 2 & 2 \\ 2 & 2 & 1 & 0 & 4 & 2 & 2 \\ 2 & 2 & 2 & 2 & 2 & 1 & 1 \\ 2 & 2 & 2 & 2 & 2 & 1 & 1 \end{pmatrix}.$$

Proof. The step follows from Steps 2, 4 and (1) after Step 1. ∎

Proof of Theorem when $p = 11$. To show the theorem, it is sufficient to determine D_{ij}'s satisfying the conditions of Step 3. We identify G with the additive group F^+ of the ring $F = \{0, 1, \dots, 10\}$ (mod 11). Changing generators of F^+ and choice of $P_1 (\in \mathfrak{P}_1)$, if necessary, we may assume that $D_{11} = \{0,1,2,4\}, \{0,1,2,5\}$ or $\{0,1,3,4\}$. We note that this procedure means a suitable choice of a permutation $x \longrightarrow kx + l$ $(k \in F^*, l \in F)$ on F; similarly, considering the permutations $x \longrightarrow x + l$ $(l \in F)$ on F, we may assume that

$$D_{12}, D_{13}, D_{16}, D_{17} \in \{\{0,k\} \mid 1 \le k \le 5\},$$

$$D_{25} = D_{66} = D_{76} = \{0\},$$

and

$$\begin{aligned} D_{ii} \in \{&\{0,1,2,4\}, \{0,1,2,5\}, \{0,1,2,6\}, \{0,1,2,7\}, \{0,1,2,8\}, \\ &\{0,1,2,9\}, \{0,1,3,4\}, \{0,1,3,5\}, \{0,1,3,6\}, \{0,1,3,7\}, \\ &\{0,1,3,8\}, \{0,1,3,9\}, \{0,1,4,5\}, \{0,1,4,6\}, \{0,1,4,7\}, \\ &\{0,1,4,9\}, \{0,1,5,7\}, \{0,1,5,8\}, \{0,1,5,9\}, \{0,1,6,8\}, \\ &\{0,1,6,9\}, \{0,1,7,9\}, \{0,2,4,7\}, \{0,2,4,8\}, \{0,2,5,7\}\} \end{aligned}$$

for each i with $2 \le i \le 5$.

Set $D_{16} = \{0,a\}$ and $D_{17} = \{0,b\}$. By interchanging $\mathfrak{P}_6$ with $\mathfrak{P}_7$, if necessary, we may assume that $a \le b$. Similarly we may assume that $a_1 \le b_1$ for $D_{61} = \{a_1, a_2\}(a_1 < a_2)$ and $D_{71} = \{b_1, b_2\}(b_1 < b_2)$.

Using a computer, we get only two possibilities for $(D_{ij})_{1\le i,j\le 7}$:

$$\begin{pmatrix} L_1 & \{0,2\} & \{0,4\} & \{0\} & \phi & \{0,5\} & \{0,5\} \\ \phi & L_2 & \{1,2\} & \{4,6\} & \{0\} & \{0,8\} & \{4,7\} \\ \{6\} & \phi & L_3 & \{1,7\} & \{4,5\} & \{0,4\} & \{2,9\} \\ \{2,7\} & \{9\} & \phi & L_4 & \{7,10\} & \{0,2\} & \{8,10\} \\ \{3,7\} & \{0,8\} & \{5\} & \phi & L_5 & \{9,10\} & \{1,2\} \\ \{1,10\} & \{4,5\} & \{3,8\} & \{2,5\} & \{2,6\} & \{0\} & \{1\} \\ \{3,8\} & \{3,6\} & \{2,9\} & \{8,10\} & \{9,10\} & \{0\} & \{1\} \end{pmatrix},$$

$$\begin{pmatrix} L_1 & \{0,4\} & \{0,5\} & \{0\} & \phi & \{0,2\} & \{0,5\} \\ \phi & L_5 & \{0,8\} & \{1,2\} & \{0\} & \{6,10\} & \{9,10\} \\ \{2\} & \phi & L_4 & \{3,8\} & \{3,5\} & \{5,8\} & \{7,9\} \\ \{6,10\} & \{8\} & \phi & L_3 & \{4,5\} & \{4,10\} & \{1,5\} \\ \{3,5\} & \{0,8\} & \{10\} & \phi & L_2 & \{8,9\} & \{3,6\} \\ \{3,9\} & \{7,8\} & \{4,6\} & \{1,5\} & \{2,10\} & \{0\} & \{7\} \\ \{4,10\} & \{5,6\} & \{8,10\} & \{4,8\} & \{4,7\} & \{0\} & \{7\} \end{pmatrix}.$$

Here $L_1 = \{0,1,3,4\}$, $L_2 = \{0,1,5,7\}$, $L_3 = \{0,1,3,9\}$, $L_4 = \{0,1,4,5\}$, $L_5 = \{0,2,5,7\}$. This implies that there are at most two biplanes of order 11 that admit automorphisms of order 11. However, as the two biplanes constructed by Aschbacher [1] of order 11 have automorphisms of order 11, this yields Theorem when $p = 11$. ∎

4. THE CASE $p = 5$

In this section we consider the case that $p = 5$. By a similar argument as in the previous section we have the following.

(i). $|F(\varphi)\cap\mathfrak{P}| = |F(\varphi)\cap\mathfrak{B}| = 4$.

(ii). G has 19 point-orbits $\mathfrak{P}_1,\ldots,\mathfrak{P}_{19}$ and 19 block-orbits $\mathfrak{B}_1,\ldots,\mathfrak{B}_{19}$. Choose a point $P_i \in \mathfrak{P}_i$ and a block $B_i \in \mathfrak{B}_i$ for each i, $1 \le i \le 19$. Then we may assume the following :

$$\begin{aligned}
&|\mathfrak{P}_1| = \cdots = |\mathfrak{P}_{15}| = 5,\\
&\mathfrak{P}_{16} = \{P_{16}\},\ldots,\mathfrak{P}_{19} = \{P_{19}\},\\
&|\mathfrak{B}_1| = \cdots = |\mathfrak{B}_{15}| = 5,\\
&\mathfrak{B}_{16} = \{B_{16}\},\ldots,\mathfrak{B}_{19} = \{B_{19}\},\\
&B_{16} = \mathfrak{P}_8 \cup \mathfrak{P}_9 \cup \{P_{17},P_{18},P_{19}\},\\
&B_{17} = \mathfrak{P}_{10} \cup \mathfrak{P}_{11} \cup \{P_{16},P_{18},P_{19}\}\\
&B_{18} = \mathfrak{P}_{12} \cup \mathfrak{P}_{13} \cup \{P_{16},P_{17},P_{19}\},\\
&B_{19} = \mathfrak{P}_{14} \cup \mathfrak{P}_{15} \cup \{P_{16},P_{17},P_{18}\},\\
&\qquad B_8, B_9 \in (P_{16}),\ B_{10}, B_{11} \in (P_{17}),\\
&\qquad B_{12}, B_{13} \in (P_{18}),\ B_{14}, B_{15} \in (P_{19})
\end{aligned}$$

(iii). Set $D_{ij} = \{r \mid P_j r \in (B_i), r \in G\}$ and $m_{ij} = |D_{ij}|$ for i, j with $1 \le i, j \le 15$. Then

$$\sum_{1 \le j \le 15} \widehat{D_{ij}} \widehat{D_{i'j}^{-1}} = \begin{cases} 13 + 2(\widehat{G} - 1) & \text{if } i = i' \text{ and } 1 \le i \le 7, \\ 12 + (\widehat{G} - 1) & \text{if } i = i' \text{ and } 8 \le i \le 15, \\ \widehat{G} & \text{if } (i, i') \in \{(8,9), (10,11), \\ & \qquad (12,13), (14,15)\}, \\ \widehat{2G} & \text{otherwise.} \end{cases}$$

(iv).

$$\sum_{1 \le j \le 15} m_{ij} = \begin{cases} 13 & \text{if } 1 \le i \le 7, \\ 12 & \text{if } 8 \le i \le 15. \end{cases}$$

$$\sum_{1 \le j \le 15} m_{ij} m_{i'j} = \begin{cases} 21 & \text{if } i = i' \text{ and } 1 \le i \le 7, \\ 16 & \text{if } i = i' \text{ and } 8 \le i \le 15, \\ 5 & \text{if } (i, i') \in \{(8,9), (10,11), (12,13), (14,15)\}, \\ 10 & \text{otherwise.} \end{cases}$$

Set $N = (m_{i,j})_{1 \le i,j \le 15}$. There are many solutions N's for the equations in (iv). However, after removing equivalent N's and their transposed by interchanging $\mathfrak{P}_i$'s (or $\mathfrak{B}_i$'s) one obtains four possibilities for N :

$$N_1 = \begin{pmatrix} 1 & 0 & 0 & 0 & 2 & 2 & 0 & 1 & 1 & 0 & 2 & 2 & 0 & 1 & 1 \\ 2 & 1 & 0 & 2 & 0 & 0 & 0 & 1 & 1 & 2 & 0 & 2 & 0 & 1 & 1 \\ 0 & 0 & 1 & 2 & 2 & 0 & 0 & 1 & 1 & 1 & 1 & 0 & 2 & 0 & 2 \\ 0 & 2 & 2 & 1 & 0 & 0 & 0 & 1 & 1 & 0 & 2 & 1 & 1 & 2 & 0 \\ 2 & 0 & 0 & 0 & 1 & 0 & 2 & 1 & 1 & 1 & 1 & 0 & 2 & 2 & 0 \\ 0 & 2 & 0 & 0 & 0 & 1 & 2 & 2 & 0 & 1 & 1 & 1 & 1 & 0 & 2 \\ 0 & 0 & 2 & 0 & 0 & 2 & 1 & 0 & 2 & 2 & 0 & 1 & 1 & 1 & 1 \\ 1 & 1 & 1 & 1 & 1 & 0 & 2 & 0 & 2 & 0 & 1 & 1 & 0 & 0 & 1 \\ 1 & 1 & 1 & 1 & 1 & 2 & 0 & 2 & 0 & 1 & 0 & 0 & 1 & 1 & 0 \\ 0 & 0 & 1 & 2 & 1 & 1 & 2 & 1 & 0 & 1 & 1 & 1 & 0 & 1 & 0 \\ 2 & 2 & 1 & 0 & 1 & 1 & 0 & 0 & 1 & 1 & 1 & 0 & 1 & 0 & 1 \\ 0 & 2 & 0 & 1 & 2 & 1 & 1 & 0 & 1 & 1 & 0 & 1 & 1 & 1 & 0 \\ 2 & 0 & 2 & 1 & 0 & 1 & 1 & 1 & 0 & 0 & 1 & 1 & 1 & 0 & 1 \\ 1 & 1 & 0 & 2 & 0 & 2 & 1 & 0 & 1 & 0 & 1 & 0 & 1 & 1 & 1 \\ 1 & 1 & 2 & 0 & 2 & 0 & 1 & 1 & 0 & 1 & 0 & 1 & 0 & 1 & 1 \end{pmatrix},$$

$$N_2 = \begin{pmatrix} 1&0&2&0&0&2&0&1&1&0&2&2&0&1&1\\ 2&1&0&0&2&0&0&1&1&2&0&2&0&1&1\\ 2&0&1&0&0&0&2&1&1&1&1&0&2&2&0\\ 0&0&0&1&0&2&2&1&1&2&0&1&1&0&2\\ 0&2&2&0&1&0&0&1&1&1&1&0&2&0&2\\ 0&2&0&2&0&1&0&0&2&1&1&1&1&2&0\\ 0&0&0&2&2&0&1&2&0&0&2&1&1&1&1\\ 1&1&1&1&1&0&2&0&2&0&1&1&0&0&1\\ 1&1&1&1&1&2&0&2&0&1&0&0&1&1&0\\ 0&2&1&0&1&1&2&1&0&1&1&1&0&1&0\\ 2&0&1&2&1&1&0&0&1&1&1&0&1&0&1\\ 0&0&2&1&2&1&1&0&1&1&0&1&1&1&0\\ 2&2&0&1&0&1&1&1&0&0&1&1&1&0&1\\ 1&1&0&0&2&2&1&0&1&0&1&0&1&1&1\\ 1&1&2&2&0&0&1&1&0&1&0&1&0&1&1 \end{pmatrix},$$

$$N_3 = \begin{pmatrix} 0&1&1&0&0&0&3&1&1&2&0&1&1&1&1\\ 1&1&1&2&0&0&0&0&2&1&1&0&2&2&0\\ 1&1&0&2&0&0&1&1&1&0&2&2&0&0&2\\ 1&0&1&0&0&2&1&2&0&0&2&1&1&2&0\\ 2&0&0&1&2&0&0&2&0&2&0&1&1&1&1\\ 0&2&0&0&1&2&0&1&1&1&1&0&2&0&2\\ 0&0&2&0&2&1&0&0&2&1&1&2&0&1&1\\ 0&1&0&2&1&2&1&1&1&1&0&1&0&1&0\\ 2&1&2&0&1&0&1&1&1&0&1&0&1&0&1\\ 0&0&2&2&1&1&1&1&0&1&1&0&1&0&1\\ 2&2&0&0&1&1&1&0&1&1&1&1&0&1&0\\ 0&2&1&1&2&0&1&1&0&0&1&1&1&1&0\\ 2&0&1&1&0&2&1&0&1&1&0&1&1&0&1\\ 1&0&0&1&2&1&2&0&1&0&1&0&1&1&1\\ 1&2&2&1&0&1&0&1&0&1&0&1&0&1&1 \end{pmatrix},$$

$$N_4 = \begin{pmatrix} 1&0&1&0&0&0&3&1&1&0&2&1&1&1&1\\ 1&1&1&2&0&0&0&0&2&1&1&0&2&2&0\\ 1&1&0&0&0&2&1&2&0&2&0&0&2&1&1\\ 0&1&1&0&2&0&1&0&2&2&0&1&1&0&2\\ 0&2&0&1&2&0&0&2&0&0&2&1&1&1&1\\ 0&0&2&0&1&2&0&1&1&1&1&2&0&2&0\\ 2&0&0&2&0&1&0&1&1&1&1&2&0&0&2\\ 0&1&0&2&1&1&2&1&1&1&0&1&0&1&0\\ 2&1&2&0&1&1&0&1&1&0&1&0&1&0&1\\ 0&0&2&2&1&1&1&1&0&1&1&0&1&0&1\\ 2&2&0&0&1&1&1&0&1&1&1&1&0&1&0\\ 0&2&1&1&0&2&1&0&1&0&1&1&1&0&1\\ 2&0&1&1&2&0&1&1&0&1&0&1&1&1&0\\ 1&0&0&1&2&2&1&0&1&0&1&0&1&1&1\\ 1&2&2&1&0&0&1&1&0&1&0&1&0&1&1 \end{pmatrix}.$$

We now regard G as the additive group of the ring $F = \{0,1,2,3,4\}$ (mod 5). Moreover, we may assume that $\mathfrak{P}$ and $\mathfrak{B}$ are represented in the following form :

$$\mathfrak{P} = \{(j,r) \mid 1 \le j \le 15, r \in F\} \cup \{P_{16}, P_{17}, P_{18}, P_{19}\},$$

$$B_{i,s} = \begin{cases} \{(j,r+s) \mid 1 \le j \le 15, r \in D_{ij}\} & \text{if } 1 \le i \le 7, \\ \{(j,r+s) \mid 1 \le j \le 15, r \in D_{ij}\} \cup \{P_{i'}\} & \text{if } 8 \le i \le 15, \end{cases}$$

$$\text{where } i' = \begin{cases} 16 & \text{if } i \in \{8,9\}, \\ 17 & \text{if } i \in \{10,11\}, \\ 18 & \text{if } i \in \{12,13\}, \\ 19 & \text{if } i \in \{14,15\}, \end{cases}$$

$$B_{16} = \{(j,r) \mid 8 \le j \le 9, r \in F\} \cup \{P_{17}, P_{18}, P_{19}\},$$
$$B_{17} = \{(j,r) \mid 10 \le j \le 11, r \in F\} \cup \{P_{16}, P_{18}, P_{19}\},$$
$$B_{18} = \{(j,r) \mid 12 \le j \le 13, r \in F\} \cup \{P_{16}, P_{17}, P_{19}\},$$
$$B_{19} = \{(j,r) \mid 14 \le j \le 15, r \in F\} \cup \{P_{16}, P_{17}, P_{18}\}.$$

Clearly $\mathfrak{D} = (\mathfrak{P}, \mathfrak{B})$ admits an automorphism φ of order 5 :

$$(j,r) \to (j,r+1), P_i \to P_i (16 \le i \le 19).$$

Proof of Theorem when $p = 5$. It is sufficient to determine D_{ij}'s, $1 \le i,j \le 15$ satisfying the conditions mentioned in (iii). We state the outline of the proof of the theorem.

CASE 1. $N = N_3$: After labelling rows and columns naturally, the incidence matrix A of $\mathfrak{D}$ satisfies the conditions of Proposition 1:

$$A = \begin{pmatrix} B & B_{12} \\ B_{21} & B_{22} \end{pmatrix},$$

$$B_{21} = {}^tB_{12} = \begin{pmatrix} O_{35} & J & J & O_5 & O_5 & O_5 & O_5 & O_5 & O_5 \\ O_{35} & O_5 & O_5 & J & J & O_5 & O_5 & O_5 & O_5 \\ O_{35} & O_5 & O_5 & O_5 & O_5 & J & J & O_5 & O_5 \\ O_{35} & O_5 & O_5 & O_5 & O_5 & O_5 & O_5 & J & J \end{pmatrix},$$

$$B_{22} = \begin{pmatrix} 0 & 1 & 1 & 1 \\ 1 & 0 & 1 & 1 \\ 1 & 1 & 0 & 1 \\ 1 & 1 & 1 & 0 \end{pmatrix}$$

$$(\text{Here } O_n = \overbrace{(0,\ldots,0)}^{n \text{ times}},\ J = (1,1,1,1,1)),$$
$$BJ_0 = J_0B = N \otimes J_5,$$
$$J_0 = I_{15} \otimes J_5,$$

$$\det(xI_{15} - N) = (x-1)^3(x^2 - 10x - 31)$$
$$\cdot (x^{10} + 3x^8 + 10x^7 - 90x^6 + 148x^5 - 990x^4 + 1210x^3 + 3993x^2 + 161051)$$

Hence, one can verify that some eigenvalue of N, say λ, has the absolute value 3.426482136.... On the other hand, as $B\,{}^tB = {}^tBB = \lambda J_{75} + 11I_{75} - B_{12}B_{21}$, $|\lambda|^2$

is equal to the absolute value of some eigenvalue of ${}^tBB$ by Proposition 1. One can check that this is a contradiction.

CASE 2. $N = N_1$: Considering the permutations $x \to kx + l (k \in F^*, l \in F)$ on F, we may assume that $D_{1,5} = \{0,1\}$. Furthermore, considering the permutations $x \to x + l (l \in F)$ on F, we may assume that

$$\begin{aligned} D_{1,1} &= D_{1,8} = D_{1,9} = D_{1,14} = D_{1,15} = D_{2,1} = D_{3,3} = D_{7,7} \\ &= D_{8,1} = D_{9,1} = D_{10,3} = D_{11,3} = D_{12,4} = D_{13,4} = D_{14,1} = D_{15,1} = \{0\} \end{aligned}$$

and

$$D_{1,6}, D_{1,11}, D_{1,12}, D_{2,1}, D_{2,4}, D_{2,10}, D_{3,4}, D_{3,13}, D_{4,2}, D_{5,1}, D_{5,7}, D_{6,2} \in \{\{0,1\},\{0,2\}\}.$$

Using a computer we get exactly two $(D_{ij})_{1 \le i,j \le 15}$'s. But the two biplanes corresponding to these are isomorphic each other. We write down one of them.

$$(D_{ij})_{1 \le i,j \le 15} = \begin{pmatrix} 0 & \phi & \phi & \phi & S_1 & S_1 & \phi & 0 & 0 & \phi & S_2 & S_2 & \phi & 0 & 0 \\ S_2 & 0 & \phi & S_2 & \phi & \phi & \phi & 4 & 1 & S_1 & \phi & S_1 & \phi & 2 & 3 \\ \phi & \phi & 0 & S_2 & S_6 & \phi & \phi & 3 & 2 & 4 & 1 & \phi & S_1 & \phi & S_4 \\ \phi & S_2 & S_2 & 3 & \phi & \phi & \phi & 0 & 3 & \phi & S_1 & 4 & 4 & S_5 & \phi \\ S_1 & \phi & \phi & \phi & 2 & \phi & S_1 & 3 & 4 & 2 & 0 & \phi & S_6 & S_9 & \phi \\ \phi & S_1 & \phi & \phi & \phi & 4 & S_6 & S_5 & \phi & 4 & 0 & 1 & 2 & \phi & S_2 \\ \phi & \phi & S_8 & \phi & \phi & S_7 & 0 & \phi & S_5 & S_3 & \phi & 4 & 3 & 3 & 0 \\ 0 & 1 & 3 & 2 & 2 & \phi & S_8 & \phi & S_3 & \phi & 3 & 4 & \phi & \phi & 4 \\ 0 & 2 & 0 & 1 & 0 & S_8 & \phi & S_7 & \phi & 4 & \phi & \phi & 3 & 3 & \phi \\ \phi & \phi & 0 & S_4 & 2 & 0 & S_6 & 0 & \phi & 0 & 4 & 3 & \phi & 3 & \phi \\ S_4 & S_7 & 0 & \phi & 3 & 4 & \phi & \phi & 1 & 2 & 2 & \phi & 4 & \phi & 1 \\ \phi & S_8 & \phi & 0 & S_2 & 4 & 2 & \phi & 2 & 2 & \phi & 0 & 2 & 1 & \phi \\ S_7 & \phi & S_8 & 0 & \phi & 3 & 3 & 4 & \phi & \phi & 0 & 2 & 0 & \phi & 1 \\ 0 & 4 & \phi & S_{10} & \phi & S_6 & 3 & \phi & 2 & \phi & 1 & \phi & 2 & 4 & 3 \\ 0 & 3 & S_9 & \phi & S_{10} & \phi & 1 & 1 & \phi & 0 & \phi & 2 & \phi & 1 & 4 \end{pmatrix}.$$

Here $S_1 = \{0,1\}$, $S_2 = \{0,2\}$, $S_3 = \{0,3\}$, $S_4 = \{0,4\}$, $S_5 = \{1,2\}$, $S_6 = \{1,3\}$, $S_7 = \{1,4\}$, $S_8 = \{2,3\}$, $S_9 = \{2,4\}$, $S_{10} = \{3,4\}$.

Since the two biplanes constructed by Aschbacher (they are dual to each other) have automorphisms of order 5, the two biplanes corresponding to N_1 coincide with them. Thus we have the theorem when $p = 5$.

CASE 3. $N = N_2$ or N_4: First suppose $N = N_2$. By a similar argument as in Case 2, we may assume that

$$D_{1,1} = D_{1,8} = D_{1,9} = D_{1,14} = D_{1,15} = D_{2,2} = D_{3,3} = D_{4,4} = \{0\},$$

$$D_{1,6} \cap D_{1,11} \cap D_{1,12} \cap D_{2,1} \cap D_{2,5} \cap D_{2,10} \cap D_{3,7} \cap D_{3,13} \cap D_{4,6} \cap D_{5,2} \cap D_{6,2} = \{0\},$$

$$D_{1,6} \cup D_{1,11} \cup D_{1,12} \cup D_{2,1} \cup D_{2,5} \cup D_{2,10} \cup D_{3,7} \cup D_{3,13} \cup D_{4,6} \cup D_{5,2} \cup D_{6,2} \subset \{0,1,2\}.$$

Then, by computer search, one can verify that there are exactly 28 possibilities for $D_{i,j}$'s, $1 \leq i \leq 6$, $1 \leq j \leq 15$ that satisfy the condition (iii). But, for each of them, one can check that there exist no $D_{7,j}$'s, $1 \leq j \leq 15$ satisfying the condition.

We now suppose $N = N_4$. Similarly we may assume that

$$\begin{aligned} D_{1,1} = D_{1,3} = D_{1,8} = D_{1,9} = D_{1,12} = D_{1,13} = D_{1,14} = D_{1,15} \\ = D_{2,1} = D_{2,2} = D_{2,10} = D_{3,1} = D_{4,2} = D_{5,4} = D_{6,5} = \{0\}, \end{aligned}$$

$$D_{1,7} = \{0, 1, 2\},$$

$$D_{1,11} \cap D_{2,4} \cap D_{3,6} \cap D_{4,5} = \{0\},$$

$$D_{1,11} \cup D_{2,4} \cup D_{3,6} \cup D_{4,5} \subset \{0, 1, 2\}.$$

Then, by computer search, one can verify that there are exactly 62 possibilities for $D_{i,j}$'s, $1 \leq i \leq 6$, $1 \leq j \leq 15$ that satisfy the condition (iii). But, for each of them, one can check that there exist no $D_{7,j}$'s, $1 \leq j \leq 15$ satisfying the condition. ∎

5. THE CASE $p = 13$

In this section we consider the case that $p = 13$. By a similar argument as in Section 3 we have the following.

(i). $|F(\varphi) \cap \mathfrak{P}| = |F(\varphi) \cap \mathfrak{B}| = 1$.

(ii). G has 7 point-orbits $\mathfrak{P}_1, \ldots, \mathfrak{P}_7$ and 7 block-orbits $\mathfrak{B}_1, \ldots, \mathfrak{B}_7$. Let $P_i \in \mathfrak{P}_i$ and $B_i \in \mathfrak{B}_i$ for each $i, 1 \leq i \leq 7$. Then we may assume that :

$$\begin{aligned} |\mathfrak{P}_1| = \cdots = |\mathfrak{P}_6| = 13, \\ \mathfrak{P}_7 = \{P_7\}, \\ |\mathfrak{B}_1| = \cdots = |\mathfrak{B}_6| = 13, \\ \mathfrak{B}_7 = \{B_7\}, \\ \mathfrak{P}_6 = (B_7), \\ \mathfrak{B}_6 = (P_7). \end{aligned}$$

(iii). Set $D_{ij} = \{r \mid P_j r \in (B_i), r \in G\}$ and $m_{ij} = |D_{ij}|$ for $i, j,\ 1 \leq i, j \leq 6$. Then

$$\sum_{1 \leq j \leq 6} \widehat{D_{ij}} \widehat{D_{i'j}^{-1}} = \begin{cases} 13 + 2(\widehat{G} - 1) & \text{if } 1 \leq i = i' \leq 6, \\ 12 + (\widehat{G} - 1) & \text{if } i = i' = 6, \\ \widehat{2G} & \text{if } i \neq i', \end{cases}$$

$$\sum_{1 \leq j \leq 6} m_{ij} m_{i'j} = \begin{cases} 37 & \text{if } 1 \leq i = i' \leq 6, \\ 24 & \text{if } i = i' = 6, \\ 26 & \text{if } i \neq i'. \end{cases}$$

Set $N = (m_{i,j})_{1 \le i,j \le 6}$. After removing equivalent N's and their transposed by suitable replacement of $\mathfrak{P}_i$'s (or $\mathfrak{B}_i$'s) one obtains exactly one possible N:

$$N = \begin{pmatrix} 4 & 3 & 2 & 2 & 0 & 2 \\ 0 & 4 & 3 & 2 & 2 & 2 \\ 2 & 0 & 4 & 3 & 2 & 2 \\ 2 & 2 & 0 & 4 & 3 & 2 \\ 3 & 2 & 2 & 0 & 4 & 2 \\ 2 & 2 & 2 & 2 & 2 & 2 \end{pmatrix}.$$

Proof of Theorem when $p = 13$. Let $A = \left(\begin{smallmatrix} B & B_{12} \\ B_{21} & B_{22} \end{smallmatrix}\right)$ be an incidence matrix of $\mathfrak{D}$. By a similar argument as in the proof of Case 1 in the previous section, we may assume that $B_{21} = (0, \ldots, 0, 1, \ldots, 1)$, $B_{12} = {}^tB_{21}$ and $B_{22} = (0)$. One can check that the eigenvalues of B^tB are 11 and $\frac{165 \pm 13\sqrt{161}}{2}$. By Proposition 1, for any eigenvalue x of N

$$|x| \in \left\{ \sqrt{11}, \sqrt{\frac{131\sqrt{161} \pm 165}{2}} \right\} = \{3.316\ldots, 20.225\ldots, 27.360\ldots\}. \qquad (2)$$

But, in fact, N has an eigenvalue x_0, which is a root of an equation

$$\det(xI - N) = (x^2 - 13x + 2)(x^4 - 9x^3 + 4x^2 - 99x + 121) = 0,$$

and $|x_0| = 2.8429\ldots$. This is contrary to (2). Thus the theorem also holds in this case. ∎

REFERENCES

[1] M. Aschbacher, On collineation groups of symmetric block designs, *J. Combin. Theory* **11** (1971), 272–281.

[2] E.F. Assmus, Jr., Biplanes in planes, *Congr. Numer.* **60** (1987), 279–294.

[3] E.S. Lander, *Symmetric Designs: an Algebraic Approach,* Cambridge University Press, Cambridge, 1983.

Arc Subgroups of Planar Singer Groups

CHAT YIN HO Department of Mathematics, University of Florida, Gainesville, Florida 32611-8105.

1. Introduction.

In this paper, projective planes and groups are of finite cardinalities. A planar *Singer group* is a collineation group acting regularly on the points of a projective plane. We use the term Singer group here to mean planar Singer group. The planar order of a Singer group is the order of the projective plane on which the group acts. The points of a projective plane admitting a Singer group S, can be identified as the elements of S. A subgroup of S is called an *arc subgroup* if it is nontrivial and its points form an arc (i.e., no three points are collinear). At the 1994 April Oberwolfach meeting, Tamás Szőnyi drew my attention to arc subgroups in his talk. Results concerning Desarguesian planes have been obtained by several authors recently. (See, for example, **[12]**,**[10]**, and **[11]**.)

Every non-trivial element s of a Singer group S can be expressed as $s_1 {s_2}^{-1}$, where s_1 and s_2 are two elements uniquely determined by s of a difference set D. Thus for $i = 1$ or 2, $\rho_i(s) = s_i$ defines a function from $S \setminus \{1\}$ to D. We present (in Theorem 2.1 below) a characterization of an arc subgroup in terms of a difference set D of the Singer group S: A subgroup H of S is an arc subgroup if and only if the restriction of one of the functions ρ_1, ρ_2 to $H \setminus \{1\}$ is injective.

Let σ be a multiplier of S. If σ is an involution , then it is known that $[S, \sigma]$ is an arc subgroup. (See, for example, [6].) We prove (in Theorem 3.3 below) that the converse of this is also true. We use this result (in Theorem 3.4 below) to characterize an involutory multiplier of S by its fixed points. In particular, if S is abelian, then $C_S(\sigma)$ is a Baer subplane if and only if σ has order 2.

An arc subgroup has at least three points as the order of a Singer group is odd. A counting argument shows that the cardinality of an arc is at most $n + 2$. We prove (in 4.3 below) that $n - \sqrt{n} + 1$ is an upper bound for an arc subgroup. This bound is attained when the planar order of an abelian Singer group is a square.

In the search for the above bound of an arc subgroup, we found a result concerning the order on a semi-regular collineation group . (A collineation group is semi-regular if the stabilizer of a point in this group is trivial. For example, a subgroup of a Singer group is semi-regular.) We show (in 4.2 below) that this order is outside the open interval $n - \sqrt{n} + 1 < x < n + \sqrt{n} + 1$. This follows from a result (in 4.1 below) on a divisor of $n^2 + n + 1$.

In 4.4, we study semi-regular collineation groups of order $n - i$, for $1 \leq i \leq 7$. We also show the non existence of arc subgroups of a cyclic Singer group of a Desarguesian plane of order 2^5, 3^5, 2^3, 3^3, 5^3.

2. A characterization of an arc-subgroup in terms of a difference set.

Let S be a Singer group with a difference set D. The points and the lines of the projective plane can be identifed with the elements of S such that the point x is incident with the line y if and only if $xy^{-1} \in D$. (See, for example, [8] or [7].)

By the definition of the difference set, every non-trivial element s can be expressed as $s_1 s_2{}^{-1}$ where s_1 and s_2 are two elements of D uniquely determined by s. Thus for $i = 1$ or 2, $\rho_i(s) = s_i$ defines a function from $S \setminus \{1\}$ to D. We use these functions to give a necessary and sufficient condition for a subgroup of S to be an arc-subgroup.

THEOREM 2.1. *A subgroup H of a Singer group S is an arc-subgroup if and only if the restriction of one of the functions ρ_1, ρ_2 to $H \setminus \{1\}$ is injective.*

PROOF: We observe that the function ρ_1 is the composite of the function ρ_2 with the function sending $s \in S$ to s^{-1}. Thus the restriction of ρ_1 to $H \setminus \{1\}$ being injective is equivalent to the restriction of ρ_2 to $H \setminus \{1\}$ being injective. Thus it suffices to prove that H is an arc subgroup if and only if the restriction of ρ_2 to $H \setminus \{1\}$ is injective. Let $\rho = \rho_2$ in the rest of this proof. Any two distinct points x, z of S define uniquely a line $l = l(x, z)$. Thus $xl^{-1} = d_x \in D$ and $zl^{-1} = d_z \in D$. Hence $1 \neq xz^{-1} = d_x l(l^{-1} d_z{}^{-1}) = d_x(d_z)^{-1}$. By the definition of the difference set, d_x, d_z are uniquely determined by the non-trivial element xz^{-1}. Note that $l = d_x{}^{-1}x = d_z{}^{-1}z$. We now apply this to $x = s = s_1 s_2{}^{-1}$, where s_1, $s_2 \in D$, and $z = 1$. From $s1^{-1} = s = s_1 s_2{}^{-1}$, we obtain $d_x = s_1$ and $d_z = s_2$. Therefore the line $l(s, 1) = s_1{}^{-1}s = s_2{}^{-1}$. We put this in the following for future reference.

(a) Suppose $1 \neq s = s_1 s_2{}^{-1}$. Then the line $l(s, 1) = s_1{}^{-1}s = s_2{}^{-1}$.

We divide the proof into two parts.

(1) Assume that H is an arc-subgroup.

Let a, c be two distinct elements of $H \setminus \{1\}$. Multiplication on the right by an element of S preserves the incidence relation, and so this multiplication is a collineation. Suppose $a_2 = c_2 = \beta$, where $a = a_1 a_2{}^{-1}$ and $c = c_1 c_2{}^{-1}$. By (a) the line $l(a, 1)$ is β^{-1} which is equal to the line $l(c, 1)$. Thus 1, a, c are points of H incident with the line β^{-1}. Since H is an arc and $a \neq 1 \neq c$, this implies $a = c$ as desired.

(2) Assume that the restriction of the function ρ to $H \setminus \{1\}$ is injective. We claim the following holds:

(b) Any line incident with the point 1 meets H in at most two points. Let t be a line incident with the point 1.

Let $h = h_1 h_2{}^{-1}$ and $k = k_1 k_2{}^{-1}$ be two points of H different from 1. Suppose h and k are incident with t. By (a), we obtain $t = l(h, 1) = h_2{}^{-1}$ and $t = l(k, 1) = k_2{}^{-1}$. Hence $h_2 = k_2$. So $h = k$ by our assumption that the restriction of the function ρ to $H \setminus \{1\}$ is injective. This establishes our claim.

Right multiplication by elements of H induces a transitive (collinear) action of H on the points of H. The number of points of H incident with a line u is equal to the number of points of H incident with the line uv for any $v \in H$. Using the last two observations we obtain, by (b), that any line meets H in at most two points. This implies that the points of H form an arc. This completes the proof for ρ_2 , and the proof of Theorem 2.1 is complete.

3. When $[S,\sigma]$ is an arc.

We begin by proving the following fact about finite groups.

LEMMA 3.1. *Let H be a finite group and $\sigma \in Aut(H)$. Then $|H| \le |C_H(\sigma)||[H,\sigma]|$. Equality holds if H is abelian.*

PROOF: Let $H = \cup_{i=1}^{h} C_H(\sigma)x_i$. We now prove that for $1 \le i,j \le h$ if $[x_i,\sigma] = [x_j,\sigma]$, then $i = j$. Suppose $[x_i,\sigma] = [x_j,\sigma]$. This implies $x_i{}^\sigma = x_i x_j{}^{-1} x_j{}^\sigma$. Hence $(x_i x_j{}^{-1})^\sigma = x_i{}^\sigma (x_j{}^\sigma)^{-1} = x_i x_j{}^{-1}$. Therefore $x_i x_j{}^{-1} \in C_H(\sigma)$ and $i = j$ as desired. So there are $h = |H : C_H(\sigma)|$ distinct elements $[x_1,\sigma], \cdots, [x_h,\sigma]$ in $[H,\sigma]$. This proves the first conclusion of the lemma.

Suppose H is abelian. For $x, y \in H$, we have $[xy,\sigma] = [x,\sigma]^y[y,\sigma] = [x,\sigma][y,\sigma]$. This shows that $[H,\sigma] = \{[z,\sigma] : z \in H\}$. Since $[C_H(\sigma)x_i,\sigma] = [x_i,\sigma]$ for $1 \le i \le h$, $|\{[z,\sigma] : z \in H\}| \le h$. Thus $|[H,\sigma]| \le h$. The second conclusion of the lemma follows from this and the first conclusion that we have proved.

3.2. Remark. It is possible that $H \ne C_H(\sigma)[H,\sigma]$. For example when H is a cyclic group of order p^2 for a prime p, and σ has order p.

THEOREM 3.3. *Let S be a Singer group, and σ be a multiplier of S. Then $[S,\sigma]$ is an arc if and only if $|\sigma| = 2$.*

PROOF: If $|\sigma| = 2$, then $[S,\sigma]$ is an arc by [5]. In the rest of the proof, assume $A = [S,\sigma]$ is an arc. Let $B = C_S(\sigma)$, and E be the group generated by σ, and n be the planar order of S. We divide the proof in the following five steps.

Step 1. $|A| < n$.

Proof. Since A is an arc, $|A| \le n+2$. Being a subgroup of S, $|A|$ divides $|S| = v(n) = n^2+n+1$ by Lagrange's Theorem. Now $(n+2, v(n))$ divides 3, which is the highest possible power of 3 dividing $v(n)$. This implies that $|A| = n+2$ if and only if $n = 1$. However $n \ge 2$. Hence $|A| \le n+1$. From $(n, v(n)) = 1 = (n+1, v(n))$, we see that $|A| < n$ as claimed.

Step 2. B is a Baer subplane.

Proof. By Lemma 3.1 and Step 1, we obtain $|B| \ge \frac{|S|}{|A|} > \frac{v(n)}{n} = n+1+\frac{1}{n}$. Since $n \ge 2$, $|B| > \frac{v(n)}{n} > 3$. So B is a subplane. Clearly $B \ne S$. Let m be the planar order of B. If $n \ne m^2$, then a result of Bruck [8] implies $n+1 \ge m^2+m+1 = |B|$, which we have shown to be bigger than $n+1+\frac{1}{n}$. This contradiction proves $n = m^2$. So B is a Baer subplane as desired.

Step 3. $A \not\le B$.

Proof. By Step 2, we obtain $|S| = v(m^2) = v(m)v(m-1)$ and $|B| = v(m)$. Since $|A| < n$ by Step 1, So $A \ne B$. Suppose $A \le B$. As $|B|$ is odd, this implies $|B| \ge 3|A|$ by Lagrange's Theorem. By Lemma 3.1, this yields $v(m) = |B| \ge 3|A| \ge 3\frac{|S|}{|B|} = 3v(m-1)$, which leads to $2m \ge m^2+1$. Thus $0 \ge (m-1)^2$. So $m = 1$ and $n = m^2 = 1$, which is impossible. This contradiction proves $A \not\le B$.

Step 4. $|\sigma|$ is a power of 2.

Proof. Let τ be an element of $<\sigma>$ of prime order p. Since $B = C_S(\sigma) \le C_S(\tau)$ and B is a Baer subplane, $B = C_S(\tau)$. By Step 3, there is a point $X \in A \setminus B$. By Step 2, X is incident with a line d of B. Let $W = X^{<\tau>}$. Then $W \le d^{<\tau>} = d$. Since $|\tau| = p$ is a prime, $|W| = 1$ or p. If $|W| = 1$, then $X \in C_S(\tau) = B$, which is impossible. Therefore W is a set of p points of the arc A incident with the line d. This implies $p = |W| \leq 2$. Hence $< \sigma >$ is a 2-group.

Step 5. $|\sigma| = 2$.

Proof. Let τ be the involution of $E =< \sigma >$. Let $Y \in [S, \tau]$. Then τ inverts every element of $[S, \tau]$ (see, for example, [6]), which is an E-invariant subgroup of A. Since E is a cyclic 2-group, this implies that the stabilizer of the point Y in E is 1. Hence $|E| = |Y^E|$. Now Y is incident with a line b of the Baer subplane B. So $Y^E \leq b^E = b$. Thus Y^E is a set of points of the arc A incident with the line b. This implies that $|\sigma| = |E| = |Y^E| \leq 2$. This completes the proof of the theorem.

We now apply Theorem 3.3 to characterize an involutory multiplier by its fixed points.

3.4. THEOREM. *Let S be a Singer group. Then the following holds.*

(1) *Suppose S has an involutory multiplier. Let σ be a multiplier of S. If $C_S(\sigma)$ is a Baer subplane, then $|\sigma| = 2$.*

(2) *If S is abelian , then a multiplier σ is an involution if and only if $C_S(\sigma)$ is a Baer subplane.*

PROOF: (1)Let i be an involutory multiplier of S. Then $S = A \cdot B$, where $B = C_S(i)$ is a Baer subplane, and $A = [S, i]$ is an abelian normal Hall (hence characteristic) arc subgroup of S, and $A \cap B = 1$. (See, [6] Theorem B.) Let $C = C_S(\sigma)$. Since C is a Baer subplane, $|C| = |B|$. So $|C|$ is prime to $|A|$. Thus $C_A(\sigma) = 1$. This implies, by Lemma 3.1, $[A, \sigma] = A$ as A is abelian and characteristic. Now $|AC| = |A||C|/(|A \cap C|) = |A||C| = |A||B|$. Hence $S = AC$. So $[S, \sigma] = [AC, \sigma] = [A, \sigma] = A$. But A is an arc. Therefore $|\sigma| = 2$ by Theorem 3.3.

(2) It is well known that if $|\sigma|$=2, then $C_S(\sigma)$ is a Baer subplane. (See, for example, [5] Lemma 3.1.)

Suppose $C_S(\sigma)$ is a Baer subplane. Then $n = m^2$. Since S is abelian, Hall's multiplier theorem [9] implies that S has an involutory multiplier, namely, the one corresponding to n^3. Hence by (1), $|\sigma| = 2$ as asserted.

3.5 Remark. Theorems 3.4 and 3.4 show that an involutory multiplier σ of an abelian Singer group S can be characterized by either $[S, \sigma]$ or $C_S(\sigma)$.

4. A bound for an arc subgroup.

Suppose the set of points (or lines) of a projective plane of order n is a disjoint union of subsets of equal size e. Then e is a divisor of $v(n) = n^2 + n + 1$. The following proposition shows that e is outside the open interval $(v(\sqrt{n} - 1) = n - \sqrt{n} + 1, v(\sqrt{n}) = n + \sqrt{n} + 1)$.

PROPOSITION 4.1. *Suppose e, n are two positive integers and e divides $n^2 + n + 1$. Then e is either $\leq n - \sqrt{n} + 1$ or $\geq n + \sqrt{n} + 1$. Further the following holds:*

(1) *If $e \leq n$, then $e \leq n - \sqrt{n} + 1$.*

(2) *If $e \geq n + 1$, then $e \geq n + \sqrt{n} + 1$.*

PROOF: Assume $e \leq n$. Let $e = n - k$. Since e divides $v(n)$, $k \geq 1$. From $n \equiv k \ (mod\ e)$, we obtain that $0 \equiv v(n) \equiv v(k) \ (mod\ e)$. Hence e divides $v(k)$. In particular, $k^2 + k + 1 \geq e = n - k$. Thus $n \leq (k+1)^2$. So $\sqrt{n} - 1 \leq k$. Therefore $n - \sqrt{n} + 1 \geq n - k = e$ and (1) holds in this case.

We now assume $e \geq n + 1$. Let $e = n + h$. Then $h \geq 1$ From $n \equiv -h \ (mod\ e)$, we obtain that $0 \equiv v(n) \equiv (-h)^2 - h + 1 \ (mod\ e)$. Hence e divides $h^2 - h + 1$. In particular, $h^2 - h + 1 \geq e = n + h$ Thus $n \leq (h-1)^2$. So $\sqrt{n} + 1 \leq h$. Therefore $e = n + h \geq n + \sqrt{n} + 1$ and (2) holds in this case. The proof is now completed.

Each orbit of points of a semi-regular collineation group of a projective plane has the same cardinality, namely, the order of the group. An application of the above proposition yields the following.

COROLLARY 4.2. *The order g of a semi-regular collineation group of a projective plane of order n is either $\leq n - \sqrt{n} + 1$ or $\geq n + \sqrt{n} + 1$. Further the following holds:*

(1) *If $g \leq n$, then $g \leq n - \sqrt{n} + 1$.*
(2) *If $g \geq n + 1$, then $g \geq n + \sqrt{n} + 1$.*

COROLLARY 4.3. *The order of an arc subgroup of a Singer group of planar order n is at most $n - \sqrt{n} + 1$.*

PROOF: Let g be the order of an arc subgroup G of a Singer group of planar order n. Being a subgroup of a Singer group, G is a semi-regular collineation group. Being an arc, $g \leq n + 2$. If $g \geq n + 1$, then, by 4.2, we see $n + 2 \geq g \geq n + \sqrt{n} + 1$. This implies $1 \geq \sqrt{n}$. So $1 \geq n$, a contradiction. This contradiction proves that $g \leq n$. Hence $g \leq n - \sqrt{n} + 1$ by 4.2 as desired.

4.4. Some Remarks. (a) Let g be the order of a semi-regular collineation group of a projective plane of order n. We study the cases $g = n - i$ with $1 \leq i \leq 7$. The congruences in the following are module g.

(1) Suppose $g = n - 1$. Then $0 \equiv v(n) \equiv 1 + 1 + 1$. Hence $g|3$. So $g = 3$ and $n = 4$.

(2) Suppose $g = n - 2$. Then $0 \equiv 2^2 + 2 + 1$. Hence $g = 7$ and $n = 9$.

(3) Suppose $g = n - 3$. Then $0 \equiv 13$. Hence $g = 13$ and $n = 13 + 3 = 16$.

(4) Suppose $g = n - 4$. Then $0 \equiv 21$. Hence $g = 3, 7$ or 21. If $g = 3$, then $n = 7$. If $g = 7$, then $n = 11$. (In the Desarguesian plane of order 11, there is a subgroup of order 7 in the cyclic Singer group. In fact this subgroup is an arc subgroup. See [**10**]. If $g = 21$, then $n = 25$.

(5) Suppose $g = n - 5$. Then $0 \equiv v(5) = 31$. Hence $g = 31$ and $n = 31 + 5 = 36$. Note that a Singer group of planar order 36 cannot exist by observing the possible multipliers 2 and 3.

(6) Suppose $g = n - 6$. Then $0 \equiv v(6) = 43$. Hence $g = 43$ and $n = 49$. In the Desarguesian plane of order 49, there is an arc subgroup of order 43 of the cyclic Singer group. (This has been confirmed by Storme.)

(7) Suppose $g = n - 7$. Then g divides $v(7) = 3 \times 19$. If $g = 3$, then $n = 10$. We know that a projective plane of order 10 does not exist by a computer search by Lam et al.

If $g = 19$, then $n = 19 + 7 = 26$. Note that there is no Singer group of planar order 26.

If $g = 57$, then $n = 57 + 7 = 64 = 8^2$.

In the above, a Desarguesian plane of a possible order has an subgroup of a cyclic Singer group of the corresponding order g.

(b) We now show the non existence of arc subgroups in some Desarguesian planes.

(1) There is no arc subgroup of the cyclic Singer group of a Desarguesian plane of order 2^5 (respectively, 3^5).

Proof. Suppose K is an arc subgroup of a cyclic Singer group of a Desarguesian plane of order 2^5 (respectively, 3^5). There is a multiplier σ of order 5 for S. Thus K is σ-invariant as S is cyclic. Since 5 does not divide $v(n)$ for any positive integer n, $(|K|, 5) = 1$. Hence $K = C_K(\sigma) \times [K, \sigma]$. Let $H = [K, \sigma]$. If $H \neq 1$, then H is an arc-subgroup as K is an arc-subgroup. By 4.3, we obtain $|H| \leq 2^5 - \sqrt{32} + 1 < 32 - 16 + 1 = 29$. (respectively, $3^5 - \sqrt{243} + 1 < 243 - 15 + 1 = 229$). But H has odd order and has an automorphism of order 5. Thus 5 divides $p - 1$ for at least one prime divisor p of $|H|$. This forces $|H| = 11$. However $32 \equiv -1$ $(mod\ 11)$, which implies $|S| = v(32)$ is prime to 11. Thus $|H|$ is also prime to 11. (Respectively, using the fact that $v(243)$ is not divisible by 3, 5, 7,11, 13, 17,19 to eliminate the possible order of H.) This contradiction proves that $H = 1$. Hence the arc-subgroup K is in the subplane $B = C_S(\sigma)$ of planar order 2. Since $|B| = v(2) = 7$ (respectively, $|B| = v(3) = 13$) is a prime and $|K| \geq 3$, $K = B$. However B is a subplane. This contradiction proves that S does not have any arc-subgroup.

(2) There is no arc subgroup of a cyclic Singer group of a Desarguesian plane of order 2^3, 3^3 or 5^3.

Proof. We use the same technique as in (1). Let $h = |H|$ and $n \in \{8, 27, 125\}$. Since $v(n)$ is a prime, it suffices to show that $h = 1$ as in the argument of (1). To prove $h = 1$, this time we have h is odd and 5 does not divide h. Also 3 divides $p - 1$ for at least one of the prime divisor of p of $v(n)$.

Suppose $n = 8$. Then $h \leq 8 - \sqrt{8} + 1 < 8 - 2 + 1 = 5$ by 4.3. The only possiblity is 3 for h. Since 3 does not divide $v(8)$, $h = 1$.

Suppose $n = 27$. Then $h \leq 27 - \sqrt{27} + 1 < 27 - 3 + 1 = 25$. Suppose $n = 125$. Then $h \leq 125 - \sqrt{125} + 1 < 125 - 11 + 1 = 115$. With the additional fact that $v(27)$ and $v(125)$ are not divisible by 3, 5, 7 ,11, 13, the possibilities for $h \neq 1$ are easily eliminated.

REFERENCES

1. L.D. Baumert, "Cyclic difference sets"," Springer Lecture Notes, New York, 1971.
2. P. Dembowski, "Finite Geometries," Springer, New York, 1968.
3. W. Feit and J.G. Thompson, *Solvability of groups of odd order*, Pacific J. of Math. **13** (1963), 755–1029.
4. D. Gorenstein, "Finite groups," Harper and Row, New York, 1968.
5. C.Y. Ho, *On bounds for groups of multipliers of planar difference sets*, J. of Algebra **148** (1992), 325–336.
6. C.Y. Ho, *Projective planes with a regular collineation group and a question about powers of a prime*, J. of Algebra a **154** (1993), 141–151.
7. C. Y. Ho, *Some basic properties of planar Singer groups*, Geom. Ded. **55** (1995), 59–70.
8. D. Hughes and F. Piper, "Projective planes," Springer, New York, 1973.

9. E. Lander, "Symmetric Designs: An Algebraic approach," London Math Soc. Lecture Notes 74, Cambridge U. Press, 1983.
10. L. Storme and H. Van Maldeghem, *Cyclic arcs in PG(2,q)*, J. of Algebraic Combinatorics **3** (1994), 113–128.
11. L. Storme and H. Van Maldeghem, *Primitive arcs in PG(2,q)*, to appear in J. Comb. Theory A.
12. Tamás Szőnyi, *On cyclic caps in Projective spaces*, manuscript.

Structure Theory for Point-Baer and Line-Baer Collineations in Affine Planes

Vikram Jha Mathematics Department, Glasgow-Caledonian University, Cowcaddens Road, Glasgow, Scotland

Norman L. Johnson Mathematics Department, University of Iowa, Iowa City, Iowa 52242

Abstract

It is well-known that the behavior of Baer collineations in finite translation planes mirrors the behaviour of central collineations. A major theme of this paper is to explore the extent to which the Baer collineation theory is analogous to central collineation theory in arbitrary projective planes, particularly in infinite planes. Special attention is given to André's classical theory of central collineations in finite projective planes: to what extent it holds in infinite translation planes, and how far does it apply to collineation groups generated by Baer collineations.

This article is dedicated to Professor T.G. Ostrom on the occasion of his eightieth birthday.

1 Introduction.

In finite affine planes all lines and Baer subplanes have the same cardinality n. Adopting a rather naive position, this suggests that Baer collineations might be somewhat analogous to central collineations — any collineation fixing $\geq n$ affine points is necessarily one of these two types — and that the proven impact of central collineations on the theory of affine planes might be extended to include a parallel theory based on Baer collineations. In particular, André's fundamental theorem on groups of central collineations fixing a line might be reflected in a corresponding theorem for appropriately defined Baer central collineations.

These goals have more or less already been achieved for finite translation planes, and resulted in tools that have a strong influence on the study of translation planes. Thus, the Hering-Ostrom theory for groups generated by elations on finite translation planes has been extended by Foulser, Walker and Dempwolff to encompass Baer collineations in translation planes of [square] order.

This approach has had a major impact on the development of finite translation planes, including, for example, the Foulser-Johnson classification of translation planes of order q^2 that admit $SL(2,q)$,[7, 8]. In particular, a frequently used tool in finite-translation plane theory is the structure theorem for Baer groups having the same fixed plane. This theorem is an exact analogue of André's theorem concerning groups of central collineations with the same axis — valid in any *finite* plane.

Thus the strong influence of Baer-group theory on finite translation planes may largely be ascribed to its resemblance to central collineation theory, which derives part of its power, in the finite case, from André's theorem.

The aim of this paper is to examine these ideas in a more general framework. Our main concerns are motivated by the desire to understand why Baer collineations have such a strong influence in finite translation planes: what is it about Baer collineations of finite translation planes that causes them to behave like central collineations, and to obey an André-type theorem? Related to this is our second theme: to what extent is André's theorem valid in infinite translation planes?

Regarding Baer structure, we introduce the notion of a Baer central collineation — or a Baer perspectivity. Basically, a Baer perspectivity is a collineation σ, of an affine plane π, such that its fixed points form a Baer subplane π_σ, called the *axis* of σ, and such that π admits a point-partition $\mathcal{C}$ such that each $\psi \in \mathcal{C}$ is a σ-invariant Baer subplane of π; $\mathcal{C}$ is a *center* for σ, and σ is called a Baer-elation if $\pi_\sigma \in \mathcal{C}$ and a Baer-homology otherwise.

Several related issues need to be addressed here: whether or when the center is unique, whether a σ might simultaneously be a Baer-elation *and* a Baer-homology, relative to distinct centers, and even how the notion of Baer subplanes ought to be generalized to infinite planes. We first assume the standard definition of a Baer subplane: the affine plane ψ is a Baer subplane, or a *proper* Baer subplane of an affine plane π if every point and line of π meets the projective closure of ψ at least once.

Our main result concerning Baer perspectivities in arbitrary affine planes, finite or infinite, implies the following:

Theorem 1 *Let σ denote a Baer-perspectivity of an affine plane π, with fixed Baer subplane π_σ, its axis. Then*

1. *the center $\mathcal{C}$ of σ is unique and consists of all the σ-invariant Baer subplanes of π.*

2. *σ is a Baer-elation (resp. Baer-homology) if and only if σ induces a translation (resp. dilation) on every σ-invariant Baer subplane.*

The above suggests that Baer-perspectivities are the natural planar analogues of ordinary central collineations. The well-known structure of Baer *collineations* in finite translation planes implies that if σ is any Baer collineation of a finite translation plane π then σ is a Baer perspectivity; this is in fact the motivating observation for this paper. Our main result for translation planes is a generalisation of this theorem to infinite translation planes.

However, in making this generalisation we have been forced to consider a putatively weaker definition of a Baer subplane, what we call *point*-Baer subplanes. An affine subplane ψ of any affine plane π is a *point-Baer* subplane of π if ψ is a Baer subplane in the net of π, defined by the slope-set of ψ: this just means every point of π lies on a line of ψ (hence our terminology). Now we may define point-Baer perspectivities, point-Baer homologies and point-Baer elations as before, but making the substitution Baer$\rightarrow$point-Baer. Our main results, concerning this possibly more general situation, are as follows:

Theorem 2 *Let π be an affine plane and σ a point-Baer perspectivity of π; thus its axis π_σ is a point-Baer subplane and π admits a point-partition by a collection C of σ-invariant point-Baer subplanes. Then the following hold:*

1. *the fixed plane π_σ, i.e. the axis of σ, is a* proper *Baer subplane of π;*
2. *σ is a point-Baer elation (i.e. $\pi_\sigma \in C$) precisely when σ induces a translation on* every *σ-invariant point-Baer affine subplane of π;*
3. *σ is a point-Baer homology (i.e. $\pi_\sigma \notin C$) precisely when σ induces a dilation on* every *σ-invariant point-Baer affine subplane of π, and C consists of all such planes;*
4. *every σ-invariant point-Baer subplane of π has slope-set contained in Δ_σ, the slope-set of π_σ, and if σ is a point-Baer homology then the slope-set of every σ-invariant point-Baer subplane of π has precisely the same slope-set as π_σ.*

Thus even in the point-Baer situation, every Baer-perspectivity σ must have as its axis a *proper* Baer subplane of π, and the concepts of point-Baer homology and point-Baer elation do not overlap: in particular the center of a point-Baer homology is *unique*. What is not clear, is whether this uniqueness extends to point-Baer elations, and whether the central planes are *proper* Baer subplanes of π: this is an open question for point-Baer elations *and* point-Baer homologies. In translation planes, we shall show that all these matters have a satisfying resolution, except the last: when do the central planes of a Baer perspectivity σ have to be proper Baer subplanes in a translation plane π? We show that if σ is any point-Baer *elation* then all its central planes are proper Baer subplanes; this is part of the structure theorem for Baer *collineations* acting on an arbitrary translation plane (c.f., theorem 37):

Theorem 3 *Let π be any translation plane admitting a point-Baer collineation σ. Then π_σ is a point-Baer perspectivity with a unique center C, and every σ-invariant point-Baer subplane of $\pi \neq \pi_\sigma$ lies in C: in fact these planes constitute* all *of C, when σ is a point-Baer homology — otherwise C includes precisely one further plane π_σ (when σ is a point-Baer elation). Moreover in all cases C is a translation orbit and hence all the central planes have precisely the same slope-set as π_σ.*

Of course, one may consider line-Baer perspectivities, line-Baer homologies and line-Baer elations as well. However, here our results are much weaker probably since for line-Baer collineations, one loses the net concepts. In spite of this, we are able to show

Theorem 4 *Let π_σ be the fixed plane of a non-trivial planar collineation σ, acting on a projective plane π. Then the following are pairwise equivalent:*

1. *π_σ is a point-Baer subplane;*
2. *π_σ is a line-Baer subplane;*
3. *π_σ is a proper Baer subplane.*

In translation planes with spreads in $PG(3, K)$ for K a skewfield, any $2-$dimensional $K-$space, in the associated vector space is either a line or a line-Baer subplane. For such translation planes, we show that the theories of Baer, point-Baer and line-Baer collineations are essentially identical.

More generally, a related consequence of the above general results is the following

Corollary 5 *Any point-Baer elation σ of a translation plane π is a proper Baer elation, i.e., all the central planes of σ are proper Baer subplanes of π; so all the σ-invariant point-Baer subplanes of π are proper Baer subplanes of π and have the same slope-set as the axis π_σ.*

It is not clear whether this corollary extends in general to point-Baer homologies. However, it does apply to Baer involutions: hence we have what amounts to a complete structure theorem for Baer involutions (see theorem 41).

Turning to André's theorem for groups of perspectivities, we note that so far this important theorem has received attention only in the finite case. Our analysis shows that the theorem cannot hold in general in infinite planes because it fails in *every* characteristic 0 translation plane. However, by exploiting the analogous structure of Baer-involutions and ordinary involutions, we are able to prove André's theorem for groups generated by [Baer]-involutory homologies that all share the same axis (or coaxis), in the case of translation planes with *prime* characteristic $p > 2$; for $p = 2$ the situation is essentially vaccuous.

One might ask if a weaker version of André might always hold: *given a group of affine perspectivities G sharing a common axis or a common coaxis (but not both), is it always the case that G contains an elation?* Section 3 briefly considers this question. For example we show that this *Little André Condition* holds in all Pappian planes (remember, the proper André fails in characteristic 0 case!), and in many other cases as well: if G contains an involution, or if G is solvable, or if G has a finite normal subgroup. However, it still remains open whether the Little André condition can ever fail.

Section 4 deals with a well-known property of finite translation planes: a Klein group cannot be a group of homologies with the same axis; this elementary fact has had a powerful bearing on finite-translation plane theory, via a trick of T. G. Ostrom, c.f. Kallaher, [17, p163, section 3]. In the section we prove a related result concerning Klein groups acting on arbitrary affine planes. This implies, for example, that in any affine plane admitting a non-trivial translation at most one involutory dilation exists: thus any finite 2-subgroup of dilations[1] is either cyclic or generalized quaternion.

In section 5 we come to the heart of the paper. We develop the basic properties of point-Baer collineations. The putatively stricter notion of an axial-Baer perspectivity is introduced, mainly to prove the structure theorem for point-Baer homologies, indicated earlier. The section ends with a Baer analogue of the 'no-Klein' theorem that was proved for ordinary involutory homologies in section 4.

Section 6 specializes to translation planes. The main structure theorem there shows that in translation planes all point-Baer *collineations* are axial-Baer perspectivities. The section ends with a brief discussion of point-Baer subplanes in (infinite)-spreads: for example it is not known whether the direct-sum of even two proper Baer-subplanes of a spread is the whole space; it is true, for example when the two Baer subplanes define the same net, as a consequence of a result in [16, Johnson and Ostrom].

Section 7, as already indicated, specializes to a version of André's theorem: for groups generated either by ordinary involutory homologies or by Baer involutory homologies. A unified treatment of both cases emphasizes the striking parallels between Baer perspectivities and ordinary perspectivities.

In section 8, line-Baer subplanes in spreads of $PG(3, K)$ and analogous results to the point-Baer situation are obtained.[2]

2 Preliminaries.

We assume familiarity with finite projective planes and related areas, particularly translation planes and their spreads, c.f., Kallaher [17]. If π^ℓ is any affine plane then ℓ will be called the ideal line, or the line at infinity, and the slope of any affine line is the point at which it meets ℓ; the slope-set of a subplane ψ of π consists of all the points on the ideal line that lie in the [projective closure] of ψ. A central collineation of π is a translation (resp. dilation) if it is an elation (resp. homology) with axis the ideal line. A collineation σ of an affine plane π

[1]We sometimes use the term to refer to homologies of an affine plane when the axis is the ideal line

[2]The authors would like to thank Bonnie Hemenover for providing a peaceful atmosphere conducive for this research.

The research for this article was intiated while the first author was visiting the University of Iowa during the summer of 1995.

is called *planar* if its fixed points and lines form a subplane of π, and we denote the fixed plane of σ by π_σ.

By André's characterisation of spreads, any translation plane may be identified with a spread $\pi = (V, \Gamma)$, where V is a vector space, over a [skew] field K, and Γ is a covering of V by a set of pairwise trivially intersecting subspaces of V such that any two distinct members of Γ direct-sum to V. The translation plane associated with V has V as its affine points and the lines are cosets of the members (or 'components') of Γ. The kern homologies of the plane (or the spread π) are the dilations with $\mathbf{O}$ as center.

It is well-known, and easily seen, that the spread $\pi = (V, \Gamma)$, coordinatising a translation plane, may be chosen so that $V = W \oplus W$, and the members of Γ include the subspaces $X = W \oplus \mathbf{O}$, $Y = \mathbf{O} \oplus \mathbf{W}$, and $\mathbf{I} = \{(\mathbf{w}, \mathbf{w}) \mid \mathbf{w} \in \mathbf{W}\}$. Now every component of Γ, other that Y, has the form

$$\text{`}y = xM\text{'} := \{(w, wM) \mid w \in W\} \exists M \in \mathrm{GL(V,+)} \cup \{\mathbf{O}\},$$

and we call such M the slope of the corresponding components.

3 Notes On The Little André Condition.

Let π^ℓ be an affine plane admitting a group of central collineations G such that the members of G all share a common affine axis X. If all $g \in G$ share the same center C then we call G a *stable* group of central collineations, otherwise G is an *unstable* group. Thus a stable group of central collineations is either an elation group (and any elation group is clearly stable — with center $y = X \cap \ell$) or a homology group.

Up to duality, André's theorem on central collineations is essentially concerned with unstable groups of central collineations:

Theorem 6 (André's Theorem For Central Collineations) *Let π^ℓ be an affine plane admitting an unstable group G of central collineations with common affine axis X. Let C_G denote the set of all the centers C_g of the non-trivial homologies $g \in G$. Then if π^ℓ is finite:*
(1) The elations in G form a non-trivial subgroup E; and
(2) E acts transitively on the set C_G.

The proof of André's theorem involves considerable counting, so it is not clear to what extent the theorem applies to infinite planes; it is conceivable, for example, that only (1) holds in all planes. We define such planes as satisfying the little André condition.

Definition 1 (Little André Condition) *An affine plane satisfies the* little André condition *if it does not admit any unstable homology groups.*

Thus the little André condition is equivalent to the assertion that unstable groups of central collineations must contain non-trivial elations (which is part (1) in André's theorem above). Alternatively, the little André condition means that the plane in question does not admit unstable homology groups.

One of the aims of this paper is to show that André's theorem does not extend unconditionally to infinite planes. This is done in the final section, corollary 50, where we show that André's theorem fails for many infinite translation planes including all the Pappian planes of characteristic 0.

The aim of this section is to consider the extent to which the weaker hypothesis, the little André condition, applies to infinite planes. We shall demonstrate that all Pappian planes satisfy the little André theorem. Thus, planes satisfying the little André condition form a strictly larger class of planes than those to which André's theorem applies.

Since failure of the little André condition is equivalent to the existence of unstable homology groups in the plane, the rest of the section is taken up mainly with the analysis of unstable homology groups. The following result shows that unstable homology groups are rather rare: so even among infinite planes it appears to be difficult to find planes that fail to satisfy the little André condition.

Proposition 7 *Let G be an unstable group of homologies acting on an affine plane π. Then*

1. *G contains no non-trivial stable subgroups N that are subnormal in G*[3].
2. *G is non-solvable and has trivial center.*
3. *G contains no element of finite even order.*
4. *All finite subgroups of G are stable.*

Thus G is an infinite non-solvable group containing no element of even order, nor any subnormal subgroup of finite order.

Proof: If G contains a stable subgroup $N \neq \mathbf{I}$ such that N is subnormal in G then G must fix the center of N forcing G to be stable, contradicting our hypothesis. Since all the elements of any abelian group of homologies with a common axis must share the same center, it now follows that G cannot contain a subnormal abelian group and so G must be non-solvable and have trivial center.

Now suppose G contains a finite subgroup of even order. Then G contains an involution α. If α is unique then it is central in G contrary to the above. Now any two distinct involutions α and β in G must fix the center of the homology $\alpha\beta$ since the abelian group $<\alpha\beta>$ is normalized by α and β. Thus the centers of α and β coincide (with that of $\alpha\beta$). Since this applies to any pair of distinct

[3] N is subnormal in G if there is a finite chain of subgroups $G = G_0 \geq G_1 \geq G_2 \dots G_n = N$ such that $G_{i+1} \lhd G_i$ for all i.

involutions in G it follows that the subgroup of G generated by its involutions is stable and normal in G, forcing the contradiction that G is stable. Hence G does not have finite elements of even order. Now as any finite group of G is solvable or has even order elements, it follows that all finite subgroups of G are stable. $\square$

We now consider the little André condition in the context of translation planes. So let $\pi = (V, \Gamma)$ be a spread associated with a translation plane: whose points are thus the elements of the vector space V, and whose lines are the cosets of the members of Γ, the set subspaces corresponding to the lines through the origin. Suppose G is a group of central collineations of π, fixing a component $X \in \Gamma$ elementwise, and not leaving invariant any other component. André's theorem implies, when π is finite, that G contains elations and that the group of all elations in G acts transitively on the set of coaxes of the non-trivial homologies in G: a coaxis of a homology g is its fixed component X_g distinct from X, the axis of g. To identify the elations in G, if any, we shall apply the following standard characterisation of elations acting on spreads.

Let the spread $\pi = (V, \Gamma)$ admit a linear group G of central collineations with axis $X \in \Gamma$, and let $g \to \hat{g}$ define the standard action of $g \in G$ on the quotient vector space V/X. Then g is an elation if and only if $\hat{g}$ is trivial. In particular, G contains an elation if and only if its natural representation on V/X, defined by $g \to \hat{g}$ is not faithful.

Proof: g is an elation iff it fixes each coset of its axis X, and this means $g(v) - v \in X$, for all $v \in V$. $\square$

Consider generalising the proof. If g fixes a proper coset of X, say $v + X$, then g fixes a line parallel to X but distinct from X. So g is clearly an elation. Thus a corollary to the above argument is:

Corollary 8 *Let the spread $\pi = (V, \Gamma)$ admit a linear group G of central collineations with axis $X \in \Gamma$, and let $g \to \hat{g}$ define the standard action of $g \in G$ on the quotient vector space V/X. Then g is an elation if and only if $\hat{g}$ fixes a non-zero element of V/X . In particular, G contains only homologies if and only if its natural representation on V/X, defined by $g \to \hat{g}$ is faithful and semiregular on the non-zero points.*

In the context of a spread θ any perspectivity with axis X has the form:

$$\theta : (x, y) \longrightarrow (xA, xB + y),$$

and hence any component $y = xM$ maps to $y = x(A^{-1}(B + M))$.

Since θ is an elation with axis X iff it acts trivially on the factor space V/X, it follows that θ is an elation precisely when $A = I$. If $A \neq I$ then θ is a homology if its coaxis is of form $y = xF$, for some component F, and now the condition above yields B in terms of A and F, and hence we may write:

$$\theta = \begin{pmatrix} A_\theta & (A_\theta - I)F_\theta \\ O & I \end{pmatrix},$$

and so the entire group G with axis X is elation-free iff the map $\theta \to A_\theta$ is faithful. In particular, the group

$$\mathcal{A} = \{\mathcal{A}_\theta \mid \theta \in \mathcal{G}\}$$

is non-solvable, semiregular and contains neither involutions nor any central elements, nor any finite group, other than homology groups that have the same center. Now if the spread is finite-dimensional over its kern then $\mathcal{A}$ is also a subgroup of some $GL(n, K)$, K a skewfield. So we need to ask if G non-solvable without center, involutions, or abelian normal subgroups can ever be semiregular subgroups of a $GL(n, K)$. In particular, if the spread is Pappian then A_θ are all members of a field and so this case cannot arise. Hence we have established the little André theorem for Pappian planes.

Theorem 9 *Let π be an affine Pappian plane admitting a group of homologies G with a common affine axis but such that G does not fix any other affine line (i.e. the group does not have a common coaxis). Then G contains non-trivial elations.*

4 Ordinary Perspectivities.

Suppose that P is an arbitrary projective plane. Can there exist two distinct involutory homologies with the same axis and coaxis? Can there exist two distinct Baer involutory homologies with the same Baer axis and Baer-coaxis? In this section we consider the case of ordinary involutory homologies, the Baer analogue will be considered in the next section. Although the emphasis here is on the involutory case, we put our results in context by first considering arbitrary homologies (and arbitrary Baer perspectivities in the next section). We begin with an elementary lemma on ordinary homologies.

Lemma 10 *Let Π be a projective plane admitting non-trivial homologies σ and τ that have the same center but have distinct axes.*

Then any non-trivial normal abelian subgroup N of $< \sigma, \tau >$ is an elation group whose axis CQ is the join of the common center C, of σ and τ, to $Q = s \cap t$, the intersection point of the axis s, of σ, with the axis t, of τ.

In particular, E the unique maximal normal subgroup of $< \sigma, \tau >$, consists of the identity and all the elations in $< \sigma, \tau >$, the common axis being CQ.

Proof: $< \sigma, \tau >$ is a group of central collineations with center C, since both σ and τ fixes every line through C. Since N is abelian it fixes the homology axis h of every non-trivial homology subgroup H that it contains, and of course h passes through Q. Since this is only possible if the group of central collineations N is a homology group with center C and axis h we have σ and τ also fixing h. But the only line not through C fixed by σ (resp. τ) is s (resp. t). This contradiction means H is trivial and so N is an elation group, with center C and axis CQ. The remark follows. □

Now we specialize the above to involutory homologies σ and τ: now not only are non-trivial elations guaranteed but we can further conclude that at most one involutory homology in $<\sigma,\tau>$ can have a given line as axis.

Theorem 11 *Let Π be a projective plane. Let σ and τ be involutory homologies with the same center C and distinct axes, say s and t respectively. Then*

(1) The product $\sigma\tau$ is an elation with center the common center C and axis CQ, the line between the center and the intersection $Q = s \cap t$ of the two distinct axes, s and t, of the involutory homologies.

(2) If ℓ is any line not incident with C, then there is at most one involutory homology in Aut π *with center C and axis ℓ. (For example, τ is the only homology of Π with center C and axis t.)*

Proof: $<\sigma\tau>$ is a cyclic abelian normal subgroup (the stem) of the dihedral group $<\sigma,\tau>$, e.g., σ normalizes $<\sigma\tau>$ because

$$\sigma(\sigma\tau)\sigma^{-1} = \tau\sigma = (\sigma\tau)^{-1} \in <\sigma\tau> .$$

Hence by the lemma above $<\sigma\tau>$ is an elation group with axis CQ. Hence (1) follows. To prove (2) it is clearly sufficient to show that if θ is also a homology with the axis t, of τ, then $\theta = \tau$. Repeating the above argument, with θ instead of τ, we have $\sigma\theta$ is also an elation with axis CQ. But now θ and τ are both homologies with center C and axis t that have the same action on a line CQ through C, viz.,

$$\theta|CQ = \sigma|CQ = \tau|CQ.$$

Hence by the semiregularity of homology groups having the same center and axis we have $\theta = \tau$, as required. □

We draw attention to some consequences of the above in the context of a dilation group of an affine plane (finite or infinite) $\mathcal{A}$. A dilation group of $\mathcal{A}$ is by definition a homology group with the ideal line as axis. Taking an affine version of the above theorem dualized, we see that if σ and τ are involutory dilations of $\mathcal{A}$ with different centers then $\mathcal{A}$ cannot admit two distinct dilations with the same center. In particular, a dilation group with a common center either has a unique involution, in which case it is not simple since the involution will be central, or all its finite subgroups will have odd order, and hence be solvable by the Feit-Thompson theorem. Summarising, we have the following result, which we shall require in the next section.

Theorem 12 *Let $\mathcal{A}$ be an affine plane admitting a non-trivial translation. Then any dilation group G of $\mathcal{A}$, with a common center, contains at most one involution. In particular, G cannot be a simple group unless it has infinite order (and has no involutions!)*

We now record another consequence of theorem 11 that applies to translation planes. Although this result is certainly known, see [19], we deduce it because we shall make extensive use of it later on.

Corollary 13 *Let $\mathcal{A}$ be an affine translation plane. Then two distinct involutory homologies of $\mathcal{A}$ cannot share the same axis and center.*

Proof: If the axis is the translation axis we may, for example, appeal to the theorem above. So assume the axis and coaxis are affine lines. Now a translation leaves the center fixed and moves the axis. Now theorem 11 directly yields the desired conclusion. □

5 Baer Perspectivities.

The aim of this section is to introduce the notion of a Baer perspectivity, or a Baer central collineation, and to study its basic properties, in completely arbitrary planes. The notion of a Baer subplane π_0 in a projective plane π is equivalent to the following two properties:

1. Every point of π is on at least one line of π_0; and
2. Every line of π is on at least one point of π_0.

We shall call a subplane of π a *point-Baer* subplane if it satisfies condition *1.* In the finite case, a well-known counting argument shows that any point-Baer subplane is necessarily a Baer subplane, i.e. property 1 implies 2. Similarly, we call a subplane a *line-Baer subplane* if it satisfies the second condition.

We recall that a spread is a dual spread in $PG(3, K)$, for K a skewfield , if and only if each plane of $PG(3, K)$ contains a line of the spread. From Bruen and Fisher [3], we see that not all spreads are dual spreads (also see Barlotti [2]). Furthermore, it is noted in Johnson [14] that a spread is a dual spread if and only if every line which is not an element of the spread becomes a Baer subplane of the associated translation plane. However, such lines become line-Baer subplanes. Hence, there are line-Baer subplanes which are not point-Baer and dualizing, there are pont-Baer subplanes which are not line-Baer subplanes.

It might be pointed out that in the event a net becomes derivable in an affine plane, this does not necessarily imply that the plane is also derivable. In Johnson [14], it is shown that there exist derivable nets which may be embedded in non-derivable planes.

So, even in the study of ordinary Baer collineations such possibilities are forced to our attention. Hence we introduce the following terminology.

Definition 2 *A planar collineation σ of an affine plane π is a* point-Baer *collineation if its fixed plane π_σ is a point-Baer subplane of π. If π_σ is a proper Baer subplane of π we shall call σ a [proper] Baer collineation. Similarly, a planar collineation is a line-Baer collineation if its fixed plane is a line-Baer subplane of π.*

However, the following result shows that it is usually not important to make such distinctions as we show that if a subplane π_0 of a projective plane π is elementwise fixed by a non-trivial collineation σ then π_0 must be a proper Baer subplane of π if it is point-Baer or line-Baer. The result is sharp in the sense that the three concepts do not coincide in arbitrary planes — even translation planes – because of the spreads admitting line-Baer subplanes that are not point-Baer, as mentioned above.

Of course, all this can be stated in terms of affine planes — our main concern — but we have chosen to work projectively to emphasize the dual nature of our work.

Theorem 14 *Let π_σ be the fixed plane of a non-trivial planar collineation σ, acting on a projective plane π. Then the following are pairwise equivalent:*

1. *π_σ is a point-Baer subplane;*
2. *π_σ is a line-Baer subplane;*
3. *π_σ is a proper Baer subplane.*

Proof: We need only check that (2) implies (1), since we can then apply duality, etc. So let π_σ be a line-Baer subplane of a projective plane π. Consider any point $P \in \pi - \pi_\sigma$. Now $P \neq P^\sigma$, so we may define the line $\lambda = PP^\sigma$. Since λ meets π_σ in a point L (by the line-Baer condition), and L is clearly distinct from P and P^σ, we now have three distinct collinear points $\{P, P^\sigma, L\}$ on the line λ. But this means λ is σ-invariant because σ maps P to P^σ and L to L. Hence λ is in π_σ, and so P lies on a line of π_σ. Since this applies to all points of π that are not in π_σ, the plane π_σ is point-Baer, and hence Baer. □

In the above definition, and throughout the paper, a Baer subplane (or 'proper' Baer subplane) has the standard meaning: i.e. a subplane satisfying both the conditions (1) and (2).

Most of this article concerns point-Baer collineations although we do consider line-Baer collineations in spreads of $PG(3, K)$.

There are several other statements concerning Baer subplanes that appear to be plausible even in the infinite case but have apparently not been established. We list two of them in the following 'conjecture', mainly to guard the reader against their use.

Conjecture 15 *Let π be an infinite plane with a point-Baer subplane π_0 whose slope-set is Φ, and suppose ψ is any proper affine subplane of π with slope-set Ψ. Then*

(a) If $\Psi \supseteq \Phi$ and π_0 is a proper Baer subplane of π then ψ is [at least] a point-Baer subplane of π_0;

(b) if Ψ is a proper subset of Φ then ψ cannot be a point-Baer subplane of π.

However, there are two elementary results on finite planes that readily extend to the infinite case, and we shall sometimes invoke them tacitly:

Lemma 16 *Let π be any affine plane and π_0 an affine subplane.*

1. *If π contains a subplane ψ that has the same slope-set as π_0, and the two subplanes share at least two distinct affine points then $\psi = \pi_0$.*
2. *If the subplane π_0 is a point-Baer subplane of π, and ψ is a proper affine subplane of π_0 then ψ cannot be a point-Baer subplane of π.*

Proof: (1) Let A and B be distinct affine points common to the two subplanes. Choose any affine point $C \in \psi$ that does not lie on the line AB. Now CA, CB meet ℓ_∞ at distinct points, say, a, b. But then the lines Aa, Bb are both in π_0, and hence so is C. So all the points of ψ not on AB are in π_0, and this is clearly only possible if $\psi \subseteq \pi_0$. Reversing the argument completes the proof.
(2) Suppose false. Since ψ is a proper subplane of π_0 its slope-set must also be a proper subset of the slope-set of π_0. Hence we may choose a point M on the ideal line of π that is a slope of π_0 but not of ψ. Now choose an affine point P of π_0, and a point Q on the line $p = PM$ such that Q is *not* in π_0. Since p is not in ψ (that would imply $M \in \psi$) Q cannot lie on any line q of ψ for then, as $q \neq p$, we have $Q = q \cap p \in \pi_0$, contradicting the fact that Q is not in π_0. Thus no line of ψ meets Q and so we have the contradiction that ψ is not point-Baer. □

The following result will form the basis for many of our arguments concerning point-Baer subplanes.

Theorem 17 (The Baer Slope Theorem) *Let π be any affine plane admitting a point-Baer collineation σ, with fixed plane π_σ. Suppose π_1 is any affine subplane of π which shares at most one affine point with π_σ and is left invariant by σ. Then the slope-set of π_1 is contained in the slope-set of π_σ.*

Proof: Consider first the case when the σ-invariant affine subplane π_1 shares no affine points with π_σ. Let Z be any affine point of π_1. Then, since π_σ is point-Baer, there exists a line ℓ of π_σ which is incident with Z. Now since $Z \in \ell$ and ℓ is fixed by σ we also have $Z^\sigma \in \ell$ and hence the line $\ell = Z^\sigma Z$ as $Z \neq Z^\sigma$. Let A be the slope of the fixed line $Z^\sigma Z$ — so A is fixed by σ. Note that Z^σ is in π_1 so that ℓ is a line of π_1.

Now take a point $W \in \pi_1$ that does not lie on $Z^\sigma Z$. Hence the σ-invariant line $W^\sigma W$ must meet $Z^\sigma Z$ in A, for otherwise σ would fix an affine point of π_1, where the two lines meet. Thus the line joining every affine point of π_1 to A is fixed by σ and σ fixes no affine point of π_1.

Hence σ induces faithfully a translation of π_1, and so the slope-set of π_1 is contained in the slope-set of π_σ, since this is precisely the set of ideal points of π that are elementwise fixed by σ.

Now consider the remaining case: when π_1 and π_σ share exactly one affine point P. Repeating the above argument, if $Z \in \pi_1$ is distinct from P then we have a line $Z^\sigma Z$ which is σ-invariant and hence meets the ideal line at a a fixed point A of σ, in π_σ and π_1. Since P is the only affine fixed point of σ in π_1, all the fixed lines of type $W^\sigma W$, $W \in \pi_1$, meets $Z^\sigma Z$ either in P or in A. Note that both types of lines, distinct from PA cannot occur, for if p is a σ-invariant line through P and a is a σ-invariant line through A then $p \cap a$ would be an affine fixed point of π_1 distinct from P unless $p = a = PA$. Thus either all the σ-invariant lines of π_1 pass through A or they all pass through P, and the union of all such σ-invariant lines form a cover of the affine points of π_1. Hence σ is a non-trivial perspectivity of π_1 with center P or A. But the latter is a contradiction, since σ fixes the affine point P, so σ is a dilation of π_1 and hence the fixed slopes of π_1 are among the fixed slopes of π_σ as required. □

It appears that not much can be said about point-Baer collineations, or indeed proper Baer collineations, of arbitrary projective planes, even when they are finite. This is in contrast to the special case of finite translation planes, where the existence of point-Baer collineations has a major impact on the structure of the plane, and such collineations display striking analogies with central collineations. It is our thesis that this is mainly due to the fact that in translation planes point-Baer collineations are point-Baer perspectivities in the following basic sense.

Definition 3 *Let π be an affine plane admitting a point-Baer collineation σ. Then σ is a* point-Baer perspectivity *with* axis π_σ *if π admits a partition of its affine points by a collection C of σ-invariant point-Baer subplanes of π such that each central plane different from π_σ meets π_σ in at most one point. C is called the* center *of the Baer perspectivity σ, and the members of C are the* central planes *of σ. If π_σ lies in C then σ is a* point-Baer elation *and is a* point-Baer homology *otherwise.*

So σ is a point-Baer elation precisely when no central planes $\xi \in C$ share any affine point with the axis π_σ. Hence σ is a point-Baer homology if and only if at least some of the central planes $\xi \in C$ are not disjoint from the affine points of the axis π_σ. We call such central planes the **coaxial subplanes** of σ. In the finite case, it will become evident that *all* the central planes of point-Baer homologies are coaxial planes, see theorem 18(e). We first summarize our terminology.

Definition 4 (Coaxis and Coaxial Homologies.) *Let π be an affine plane admitting a point-Baer homology σ with* axis π_σ *and center C. Then a central plane ψ is called a* coaxis *if it shares an affine point Q with π_σ, and the set $\mathcal{H}$ of all coaxes is called the* cocenter *of σ. The (clearly unique) coaxial plane through each affine point $Q \in \pi_\sigma$ is called the Q-*coaxis *of σ. If furthermore all the central planes of the point-Baer homology σ are coaxial then σ is a* coaxial *point-Baer homology.*

It will eventually turn out that any point-Baer homology is coaxial (c.f. corollary 28).

We now list some consequences of the above definitions for arbitrary finite affine planes. One of our main goals in this section is to examine the extent to which these properties generalize to the infinite case.

Theorem 18 *If σ is a point-Baer perspectivity of a finite plane π with axis π_σ and center C then all the following hold.*

(a) π_σ and every plane in the center C are proper Baer subplanes of π.

(b) Every affine σ-invariant point-Baer subplane (i.e., a proper Baer subplane in the finite case) coincides with the axis π_σ or coincides with one of the planes in the center C.

(c) The center C of σ is unique, and consists of all 'point-Baer' (= Baer) subplanes of π that are σ-invariant, but excluding the axis π_σ when σ is a point-Baer homology, and otherwise, when σ is a point-Baer elation, the center includes π_σ. In both situations the slope-set of each of the central planes in C coincides with the slope-set of π_σ.

(d) σ is a point-Baer elation if and only if it induces a translation on every central plane distinct from π_σ.

(e) σ is a point-Baer homology if and only if it induces a homology on every central plane.

(f) The center of σ coincides with its cocenter: so every central plane ψ meets the axis π_σ at a unique affine point Q_ψ and $\psi \mapsto Q_\psi$ is a bijection from C to the affine points π_σ

Proof: (a) is immediate since in the finite case all point-Baer subplanes are proper Baer subplanes. Consider (b). By the Baer-slope theorem, theorem 17, if σ is a point-Baer perspectivity then all the point-Baer subplanes in the center C have their slope-set contained in the slope-set of the axis π_σ. Thus in the finite case the slopes of the central subplanes are precisely the slopes of the axis π_σ. Now if ψ is any σ-invariant Baer subplane distinct from π_σ then it contains an affine point X such that $X^\sigma \neq X$. Again, by the Baer-slope theorem the slopes of ψ coincide with the slopes of π_σ and ψ shares the two affine points X and X^σ with the unique central plane containing X. Hence ψ coincides with the central plane containing X, and (b) follows. (c) is an immediate consequence.

Consider (d) and (e) together. Since any central plane ξ meets π_σ in at most one affine point, and has the same slope-set as σ, σ induces faithfully on each central plane ξ, distinct from the axis π_σ, either a dilation or a translation of order $k > 1$. Now it is impossible for σ to induce (faithfully and simultaneously) a homology on a finite plane ξ_1 of order n while inducing an elation on another plane ξ_2 of the same order n, for then $k = |\sigma|$ would be forced to divide $\gcd(n-1, n) = 1$. So (d) and (e) follow.

Failiure of part (f) would imply that σ induces translations on some central planes and homologies on others, contradicting (d) or (e). □

It is worth stressing that part (c) above implies that in the finite case a non-trivial point-Baer perspectivity cannot simultaneously be a point-Baer elation and a point-Baer homology (relative to different choices for the center). One of our goals here is to show that this also holds in arbitrary infinite planes; in fact we shall see that the whole of part (c) carries over to arbirary infinite planes for point-Baer *homologies.* In the next section we show that for infinite *translation* planes part (c) holds for all point-Baer perspectivities, and that point-Baer elations satisfies the additional requirement that all its central planes, as well as its axis, are *proper* Baer subplanes: the axis must be proper-Baer by theorem 14.

We have seen that in the finite case the point-Baer subplanes ψ that are left-invariant by a point-Baer perspectivity σ have precisely the same slope-set as π_σ. We now show this continues to hold in the infinite case whenever σ induces a homology on ψ.

Theorem 19 *Let π be an affine plane that admits a point-Baer perspectivity σ that leaves invariant a point-Baer subplane ψ, distinct from π_σ. Assume that σ induces a homology on ψ. Then the slopes-set of ψ is precisely the slope-set of π_σ.*

Proof: By the Baer-slope theorem, the homology induced on ψ by σ must have the ideal line as axis. So let F be the unique affine point of ψ that is fixed by σ: note that σ must induce a non-trivial collineation on ψ (apply lemma 16). Now assume, if possible, that M is a slope of the axis π_σ of σ such that $M \notin \psi$. Then the line FM, which is clearly in π_σ, has at least a third affine point $G \in \pi_\sigma$, distinct from F and M. Since ψ is point-Baer, G lies on a line $g \in \psi$, and such a g is unique (otherwise F and $G \in \psi$, so ψ becomes a subplane of π_σ). Thus g is the only line in ψ that meets G. Hence the σ-invariance of G and ψ forces g also to be σ invariant. But now g is a σ-invariant affine line of ψ which does not pass through F (or FG is a line of ψ containing the slope M), and so σ cannot induce a non-trivial homology on ψ with center F and axis the ideal line. The contradiction implies the desired result. □

We shall eventually show, in theorem 26, that the center and axis of any point-Baer homology σ are unique, and all the central planes have precisely the same slope-set as π_σ. The first step is to establish the result for coaxial point-Baer homologies.

Corollary 20 *If σ is a coaxial homology then the slope-set of π_σ is precisely the slope-set of every central plane. Moreover the center of σ is unique, and consists of all σ-invariant point-Baer subplanes distinct from π_σ.*

Proof: By theorem 19 above, the slopes of the central planes coincide with the slopes of π_σ. Now if the center C of σ were not unique there would be a

σ-invariant point-Baer subplane $\psi \notin \mathcal{C}$. Now choose an affine point $X \in \psi$ that is not fixed by σ: we may assume that X exists since clearly we may ignore the case when ψ is a subplane of the axis π_σ — for this contradicts lemma 16. Now consider the σ-invariant plane generated by the slopes of ψ and the two points $\{X, X^\sigma\}$. This is clearly contained in the member ξ of $\mathcal{C}$ that contains $\{X, X^\sigma\}$, and so by lemma 16 we have the contradiction that ψ is not point-Baer unless ψ coincides with ξ. The desired result follows. □

To extend the above corollary to arbitrary point-Baer homologies, we first need to examine the possibility of a point-Baer elation σ having a central plane whose slope-set is not coincident with that of π_σ. We prove that either a central plane ψ has the same slope-set as π_σ or a sort of dual holds. Thus, instead of having a 'full range of points' of π_σ in ψ we might (conceivably) have the dual condition that a 'full pencil' of π_σ contains ψ.

Lemma 21 *Let σ be a point-Baer collineation of an affine plane π such that σ induces a translation on a σ-invariant point-Baer subplane ψ of π. Then either the slopes-set of ψ is precisely the slope-set of π_σ or* both *the following conditions hold:*

(1) π_σ has a unique slope N such that the set $[N]_\sigma$, consisting of all the affine lines of π_σ through N, contains all the points of ψ and each affine point of ψ lies in precisely one line of $[N]_\sigma$; so the pencil $[N]_\sigma$ induces a partition on the points of ψ, and the partition is minimal in the sense that all the lines of $[N]_\sigma$ intersect ψ non-trivially. The point N is in fact the center of the translation that σ induces on ψ.

(2) If K is any slope of π_σ that is not also a slope of ψ then the pencil $[K]_\sigma$, consisting of the π_σ-lines through K, does not contain any affine point of ψ.

Proof: We may assume throughout that there is a slope K of π_σ that does not lie in ψ. Now if a line k of π_σ through K contains two points of ψ then we obtain the contradiction that $K \in \psi$. If such a k contains one affine point P of ψ then P would clearly be fixed by σ and so σ would not be able to induce a translation on ψ — contrary to our hypothesis. Hence (2) holds.

Now let N be the center of the translation that σ induces on ψ. It is immediate that if P is any affine point of ψ then PP^σ is a fixed line of π_σ passing through N. Thus the lines of π_σ through N that meet ψ form a cover of the points of ψ. We are still required to verify that every affine line of π_σ that passes through N must meet ψ in an affine point. So let P be any affine point of π_σ. Now $P \notin \psi$ and so at most one line f of ψ meets P. But by the point-Baer condition on ψ such an f does exist. But the uniqueness of f forces it to be σ-invariant, so f must pass through the center of the translation that σ induces on π_σ, i.e., f lies on N. But since f is σ-invariant it is also a line of π_σ. The desired result follows. □

We shall also need to force the configuration arising in part (1) of lemma 21 above, *without assuming* the existence of any collineations, but using *instead* the configurational assumption that the slope-sets of the planes involved coincide. This will also prove important also when we deal with infinite translation planes. The first step is a general lemma for affine planes.

Lemma 22 *Let α and β denote point-Baer subplanes of an affine plane π such that α and β have the* same *slope-set Δ but are affine-point disjoint. Then there is a point $M \in \Delta$ such that α and β have precisely the same set of lines through M and no other affine line is common to α and β.*

Proof: Let A be any affine point of α. Then since β is a point-Baer subplane of π, one of its affine lines b passes through A. Let M be the slope of b. So $M \in \alpha$ and hence b is a line of α also. Now if a line $c \in \beta$ does not pass through M then it cannot meet α for otherwise the intersection of $c \cap b$ would be a common affine point of both α and β, contrary to our hypothesis. Thus clearly every point of α lies on a unique line of β and this line passes through M. By reversing the argument we see that the lines of β through M coincide with the lines of α through M. □

The following consequence of the last two lemmas will yield the structure of Baer-homologies.

Lemma 23 *Let π_0 be a point-Baer subplane of an affine plane π and let π_1 be another subplane of π that has precisely the same slope-set Δ as π_0, and such that π_1 and π_0 share exactly one affine point O. Suppose ψ is a point-Baer subplane of π with slope-set $\Delta_\psi \subseteq \Delta$ and such that ψ is affine-point disjoint from π_0.*

Assume ψ and π_0 satisfy at least one of the following conditions:

1. *ψ and π_0 have the same slope-set, i.e., $\Delta_\psi = \Delta$;*
2. *if ψ and π_0 have precisely the same set of lines through a point $A \in \Delta$ (the 'dual condition').*

Then ψ meets π_1 in precisely one affine point.

Proof: By lemma 22, (1) $\Rightarrow$ (2). Hence in all cases we may assume that there is a (clearly unique) point $A \in \Delta$ such that the lines of ψ and π_0 through A are the same and the two planes share no other lines.

Choose an affine point $T \in \pi_1$ that does not lie on the line OA. Since ψ is a point-Baer subplane of π, a line $t \in \psi$ passes through T. Now let the intersection of t with Δ be t_Δ. Now as the line t is also clearly in π_1 and distinct from OA, the unique point $P{:=}OA \cap t$ lies in both the planes ψ and π_1. If P is an affine point then it must be the unique intersection of ψ and π_1: because if the two planes shared a second affine point they would coincide (e.g. by lemma 16.1), forcing ψ and π_0 to share the affine point O, contrary to our hypothesis. So assume OA

and t are parallel lines, meeting at $A \in \Delta$. But since π_0 and ψ share all their lines through A, it follows $t(\neq OA)$ lies in π_0 and π_1. But these planes share all their lines through the affine point O and hence cannot share any further affine lines including t. This contradiction implies the result. □

We record the following corollary for future reference.

Corollary 24 *Let π be an affine plane admitting a point-Baer subplane π_0 and another subplane π_1 that shares with π_0 the same slope-set and an affine point O. Suppose π admits a translation τ such that $\psi = \pi_0^{\tau}$ is a plane $\neq \pi_0$. Then ψ meets π_1 in a unique affine point.*

Proof: A well-known elementary argument shows that the image of any subplane π_0 under a non-trivial translation either coincides with π_0 or shares no affine points with it. Now apply the lemma. □

Theorem 25 *If a point-Baer perspectivity σ, of an affine plane π, induces a translation on a σ-invariant point-Baer subplane ψ of π then σ induces a translation on every point-Baer subplane $\neq \pi_\sigma$ that it leaves invariant.*

Proof: To obtain a contradiction, throughout the proof we assume that there exists a point-Baer subplane π_1 on which σ induces a non-trivial homology whose center we denote by O. By theorem 19 we may assume that π_1 and π_σ have precisely the same slope-set, which we denote by Δ. Also by the Baer slope theorem the slope-set ψ_Δ of ψ satisfies the condition $\psi_\Delta \subseteq \Delta$.

If $\psi_\Delta = \Delta$ then corollary 24 implies that ψ meets π_1 in a unique affine point T and, since both these planes are invariant under σ, T is fixed by σ, and hence σ cannot induce a translation on ψ, contrary to hypothesis. So assume that we have a strict containment $\psi_\Delta \subset \Delta$. Now the conditions (1) and (2) of lemma 21 both hold; so all the lines of ψ through one of its slopes $A \in \Delta$ are precisely the lines of π_σ through A. Thus hypothesis 2 of lemma 23 above applies and we again have the contradiction that ψ and π_0 share a σ-invariant affine point. Hence ψ cannot induce a non-trivial homology on any point-Baer subplane of π, and the result follows. □

The above theorem implies that point-Baer homologies and point-Baer elations are disjoint concepts, and that the center (and axis) of point-Baer homologies are unique. We record this explicitly.

Theorem 26 (Structure Of Point-Baer Homologies) *A point-Baer perspectivity σ cannot simultaneously be a point-Baer elation and a point-Baer homology. In fact, σ cannot simultaneously induce a non-trivial homology and a non-trivial elation on distinct σ-invariant point-Baer subplanes.*

If σ is a point-Baer homology it has a unique center and this consists of all the point-Baer subplanes ($\neq \pi_\sigma$) that it leaves invariant; all these planes have precisely the same slope-set as π_σ. In particular, every central plane does *meet the axis π_σ in a unique point.*

Proof: Take the union of the last theorem and corollary 20. □

Corollary 27 *A point-Baer perspectivity σ is a point-Baer elation [homology] if and only if it induces a non-trivial translation [homology] on some and hence every σ-invariant point-Baer subplane ($\neq \pi_\sigma$).*

It should also be stressed that the concept of a 'coaxial' homology is no longer necessary because of the above corollary.

Corollary 28 *Any point-Baer homology is a coaxial homology and vice-versa.*

The above structure theorem for point-Baer homologies shows that they have unique centers and their slope-set coincides with the slope-set of their axis. Unfortunately, it is not clear to us whether either of these properties are universally valid for point-Baer *elations.* However, we shall see that these properties hold when we restrict ourselves to what we call *axial*-Baer perspectivity: this means that all the central planes of the point-Baer perspectivity σ are required to be 'Baer in the net' defined by the axis π_σ. In translation planes, finite or infinite, we shall show that the notion of a point-Baer perspectivity and an axial-Baer perspectivity coincide.

Definition 5 (Axial-Baer Perspectivity.) *Let σ be a point-Baer perspectivity of an affine plane π relative to a center C. Then σ is called an* axial-*Baer perspectivity if each line of π_σ is incident with at least one point of each [projective] plane ψ in the center C. An axial-Baer perspectivity is an* axial-Baer elation *or an* axial-Baer homology *according to whether σ is a point-Baer elation or a point-Baer homology.*

In particular, a point-Baer perspectivity, of an affine plane π, is an axial-Baer perspectivity if all its central planes are proper Baer subplanes of π. In translation planes, finite or infinite, we shall see that all point Baer-elations are of the latter type and hence axial-Baer perspectivities. For point-Baer homologies the situation is somewhat different. In *any* plane, point-Baer homologies are axial, but even in translation planes (of infinite order) it is not clear whether their central planes are proper Baer subplanes of π.

All this follows immediately from what amounts to an equivalent definition of an axial-Baer perspectivity: a point-Baer perspectivity is axial if the slope-sets of its central planes coincide with the slope-set of its axis (explaining our terminology).

Theorem 29 *A point-Baer perspectivity σ, with a center C, is an axial-Baer perspectivity relative to C if and only if the slope-set of all the planes in the center C coincide with the slope-set of the axis π_σ.*

Proof: If the central planes have the slopes of π_σ as their slope-sets then every line of π_σ meets the central planes in one of their slopes. So consider the converse situation: when every central plane, viewed projectively, meets each line of π_σ at least once. Now, by the Baer slope theorem, the converse cannot fail unless some central plane ψ has a slope-set which is strictly contained in F_σ the slope-set of π_σ. By corollary 20 this is impossible when σ induces a homology on ψ, and so we may assume that σ induces faithfully a translation on ψ. But now, if possible, let M be one of the slopes of π_σ that is not included among the slope-set of ψ. So, since we are assuming that σ is an axial-Baer perspectivity, each affine line k of π_σ through M meets ψ in a unique affine point K and σ clearly fixes K. Hence we have the contradiction that σ does not induce a translation on ψ. The desired result folows. □

The above taken together with the structure of point-Baer homologies, theorem 26, implies that:

Corollary 30 *σ is a point-Baer homology if and only if it is an axial-Baer homology; so every axial-Baer homology has a unique center and axis.*

We now extend the corollary to axial-Baer elations:

Theorem 31 *Let π be an affine plane admitting an axial-Baer perspectivity σ. Then the center of σ is unique, and all the central planes have the same slope-set as the axis π_σ.*

Proof: Through any affine point P of $\pi - \pi_\sigma$ let β and γ be respectively the subplanes of π in the centers $\mathcal{B}$ and $\mathcal{C}$ of σ. Now since β and γ are both σ-invariant they have, by theorem 29 above the same slope-set as σ and they additionally share at least two affine points in common, viz., P and its image $\sigma(P)$: but two affine planes with the same slope-set and two affine points in common must coincide. Thus the class of central planes in $\mathcal{B}$ containing at least one point not in the axis π_σ coincides with the class of central planes in $\mathcal{C}$ containing at least one point not in the axis π_σ. The desired result clearly follows. □

Recall that the unique center of a point-Baer homology σ consists of all the point-Baer subplanes that are σ-invariant. The same result is obtained for general axial-Baer perspectivities by slightly modifying the proof above.

Corollary 32 *Let σ be an axial-Baer perspectivity of an affine plane π and suppose ψ is a σ-invariant affine subplane of π, distinct from the axis π_σ. Then the following are pairwise equivalent.*

(a) ψ is a point-Baer subplane of π;

(b) ψ has the same slope-set as σ;

(c) ψ is a plane in the (unique) center of σ.

Proof: By the argument in the theorem above, ψ is contained in some subplane Ψ in the center C of σ. So by lemma 16 ψ is a point-Baer subplane of π only when it coincides with Ψ, and so (a)$\Rightarrow$(b) holds. If ψ satisfies (*b*) then since it has the same slope-set as Ψ and shares two affine points with it (by the argument in the theorem above) we have $\Psi = \psi$. Thus (b) clearly implies (c), and (c) implies (a) by the very definition of an axial-Baer perspectivity. □

Note that the above corollary implies that an *axial*-Baer perspectivity σ has a unique center even when it is regarded as a *point*-Baer perspectivity; the central planes of σ consists of all σ-invariant planes that have the same slope-set as the axis π_σ — with due allowance made for the fact that π_σ is not always in the center. We record this for reference.

Corollary 33 *Let π be an affine plane admitting an axial-Baer perspectivity σ, and let Φ denote the slope-set of π_σ. Also write $C^* = C - \{\pi_\sigma\}$, where C denotes the center of σ. Then C^* consists of all the subplanes of π that are generated by point-sets of type $\{X, X^\sigma\} \cup \Phi$, where $X \neq X^\sigma$, or equivalently all the σ-invariant subplanes $\psi \neq \pi_\sigma$ that have Φ as their slope-set.*

By definition, a point-Baer perspectivity σ, of an affine plane π, is a *proper* Baer-perspectivity if all the central planes of σ are proper Baer subplanes of π; a fortiori σ is an axial Baer collineation. Thus theorem 1 of the introduction follows from corollary 33 above and theorem 26.

The product of two (ordinary) perspectivities with the same center (resp. axis) is again a perspectivity with the same center (resp. axis). To what extent is this true for *Baer* perspectivities? We now show this applies to axial-Baer perspectivities with a common center. In particular, this means that the product of any two point-Baer homologies [hence axial-Baer homologies] with a common center is again a point-Baer-perspectivity [in fact an axial-Baer perspectivity] with the same center.

Theorem 34 *Let σ and τ be axial-Baer perspectivities of an affine plane π such that they both share a common center C and their axes share at least one common affine point Q. Then*

(1) $\sigma\tau$ is also an axial-Baer perspectivity, and its unique center is C.

(2) $\sigma\tau$ is an axial-Baer elation if and only if it acts unfaithfully on the unique central plane π_Q through Q (i.e. on the common Q-Baer coaxis), and now π_Q is the axis of $\sigma\tau$.

Proof: Since (1) easily yields (2), we shall only consider (1). By the characterisation of the axial condition, theorem 29, we may let Φ be the slope-set of all the central planes of σ and τ and also of their axes, π_σ and π_τ. Also note that we are done if π_σ and π_τ share a second affine point $R \neq Q$, for then the plane π_0 generated by $\Phi \cup \{Q, R\}$ satisfies $\pi_0 = \pi_\sigma = \pi_\tau$ and $\sigma\tau$ is an axial Baer-elation, with axis π_0. So assume Q is the only affine point common to π_σ and π_τ.

Now to show $\sigma\tau$ is an axial-Baer collineation, with center $\mathcal{C}$, it is clearly sufficient to show that it fixes some point-Baer subplane π_0 elementwise, and that the slope-set of π_0 is *not* strictly larger than Φ.

The main step is to show that $\sigma\tau$ is a point-Baer collineation. Deny. Then $\sigma\tau$ must induce faithfully a translation or a dilation on each central plane, as they are all point-Baer, and hence must induce a dilation on $\pi_Q \in \mathcal{C}$ — the unique central plane containing Q.

Consider first the case when $\sigma\tau$ induces a translation on a central plane $\psi \in \mathcal{C}$. Now $\sigma\tau$ clearly fixes all the lines of ψ through a slope $F \in \Phi$ and at the same time it fixes all the lines through Q of π_Q. So $\sigma\tau$ certainly fixes at least two affine points in addition to the Baer slope-set Φ. Hence $\sigma\tau$ is a planar collineation of π such that the slopes of $\pi_{\sigma\tau}$ contains the set Φ. On the other hand, if $\sigma\tau$ induces a homology on every central plane the same conclusion follows a fortiori, since $\sigma\tau$ fixes affine points of every central plane. Thus in all cases $\sigma\tau$ is planar.

Next we must establish that the plane $\pi_{\sigma\tau}$ is a point-Baer subplane of π. So we need to show that every affine point X lies on a fixed line of $\sigma\tau$. But X lies in a central plane $\xi \in \mathcal{C}$ and $\sigma\tau$ induces a perspectivity on this plane (as it fixes Φ the common slope-set of all the central planes). But in any plane admitting a perspectivity every point must lie on a fixed line of the perspectivity. Thus X lies on a fixed line of $\sigma\tau$, and hence $\pi_{\sigma\tau}$ is a point-Baer subplane, whose slope-set includes Φ.

This means that $\sigma\tau$ is an axial-Baer perspectivity, as required, provided we can exclude the putative possiblity that the slope-set of the point-Baer subplane $\pi_{\sigma\tau}$ strictly contains Φ (c.f., conjecture 15(b)). Note that by the structure of point-Baer homologies, theorem 26, we may assume further that $\sigma\tau$ is a point-Baer-elation and hence it does not induce a non-trivial homology on any of its central planes in $\mathcal{C}$, c.f. theorem 25.

But since $\sigma\tau$ fixes Q it must induce a homology on the central plane π_Q through Q. Hence we have a contradiction unless $\sigma\tau$ fixes π_Q elementwise. But a central plane is point-Baer and so cannot, by lemma 16.2, be a subplane of any larger plane containing its slope-set Φ. Hence the axis of the point-Baer-elation σ must coincide with π_Q, and so its slope-set can be no larger than Φ. The result follows. □

It will be convenient to restate the first part of the theorem in terms of the group generated by σ and τ:

Corollary 35 *Let σ and τ be axial-Baer perspectivities of an affine plane π such that they both share a common center $\mathcal{C}$ and their axes share at least one common point Q. Then the group $<\sigma,\tau>$ that they generate is a group of axial-Baer perspectivities sharing the common point Q and the common center $\mathcal{C}$.*

We remark briefly on what might be called the dual situation, when two proper axial-Baer homologies have the same axis but distinct centers. In general, it is not even clear whether such a product is a point-Baer perspectivity, so we focus mainly on the abelian case.

Remark 1 *Let π be an affine plane admitting point-Baer homologies σ and τ that have the same axis π_0 $(= \pi_\sigma = \pi_\tau)$. Denote the centers of σ and τ by $\mathcal{C}_\sigma$ and $\mathcal{C}_\tau$ resp. Then*

1. *σ and τ cannot commute unless $\mathcal{C}_\sigma = \mathcal{C}_\tau$;*
2. *If σ and τ have the same center $\mathcal{C}$ $(= \mathcal{C}_\sigma = \mathcal{C}_\tau)$ then the group $\langle \sigma, \tau \rangle$ is a point-Baer homology group, with axis π_0 and center $\mathcal{C}$.*
3. *A commuting set S of point-Baer perspectivities with a common axis, containing at least one non-trivial point-Baer homology σ, generates an abelian group of point-Baer homologies with the same center and axis as σ.*

Proof: If σ and τ commute then σ permutes the central planes of τ and hence must leave invariant every central plane on which τ induces a homology: since exactly one such plane τ_Q passes through any affine point Q of π_τ, and Q is also fixed by σ since this has the same axis as τ. But as all the central planes of any point-Baer homology meets the axis (the 'coaxial property'), σ and τ fix each of the central planes in $\mathcal{C}_\sigma$ and $\mathcal{C}_\tau$. But any point-Baer homology has a unique center, and hence $\mathcal{C}_\sigma = \mathcal{C}_\tau$, yielding (1).

Now (2) follows immediately by appealing again to the the coaxial-property and the uniqueness-of-center-property of all point-Baer homologies. (3) is an immediate consequence 1 and 2. □

We now turn to the Baer analogue of the theorem 11 concerning dihedral groups of homologies with a common center.

Theorem 36 *Let σ and τ be point-Baer involutory homologies of an affine plane π. Suppose σ and τ have the same center $\mathcal{C}$, and also have a common fixed affine point $Q \in \pi_\sigma \cap \pi_\tau$. Then*

(a) $\sigma\tau$ is an axial-Baer elation unless $\pi_\sigma = \pi_\tau$, (in which case $< \sigma, \tau >$ is just a group of axial-Baer homologies, sharing the common center and axis of σ and τ).

(b) Let $\mathcal{G}$ be the full *collineation group of the affine plane π that leaves invariant the axis of τ and also all the planes in the center $\mathcal{C}$ of τ. Then either σ and τ have the same center and axis (i.e. $\mathcal{C}$ and π_τ), or the only involution in the group $\mathcal{G}$ is τ itself; so the finite groups in $\mathcal{G}$ have cyclic or dihedral Sylow 2-subgroups.*

Proof: (a) Since point-Baer homologies are always axial, theorem 34 implies $\sigma\tau$ is an axial-Baer perspectivity with unique center $\mathcal{C}$. So if $\sigma\tau$ is not an axial Baer-elation then $\sigma\tau$ must be an axial-Baer homology whose unique center is also $\mathcal{C}$, and so $\sigma\tau$ fixes a unique point of each affine plane $C \in \mathcal{C}$. But the normality of $\sigma\tau$ in the dihedral group $< \sigma, \tau >$ now means that both σ and τ must fix the

unique center of the homology that $\sigma\tau$ induces on each central plane: but this means that σ and τ also share the same axis, and (a) follows.
(b) Note that $\mathcal{G}$ must fix π_τ elementwise since each central plane of τ meets π_τ at a single affine point; thus $\mathcal{G}$ is a group of axial-Baer perspectivities with the same center and axis as π_τ.

Assume $\pi_\sigma \neq \pi_\tau$, and suppose an involution $\rho \in \mathcal{G}$. Now consider any central plane $\xi \in \mathcal{C} - \{\pi_{\sigma\tau}\}$ — so it meets π_τ in a (unique) affine point G. Since by (a) $\sigma\tau$ is an axial-Baer elation, it must induce a non-trivial translation on ξ, and now by theorem 12, ξ cannot admit distinct involutory dilations with center G. Hence the action of ρ and τ agree on ξ, and since the ξ's cover all the affine points of π in the complement of $\pi_{\sigma\tau}$ we clearly have $\rho = \tau$, as required. □

6 Baer Collineations of Translation Planes.

Up until now, we have been considering point-Baer and axial Baer perspectivities in arbitrary affine planes. We now turn our attention to translation planes. One of the main differences from the general case is that in translation planes we no longer need to explicitly impose a Baer-*perspectivity* type of condition on Baer collineations. Thus we shall show that if σ is any point-Baer *collineation*, of a translation plane π, then σ is an axial-Baer *perspectivity* — its axis π_σ will of course be a *proper* Baer subplane.

it should also be noted that point-Baer elations appear to behave better in translation planes than do point-Baer homologies: reversing the situation in the general case where point-Baer *homologies* are always *axial* Baer homologies. In a translation plane π, not only are the point-Baer elations σ *axial*, as indicated above, but they satisfy what may be a strictly stronger condition: all the central planes of σ, as well as its axis, are proper Baer subplanes of the ambient translation plane π.

Note that in the finite case the distinction between point-Baer subplanes and proper Baer subplanes is vaccuous. Thus the aim here is to extend the existing theory of finite translation planes to infinite translation planes. In particular, part of our argument is based on Walker's work in the finite case [21].

Theorem 37 (Baer-Collineation Structure For Translation Planes) *Let* Π *be a translation plane and let* σ *be a collineation which fixes a point-Baer subplane elementwise. Then* σ *is either a point-Baer homology (and hence an axial-Baer homology) or* σ *is an axial-Baer elation and in this case all the central planes are proper Baer subplanes. In particular, in all cases the axis* π_σ *is a proper Baer subplane of* Π*, and* σ *has a unique center.*

Proof: We shall identify the points of the affine translation plane Π with a vector space π, over the prime field in the kern of Π, and we let Γ be a spread on π whose members and their cosets are precisely the affine lines of the translation plane Π. By the fundamental André characterisation of spreads, [19], we may regard σ as being in $GL(\pi, +)$ and permuting the members of Γ. We let π_0 denote the subspace of the vector space π associated with the points of π_σ, and $\pi_1 := \pi(\sigma - 1)$.

We want to first establish that π_1 is a spread with component set

$$\Gamma_1 = \{F \cap \pi_1 \mid F \in \Gamma_0\},$$

where $\Gamma_0 \subset \Gamma$ denotes the components of π_0, i.e., all the components $\gamma \in \Gamma$ that meet the subspace π_0 non-trivially.

Now consider any affine point T not in π_0. By our point-Baer hypothesis T lies on a line t of π_0, and by the σ-invariance of t we clearly have $t = T\sigma T$. Since $T\sigma T$ is a line of π_0 it must be a translate of a component of π_0. Hence, the difference $T\sigma - T = T(\sigma - 1)$ is in a component of π_0. Thus, the vector subspace $\pi_1 := \pi(\sigma - 1)$ of π is contained in the union of the set of points of π that lie on components of π_0, i.e., π_1 is covered by the partial spread Γ_0, defined by the components of π_0. In fact, every component of Γ_0 meets π_1 non-trivially because if $F \in \Gamma_0$ then $F(\sigma - 1)$ is clearly a non-trivial subspace of F.

Now π_1 is partitioned by the following collection of non-trivial subspaces of π:

$$\Gamma_1 := \{F \cap \pi_1 \mid F \in \Gamma_0\},$$

and so to demonstrate that this defines a spread on π_1 (with component-set precisely those of Γ_0 because each of the intersections $F \cap \pi_1$ above are clearly non-trivial) it is sufficient to establish that if C and F are arbitrary but distinct members of Γ_0 then $C_1 \oplus F_1 = \pi_1$, where $C_1 = C \cap \pi_1$ and $F_1 = F \cap \pi_1$. This holds because since $\pi = C \oplus F$, and σ leaves C and F invariant, we have

$$\pi_1 = \pi(\sigma - 1) = C(\sigma - 1) \oplus F(\sigma - 1) \subseteq C_1 \oplus F_1 \subseteq \pi_1,$$

yielding the desired result. Note that π maps to π_1 under the mapping $\sigma - 1$ so that $\pi_1(\sigma - 1)$ is contained in π_1 which implies that π_1 is $\sigma-$invariant.

Summarizing, we have now established that $\pi_1 := \pi(\sigma - 1)$ is an affine translation subplane of Π that has precisely the same components as the point-Baer subplane π_0.

Hence either $\pi_0 = \pi_1$ or the only affine vector they share is $\mathbf{O}$. We assert that in either case σ is an axial-Baer perspectivity.

First assume that the two subplanes are identical. Then $\pi_0(\sigma - 1) = \pi(\sigma - 1)^2$ is the zero space so that $(\sigma - 1)^2 \equiv 0$. We assert that σ is an axial-Baer elation. To prove this, we need only show that σ leaves each translate of π_0 invariant — since the translates of π_0 are clearly point-Baer subplanes, and partition the affine points of π, and share the same slopes as π_0. This is equivalent to the collineation σ acting trivially on the quotient vector space π/π_0 and this is implied by the fact that $(\sigma - 1)^2 \equiv 0$.

Hence we may assume that the subplanes π_0 and π_1 are distinct and have have trivial intersection. Hence $\pi \geq \pi_1 \oplus \pi_0$. But by corollary 24, for any $t \in \pi - \pi_1$

the plane $t + \pi_0$ meets π_1 at a unique point t_1, and hence $t = t_0 + t_1$, for some $t_0 \in \pi_0$ and so we have:

$$\pi = \pi_1 \oplus \pi_0.$$

Now let $\mathcal{C}$ be the set of subplanes of π that are translates of π_1. Because $\pi_1 \oplus \pi_0 = \pi$,then every plane $\psi \in \mathcal{C}$ is of the form $\pi_1 + p_0$, where $p_0 \in \pi_0$, and so ψ is σ-invariant as π_1 is invariant. Thus $\mathcal{C}$ is a class of σ-invariant subplanes that are translates of π_1, and since distinct cosets of π_1 are pairwise disjoint planes, sharing precisely the same slopes as π_1, we see that the planes ψ partition the affine points of the plane π and all the (affine) planes of $\mathcal{C}$ have precisely the same set of ideal points, viz., those of the point-Baer subplane π_0.

To show that σ is a point-Baer homology (and hence of course an axial-Baer homology) it only remains to check that the planes in $\mathcal{C}$ are point-Baer, or equivalently, that π_1 is point-Baer. Thus we must check that if T is any affine point of π outside π_1 then T lies on a (unique) line of π_1. Since we have already seen that the components of π_0 and π_1 coincide, we may assume without loss of generality that T is not in $\pi_0 \cup \pi_1$.

Now, since $\pi = \pi_0 \oplus \pi_1$, the coset $\pi_0 + T$ must meet the subspace π_1 at a unique point S_1. For example, if $T = S_o + S_1$ for S_i in π_i then $(\pi_o + T) \cap \pi_1$ contains $\{S_1\}$ But since the slope-set of $\pi_0 + T$ is the same as the slope-set of π_0 and of π_1, S_1T meets the ideal line at a point δ of the affine plane π_1, and S_1T shares a second point with π_1, viz., S_1. Hence the line $S_1T \in \pi_1$ and certainly meets T. It follows that π_1, and hence every member of $\mathcal{C}$ is a point-Baer subplane of π.

Thus in all cases σ is an axial-Baer perspectivity and so, by theorem 14, π_0 is a strict Baer subplane of Π. Hence, since the center is the set of translates of π_1, we further have that all the central planes are strictly Baer when $\pi_1 = \pi_0$. This completes the proof of all parts of the theorem. □

For future reference we note the following immediate consequence of the theorem:

Corollary 38 *Let Π be a translation plane admitting a point-Baer collineation σ. Then σ is a point-Baer perspectivity such that the net determined by each of its central planes is precisely the net determined by the slopes of the axis π_σ.*

The structure of Baer-groups in finite translation planes are completely known. Thus if G is such a group, fixing elementwise a Baer subplane π_0, then G has the structure of a Frobenius group with the Baer elations (in our terminology) constituting the Frobenius kernel E, and the complement $G - E$ being partitioned by the maximal Baer homology groups H_i in G. Each H_i corresponds to the global stabilizer of a unique second Baer subplane π_i, through the origin, with the same slope-set as π_0 and H_i is clearly isomorphic to a subgroup of the kern of π_i. Moreover all the H_i are all conjugate in G. With the important exception of this conjugacy property, we may apply the theorem above to show that the situation is fairly similar for infinite translation planes.

Corollary 39 *Let π be a translation plane and π_0 a point-Baer subplane of π. Let G be any group of Baer collineations of π that fixes π_0 elementwise.*

Let $P_i = \{\pi_i \mid \pi_i \text{ in } I\}$ denote the set of distinct subplanes $\pi_i(\neq \pi_0)$ through the origin with the same slope-set as π_0 and satisfying the additional condition that its global stabilizer H_i, in G, is non-trivial: thus each H_i is a non-trivial group of axial-Baer homologies with axis π_0, and its center consists of all the additive cosets of π_o.

Then the following apply:

(1) π_0 is a proper Baer subplane of π.

(2) G is a group of axial-Baer perspectivities with axis π_0.

(3) The subset E of Baer elations in G is a normal subgroup of G; we call this the Frobenius kernel of G, (c.f., (6) below).

(4) The Frobenius kernel E is isomorphic to a translation subgroup of the translation plane π_0, and hence E is elementary abelian in the sense that it is an additive subgroup of a vector space.

(5) The global stabilizers H_i of π_i, for $i \in I$, are pairwise trivially-intersecting maximal subgroups of G such that $G - E = \cup_{i \in I} H_i$. (So each of these maximal axial-Baer homology groups H_i are isomorphic to a multiplicative subgroup of a skewfield in the kern of the plane π_i.)

(6) If $G - E \neq \emptyset$ then the action of G induced on the planes $\mathcal{P}_I$ is such that on each G-orbit in $\mathcal{P}_I$ the stabilizer of any two planes in $\mathcal{P}_I$ is trivial.

Proof: Theorem 37 implies that all Baer collineations are Baer perspectivities, so the first two parts follow. Theorem 37 also implies that the Baer-elations with axis π_0 have as their center precisely the additive cosets of π_0. Thus the product of two Baer elations is again a Baer elation. Hence (3) follows. Now (4) holds because theorem 37 implies that E induces faithfully a translation group on the coset of π_0 (excluding π_0). Parts (5) and (6) follow readily from the observation that a non-trivial $h \in H_i$ cannot leave invariant more than one subplane with the same slope-set as π_0 while sharing with it an affine point $\mathbf{O}$ (c.f., corollary 33). □

In the above result, several questions are left open. First of all, we mentioned the question of the conjugacy of the groups H_i but it may also be the case that there are no other Baer subplanes other than π_σ which share its infinite points and the zero vector and hence the set I could be empty. Or there could be exactly one subplane indexed by I. That is, I could have 0,1 or at least 2 elements.

In sections 8 and 9, we consider the structure of vector space nets which admit point-Baer subplanes. In particular, when there are at least three such point-Baer subplanes of the net which share an affine point, we have the situation above when I has cardinality at least two. We are able to prove that conjugacy exists between the groups mentioned and furthermore are able to completely determine the Baer group when I is non-empty. If I is empty, and we have a particular decomposition of the vector space relative to the point-Baer subplane then we also may completely determine the Baer collineation group.

Recall that if an arbitrary affine plane π admits a point-Baer collineation σ, then π_σ is a *proper* Baer subplane of π. In this section we have seen that if π is a translation plane then all the central planes of σ have the same slope-set as the proper Baer subplane π_σ. However, it appears conceivable that the central planes associated with σ are not necessarily proper Baer subplanes of π, c.f., conjecture 15(a). To discuss this possibility we introduce the following terminology.

Definition 6 *A point-Baer perspectivity of any affine plane π is* proper *if all its central planes are [*proper*] Baer subplanes of π.*

We shall make the convention that an unprefixed 'Baer' always means '*proper* Baer'. In terms of this convention, the structure theorem 37 implies:

Corollary 40 *A point-Baer elation of an affine translation plane is a Baer elation, with unique axis and center.*

The above extends to arbitrary Baer involutions acting on translation planes.

Theorem 41 (Structure of Involutions In Translation Planes) *Let π be a translation plane of characteristic $p \geq 0$. Then any involutory collineation σ is either a central collineation or both the following apply:*

(a) The axis of σ as well as all its central planes are proper Baer subplanes of π; thus σ is a proper Baer perspectivity.

(b) σ is a Baer involutory elation or Baer involutory homology according to whether the characteristic $p = 2$ or $p \neq 2$.

Proof: In general, any involution of an affine or projective plane is either a central collineation or a Baer collineation. Thus we need only consider the Baer situation and now by the structure theorem for Baer collineations σ is an axial Baer perspectivity, and so π_σ is a proper Baer subplane of π, i.e., a Baer elation or an axial-Baer homology. We first establish part (b).
The proof of theorem 37 shows that when σ is an axial-Baer homology then at at least one central plane ψ of σ is a σ-invariant complement of the axis π_σ. Now in the characteristic 2 case the restriction $\sigma|\psi$ has 1 as an eigenvalue and so it must fix a non-zero point of ψ, a contradiction. ($\neq \psi$ intersects π_σ nontrivially so that $\sigma \mid \psi$ is a dilation. But, $\sigma^2 = 1$ and the characteristic is two so that $(\alpha - 1)^2 = 0$ so that $\pi(\sigma - 1) = \pi_\sigma$. Thus, σ is an axial Baer elation.) Hence σ cannot be an axial-Baer homology when $p = 2$, and when $p \neq 2$ we clearly have, for all $X \in \psi$, that $\sigma: X \mapsto -X$, and so σ induces a kern homology on ψ, yielding (b).
For (a) we may assume σ is a point-Baer homology (as the point-Baer elation

case is covered in theorem 37), and hence $p \neq 2$. Now σ and the kern involution both induce -1 on each central plane ψ. Hence $-\sigma$ fixes the point-Baer subplane ψ elementwise and theorem 37 forces ψ to be a proper Baer subplane. This completes the proof of (a). □

Note that the above argument shows that the central planes of a point-Baer homology will be proper Baer subplanes of the ambient translation plane π whenever the co-axis subplane incident with the zero vector is left invariant by a kern homology inducing the same action as the point-Baer homology. In section 9, we note several situations where this situation occurs.

Remark 2 *In any Desarguesian plane π any point-Baer collineation is a proper Baer perspectivity.*

6.1 Point-Baer Subplanes Of Translation Planes

In this subsection we characterize Baer-type subplanes of a spread π, *without* reference to any collineations stabilising them. Roughly speaking, we show that if a subplane π_0 of a spread π has certain complements in the vector space associated with π, then π_0 has a tendency to be a certain type of Baer subplane of π. The results we describe are concerned with generalising known facts about Baer subplanes of finite spreads. We begin by characterising proper Baer subplanes among point-Baer subplanes.

Proposition 42 *Let π be a translation plane, associated with a spread Γ on a vector space V, and let π_0 be a point-Baer subplane of π through the zero-vector $\mathbf{O} \in \mathbf{V}$. Then π_0 is a proper Baer subplane if and only if:*

$$M \in \Gamma \quad \textit{such that} \quad M \cap \pi_0 = \mathbf{O} \Longrightarrow \mathbf{V} = \pi_0 \oplus \mathbf{M} \qquad (*)$$

Proof: Assume the direct-sum hypothesis $(*)$. To show that π_0 is a proper Baer subplane of π, it is clearly sufficient to show that $(*)$ implies that every affine line of π meets π_0 at least once. So we need only consider lines of type $M + t$, where M is a component of the spread that trivially intersects π_0. Hence, by the direct-sum hypothesis, $t \in M \oplus \pi_0$, and this implies the line $t + M$ meets π_0. Thus π_0 is a proper Baer subplane of π. The converse follows by reversing the argument. □

We now prove another recognition theorem — for a subplane of a spread to be point-Baer; if subplanes π_0 and π_1, of a spread π, share the same components and direct-sum to π then both are point-Baer subplanes of π.

Proposition 43 *Let π_0 and π_1 denote two subplanes of a translation plane π with the same slope-set and sharing an affine point $\mathbf{O}$ such that $\pi = \pi_0 \oplus \pi_1$, when π is viewed as a vector space and π_0 and π_1 are subspaces. Then π_0 and π_1 are both point-Baer subplanes of π.*

Proof: We need to show that any affine point $T \in \pi$ lies on at least one line of each of the subplanes π_0 and π_1. So assume T does not lie in π_0 or π_1.

Since $T \in \pi_0 \oplus \pi_1$, the coset $\pi_0 + T$ must meet the subspace π_1 at a unique point S, and $S \neq T$ — otherwise $T = S = \mathbf{O}$, contradicting $T \notin \pi_0$. But since the slope-set of $\pi_0 + T$ is the same as the common slope-set of π_0 and π_1, the line ST meets the ideal line at a point of the affine plane π_1, and ST shares a second point with π_1, viz., S. Hence the line $ST \in \pi_1$ and certainly meets T. It follows that π_1 is point-Baer. Interchanging the roles of π_1 and π_0 above shows that π_0 is also point-Baer. □

The converse of the above is also valid, whenever π_0 and π_1 are *proper* Baer subplanes of π. This follows from the analysis of the direct-product construction for nets introduced in [16, Johnson-Ostrom], and will be necessary in our extension of André's theorem in the next section.

Proposition 44 (Johnson-Ostrom [16].) *Let π_0 and π_1 denote two subplanes of a spread π, associated with a translation plane, such that π_0 and π_1 have the same slope-set and share an affine point $\mathbf{O}$. If π_0 and π_1 are point-Baer subplanes of π then $\pi = \pi_0 \oplus \pi_1$.*

Proof: The direct product of subplanes, in the sense of [16] implies that if two Baer subplanes have a common point and define the same net then their direct product is the whole space, [16, corollary (4.18)] . In the context of translation planes, the direct product of π_0 and π_1 is the direct sum $\pi_0 \oplus \pi_1$, yielding the desired conclusion. □

We have already noted that a partial converse of the above proposition is true, viz., if two subplanes π_0 and π_1 of a spread (V, Γ) share the same components and direct-sum to V, then both the subplanes are *point*-Baer subplanes; the complete converse requires the subplanes to be *proper* Baer subplanes of the given spread. However, the following configuration appears to suggest that this might not always be the case.

Example 1 *Let $V = M \oplus B$ where M and B are infinite-dimensional K-vector spaces, over a [skew] field K, such that both M and B have K-bases of the same cardinality. Let $B = \{b_i \mid i \text{ in } \lambda\}$, and $M = \{m_i \mid i \text{ in } \lambda\}$ respectively be bases of the vector spaces B and M, bijectively indexed by the same set λ. Since λ is infinite, we may choose a proper subset B^* of B also bijectively indexed by λ, and so we may write: $B^* = \{b_i^* \mid i \text{ in } \lambda\}$ where the b_i^*'s are distinct elements of B.*

Define $C = \{m_i + b_i^* \mid i \text{ in } \lambda\}$ and let C be the subspace with basis C. Then $C \oplus B = V = M \oplus B$ but $M \oplus C$ is a proper subspace of V.

Now if a spread π contains M as a component and has $\pi_0 = B$ as a proper Baer subplane such that $\pi_1 = C$ is a subplane with the same slope-set as π_0, then π_1 will be a point-Baer subplane, by proposition 43, but not a proper Baer subplane of π — by proposition 42.

It is possible to construct derivable nets in non-derivable planes. However, the planes in question are all dual-translation planes. It may be possible to show that there are derivable nets in non-derivable translation planes in which case there would be point-Baer subplanes which are not line-Baer. However, none of these could admit point-Baer collineations.

7 André's Theorem on Semi-Perspectivities in Translation Planes.

A well-known counting argument shows that a finite plane cannot admit simultaneously a non-trivial homology and elation of the same order. In general this is false for infinite planes if only because translation planes with characteristic 0 simultaneously admit elations and homologies of order $\aleph_0$. However, for involutions we have seen previously that the situation is the same for finite and infinite translation planes at least for non-central collineations.

We now verify and complete this by listing all the possibilities for an involution.

Remark 3 *Let π be a translation plane with characteristic $p \geq 0$, and assume π admits an involutory collineation α. Then exactly one of the following applies:*

1. *If $p = 2$ then α is a Baer-elation or an ordinary elation;*
2. *If $p \neq 2$ then α is a Baer-homology or an ordinary homology.*

Proof: By the previous structure theorem, we may assume that α is a central collineation with the axis the ideal line then either α is a translation with order $p = 2$, or α is the kern homology -1, with order 2 when $p \neq 2$. . Since any involution is either a Baer involution (in which case it is a Baer-perspectivity) or a central collineation, it remains to consider the case when α is a central collineation whose axis is a component F of the spread associated with π. Since α is an involution, its eigenvalues are precisely ± 1, and so α must leave invariant all the rank-one subspaces of any subspace that it leaves invariant. Now, in the case $p = 2$, α clearly cannot fix any non-trivial space disjoint from F, and hence α cannot be a homology. Thus if $p = 2$, α is always an elation. Conversely if α is not an elation it acts faithfully as -1 on a coaxis, and so $p \neq 2$. Hence the remark always holds. □

Note that the above remark implies that André's theorem when applied to groups generated by involutory homologies does not have a meaning unless $p \neq 2$. The following asserts that André's theorem does hold for *finite* characteristic $p > 2$, and moreover, our argument shows the theorem must fail for $p = 0$.

Theorem 45 *Let π be an arbitrary translation plane with* finite *characteristic $p > 2$. Let G be a group of collineations in the translation complement of π*

generated by a set of affine involutory homologies that share a common coaxis (resp. common axis).

Then G contains a normal elementary abelian p-group E of elations which acts regularly on the axes (resp. coaxes) of all the involutory central collineations (i.e. involutory homologies) in G.

We shall also establish the Baer-analogue of the theorem above.

Theorem 46 *Let π be an arbitrary translation plane with finite characteristic $p > 2$. Let G be a group of collineations in the translation complement of π generated by a set of affine Baer involutory homologies that share a common coaxis (resp., common axis).*

Then G contains a normal elementary abelian p-group E of affine Baer elations which acts regularly on the axes (resp. coaxes) of all the involutory Baer central collineations (i.e. involutory Baer homologies) in G.

We introduce the following terminology to provide a uniform treatment of the two theorems and to stress the analogy between Baer-perspectivities and ordinary perspectivities.

Definition 7 *Let $\mathcal{I}$ denote a set of involutory automorphisms of a spread π, associated with an affine translation plane. The involutions in $\mathcal{I}$ are all of the same* type *if and only if $\mathcal{I}$ consists entirely of affine homologies or entirely of Baer homologies.*

An automorphism σ of π is called a semi-elation *with the* same type *as $\mathcal{I}$ if σ is an ordinary elation (resp. Baer-elation) when the involutions in $\mathcal{I}$ are all ordinary affine homologies (resp. Baer homologies).*

Since by remark 3 all involutions are either ordinary or Baer central collineations, the two theorems above may be viewed as concerning sets of involutions of a given type — homology-type (theorem 45) or Baer-homology type (theorem 46). We now state a unified version of these theorems.

Theorem 47 (André's Involutory Theorem For Translation Planes) *Let π be an arbitrary translation plane of odd (prime) characteristic p. Let G be a group of collineations in the translation complement of π generated by a set $\mathcal{I}$ of involutions of the same type such that all the involutions in $\mathcal{I}$ share a common coaxis (resp. common axis).*

Then G contains a normal elementary abelian p-group E consisting of semi-elations of the same type as $\mathcal{I}$ which acts regularly on the axes (resp. coaxes) of all *the involutions in G.*

Note: Since all involutions in the theorem above fix the origin, all the coaxes considered are necessarily **O**-coaxis.

Proof: This is immediate if the involutions involved are central collineations, so consider the case when the involutions involved include distinct Baer homologies α and β that share a common coaxis ψ — not through the origin. Now the two involution share at least two fixed affine points, viz., the origin and the point(s) where ψ meets the axes π_α and π_β. So $\pi_\alpha = \pi_\beta$, since these planes have the same slope-set as ψ. Thus $\alpha\beta$ is the identity on a set $\psi \cup \pi_\alpha$ strictly containing the Baer subplane π_α, forcing $\alpha = \beta$. □

To prove the theorem, we generalize the basic coordinatisation principle (c.f. section 2) for arbitrary spreads: *relative to a suitable basis any three components may be chosen to be $x = 0$, $y = 0$, and $x = y$,* . We require this choice to be valid if the components are replaced by appropriate Baer subplanes.

Lemma 48 (Triad-Lemma) *Let Σ be the spread, of an affine translation plane π, defined on some vector space U over the prime field F in kern π. Suppose X, Y and Z are three Baer subsplanes (or three components) of U through $\mathbf{O}$ such that any two of them intersect precisely in $\mathbf{O}$, and direct-sum to U, viewed as an F-space. Then there is an F-vector space W and a spread Γ on the vector space $V = W \oplus W$, such that some affine-plane isomorphism Φ from π to the translation plane associated with Γ, maps O to the zero vector, and the images of X, Y and Z onto, respectively, the subspaces '$y = \mathbf{O}$', '$x = \mathbf{O}$' and '$y = x$' of V, defined in the natural way.*

Proof: The condition that any two of X, Y and Z direct-sum to U implies that we may choose a basis of Z of type $(x_i + y_i)_{i \in I}$ such that $(x_i)_{i \in I}$ and $(y_i)_{i \in I}$ are bases of X and Y resp. The desired identification can now easily be specified in terms of these bases: define the linear bijection Φ in the obvious way, and let Γ consist of the images of the components of Σ under Φ. □

Note that when the three spaces X, Y and Z above are *components* (rather than Baer subplanes) of the spread Σ there is no need to explicitly assume in the lemma that the spaces direct-sum pairwise to the whole space U. It is not clear if the assumption is ever necessary when X, Y and Z are Baer subplanes of the spread Σ.

Lemma 49 *Let (V, Γ) be a spread on a vector space V, with component set Γ admitting distinct involutions $\rho, \sigma \in \mathrm{GL}(V, +)$, that permute the components in Γ. If ρ and σ are involutions of the same type sharing the same $\mathbf{O}$-coaxis (resp., axis), but with distinct axes (resp., $\mathbf{O}$-coaxes), then V may be assigned a basis so that the common $\mathbf{O}$-coaxis (resp., axis) of ρ and σ becomes $x = 0$, the axis (resp., $\mathbf{O}$-coaxis) of σ becomes $y = 0$, and the axis (resp. $\mathbf{O}$-coaxis) of ρ becomes $y = x$.*

Proof: First consider the case when the involutions share the same coaxis. When ρ and σ are ordinary homologies then the axes and coaxes are, by definition, components of the spread Γ, and since any two distinct components of a spread on V must direct-sum to V, the lemma above yields the desired result.

It remains to consider the case when ρ and σ are involutory Baer homologies. Since by corollary 38 the slope-set of the central planes of a Baer perspectivity, of any translation plane, determine the same net, it follows that each of the central planes, as well as the axis of σ and ρ, determine precisely the same net, and the planes among these through **O** are pairwise disjoint proper Baer subplanes.

Thus we are dealing with three pairwise disjoint[4] Baer subplanes defining the same net. Now the lemma follows, for example, by proposition 44. □

We can now complete the proof of theorem 47, André's theorem for odd characteristic p; the proof will also show that André must fail in characteristic 0.

Proof: Let ρ and σ denote distinct involutions in $\mathcal{I}$. First consider the case when all the involutions in $\mathcal{I}$ share a common **O**-coaxis $\pi_{\mathcal{I}}$ and so by lemma 49 we may choose a basis of the spread so that $\pi_{\mathcal{I}}$ becomes $x = 0$, the axis of σ becomes $y = 0$, and the axis of ρ becomes $y = x$. Thus ρ maps (x, x) to (x, x), and induces -1 on $x = 0$ (this being the **O**-coaxis of an involution), and so the linear map ρ must be the mapping $(x, y) \to (x, 2x - y)$, and by another easy computation σ is the mapping $(x, y) \to (x, -y)$.

Now consider the dihedral group $D =< \rho, \sigma >$, whose stem is clearly the group:

$$E =< \rho\sigma >= \{(x, y) \to (x, 2kx + y) \mid k \in Z(\text{integers})\},$$

and this group fixes $x = 0$ elementwise and leaves invariant each of its cosets $x = c$. Hence E is a [Baer] elation group with axis $x = 0$ and center $\{x = c \mid c \in V\}$ (i.e., the center consists of the cosets of the Baer subplane $x = 0$, in the Baer case, and is otherwise the ideal point on the parallel class defined by the lines $x = c$.)

Now the orbit of $y = 0$ under E is clearly the set of subspaces of type $y = 2kx$, where k ranges over the integers. Hence we may map $y = 0$ onto $y = x$, using E, if and only if there is an integer k such that $2k = 1$. Note that this holds for some k if and only if the integer p is an odd prime.

Hence, we have established that there is a [Baer]-elation within $< \rho, \sigma >$ which maps one [Baer]-axis to the other if and only if the characteristic of the translation plane is an odd prime p[5]. By varying ρ and σ we see that the full [Baer] elation group $\mathcal{E}$ in G acts transitively on the [Baer]-axes of all the ρ and σ. To establish the regularity of $\mathcal{E}$ on these axes, note that since the [Baer]-elation in G must leave invariant $\pi_{\mathcal{I}}$, the common **O**-coaxis of the involutions in $\mathcal{I}$, $\pi_{\mathcal{I}}$ must in fact be the common axis of all the [Baer]-elations in G, since [Baer]-elations do not leave invariant any [Baer subplane] component distinct from their axes and passing through **O**. This also shows that the non-trivial elements of $\mathcal{E}$ cannot hope to fix any axes of the involutions of type $\mathcal{I}$ in G. Thus the regularity of $\mathcal{E}$ follows, and so does the theorem for the case when the involutions in $\mathcal{I}$ share a common **O**-coaxis.

---

[4]By abuse of notation trivially intersecting spaces will be called pairwise-disjoint.

[5]The argument shows that *any* dihedral subgroup of type $< \rho, \sigma >$ *fails* to satisfy André's theorem when the characteristic $p = 0$.

Now we turn to the remaining case, when the involutions in $\mathcal{I}$ share a common axis. Although duality applies in the central collineation case, we use a direct argument to accommodate the Baer case.

We may assume that the [Baer]-homologies σ and ρ are of the same type as all the involutions in $\mathcal{I}$, and hence have the same axis and distinct coaxes.

Following the coordinatisation principle used in the previous case, we choose the common axis as $x = 0$, the coaxis of σ as $y = 0$ and the coaxis of ρ as $y = x$. Arguing as before, the involution ρ induces -1 on $y = x$ and the identity on $x = 0$, and hence it can only be the linear map $(x, y) \to (-x, -2x + y)$, and σ becomes the map $(x, y) \to (-x, y)$. The stem E of $\langle \rho, \sigma \rangle$ may now be expressed as:

$$E = < \rho\sigma >= \{(x, y) \to (x, 2kx + y) \mid k \in Z(\text{integers})\},$$

and again E is a [Baer] elation subgroup of $< \rho, \sigma >$ that has the axis of σ and ρ in the same orbit precisely when p is an odd prime. The proof can now be completed as before. □

Corollary 50 *Let π be a translation plane of characteristic* 0. *Then André's theorem for involutions always fails*[6].

Proof: The previous proof shows that when there are either two semi-involutions with the same axis and distinct coaxes or the same coaxis and distinct axes then there can never be a semi-elation in the group generated by the two semi-involutions which maps one coaxis (axis) onto the other since the equation $2k = 1$ for k an integer never has a solution over fields of characteristic zero. □

8 Line-Baer subplanes in $PG(3, K)$ spreads.

Let S be a spread in $PG(3, K)$ for K a skewfield. Let π_o be any 2−dimensional K-subspace. Then it follows from Johnson [14] that π_o is a line-Baer affine subplane, if it's not a component.

Theorem 51 *Let S be a spread in $PG(3, K)$ for K a skewfield and let π_o and π_1 be* 2−*dimensional K−subspaces which as line-Baer affine plane share an affine point. If the set of parallel classes for π_o is contained in the set of parallel classes for π_1 then the two affine plane share all of their parallel classes.*

Proof: Decompose the space as (x_1, x_2, y_1, y_2) so that $x = 0, y = 0, y = x$ are common components of the two line-Baer subplanes. Assume that $x_2 = 0 = y_2$ defines π_o. It is possible to choose coordinates so that π_1 has equation $x_1 = 0 = y_1$. The components of π_o have the form $x = 0, y = x \begin{bmatrix} a & 0 \\ 0 & b(a) \end{bmatrix}$ for all

[6] Provided of course that π admits distinct affine involutions of the same type and sharing a common axis [coaxis] but not both; without this tacit assumption André's theorem is arguably true in a pathological sense.

a in K where b is a function on K. The components of π_1 have the general form $x = 0, y = x \begin{bmatrix} d & 0 \\ e & c \end{bmatrix}$ for elements d, e, c of $\dot{K}$. Since the difference of the corresponding matrices is nonsingular or identically zero, it follows that when $a = -d$ then $e = 0$. That is, all components of π_1 are common to π_o

Theorem 52 *Let S be a spread in $PG(3, K)$ for K a skewfield. Let π denote the corresponding translation plane.*

Let σ be a collineation of π which fixes a 2−dimensional $K-$ space which is not a component. Let $\pi_o = Fix\sigma$.

(1) Then σ is in $GL(2, K)$ and $\pi_o(\sigma - 1)$ is a 2−dimensional K-subspace such that π_o and $\pi_o(\sigma - 1)$ as line-Baer subplanes share all parallel classes.

(2) π_o and $\pi_o(\sigma - 1)$ are point-Baer subplanes so hence Baer subplanes.

Proof: Since σ is a collineation of π, we may assume that σ is in $\Gamma L(4, K)$. However, writing the elements of π as (x_1, x_2, y_1, y_2) and the equation of π_o as $x_1 = y_1 = 0$, if σ has the general form $(x_1, x_2, y_1, y_2) \to (x_1^\rho, x_2^\rho, y_1^\rho, y_2^\rho) \begin{bmatrix} a & b \\ c & d \end{bmatrix}$ for a, b, c, d in K and considering π as a left $K-$space for ρ an automorphism of K then it follow that $c = 0$ and $x_2^\rho a = x_2$ for all x_2 in K. That is, $a = 1$ and ρ is the identity mapping.

It is immediate that then $\pi_o(\sigma - 1)$ is a left $K-$subspace. Furthermore, since π/π_o is now $K-$isomorphic to $\pi_o(\sigma-1)$ it follows that $\pi_o(\sigma-1)$ is a 2−dimensional $K-$subspace.

Now let F be any component of π_o. Clearly, $F(\sigma - 1)$ is a component of $\pi_o(\sigma - 1)$ so that all components of π_o are components of $\pi_o(\sigma - 1)$. By the above results, the two line-Baer subplanes share all of their parallel classes.

Hence, it follows that either $\pi_o = \pi_o(\sigma - 1)$ or the two line-Baer subplanes share exactly one common affine point.

Assume first that $\pi_o = \pi_o(\sigma - 1)$. Hence, we must have $(\sigma - 1)^2 = 0$ which means also that σ fixes π/π_o pointwise. Let P be a point of $\pi - \pi_o$. Then $(P + \pi_o)\sigma = P + \pi_o$ so that $P\sigma - P$ is in π_o.

Form the line $PP\sigma$ and assume it is the line $V_\alpha + c$ where V_α is in the spread of π. Hence, $P\sigma - P$ must be in V_α which means that V_α intersects π_o nontrivially so that V_α is a component of π_o. To show that $V_\alpha + c$ is a line of π_o, it must be shown that $c = v_o + d$ for v_o in V_α and d in π_o.

However, we may assume that $c = P\sigma$. Now $P\sigma + V_\alpha$ is a line of π_o if and only if it is fixed by σ if and only if $P\sigma^2 - P\sigma$ is in V_α. Since $(\sigma - 1)^2 = 0$, it follows that $P\sigma^2 - P\sigma = P(\sigma^2 - \sigma) = P(\sigma - 1) = P\sigma - P$ which is in V_α.

Hence, each point of π is incident with a line of π_o so that π_o is a point-Baer and hence a Baer subplane.

Now assume that $\pi_o \neq \pi_o(\sigma - 1)$. Then, we can use (6.8) to show that both subplanes are point-Baer subplanes and hence both are Baer subplanes.

Theorem 53 *Let S be any spread in $PG(3, K)$ for K a skewfield. Let L be any line in $PG(3, K)$ which is not a line of S.*

If L is pointwise fixed by a collineation of S then L is a Baer subplane of the corresponding translation planeπ.

If each plane of $PG(3, K)$ contains a line not in S which is fixed pointwise by a collineation of π then S is a dual spread.

Proof: The first statement is simply a restatement of part of the previous result since L becomes a $2-$dimensional $K-$subspace of the corresponding vector space and hence a line-Baer subplane of π.

Let Σ be an plane and let L denote a line of Σ which not in S and which is fixed pointwise by a collineation of Σ. Then L is Baer and from Johnson [14], it follows that Σ must contain a line of the spread.

References

[1] André, J. Über Perspektivitäten in endlichen projektiven Ebenen., Arch. Math. 6, 29-32.

[2] Barlotti, A. On the definition of Baer subplanes of infinite planes. J. Geom. 3 (1973), 87-92.

[3] Bruen, A. and Fisher, J.C. Spreads which are not dual spreads. Canad. Math. Bull. 12 (1969), 801-803.

[4] Cohn, P.M. Algebra, Vol. 2. John Wiley and Sons, London-New York-Sydney-Toronto, 1977.

[5] Dembowski, P. **Finite Geometries.** Ergebnisse der Mathematik und Ihrer Grenzgebiete. Band 41, Springer-Verlag, New York, Inc. 1968.

[6] Foulser, D.A. Baer p-elements in translation planes. J. Alg. 31 (1974), 354-366.

[7] Foulser, D.A. and Johnson, N.L. The translation planes of order q^2 that admit $SL(2, q)$ as a colineation group. I. Even Order. J. Alg. 86 (1984), 385-406.

[8] Foulser, D.A. and Johnson, N.L. The translation planes of order q^2 that admit $SL(2, q)$ as a colineation group. II. Odd Order. J. Geom. 18 (1982), 122-139.

[9] Grundhofer, T. Eine Charakterisierung ableitbarer Translationsebenen. Geom. Dedicata 11, (1981), 177-185.

[10] Jha, V. and Johnson, N.L. Infinite flocks of a quadratic cone. J. Geom. (to appear).

[11] Jha, V. and Johnson, N.L. Quasifibrations. Bull. Belgian Math. Soc./Simon Stevin 3 (1996), 313-324.

[12] Johnson, N.L. Flocks of infinite hyperbolic quadrics. J. Alg. Comb. (to appear).

[13] Johnson, N.L. Derivable nets and three dimensional projective spaces and II. The Structure. Abh. Math. Sem. Univ. Hamburg 58 (1988), 245-25 and Arch. Math. (Basel) 55 (1990), 84-104 respectively.

[14] Johnson, N.L. Derivable nets in nonderivable planes. Groups and Geometries (Siena) (submitted).

[15] Johnson, N.L. The characterization of subplane covered nets. Bull. Belgian Math. Soc./Simon Stevin 2 (1995), 487-507.

[16] Johnson, N.L. and Ostrom, T.G. Direct products of affine partial linear spaces. J. Comb. Theory (Series A), vol. 75 (1996), 99-140.

[17] M. Kallaher, Translation Planes, **Handbook of Incidence Geometry**, ed. F. Buekenhout, Elsevier Science B.V., 1995.

[18] Pickert, G. Projektive Ebenen. Math. Wissenschaften, Springer-Verlag. Berling-Göttingen-Heidelberg, 1955.

[19] Lüneburg, H. **Translation planes**. Springer-Verlag. Berlin, Heidelberg, New York, 1980.

[20] Yaqub, Jill C.D.S. Translation ovoids over skew fields. Geom. Ded. 36 (1990), 261-271.

[21] Walker, M. On translation planes and their collineation groups. Thesis, Univ. London, 1973.

Flocks of Quadratic and Semi-Elliptic Cones

NORMAN L. JOHNSON
and
XIAOBIN LIU
Department of Mathematics
The University of Iowa
Iowa City, IA 52242

Abstract

In this paper, we study mostly infinite flocks of quadratic cones in $PG(3, \mathbb{K})$. Furthermore, the notion of a semi-elliptic flock is introduced. Many new classes of infinite conical flocks and infinite semi-elliptic flocks in $PG(3, \mathbb{K})$ will be constructed.

1 Introduction

Flocks of quadratic cones in three dimensional finite projective spaces have been studied for roughly twenty years. Recently, some types of finite flocks of quadratic cones were generalized to those of infinite ones, due to the efforts by several mathematicians: M. Biliotti, F. De Clerck, V. Jha, N. L. Johnson, H. Van Maldeghem and others. (see [2], [4] and [6].)

Let $\mathbb{K}$ be a field. Let $\mathcal{Q}_C$: $x_0x_1 - x_2^2 = 0$ be the quadratic cone in $PG(3, \mathbb{K})$, whose vertex is $\mathbf{v} = (0, 0, 0, 1)$. $\mathcal{Q}_C$ can then be considered as a set of *generating lines* $\{\ell\} \cup \{\ell_s \mid s \in \mathbb{K}\}$, where $\ell = \langle (0, 0, 0, 1), (1, 0, 0, 0) \rangle$ and $\ell_s = \langle (0, 0, 0, 1), (s^2, 1, s, 0) \rangle$, such that
(1) each plane not containing $\mathbf{v}$ in $PG(3, \mathbb{K})$ will intersect the lines, and
(2) the intersection points form a conic.

Let $\mathcal{P} = \{\pi_t : tx_0 + f(t)x_1 + g(t)x_2 + x_3 = 0 \mid t \in \mathbb{K}\}$ be a set of planes which do not contain $\mathbf{v}$. A *conical flock* of $\mathcal{Q}_C$ in $PG(3, \mathbb{K})$ is just a set of

This article is dedicated to Professor T. G. Ostrom on his eightieth birthday.

conics partitioning the points of $\mathcal{Q}_C - \mathbf{v}$. A flock $\mathcal{F}$ is called *linear* if the associated planes of $\mathcal{F}$ are collinear; a flock $\mathcal{F}$ is called *concurrent* if the associated planes of $\mathcal{F}$ are concurrent; and a flock $\mathcal{F}$ is called *monomial* if f and g in $\mathcal{P}$ are monomials in $\mathbb{K}$. Two flocks are said to be *projectively equivalent* if there exists a semilinear transformation that maps the associated planes of one flock to the associated planes of the other.

The planes $\pi_t : tx_0 + mtx_1 + tx_2 + x_3 = 0,\ t \in \mathbb{K}$, where m satisfies the condition that $x^2 + x + m \neq 0,\ \forall x \in \mathbb{K}$, will define a linear conical flock in $PG(3, \mathbb{K})$. If $\mathbb{K} = GF(q)$, q odd, then any m such that $1 - 4m$ is nonsquare will be sufficient. If $\mathbb{K} = GF(q)$, q even, then m can be taken as 1. If $\mathbb{K} \subset \mathbb{R}$, then any m such that $m > \frac{1}{4}$ can be chosen to obtain a linear flock. However, if $\mathbb{K}$ is a field of characteristic $\neq 2$ and $1 - 4m$ is always a square for all $m \in \mathbb{K}$, then no linear flock can exist in $PG(3, \mathbb{K})$. More generally, if $\mathbb{K}$ is an algebraically closed field, any pair of the planes will intersect with the quadratic cone $\mathcal{Q}_C$ in $PG(3, \mathbb{K})$ nontrivially. Therefore, no partial flock of two conics can occur in this case.

The first example of a nonlinear infinite conical flock is the Betten flock in $PG(3, \mathbb{R})$, which has been generalized to a flock in $PG(3, \mathbb{K})$, for any field with characteristic $\neq 3$ over which x^3 is bijective, by V. Jha and N. L. Johnson [6]. This flock is called a *generalized Betten flock.* This flock was also studied by F. De Clerck and H. Van Maldeghem in [4]. We shall see in section 5 that this type of flock can be generalized to an infinite class of conical flocks in $PG(3, \mathbb{R})$.

Using the technique of nest replacements, V. Jha and N. L. Johnson [6] obtained a class of translation planes which correspond to the *generalized Fisher flocks* in $PG(3, \mathbb{K})$. When $\mathbb{K}$ is finite, the flocks are *Fisher flocks.* The algebraic construction to this class of flocks is still unknown in the infinite case.

In [2], M. Biliotti and N. L. Johnson constructed a new flock whose planes share one of two fixed lines which are called the *bilinear flock.* (Note that there is no finite bilinear flock.) This idea has been generalized to obtain an *n-linear flock*, which consists of n linear partial flocks and the n associated lines are distinct. The reader is referred to the article of Biliotti and Johnson in this volume.

The Kantor-Knuth flocks in $PG(3, q)$ has various infinite analogues which we call the *generalized Kantor-Knutn flocks.* We mention these flocks briefly in this article and determine their structure more generally in a paper in this volume on the generalized Kantor-Knuth flocks.

In section 2, we shall prove the fundamental theorem of flocks of the quadratic cone in $PG(3, \mathbb{K})$. Sections 3 and 4 will be devoted to revisiting some of the well-known (finite) conical flocks in $PG(3, q)$, namely, the monomial flocks, for q being odd and even. Most of these flocks have infinite analogues which we construct. For a more complete treatment of the examples in the finite case, the reader is referred to the survey article by Johnson and Payne in this volume.

In section 5, we construct many classes of conical flocks in $PG(3, \mathbb{R})$, from smooth flocks to discontinuous flocks. The n-linear flocks will be constructed in section 6 for any ordered field. In section 7, we shall study the concurrent flocks whose associated planes share the point $\mathbf{u}=(0,0,1,0)$. Finally in section 8, semi-elliptic flocks will be studied.

2 The Fundamental Theorem

The following result is the fundamental theorem of flocks of quadratic cones in $PG(3, \mathbb{K})$: (See F. De Clerck and H. Van Maldeghem in [4] for this result stated in a slightly different manner.)

Theorem 1 *$\mathcal{P}$ is a set of associated planes of a flock of $\mathcal{Q}_C$ in $PG(3, \mathbb{K})$ iff $\phi_s(t) = ts^2 + g(t)s + f(t)$ is a bijective mapping from $\mathbb{K}$ to $\mathbb{K}$, for each $s \in \mathbb{K}$.*

Proof. $\mathcal{P}$ defines a partial flock of $\mathcal{Q}_C$ iff $\pi_t \cap \pi_{\overline{t}} \cap \mathcal{Q}_C = \emptyset$, iff $x_1 \neq 0$ and

$$(t-\overline{t})x_2^2 + (f(t) - f(\overline{t}))x_1^2 + (g(t) - g(\overline{t}))x_2x_1 = 0$$

has no nontrivial solution, iff

$$(t-\overline{t})x^2 + (g(t) - g(\overline{t}))x + (f(t) - f(\overline{t}))$$

is nonzero for all $x \in \mathbb{K}$, iff $ts^2 + g(t)s + f(t)$ is one-to-one, $\forall s \in \mathbb{K}$.

Consider the lines

$$\ell = \langle(0,0,0,1),(1,0,0,0)\rangle$$

and

$$\ell_s = \langle(0,0,0,1),(s^2,1,s,0)\rangle$$

that generate $\mathcal{Q}_C$. ℓ is covered by $\mathcal{P}$. In fact, $\forall\beta$, $\forall\alpha \neq 0$, $(\alpha,0,0,\beta) \in \pi_{-\alpha^{-1}\beta}$. On the other hand, ℓ_s is covered by $\mathcal{P}$ iff

$$\forall\beta,\ \forall\alpha \neq 0,\ (\alpha s^2, \alpha, \alpha s, \beta) \in \pi_t \text{ for some } t \in \mathbb{K} \text{ iff}$$

$$\alpha s^2 t + \alpha f(t) + \alpha s g(t) + \beta = 0 \text{ iff}$$

$$ts^2 + g(t)s + f(t) = -\alpha^{-1}\beta \text{ iff}$$
$$ts^2 + g(t)s + f(t) \text{ is onto.}$$

Therefore, $\{\pi_t \cap \mathcal{Q}_C \mid t \in \mathbb{K}\}$ is a flock iff $\phi_s(t) = ts^2 + g(t)s + f(t)$ is bijective from $\mathbb{K}$ to $\mathbb{K}$ for each $s \in \mathbb{K}$. Q.E.D.

As a simple observation, we have

Theorem 2 *Let k be a fixed element in $\mathbb{K}$. If $(f(t), g(t))$ defines a conical flock in $PG(3,\mathbb{K})$, then $(f(t)+k, g(t))$ will also define a conical flock. Consequently, f can be taken such that $f(0) = 0$.*

Proof. It follows from the fact that $\phi_s(t) = ts^2 + g(t)s + f(t)$ is bijective iff $\overline{\phi}_s(t) = ts^2 + g(t)s + f(t) + k$ is bijective. Q.E.D.

As the first example of applying Theorem 1, we have the following result.

Theorem 3 *Let $\mathbb{K}$ be a field with characteristic $\neq 3$ such that x^3 is bijective. Then the set of planes*

$$\mathcal{P} = \{\pi_t : tx_0 + \frac{1}{3}c^2t^3x_1 + ct^2x_2 + x_3 = 0 \mid t \in \mathbb{K}\}$$

defines a conical flock in $PG(3,\mathbb{K})$, where $0 \neq c \in \mathbb{K}$. Moreover, any two such flocks are projectively equivalent.

Proof. Notice that

$$\begin{aligned}
\phi_s(t) &= ts^2 + g(t)s + f(t) \\
&= ts^2 + ct^2s + \frac{1}{3}c^2t^3 \\
&= \frac{1}{3}c^2\left[\left(t + \frac{s}{c}\right)^3 - \left(\frac{s}{c}\right)^3\right] \\
&= \frac{1}{3}c^2\left(t + \frac{s}{c}\right)^3 - \frac{s^3}{3c}.
\end{aligned}$$

Since x^3 is bijective on $\mathbb{K}$, $\phi_s(t) = \frac{1}{3}c^2\left(t + \frac{s}{c}\right)^3 - \frac{s^3}{3c}$ is bijective on $\mathbb{K}$, for each fixed $s \in \mathbb{K}$. It follows from Theorem 1 that $\mathcal{P}$ defines a conical flock in $PG(3,\mathbb{K})$.

Let π'_t have equation: $tx_0 + \frac{1}{3}c'^2t^3x_1 + c't^2x_2 + x_3 = 0$. Then it is easy to see that there is a mapping $\sigma \in GL(4, \mathbb{K})$ which maps $\pi_t : tx_0 + \frac{1}{3}c^2t^3x_1 + ct^2x_2 + x_3 = 0$ to $\pi'_t : tx_0 + \frac{1}{3}c'^2t^3x_1 + c't^2x_2 + x_3 = 0$ and preserves $\mathcal{Q}_C$:

$$\sigma = \begin{bmatrix} 1 & 0 & 0 & 0 \\ 0 & (\frac{c}{c'})^2 & 0 & 0 \\ 0 & 0 & \frac{c}{c'} & 0 \\ 0 & 0 & 0 & 1 \end{bmatrix}.$$

Therefore, the flock is unique up to isomorphism. Q.E.D.

A *generalized Betten flock* is a flock in $PG(3, \mathbb{K})$ such that $f(t) = at^3$ and $g(t) = bt^2$ for some $a, b \in \mathbb{K}$. Note that Theorem 3 gives us an example of generalized Betten flocks.

Remark 4 *When* $\mathbb{K} = GF(q)$, *where* $q \equiv -1 \pmod 3$, *this finite flock is the* Fisher-Thas-Walker flock. *When* $\mathbb{K} = \mathbb{R}$, *then the flock is called the* Betten flock, *as this flock is obtainable from a spread constructed by D. Betten in [1]. When* $c = -1$, *the result is due to V. Jha and N. L. Johnson [6]. When* $c = 3$, *the result is due to F. De Clerck and H. Van Maldeghem [4].*

3 Monomial Flocks in $PG(3, p^n)$, p odd

Let $q = p^n$ be an odd prime power. To obtain a conical flock, we need to find $f(t)$ which is bijective over $GF(q)$, and $g(t)$ which makes $\phi_s(t) = ts^2 + g(t)s + f(t)$ to be bijective for each $s \in GF(q)$. Notice that in a finite field, bijectivity is equivalent to injectivity.

$\phi_s(t)$ is injective iff $(t - \bar{t})s^2 + (g(t) - g(\bar{t}))s + (f(t) - f(\bar{t})) \neq 0, \forall t \neq \bar{t}$, iff $(g(t) - g(\bar{t}))^2 - 4(t - \bar{t})(f(t) - f(\bar{t}))$ is a nonsquare, which is a well-known result. In this section, we shall consider the case when f and g are monomials which give rise to conical flocks in $PG(3, p^n)$, p odd.

3.1 The Linear Flocks

Let $f(t) = at$ and $g(t) = bt$. Then $\phi_s(t) = ts^2 + g(t)s + f(t) = t(s^2 + bs + a)$ is bijective iff $(s^2 + bs + a) \neq 0$ for all $s \in GF(q)$ iff $b^2 - 4a$ is a nonsquare. The flock defined by such f and g is linear. The associated line has equation: $x_0 + ax_1 + bx_2 = 0 = x_3$. In particular, if $g(t) \equiv 0$, then $\phi_s(t)$ is bijective iff $-a$ is nonsquare.

3.2 The Kantor-Knuth Flocks

J. A. Thas [11] has completely determined finite concurrent flocks. A nonlinear concurrent flock must be a Kantor-Knuth flock which has the following functions f and g. This flock is equivalent to the semifield translation plane of Knuth type [8] and the generalized quadrangle of Kantor type [7].

Let $f(t) = -mt^\sigma$ and $g(t) \equiv 0$, where σ is a nontrivial automorphism of $GF(q)$ (that is, $t^\sigma = t^{p^r}, 1 \leq r \leq n-1$).

Then $\phi_s(t) = ts^2 + g(t)s + f(t) = ts^2 - mt^\sigma$ is bijective, iff $ts^2 - mt^\sigma$ is injective, iff $ts^2 - mt^\sigma = 0$ has no nontrivial solution of t, iff $s^2 - mt^{\sigma-1} \neq 0$, iff $m \neq s^2 t^{-(p^r-1)}$, iff m is a nonsquare.

On the other hand, for two nonsquare m_1 and m_2 in $GF(q)$, we have two Kantor-Knuth flocks which are denoted by $\mathcal{F}(m_1, \sigma)$ and $\mathcal{F}(m_2, \sigma)$, respectively. Then it is ready to see that $\mathcal{F}(m_1, \sigma) \cong \mathcal{F}(m_2, \sigma)$ by the following mapping τ in $GL(4, q)$:

$$\tau = \begin{bmatrix} 1 & 0 & 0 & 0 \\ 0 & \frac{m_1}{m_2} & 0 & 0 \\ 0 & 0 & \sqrt{\frac{m_1}{m_2}} & 0 \\ 0 & 0 & 0 & 1 \end{bmatrix},$$

where $\frac{m_1}{m_2}$ is a square and so $\sqrt{\frac{m_1}{m_2}} \in GF(q)$.

3.3 The Fisher-Thas-Walker Flocks

Since x^3 is bijective on $GF(q)$ iff $q \equiv -1 \pmod 3$ and q is odd, the finite generalized Betten flock—the Fisher-Thas-Walker flock [5], can exist only when $q \equiv -1 \pmod 3$. Functions $f(t) = \frac{1}{3}t^3$ and $g(t) = t^2$ will define a Fisher-Thas-Walker flock, as is shown in Theorem 3.

3.4 The Kantor Flocks

Suppose that $q \equiv \pm 3 \pmod{10}$. Let $f(t) = 5t^5$ and $g(t) = 5t^3$. Then f and g will define a conical flock in $PG(3, q)$, which is equivalent to the generalized quadrangle discovered by W. M. Kantor [7]. In fact, $\phi_s(t) = ts^2 + g(t)s + f(t) = ts^2 + 5t^3 s + 5t^5$. The discriminant is $(5t^3)^2 - 4(t)(5t^5) = 5t^6$, which is a nonsquare for $t \neq 0$, since 5 is a nonsquare in $GF(q)$ for $q \equiv \pm 3 \pmod{10}$. Therefore, $\phi_s(t)$ is injective and hence it is bijective. It follows from Theorem 1 that f and g indeed define a conical flock.

These four classes of conical flocks, namely, the linear flocks, the Kantor-Knuth flocks, the Fisher-Thas-Walker flocks and the Kantor flocks, are the only known monomial flocks in $PG(3,q)$ for q odd.

4 Monomial Flocks in $PG(3, 2^n)$

Define the function *trace* in $GF(2^n)$ by $Tr : GF(2^n) \to GF(2)$ such that

$$Tr(x) = x + x^2 + x^{2^2} + x^{2^3} + \ldots + x^{2^{n-1}}.$$

Since $x^{2^n} = x$, $Tr(x^2) = Tr(x)$, or $Tr(x^2 + x) = 0$. In particular, $Tr(1) = 1$ in $GF(2^{2m+1})$.

Note that there is no nonsquare element in $GF(2^n)$. $\phi_s(t)$ is bijective iff $\phi_s(t)$ is injective, iff

$$(t + \bar{t})s^2 + (g(t) + g(\bar{t}))s + (f(t) + f(\bar{t})) \neq 0, \ \forall t \neq \bar{t}, \quad \text{iff}$$

$$Tr\left(\frac{(t+\bar{t})(f(t) + f(\bar{t}))}{(g(t) + g(\bar{t}))^2}\right) = 1,$$

which is also a well-known result. In this section, we shall consider the case when f and g are monomials which give rise to conical flocks in $PG(3, 2^n)$.

4.1 The Linear Flocks

Let $f(t) = at$ and $g(t) = bt$. Then

$$\begin{aligned} & Tr\left(\frac{(t+\bar{t})(f(t) + f(\bar{t}))}{(g(t) + g(\bar{t}))^2}\right) \\ = \ & Tr\left(\frac{(t+\bar{t})(at + a\bar{t})}{(bt + b\bar{t})^2}\right) \\ = \ & Tr\left(\frac{a}{b^2}\right). \end{aligned}$$

Therefore, $f(t) = at$ and $g(t) = bt$ will define a linear flock with respect to the line $\ell : x_0 + ax_1 + bx_2 = 0 = x_3$ iff $Tr\left(\frac{a}{b^2}\right) = 1$.

Notice that if a or b is 0 then $\phi_s(t)$ will not be bijective for some s.

4.2 The Fisher-Thas-Walker Flocks

Again x^3 is bijective on $GF(2^n)$ iff $2^n \equiv -1 \pmod 3$, iff n is odd. Hence $f(t) = x^3$ and $g(t) = x^2$ define a Fisher-Thas-Walker flock (see [5]) in $PG(3, 2^{2m+1})$. We shall use the trace argument to verify that such functions indeed give rise to a flock.

$$\begin{aligned}
& Tr\left(\frac{(t+\bar{t})(f(t)+f(\bar{t}))}{(g(t)+g(\bar{t}))^2}\right) \\
= \;& Tr\left(\frac{(t+\bar{t})(t^3+\bar{t}^3)}{(t^2+\bar{t}^2)^2}\right) \\
= \;& Tr\left(1+\frac{t}{t+\bar{t}}+\frac{t^2}{(t+\bar{t})^2}\right) = 1,
\end{aligned}$$

as desired.

4.3 The Payne Flocks

Consider the Galois field $GF(2^{2m+1})$, in which $Tr(1) = 1$. Let $f(t) = t^5$ and $g(t) = t^3$. Then

$$\begin{aligned}
& Tr\left(\frac{(t+\bar{t})(f(t)+f(\bar{t}))}{(g(t)+g(\bar{t}))^2}\right) \\
= \;& Tr\left(\frac{(t+\bar{t})(t^5+\bar{t}^5)}{(t^3+\bar{t}^3)^2}\right) \\
= \;& Tr\left(1+\frac{t\bar{t}}{t^2+t\bar{t}+\bar{t}^2}+\frac{(t\bar{t})^2}{(t^2+t\bar{t}+\bar{t}^2)^2}\right) = 1.
\end{aligned}$$

Therefore, $f(t) = t^5$ and $g(t) = t^3$ define a flock in $PG(3, 2^{2m+1})$. This flock is called the Payne flock, as it is equivalent to the generalized quadrangle found by S. E. Payne [9].

One may wish to find all the pairs of monomial functions f and g that define a conical flock in $PG(3, 2^n)$. T. Penttila and L. Storme [10] have shown that the Fisher-Thas-Walker flock and the Payne flock are the only two nonlinear monomial flocks in $PG(3, 2^n)$, and n must be odd. This beautiful result completes the search of monomial flocks in $PG(3, 2^n)$.

5 Flocks in the Real Projective Space

In this section, we shall construct the smooth (continuous, discontinuous, resp.) flocks defined by smooth (continuous, discontinuous, resp.) functions f and g over $\mathbb{R}$. Various of our examples were inspired by results in F. De Clerck and H. Van Maldeghem in [4]. First, we have the following fundamental result:

Theorem 5 *Let f_1, g_1 and f_2, g_2 define two conical flocks $\mathcal{F}_1$ and $\mathcal{F}_2$ in $PG(3, \mathbb{R})$, respectively. Then $\mathcal{F}_1 \cong \mathcal{F}_2$ iff there is a bijective function $h : \mathbb{R} \to \mathbb{R}$ and a 3×3 nonsingular matrix over $\mathbb{R}$ and $d_1, d_2, d_3 \in \mathbb{R}$ such that*

$$\begin{bmatrix} a_1 & b_1 & c_1 \\ a_2 & b_2 & c_2 \\ a_3 & b_3 & c_3 \end{bmatrix} \cdot \begin{bmatrix} h(t) \\ f_2(h(t)) \\ g_2(h(t)) \end{bmatrix} + \begin{bmatrix} d_1 \\ d_2 \\ d_3 \end{bmatrix} = \begin{bmatrix} t \\ f_1(t) \\ g_1(t) \end{bmatrix}, \tag{5.1}$$

where

$$c_1^2 = a_1 b_1, \quad c_2^2 = a_2 b_2,$$

and

$$c_3 = \pm\sqrt{(a_1 + a_2 + a_3)(b_1 + b_2 + b_3)} - c_1 - c_2.$$

Proof. Notice that $Aut(\mathbb{R}) = 1$ and a semilinear transformation is linear. Then (5.1) holds iff

$$\begin{aligned} a_1 h(t) + b_1 f_2(h(t)) + c_1 g_2(h(t)) + d_1 &= t, \\ a_2 h(t) + b_2 f_2(h(t)) + c_2 g_2(h(t)) + d_2 &= f_1(t), \\ a_3 h(t) + b_3 f_2(h(t)) + c_3 g_2(h(t)) + d_3 &= g_1(t), \end{aligned}$$

iff

$$\begin{aligned} h(t)(a_1 x_0 + a_2 x_1 + a_3 x_2) + f_2(h(t))(b_1 x_0 + b_2 x_1 + b_3 x_2) + \\ g_2(h(t))(c_1 x_0 + c_2 x_1 + c_3 x_2) + (d_1 x_0 + d_2 x_1 + d_3 x_2) \\ = t x_0 + f_1(t) x_1 + g_1(t) x_2, \end{aligned}$$

iff there is a linear transformation $\tau \in PG(4, \mathbb{R})$ that maps $\pi_t^1 : t x_0 + f_1(t) x_1 + g_1(t) x_2 + x_3 = 0$ to $\pi_{h(t)}^2 : h(t) x_0 + f_2(h(t)) x_1 + g_2(h(t)) x_2 + x_3 = 0$ and preserves the quadratic cone as follows:

$$\tau = \begin{bmatrix} a_1 & b_1 & c_1 & d_1 \\ a_2 & b_2 & c_2 & d_2 \\ a_3 & b_3 & c_3 & d_3 \\ 0 & 0 & 0 & 1 \end{bmatrix}.$$

Since τ preserves the cone: $x_0x_1 - x_2^2 = 0$, $(a_1x_0 + a_2x_1 + a_3x_2)(b_1x_0 + b_2x_1 + b_3x_2) = (c_1x_0 + c_2x_1 + c_3x_2)^2$. Take three points on the cone, namely, $(1,0,0,0), (0,1,0,0), (1,1,1,0)$. Then $c_1^2 = a_1b_1$, $c_2^2 = a_2b_2$ and $(c_1 + c_2 + c_3)^2 = (a_1 + a_2 + a_3)(b_1 + b_2 + b_3)$. Therefore,

$$c_3 = \pm\sqrt{(a_1 + a_2 + a_3)(b_1 + b_2 + b_3)} - c_1 - c_2,$$

as desired. Q.E.D.

We are now in a position to construct various new and inequivalent classes of conical flocks in $PG(3,\mathbb{R})$ in the rest of this section.

5.1 The Smooth Flocks

Smooth flocks in $PG(3,\mathbb{R})$ can easily be obtained by choosing appropriate smooth functions f and g. In general, f and g can be chosen to be only piecewise smooth or even discontinuous on $\mathbb{R}$. The following result is a criterion for a set of planes to define a smooth conical flocks in $PG(3,\mathbb{R})$.

Theorem 6 *Let f and g be smooth functions on $\mathbb{R}$. Suppose that there does not exist s in $\mathbb{R}$ such that $s^2 + g'(t)s + f'(t) \equiv 0$. Then the set of planes*

$$\mathcal{P} = \{\pi_t : tx_0 + f(t)x_1 + g(t)x_2 + x_3 = 0 \mid t \in \mathbb{R}\}$$

defines a flock of $\mathcal{Q}_C : x_0x_1 - x_2^2 = 0$ in $PG(3,\mathbb{R})$ iff

$$f'(t) \geq \left(\frac{g'(t)}{2}\right)^2 \tag{5.2}$$

and

$$\lim_{t\to\pm\infty} \phi_s(t) = \pm\infty. \tag{5.3}$$

Proof. Note that $\phi_s(t) = ts^2 + g(t)s + f(t)$, for each fixed $s \in \mathbb{R}$. Obviously $\phi_s(t)$ is smooth on $\mathbb{R}$. Then $\frac{d}{dt}(\phi_s(t)) = s^2 + g'(t)s + f'(t)$ is also smooth. Consider $\frac{d}{dt}(\phi_s(t))$ as a quadratic polynomial in s. Then $\phi_s(t)$ is one-to-one iff $\frac{d}{dt}(\phi_s(t)) \not\equiv 0$ is nonnegative (since the coefficient of s^2 is positive), iff the discriminant $(g'(t))^2 - 4(f'(t)) \leq 0$ iff $f'(t) \geq \left(\frac{g'(t)}{2}\right)^2$, which is (5.2) but the equality cannot always hold. Since $\frac{d}{dt}(\phi_s(t)) \geq 0$, $\phi_s(t)$ is onto iff (5.3) holds. It follows from Theorem 1 that $\mathcal{P}$ defines a conical flock iff ϕ_s is bijective for each $s \in \mathbb{R}$, iff (5.2) and (5.3) hold and the equality in (5.2) cannot hold for all $t \in \mathbb{R}$. Q.E.D.

The first immediate corollary is the simple but important flock—the linear flock in $PG(3,\mathbb{R})$.

Corollary 7 *The set of planes*

$$\mathcal{P}(\ell) = \{\pi_t : tx_0 + atx_1 + btx_2 + x_3 = 0 \mid t \in \mathbb{R}\},$$

where $0 \leq \frac{b^2}{4a} < 1$, *defines a linear flock whose associated planes share the line* $\ell : x_0 + ax_1 + bx_2 = 0 = x_3$.

Proof: Notice that $f(t) = at$ and $g(t) = bt$ are both smooth functions on $\mathbb{R}$ and $a > 0$. Since $b^2 - 4a < 0$, $s^2 + bs + a > 0$, for all $s \in \mathbb{R}$.

$$\begin{aligned} & \lim_{t\to\pm\infty} \phi_s(t) \\ = \ & \lim_{t\to\pm\infty} (ts^2 + bts + at) = \lim_{t\to\pm\infty} t(s^2 + bs + a) \\ = \ & \pm\infty, \ \forall s \in \mathbb{R}. \end{aligned}$$

$f'(t) = a$ and $\left(\frac{g'(t)}{2}\right)^2 = \frac{b^2}{4}$. Then $f'(t) \geq \left(\frac{g'(t)}{2}\right)^2$ iff $\frac{b^2}{4a} < 1$. It follows from Theorem 6 that $\mathcal{P}$ defines a conical flock of $\mathcal{Q}_C$ iff $0 \leq \frac{b^2}{4a} < 1$.

Finally, if $\frac{b^2}{4a} = 1$, then $\frac{d}{dt}(\phi_s(t)) \equiv 0$, so $\phi_s(t)$ is constant which implies that it is not bijective. Q.E.D.

Corollary 8 *The set of planes*

$$\mathcal{P} = \{\pi_t : tx_0 + at^3x_1 + bt^2x_2 + x_3 = 0 \mid t \in \mathbb{R}\}$$

defines a conical flock of $\mathcal{Q}_C : x_0x_1 - x_2^2 = 0$ *in* $PG(3,\mathbb{R})$ *iff* $0 \leq \frac{b^2}{3a} \leq 1$. *Moreover, if* $a = b^2$, *then the generalized Betten flock defined by* $\mathcal{P}$ *is unique up to isomorphism.*

Proof. Notice that $f(t) = at^3$ and $g(t) = bt^2$ are both smooth functions on $\mathbb{R}$ and $a > 0$. Obviously $s^2 + g'(t)s + f'(t) = s^2 + 2bts + 3at^2 \not\equiv 0$.

$$\lim_{t\to\pm\infty} \phi_s(t) = \lim_{t\to\pm\infty} \left(ts^2 + bt^2s + at^3\right) = \pm\infty, \ \forall s \in \mathbb{R}.$$

$f'(t) = 3at^2$ and $\left(\frac{g'(t)}{2}\right)^2 = b^2t^2$. Then $f'(t) \geq \left(\frac{g'(t)}{2}\right)^2$ iff $3a \geq b^2$ iff $\frac{b^2}{3a} \leq 1$. It follows from Theorem 6 that $\mathcal{P}$ defines a conical flock of $\mathcal{Q}_C$ iff $0 \leq \frac{b^2}{3a} \leq 1$.

Suppose that $a = b^2$, $a' = b'^2$ and $\mathcal{P}' = \{\pi'_t : tx_0 + a't^3x_1 + b't^2x_2 + x_3 = 0 \mid t \in \mathbb{R}\}$. Then τ defined as below will map $\mathcal{P}$ to $\mathcal{P}'$ and preserve the cone $\mathcal{Q}_C$:

$$\tau = \begin{bmatrix} 1 & 0 & 0 & 0 \\ 0 & \frac{a}{a'} & 0 & 0 \\ 0 & 0 & \frac{b}{b'} & 0 \\ 0 & 0 & 0 & 1 \end{bmatrix}.$$

Therefore, when $a = b^2$, the generalized Betten flock in $PG(3, \mathbb{R})$ is unique. Q.E.D.

We give two examples of proper subfields of $\mathbb{R}$ over which x^3 is bijective as follows.

Example 9 *The subfield of $\mathbb{R}$ that contains all the algebraic numbers.*

Example 10 *Define $(\mathbb{K})^{\frac{1}{3}}$ to be the extension field of $\mathbb{K}$ by joining all the cubed roots of itself. Then the union of the ascending chain of fields:*

$$\mathbb{Q} \ \cup \ (\mathbb{Q})^{\frac{1}{3}} \ \cup \ ((\mathbb{Q})^{\frac{1}{3}})^{\frac{1}{3}} \ \cup \ (((\mathbb{Q})^{\frac{1}{3}})^{\frac{1}{3}})^{\frac{1}{3}} \ \cup \ \cdots$$

will define the smallest subfield of $\mathbb{R}$ satisfying the condition that x^3 is bijective.

The following result will provide infinitely many classes of conical flocks in $PG(3, \mathbb{R})$.

Theorem 11 *Let n be a positive integer. Let $0 < |\delta| \leq 1$. Then*

$$\mathcal{P}_1^n = \{\pi_t : tx_0 + \left(\frac{1}{2n+1}\right) t^{2n+1}x_1 + \delta \left(\frac{2}{n+1}\right) t^{n+1}x_2 + x_3 = 0 \mid t \in \mathbb{R}\}$$

defines a nonlinear flock of $\mathcal{Q}_C : x_0x_1 - x_2^2 = 0$ in $PG(3, \mathbb{R})$.

Proof. Notice that $f(t) = (\frac{1}{2n+1})t^{2n+1}$ and $g(t) = \delta(\frac{2}{n+1})t^{n+1}$ are smooth functions on $\mathbb{R}$. $\phi_s(t) = ts^2 + g(t)s + f(t) = ts^2 + \delta(\frac{2}{n+1})t^{n+1}s + (\frac{1}{2n+1})t^{2n+1}$. Obviously $s^2 + g'(t)s + f'(t) \not\equiv 0$.

$$\begin{aligned} &\lim_{t \to \pm\infty} \phi_s(t) \\ = &\lim_{t \to \pm\infty} \left(ts^2 + \delta \left(\frac{2}{n+1}\right) t^{n+1}s + \left(\frac{1}{2n+1}\right) t^{2n+1}\right) \\ = &\pm\infty, \ \forall s \in \mathbb{R}. \end{aligned}$$

Also, $f'(t) = t^{2n}$ and $g'(t) = 2\delta t^n$, hence $f'(t) \geq \left(\frac{g'(t)}{2}\right)^2$. It follows from Theorem 6 that $\mathcal{P}_1^n$ defines a (nonlinear) conical flock. Q.E.D.

When $\delta = 0$, the planes of $\mathcal{P}_1^n$ also define a flock, as is shown below. Such flocks will form another class of conical flocks—the concurrent flocks. This class will be studied extensively later in section 7.

Theorem 12 *Let n be a positive integer. Let γ be a positive real number. Then*

$$\mathcal{P}_2^n = \{\pi_t : tx_0 + \gamma t^{2n+1}x_1 + x_3 = 0 \mid t \in \mathbb{R}\}$$

defines a nonlinear flock of $\mathcal{Q}_C : x_0x_1 - x_2^2 = 0$ in $PG(3, \mathbb{R})$.

Proof. Notice that $f(t) = \gamma t^{2n+1}$ and $g(t) \equiv 0$ are smooth functions on $\mathbb{R}$. $\phi_s(t) = ts^2 + g(t)s + f(t) = ts^2 + \gamma t^{2n+1}$. Obviously $s^2 + g'(t)s + f'(t) \not\equiv 0$.

$$\lim_{t\to\pm\infty} \phi_s(t) = \lim_{t\to\pm\infty} \left(ts^2 + \gamma t^{2n+1}\right) = \pm\infty, \ \forall s \in \mathbb{R}.$$

Also, $f'(t) = \gamma(2n+1)t^{2n} \geq 0$ and $g'(t) = 0$, hence $f'(t) \geq \left(\frac{g'(t)}{2}\right)^2$. It follows from Theorem 6 that $\mathcal{P}_2^n$ defines a (nonlinear) conical flock. Q.E.D.

Remark 13 *When $n = 1$, $\mathcal{P}_1^1$ defines the Betten flock. Since x^{2n+1} is a one-to-one but not onto function on some proper subfield of $\mathbb{R}$ such as the field of rationals $\mathbb{Q}$, one can immediately obtain infinitely many classes of maximal partial conical flocks in $PG(3, \mathbb{Q})$.*

More can be said about $\mathcal{P}_1^n$ and $\mathcal{P}_2^n$. In fact, the following is true.

Theorem 14 *Let a and b be positive integers. Let $c_1 > 0$ and c_2 be two fixed real numbers. Suppose that*

$$\mathcal{P}^{(a,b)} = \{\pi_t : tx_0 + c_1t^a x_1 + c_2t^b x_2 + x_3 = 0 \mid t \in \mathbb{R}\}$$

defines a nonlinear monomial conical flock in $PG(3, \mathbb{R})$. Then $\mathcal{P}$ is either $\mathcal{P}_1^n$ and $\frac{c_2^2}{4c_1} \leq 1$, or $\mathcal{P}_2^n$ for some positive integer n.

Proof. First, when $s = 0$, obviously a must be odd and $c_1 > 0$ in order for (5.3) to hold. Next, $f'(t) = c_1at^{a-1}$ and $g'(t) = c_2bt^{b-1}$, obviously $s^2 + g'(t)s + f'(t) \not\equiv 0$. By (5.2), $c_1at^{a-1} \geq \frac{(c_2b)^2}{4}t^{2b-2}$ must hold for all $t \in \mathbb{R}$. Hence $t^{a-2b+1} \geq \frac{(c_2b)^2}{4c_1a} \geq 0$.

If $c_2 \neq 0$, then $t^{a-2b+1} \geq \frac{(c_2 b)^2}{4c_1 a} > 0$ for all $t \in \mathbb{R}$. If $a - 2b + 1 < 0$, then $t^{-(a-2b+1)} \leq \frac{4c_1 a}{(c_2 b)^2}$, which means $t^{-(a-2b+1)}$ is bounded on $\mathbb{R}$, a contradiction. If $a - 2b + 1 > 0$, then a sufficiently small t will contradict $t^{a-2b+1} \geq \frac{(c_2 b)^2}{4c_1 a}$. Therefore, $a - 2b + 1 = 0$, that is, $a = 2b - 1$.

If $b = 1$, then $a = 2b - 1 = 1$. Hence $\mathcal{P}^{(1,1)}$ defines a linear flock. Let $b = n + 1$ for some positive integer n. Then $a = 2b - 1 = 2n + 1$. So $\mathcal{P}^{(a,b)} = \{tx_0 + c_1 t^{2n+1} x_1 + c_2 t^{n+1} x_2 + x_3 = 0 \mid t \in \mathbb{R}\}$.

Again $t^{a-2b+1} \geq \frac{(c_2 b)^2}{4c_1 a}$, for all $t \in \mathbb{R}$. Since $a = 2n + 1$, $b = n + 1$ and $a - 2b + 1 = 0$,

$$\frac{(c_2(n+1))^2}{4c_1(2n+1)} \leq 1. \tag{5.4}$$

Then

$$n^2 + 2\left(1 - \frac{4c_1}{c_2^2}\right) n + \left(1 - \frac{4c_1}{c_2^2}\right) \leq 0.$$

This is equivalent to

$$\left[n + \left(1 - \frac{4c_1}{c_2^2}\right)\right]^2 \leq \left(1 - \frac{4c_1}{c_2^2}\right)^2 - \left(1 - \frac{4c_1}{c_2^2}\right),$$

or

$$0 \leq \left[n + \left(1 - \frac{4c_1}{c_2^2}\right)\right]^2 \leq \left(1 - \frac{4c_1}{c_2^2}\right)\left(-\frac{4c_1}{c_2^2}\right),$$

which implies that $\frac{c_2^2}{4c_1} \leq 1$.

Since (5.4) holds, let $\delta^2 = \frac{(c_2(n+1))^2}{4c_1(2n+1)}$. Then $0 < |\delta| \leq 1$, and $\left(\frac{(n+1)c_2}{2\delta}\right)^2 = (2n+1)c_1$. Take $\sigma \in GL(4, \mathbb{R})$ as follows:

$$\sigma = \begin{bmatrix} 1 & 0 & 0 & 0 \\ 0 & (2n+1)c_1 & 0 & 0 \\ 0 & 0 & \frac{(n+1)c_2}{2\delta} & 0 \\ 0 & 0 & 0 & 1 \end{bmatrix}.$$

Then for any point $(x_0, x_1, x_2, -(tx_0 + c_1 t^{2n+1} x_1 + c_2 t^{n+1} x_2))$ on the plane with equation $tx_0 + c_1 t^{2n+1} x_1 + c_2 t^{n+1} x_2 + x_3 = 0$,

$$\begin{aligned} & (x_0, x_1, x_2, -(tx_0 + c_1 t^{2n+1} x_1 + c_2 t^{n+1} x_2))^\sigma \\ = \; & \left(x_0, (2n+1)c_1 x_1, \frac{(n+1)c_2}{2\delta} x_2, -(tx_0 + c_1 t^{2n+1} x_1 + c_2 t^{n+1} x_2)\right), \end{aligned}$$

which is on the plane with equation $tx_0 + \left(\frac{1}{2n+1}\right) t^{2n+1}x_1 + \delta \left(\frac{2}{n+1}\right) t^{n+1}x_2 + x_3 = 0$, that is,

$$\begin{aligned} & (tx_0 + c_1 t^{2n+1}x_1 + c_2 t^{n+1}x_2 + x_3 = 0)^\sigma \\ = \ & (tx_0 + \left(\frac{1}{2n+1}\right) t^{2n+1}x_1 + \delta \left(\frac{2}{n+1}\right) t^{n+1}x_2 + x_3 = 0), \end{aligned}$$

and σ fixes $\mathcal{Q}_C$ and $\mathbf{v}$. Therefore, in this case, $\mathcal{P}^{(a,b)}$ is projectively equivalent to $\mathcal{P}_1^n$ for some positive integer n.

If $c_2 = 0$, then $\mathcal{P}^{(a,b)} = \{tx_0 + c_1 t^a x_1 + x_3 = 0 \mid t \in \mathbb{R}\}$. Notice that a is odd. $a = 1$ again will produce a linear conical flock. Therefore, $a = 2n + 1$ for some positive integer n and $\mathcal{P}^{(a,b)} = \{tx_0 + c_1 t^{2n+1}x_1 + x_3 = 0 \mid t \in \mathbb{R}\}$.

Finally, for any $\gamma > 0$, take $\sigma' \in GL(4, \mathbb{R})$ as follows:

$$\sigma' = \begin{bmatrix} 1 & 0 & 0 & 0 \\ 0 & \frac{c_1}{\gamma} & 0 & 0 \\ 0 & 0 & \sqrt{\frac{c_1}{\gamma}} & 0 \\ 0 & 0 & 0 & 1 \end{bmatrix}.$$

Then it is easy to see that

$$\begin{aligned} & (tx_0 + c_1 t^{2n+1}x_1 + x_3 = 0)^{\sigma'} \\ = \ & (tx_0 + \gamma t^{2n+1}x_1 + x_3 = 0). \end{aligned}$$

and σ' fixes $\mathcal{Q}_C$ and $\mathbf{v}$. Therefore, in this case, $\mathcal{P}^{(a,b)}$ is projectively equivalent to $\mathcal{P}_2^n$ for some positive integer n. Q.E.D.

The following result provides another two classes of smooth flocks in $PG(3, \mathbb{R})$.

Theorem 15 *Let $\gamma > 0$. Let $|\delta| \leq 1$. Then the following two sets of planes define nonlinear flocks of $\mathcal{Q}_C : x_0x_1 - x_2^2 = 0$ in $PG(3, \mathbb{R})$, respectively.*

$$tx_0 + (\gamma t + \sin(\gamma t))x_1 + 4\delta\sqrt{\frac{2}{\gamma}} \sin\left(\frac{\gamma}{2}t\right) x_2 + x_3 = 0, \quad t \in \mathbb{R}, \tag{5.5}$$

$$tx_0 + (\gamma t - \cos(\gamma t))x_1 + \frac{4\delta}{\sqrt{\gamma}} \left(\sin\left(\frac{\gamma}{2}t\right) - \cos\left(\frac{\gamma}{2}t\right)\right) x_2 + x_3 = 0, \quad t \in \mathbb{R}. \tag{5.6}$$

Proof. Note that $s^2 + g'(t)s + f'(t) \not\equiv 0$. (1) $f(t) = \gamma t + \sin(\gamma t)$ and $g(t) = 4\delta\sqrt{\frac{2}{\gamma}}\sin\left(\frac{\gamma}{2}t\right)$ are both smooth functions on $\mathbb{R}$ and obviously

$$\begin{aligned} & \lim_{t\to\pm\infty} \phi_s(t) \\ = & \lim_{t\to\pm\infty}\left(ts^2 + \left(4\delta\sqrt{\frac{2}{\gamma}}\sin\left(\frac{\gamma}{2}t\right)\right)s + (\gamma t + \sin(\gamma t))\right) \\ = & \pm\infty, \ \forall s \in \mathbb{R}. \end{aligned}$$

$f'(t) = \gamma(1+\cos(\gamma t))$, $g'(t) = 2\delta\sqrt{2\gamma}\cos\left(\frac{\gamma}{2}t\right)$. Then $f'(t) \geq \left(\frac{g'(t)}{2}\right)^2$. It follows from Theorem 6 that (5.5) defines a (nonlinear) flock of $\mathcal{Q}_C$.
(2) $f(t) = \gamma t - \cos(\gamma t)$ and $g(t) = \frac{4\delta}{\sqrt{\gamma}}\left(\sin\left(\frac{\gamma}{2}t\right) - \cos\left(\frac{\gamma}{2}t\right)\right)$ are both smooth functions on $\mathbb{R}$ and obviously for each $s \in \mathbb{R}$,

$$\begin{aligned} & \lim_{t\to\pm\infty} \phi_s(t) \\ = & \lim_{t\to\pm\infty}\left(ts^2 + \frac{4\delta}{\sqrt{\gamma}}\left(\sin\left(\frac{\gamma}{2}t\right) - \cos\left(\frac{\gamma}{2}t\right)\right)s + (\gamma t - \cos(\gamma t))\right) \\ = & \pm\infty. \end{aligned}$$

$f'(t) = \gamma(1+\sin(\gamma t))$, $g'(t) = 2\delta\sqrt{\gamma}\left(\cos\left(\frac{\gamma}{2}t\right) + \sin\left(\frac{\gamma}{2}t\right)\right)$. Then $f'(t) \geq \left(\frac{g'(t)}{2}\right)^2$. It follows from Theorem 6 that (5.6) defines a (nonlinear) flock of $\mathcal{Q}_C$. Q.E.D.

5.2 The Continuous Flocks

Let $g(t) = 0$. Then we can easily construct various continuous flocks in $PG(3, \mathbb{R})$. Note that such flocks are concurrent flocks with respect to the point $(0, 0, 1, 0)$. We shall study these special concurrent flocks in a general field $\mathbb{K}$ in section 7.

Theorem 16 *Let f be a continuous, strictly increasing function on $\mathbb{R}$. If $\lim_{t\to\pm\infty} f(t) = \pm\infty$, then $\mathcal{P}^f = \{\pi_t : tx_0 + f(t)x_1 + x_3 = 0 \mid t \in \mathbb{R}\}$ defines a flock of $\mathcal{Q}_C : x_0x_1 - x_2^2 = 0$ in $PG(3, \mathbb{R})$.*

Proof. Since $f(t)$ is strictly increasing, so is $\phi_s(t) = ts^2 + f(t)$ for each fixed $s \in \mathbb{R}$. In fact, if $t_1 < t_2$, then $\phi_s(t_1) = t_1s^2 + f(t_1) < t_2s^2 + f(t_2) = \phi_s(t_2)$. This implies that ϕ_s is one-to-one.

Since $\lim_{t\to\pm\infty} f(t) = \pm\infty$, then $\lim_{t\to\pm\infty}\left(ts^2 + f(t)\right) = \pm\infty$. Thus, ϕ_s is onto and consequently it is bijective. It then follows from Theorem 1 that $\mathcal{P}^f$ will define a conical flock. Q.E.D.

As an immediate corollary to Theorem 16, we have

Corollary 17 *A continuous function f consisting of piecewise positive odd power functions with positive coefficients will give rise to $\mathcal{P}^f$ which defines a concurrent conical flock in $PG(3,\mathbb{R})$.*

In general, we have the following construction to continuous (piecewise smooth) flocks in $PG(3,\mathbb{R})$.

Theorem 18 *Suppose that (f_1, g_1) and (f_2, g_2) define two smooth flocks in $PG(3,\mathbb{R})$. If $f_1(0) = f_2(0)$ and $g_1(0) = g_2(0)$, then (f, g) will define a continuous flock in $PG(3,\mathbb{R})$, where*

$$f(t) = \begin{cases} f_1(t), & t \geq 0, \\ f_2(t), & t < 0, \end{cases}$$

and

$$g(t) = \begin{cases} g_1(t), & t \geq 0, \\ g_2(t), & t < 0. \end{cases}$$

Proof. $\phi_s(t)$ is a continuous function on $\mathbb{R}$. In fact,

$$\begin{aligned} \phi_s(t) &= ts^2 + g(t)s + f(t) \\ &= \begin{cases} ts^2 + g_1(t)s + f_1(t), & t \geq 0, \\ ts^2 + g_2(t)s + f_2(t), & t < 0. \end{cases} \end{aligned}$$

Since

$$\lim_{t\to 0^+} \phi_s(t) = g_1(0)s + f_1(0) = g_2(0)s + f_2(0) = \lim_{t\to 0^-} \phi_s(t),$$

$\phi_s(t)$ is continuous. Note that $ts^2 + g_1(t)s + f_1(t)$ and $ts^2 + g_2(t)s + f_2(t)$ are increasing and

$$\lim_{t\to\infty}\left(ts^2 + g_1(t)s + f_1(t)\right) = \infty,$$

$$\lim_{t\to-\infty}\left(ts^2 + g_2(t)s + f_2(t)\right) = -\infty.$$

Therefore, $\phi_s(t)$ is bijective on $\mathbb{R}$ for each $s \in \mathbb{R}$. It follows from Theorem 1 that f and g define a conical flock in $PG(3,\mathbb{R})$. Q.E.D.

5.3 The Discontinuous Flocks

We first complete the list of conical flocks defined by power functions f and g in $\mathbb{R}$.

Theorem 19 *Let m be a negative integers. Let $|\delta| \leq 1$. Then*

$$\{\pi_t : tx_0 + \left(\frac{1}{2m-1}\right) t^{2m-1} x_1 + \delta\left(\frac{2}{m}\right) t^m x_2 + x_3 = 0 \mid t \neq 0\} \cup \{\pi_0 : x_3 = 0\} \tag{5.7}$$

defines a (nonlinear) flock of $\mathcal{Q}_C : x_0x_1 - x_2^2 = 0$ in $PG(3, \mathbb{R})$.

Proof. Note that

$$f(t) = \begin{cases} \left(\frac{1}{2m-1}\right) t^{2m-1}, & \text{if } t \neq 0, \\ 0, & \text{if } t = 0 \end{cases}$$

and

$$g(t) = \begin{cases} \delta\left(\frac{2}{m}\right) t^m, & \text{if } t \neq 0, \\ 0, & \text{if } t = 0. \end{cases}$$

Then

$$\phi_s(t) = \begin{cases} ts^2 + \delta\left(\frac{2}{m}\right) t^m s + \left(\frac{1}{2m-1}\right) t^{2m-1}, & \text{if } t \neq 0, \\ 0, & \text{if } t = 0. \end{cases}$$

Hence when $t \neq 0$,

$$\begin{aligned} \frac{d}{dt}(\phi_s(t)) &= s^2 + 2\delta t^{m-1}s + t^{2m-2} \\ &\geq s^2 + 2\delta t^{m-1}s + \delta^2 t^{2m-2} \\ &= \left(s + \delta t^{m-1}\right)^2 \geq 0. \end{aligned}$$

That is, $\frac{d}{dt}(\phi_s(t)) \geq 0$, $\forall t \neq 0$, which implies that $\phi_s(t)$ is increasing on $(0, \infty)$ and $(-\infty, 0)$. On the other hand,

$$\lim_{t \to 0^{\pm}} \phi_s(t) = \lim_{t \to 0^{\pm}} \left(ts^2 + \delta\left(\frac{2}{m}\right) t^m s + \left(\frac{1}{2m-1}\right) t^{2m-1}\right) = \mp\infty, \ \forall s \in \mathbb{R},$$

and

$$\lim_{t \to \pm\infty} \phi_s(t) = \lim_{t \to \pm\infty} \left(ts^2 + \delta\left(\frac{2}{m}\right) t^m s + \left(\frac{1}{2m-1}\right) t^{2m-1}\right) = 0.$$

Therefore, $\phi_s(t)$ is onto and one-to-one on $\mathbb{R}$. It follows from Theorem 1 that (5.7) defines a (nonlinear) conical flock in $PG(3, \mathbb{R})$. Q.E.D.

Note that in Theorem 11 and Theorem 19, $f(t) = \frac{1}{2n-1}t^{2n-1}$, where $n = \pm 1, \pm 2, \ldots$. In fact, when $n = 0$, we have the following result.

Theorem 20 *Let $|\delta| \leq 1$. Then*

$$\{\pi_t : tx_0 - \frac{1}{t}x_1 + \delta \ln(t^2)x_2 + x_3 = 0 \mid t \neq 0\} \cup \{\pi_0 : x_3 = 0\} \qquad (5.8)$$

defines a nonlinear flock of $\mathcal{Q}_C : x_0x_1 - x_2^2 = 0$ in $PG(3, \mathbb{R})$.

Proof. Again note that

$$f(t) = \begin{cases} -\frac{1}{t}, & \text{if } t \neq 0, \\ 0, & \text{if } t = 0, \end{cases}$$

$$g(t) = \begin{cases} \delta \ln(t^2), & \text{if } t \neq 0, \\ 0, & \text{if } t = 0, \end{cases}$$

$$\phi_s(t) = \begin{cases} ts^2 + \delta \ln(t^2)s - \frac{1}{t}, & \text{if } t \neq 0, \\ 0, & \text{if } t = 0. \end{cases}$$

Then when $t \neq 0$,

$$\begin{aligned} \frac{d}{dt}(\phi_s(t)) &= s^2 + 2\delta\frac{1}{t} + \frac{1}{t^2} \\ &\geq s^2 + 2\delta\frac{1}{t} + \delta^2\frac{1}{t^2} \\ &= \left(s + \delta\frac{1}{t}\right)^2 \geq 0. \end{aligned}$$

Hence $\phi_s(t)$ is a strictly increasing function on $(0, \infty)$ and $(-\infty, 0)$. On the other hand,

$$\lim_{t \to 0^\pm} \phi_s(t) = \lim_{t \to 0^\pm} \left(ts^2 + \delta \ln(t^2)s - \frac{1}{t}\right) = \mp\infty, \; \forall s \in \mathbb{R},$$

and

$$\lim_{t \to \pm\infty} \phi_s(t) = \lim_{t \to \pm\infty} \left(ts^2 + \delta \ln(t^2)s - \frac{1}{t}\right) = 0.$$

Therefore, $\phi_s(t)$ is onto and one-to-one on $\mathbb{R}$. It follows from Theorem 1 that (5.8) defines a (nonlinear) conical flock in $PG(3, \mathbb{R})$. Q.E.D.

The following result will show the structure of some discontinuous flocks in $PG(3, \mathbb{R})$.

Theorem 21 *A bijective discontinuous function f on $\mathbb{R}$ gives rise to $\mathcal{P}^f$ which defines a concurrent conical flock in $PG(3,\mathbb{R})$ only if f has the form:*

$$f(t) = \begin{cases} f_1(t), & t < r, \\ c, & t = r, \\ f_2(t), & t > r, \end{cases}$$

where $c, r \in \mathbb{R}$, f_1 and f_2 are continuous, increasing functions. Moreover,

$$\lim_{t \to r^-} f_1(t) = \infty, \qquad \lim_{t \to r^+} f_2(t) = -\infty,$$

and

$$\lim_{t \to -\infty} f_1(t) = c, \qquad \lim_{t \to \infty} f_2(t) = c.$$

Proof. If f consists of n pieces, i.e., there is an increasing sequence of real numbers $r_1 < r_2 < \ldots < r_{n-1}$ and $r_0 = -\infty, r_n = \infty$ such that

$$f(t) = \begin{cases} f_1(t), & t < r_1, \\ c_1, & t = r_1, \\ f_2(t), & r_1 < t < r_2, \\ c_2, & t = r_2, \\ \vdots & \\ f_{n-1}(t), & r_{n-2} < t < r_{n-1}, \\ c_{n-1}, & t = r_{n-1}, \\ f_n(t), & t > r_{n-1}, \end{cases}$$

then $f_i(t)$ must be continuous, increasing on (r_{i-1}, r_i) for $1 \le i \le n$ so that $ts^2 + f_i(t)$ is also continuous and increasing on (r_{i-1}, r_i) for $1 \le i \le n$.

If $n \ge 3$, then without loss of generality, we may assume that

$$\lim_{t \to r_1^-} f_1(t) = \lim_{t \to r_2^+} f_3(t).$$

On the other hand, since $r_1 s^2 < r_2 s^2$ if $s \ne 0$,

$$\lim_{t \to r_1^-} (ts^2 + f_1(t)) \ne \lim_{t \to r_2^+} (ts^2 + f_3(t)).$$

This implies that $\phi_s(t)$ is not onto, a contradiction. Therefore, $n = 2$, and

$$f(t) = \begin{cases} f_1(t), & t < r, \\ c, & t = r, \\ f_2(t), & t > r, \end{cases}$$

for some $c, r \in \mathbb{R}$. Finally, since $f(t)$ is bijective,

$$\lim_{t \to r^-} f_1(t) = \infty, \qquad \lim_{t \to r^+} f_2(t) = -\infty,$$

and

$$\lim_{t \to -\infty} f_1(t) = c, \qquad \lim_{t \to \infty} f_2(t) = c,$$

as desired. Q.E.D.

6 The Bilinear Flocks

In [2], M. Biliotti and N. L. Johnson studied bilinear flocks of quadratic cones in $PG(3, \mathbb{K})$, where $\mathbb{K}$ is an infinite ordered field. The corresponding translation planes admit $SL(2, \mathbb{K})$ as a collineation group.

Theorem 22 (Biliotti-Johnson [2]) *Let $\mathbb{K}$ be an ordered field in which there exist $\alpha_1 > 0, \alpha_2 > 0, \beta_1, \beta_2 \in \mathbb{K}$, such that*

$$F_1(x) = x^2 + \beta_1 x + \alpha_1 > 0 \text{ and } F_2(x) = x^2 + \beta_2 x + \alpha_2 > 0, \ \forall x \in \mathbb{K}.$$

Then the set of planes

$$\{\pi_t^+ : tx_0 + \alpha_1 tx_1 + \beta_1 tx_2 + x_3 = 0 \mid t > 0\} \cup$$

$$\{\pi_t^- : tx_0 + \alpha_2 tx_1 + \beta_2 tx_2 + x_3 = 0 \mid t < 0\} \cup \{\pi_0 : x_3 = 0\} \tag{6.9}$$

will define a bilinear flock of $\mathcal{Q}_C : x_0x_1 - x_2^2 = 0$ in $PG(3, \mathbb{K})$, provided that $(\alpha_1, \beta_1) \neq (\alpha_2, \beta_2)$.

Proof. Notice that

$$\begin{aligned} \phi_s(t) &= \begin{cases} ts^2 + \beta_1 ts + \alpha_1 t, & \text{if } t > 0, \\ ts^2 + \beta_2 ts + \alpha_2 t, & \text{if } t < 0, \\ 0, & \text{if } t = 0, \end{cases} \\ &= \begin{cases} t(s^2 + \beta_1 s + \alpha_1), & \text{if } t > 0, \\ t(s^2 + \beta_2 s + \alpha_2), & \text{if } t < 0, \\ 0, & \text{if } t = 0. \end{cases} \end{aligned}$$

Since $s^2 + \beta_1 s + \alpha_1 > 0$, and $s^2 + \beta_1 s + \alpha_1 > 0$ are fixed for each fixed $s \in \mathbb{K}$, ϕ_s is bijective from $\mathbb{K}$ to $\mathbb{K}$. It follows from Theorem 1 that the planes in (6.9) define a conical flock. Furthermore, this flock is bilinear: its associated planes share exactly two distinct lines, namely, $\ell^+ : x_0+\alpha_1x_1+\beta_1x_2 = 0 = x_3$ and $\ell^- : x_0 + \alpha_2x_1 + \beta_2x_2 = 0 = x_3$. Q.E.D.

This construction can be generalized to that of n-linear flocks as follows. In Biliotti, Johnson [3], a construction is given of $n-$linear flocks over $PG(3, K)$ which requires that K is a subfield of the reals. Here, we show that at least some $n-$linear flocks may be defined over any ordered field.

We first require a lemma.

Lemma 23 *Let K be an ordered field, let K^+ denote the nonnegative elements of K and let g be a function acting in $AG(2, K)$ which is composed of n linear functions and defined as follows:*

Let $a_i > 0$ for $i = 1, 2, ..., n+1$ and let $\{0 = d_0, d_1, ..., d_{n-1}, d_n = \infty\}$ where $d_i < d_{i+1}$ for $i = 0, 2, ..., n-1$ and where a_i, d_i are in $K \cup \{\infty\}$

Let g be defined as follows:

$g(x) = a_1x$ for $d_o = 0 \leq x < d_1$, $g(x) = a_k(x-d_{k-1}) + \Sigma_{i=1}^{k-1} a_i(d_i - d_{i-1})$ for $d_{k-1} \leq x < d_k$ for $k = 2, 3, ..., n$.

(1) g is a bijective function K^+.

(2) If r in K is positive the function $r+g$ defined by $(r+g)(x) = rx+g(x)$ for all x in K^+ is a bijective function on K^+.

Proof: Since g is strictly increasing , it follows immediately that g is bijective so that (1) is clear.

Let $\{m_0 = 0, m_i = (d_i + r)$ for $i = 1, 2, ..., n\}$

Then, the function h defined by $h(x) = a_1x$ for $0 \leq x < m_1$, $h(x) = a_k(x - m_{k-1}) + \Sigma_{i=1}^{k-1} a_i(m_i - m_{i-1})$ for $m_{k-1} \leq x < m_k$ for $k = 2, 3, ..., n$ is bijective by (1).

Since, $h = r + g$, (2) is proved.

Theorem 24 *Let K be an ordered field. Let $b_i < 0$ for $i = 1, 2, ..., n+1$ and let $\{0 = d_0, d_1, ..., d_{n-1}, d_n = \infty\}$ where $d_i < d_{i+1}$ for $i = 0, 2, ..., n-1$ and where a_i, d_i are in $K \cup \{\infty\}$. Let C denote the quadratic cone in $PG(3, K)$ defined by $x_0x_1 = x_2^2$ with vertex $(0, 0, 0, 1)$ and points are given in homogeneous coordinates.*

Define the planes π_t^k as follows:

$\pi_t^1 : tx_o - b_1tx_1 + x_3 = 0$ for $0 \leq t < d_1$,

$\pi_t^k : tx_o - (b_k(t - d_{k-1}) + \Sigma_{i=1}^{k-1} b_i(d_i - d_{i-1}))x_1 + x_3$ for $d_{k-1} \leq t < d_k$ for $k = 2, 3, ..., n$.

Then $\{\pi_t^k\}$ is a partial flock of the quadratic cone C which covers $(x_3 = 0) \cap C$ and all points $(\alpha/s^2, \alpha, \alpha/s, 1)$ for all s in $K - \{0\}$ and for all $\alpha > 0$.

Proof: Since all elements b_i are negative, replace $-b_i$ by $a_i > 0$. Let g be a function defined by

$g(x) = a_1 x$ for $d_o = 0 \leq x < d_1$, $g(x) = a_k(x - d_{k-1}) + \Sigma_{i=1}^{k-1} a_i(d_i - d_{i-1})$ for $d_{k-1} \leq x < d_k$ for $k = 2, 3, ..., n$.

Hence, we may represent the planes of the partial flock in the form $tx_o + g(t)x_1 + x_2 = 0$ for all t in K.

We note that $x_3 = 0$ is π_0^1. To cover the points $(\alpha/s^2, \alpha, \alpha/s, 1)$ where $\alpha > 0$, we require that $t\alpha/s^2 + g(t)\alpha + 1 = 0$ for some t. Let $r = 1/s^2$ so that it suffices to show that

$rt + g(t) = 1/\alpha$ for all α in $K - \{0\}$. By the previous lemma, this follows immediately and we have the proof to the theorem.

Corollary 25 *For any ordered field K and any positive integer m, there exists $m-$linear flocks of a quadratic cone in $PG(3, K)$.*

Proof: Define the planes $\rho_t : tx_0 - c_1 tx_1 + c_2 tx_2 + x_3 = 0$ where $c_2^2 - 4c_1 < 0$ for $-\infty < t < 0\}$. Then the points $(\alpha/s^2, \alpha, \alpha/s, 1)$ for $\alpha < 0$ are covered by a plane ρ_t by results of Biliotti and Johnson. Hence, if we assume that the line $\cap \rho_t$ is not L_1 of the previous theorem, then we follows immediately that $\{\pi_t^i : i = 1, 2, ..., n\} \cup \rho_t$ is a $m = n + 1-$linear flock of the quadratic cone C.

Actually, using the previously construction but for infinitely many lines, we may obtain a maximal partial flock of a quadratic cone. We illustrate this in the following result:

Theorem 26 *Let K be any ordered field.*

$\omega_t^1 : tx_o + tx_1 + x_3 = 0$ *for* $0 \leq t < 1$,

$\omega_t^k : tx_o + ((t - (k+1))/2^{k-1}) + 2)x_1 + x_3 = 0$ *for* $k - 1 \leq t < k$.

Then $\cup_{i=1}^{\infty} \omega_t^i \cup \{ \rho_t : tx_0 - c_1 tx_1 + c_2 tx_2 + x_3 = 0$ *where* $c_2^2 - 4c_1 < 0$ *for* $-\infty < t < 0\}$ *is a maximal partial flock of a quadratic cone in $PG(3, K)$ which is an $\infty-$linear partial flock but is not a flock.*

Proof: Clearly, we have at least a partial flock by the previous theorem.

Note that the points on the line $L_\infty =< (1,0,0,0), (0,0,0,1) >$ are clearly covered by the planes of the partial flock so we must have a maximal partial flock which perhaps is a flock. However, we see that the function g defined by $g(t) = t$ for $0 \leq t < 1$, $g(t) = (t-(k+1))/2^{k-1} + 2$ for $k-1 \leq t < k$ is bouned by 2. Thus, the points on the line $L_o =< (0,1,0,0), (0,0,0,1) >$ cannot all be covered by the partial flock.

Note that in the previous theorem using a finite number of lines, if $\mathbb{K} = \mathbb{R}$, then we obtain a piecewise smooth flock, which is a continuous flock in $PG(3, \mathbb{R})$.

7 The Concurrent Flocks

It is natural to consider the infinite concurrent flocks of quadratic cones. In this section, the concurrent flock in $PG(3,\mathbb{K}), \mathbb{K} \subset \mathbb{R}$, whose associated planes share the common point $(0,0,1,0)$, which is equivalent to $g(t) \equiv 0$, will be studied. In this case, π_t has equation: $tx_0 + f(t)x_1 + x_3 = 0, \ \ t \in \mathbb{K}$, and $\phi_s(t) = ts^2 + f(t)$ is bijective for each $s \in \mathbb{K}$. In particular, f must be bijective as well. In order to obtain a partial flock, ϕ_s must be one-to-one. Notice that if ϕ_s is not onto, then $\{\pi_t \cap \mathcal{Q}_C \mid t \in \mathbb{K}\}$ forms a maximal partial flock, since $\ell = \langle (0,0,0,1), (1,0,0,0) \rangle$ is always be covered by $\mathcal{P}^f = \{\pi_t : tx_0 + f(t)x_1 + x_3 = 0 \mid t \in \mathbb{K}\}$.

The Kantor-Knuth flocks are the only finite nonlinear concurrent flocks. In Theorem 12, $\mathcal{P}_2^n$ defines the first example of infinite nonlinear concurrent flocks in $PG(3,\mathbb{R})$. In (5.5) and (5.6) of Theorem 15, $\delta = 0$ will give rise to other concurrent flocks in $PG(3,\mathbb{R})$.

By our previous results on n-linear flocks, when $\mathbb{K} = \mathbb{R}$ and $\beta_i = 0, \ \forall i = 1,2,\ldots,n$, an n-linear concurrent flock is obtained. This is an immediate corollary to Theorem 16, in which f is a piecewise linear function with n different positive slopes.

The following result follows immediately from Theorem 2.

Corollary 27 *Suppose that $f(t)$ defines a concurrent flock in $PG(3,\mathbb{K})$. Then for any fixed element $k \in \mathbb{K}$, $f(t) + k$ also defines a concurrent flock in $PG(3,\mathbb{K})$. Consequently, f can be taken such that $f(0) = 0$.*

We have the following sufficient condition for two concurrent flocks to be projectively equivalent.

Theorem 28 *Let two concurrent flocks in $PG(3,\mathbb{K})$ be defined by $\mathcal{P}^{f_1}$ and $\mathcal{P}^{f_2}$, respectively. Then the flock defined by $\mathcal{P}^{f_1}$ is projectively equivalent to the flock defined by $\mathcal{P}^{f_2}$ if $f_1(t) = c^2 f_2(t)$.*

Proof. Let $\sigma \in GL(4,\mathbb{R})$ be defined as follows:

$$\sigma = \begin{bmatrix} 1 & 0 & 0 & 0 \\ 0 & c^2 & 0 & 0 \\ 0 & 0 & c & 0 \\ 0 & 0 & 0 & 1 \end{bmatrix}.$$

Then σ maps a plane π_t^1 of $\mathcal{P}^{f_1}$ to a plane π_t^2 of $\mathcal{P}^{f_2}$ and fixes the quadratic cone: $x_0x_1 - x_2^2 = 0$ and its vertex $\mathbf{v} = (0,0,0,1)$, as desired. Q.E.D.

Another important class of concurrent flocks in $PG(3,\mathbb{K})$ is the *generalized Kantor-Knuth flocks*, denoted by $\mathcal{F}(m,\sigma)$, whose associated planes are:

$$tx_0 - mt^\sigma x_1 + x_3 = 0,\ t \in \mathbb{K},$$

where m is a fixed element of $\mathbb{K}$ such that $ts^2 - mt^\sigma$ is bijective for each $s \in \mathbb{K}$, and $1 \neq \sigma \in Aut(\mathbb{K})$. Notice that in $PG(3,\mathbb{R})$, such a flock does not exist, since $Aut(\mathbb{K})$ is trivial. When $\mathbb{K} = GF(q)$, if $q = 2^n$, then $ts^2 - mt^\sigma$ cannot be bijective, since $s^2 - mt^{\sigma-1} = 0$ has a solution of t. Hence there are no generalized Kantor-Knuth flocks in $PG(3,2^n)$. More generally, if $\mathbb{K}$ is a perfect field with characteristic 2, then there are no generalized Kantor-Knuth flocks in $PG(3,\mathbb{K})$. On the other hand, when $\mathbb{K} = GF(q)$, if $q = p^e$, p an odd prime, then m can be taken as a nonsquare and the flock is the Kantor-Knuth flock. Infinite generalized Kantor-Knuth flocks in $PG(3,\mathbb{K})$ may exist if $Aut(\mathbb{K})$ is nontrivial.

Let $\mathbb{F}$ be any field. Let $\mathbb{K} = \mathbb{F}((X)) = \left\{ \sum_{j=m}^{\infty} c_j X^j \mid c_j \in \mathbb{F}, m \in \mathbb{Z} \right\}$, the field of all Laurent series over $\mathbb{F}$ in the indeterminate X. Define the *valuation* of $\sum_{j=m}^{\infty} c_j X^j$ by $v\left(\sum_{j=m}^{\infty} c_j X^j\right) = m$ if $c_m \neq 0$.

Theorem 29 *Let $\mathbb{K} = \mathbb{F}((X))$. Let $f(t) = \sum_{i=1}^{N} a_i X^i t^{\sigma_i}$ for N finite where $a_1 \neq 0$, $\sigma_i \in Aut(\mathbb{K})$ such that σ_i fixes every element of $\mathbb{F}$ and preserves the valuation of X, for all $i \geq 1$. Then $\mathcal{P}^f$ defines a conical flock in $PG(3,\mathbb{K})$.*

Proof. $\phi_s(t) = ts^2 + f(t) = ts^2 + \sum_{i=1}^{N} a_i X^i t^{\sigma_i}$. Notice that f is additive, so is ϕ_s. If $t \neq 0$, then $ts^2 + \sum_{i=1}^{N} a_i X^i t^{\sigma_i} = 0$, iff $\sum_{i=1}^{N} a_i X^i t^{\sigma_i} = -ts^2$. Hence $v\left(\sum_{i=1}^{N} a_i X^i t^{\sigma_i}\right) = v(t) + 2v(s)$, or $1 + v(t^{\sigma_1}) = v(t) + 2v(s)$. Since $v(t^{\sigma_1}) = v(t)$, $1 = 2v(s)$, which is a contradiction. Therefore, $\phi_s(t)$ is one-to-one on $\mathbb{K}$.

On the other hand, $\phi_s(t)$ is onto. In fact, $s^2 \in \mathbb{K}$ can be represented as $s^2 = X^{2m}\sum_{i=0}^{N} s_i X^i$. For any $r \in \mathbb{K}$, write $r = X^k \sum_{i=0}^{N} r_i X^i$. It suffices to show that the following equation has a solution of t in $\mathbb{K}$:

$$X^{2m}\sum_{i=0}^{\infty} s_i X^i t + \sum_{i=1}^{N} a_i X^i t^{\sigma_i} = X^k \sum_{i=0}^{\infty} r_i X^i.$$

Let $t = X^n \sum_{j=0}^{\infty} t_j X^j$. Then the above equation will become

$$X^{2m+n}\left(\sum_{i=0}^{\infty} s_i X^i\right)\left(\sum_{j=0}^{\infty} t_j X^j\right) + \sum_{i=1}^{N} a_i X^i \left(X^n \sum_{j=0}^{\infty} t_j X^j\right)^{\sigma_i} = X^k \sum_{i=0}^{\infty} r_i X^i, \tag{7.10}$$

where $n \in \mathbb{Z}$ and those $t_j \in \mathbb{F}$ are to be determined. Notice that σ_i preserves the valuation of X. Let

$$X^{\sigma_i} = \sum_{j=1}^{\infty} c_j^{(i)} X^j, \quad \text{and} \quad c_1^{(i)} \neq 0, \ i = 1, 2, 3, \ldots.$$

Comparing the first term of both sides in (7.10), it is ready to see that $n = max\{k-1, k-2m\}$, and

$$t_0 = \begin{cases} \dfrac{r_0}{a_1\left(c_1^{(1)}\right)^n}, & \text{if } n = k-1, \\ \dfrac{r_0}{s_0}, & \text{if } n = k-2m. \end{cases}$$

Subtracting the first term $r_0 X^k$ from both sides in (7.10), then removing t_0 by substitution, an equation similar to (7.10) can be obtained. Repeat the procedure. $t_1, t_2, \ldots$ can be solved recursively. Consequently, a solution $t \in \mathbb{K}$ can be found. Therefore, $\phi_s(t)$ is bijective on $\mathbb{K}$. It follows from Theorem 1 that $\mathcal{P}^f$ defines a conical flock in $PG(3, \mathbb{K})$. Q.E.D.

The following result follows immediately from Theorem 29.

Corollary 30 *Let $\mathbb{K} = \mathbb{F}((X))$. Let $f(t) = -X^{2n-1}t^{\sigma}$ where $n \in \mathbb{Z}$, $\sigma \in Aut(\mathbb{K})$ such that σ fixes every element of $\mathbb{F}$ and $v(X^{\sigma}) = v(X)$. Then $\mathcal{P}^f$ defines a generalized Kantor-Knuth flock in $PG(3, \mathbb{K})$.*

The simplest case in Corollary 30 is: $\mathbb{F} = \mathbb{R}$, $f(t) = -Xt^{\sigma}$, where σ fixes every real number and $v(X^{\sigma}) = v(X)$. This is the first infinite generalized Kantor-Knuth flock given by F. De Clerck and H. Van Maldeghem in [4].

8 Semi-elliptic flocks.

Let $\mathbb{R}$ denote the field of real numbers. Cameron (p. 45) notices that there are ovals in $PG(2,\mathbb{R})$ which are not conics. Such an oval may be obtained by gluing two semi-ellipses with the same semi-axis along this line. More generally, there are examples of ovals obtained by gluing various pieces of semi-ellipses.

For example, one may take the following oval: The set of points (x,y), where

(i) $x^2+y^2/2^2=1$ for (x,y) in quadrants I and II and

(ii) $x^2+y^2=1$ for (x,y) in quadrants III and IV.

If we take say the line $x=1$ as the infinite line, and recoordinatize so that $(-1,0)$ is $(0,0)$, we obtain a parabolic version of this oval.

More generally, if we take (x,y) such that $x=y^2$ for $y\geq 0$ and $x=ay^2$ for $y<0$ where a is a fixed positive number and adjoin the infinite point (0), an oval is obtained. Note that $x=t^2$ contains the oval points (t^2,t) and (as^2,s) where $s^2=t^2/a$ and $t>0,\ s<0$. And, $y=w$ contains the oval points (w^2,w) and (0) or (aw^2,w) and (0). It follows that the affine tangent line to the curve is the unique tangent line of the oval in the projective plane.

Another example can be obtained by gluing twice but using only a quarter of one ellipse or a semi-semi-ellipse.

For example, the following set of points (x,y) forms an oval:

(i) $x^2+y^2=1$ for (x,y) in quadrant III,

(ii) $x^2+y^2/3^2=1$ for (x,y) in quadrant II, and

(iii) $x^2/2^2+(y-1)^2/2^2=1$ for (x,y) in quadrants I and IV.

Definition 31 *Any such oval obtained by gluing portions of ellipses together shall be called a* semi-elliptic oval. *Embedding a semi-elliptic oval into a plane in $PG(3,\mathbb{R})$, we may form the corresponding* semi-elliptic cone. *Any flock of a semi-elliptic cone shall be called a* semi-elliptic flock.

In this section, we find infinitely many classes of nonlinear flocks of semi-elliptic cones. The manner of construction is simple. Find a nonlinear flock of a quadratic cone in $PG(3,\mathbb{R})$ considered as an elliptic cone. Glue together portions of ellipses to form a corresponding semi-elliptic cone. Show that the planes of the original flock also define a semi-elliptic cone.

We consider only the semi-elliptic oval obtained by gluing two semi-ellipses along their common semi-axis.

Definition 32 *The set of points*

$$O_a = \{(1,t^2,t,0) \mid t \geq 0\} \cup \{(1,at^2,t,0) \mid t < 0\} \cup \{(0,1,0,0)\}$$

for $a > 0$ *is an oval in* $PG(3,\mathbb{R})$ *which shall be called the* $\{a\}-$ semi-elliptic oval. *The cone obtained by projecting from* $(0,0,0,1)$ *shall be called the* $\{a\}-$semi-elliptic oval cone. *We represent* $PG(3,\mathbb{R})$ *by homogeneous coordinates* (x_0,x_1,x_2,x_3) *for* x_i *in* $\mathbb{R}$ *for* $i = 0,1,2,3$.

Theorem 33 *The planes* π_t: $tx_0 + f(t)x_1 + g(t)x_2 + x_3 = 0$ *form a flock of the* $\{a\}-$*semi-elliptic oval cone if and only if*

(*i*) *for* $s \geq 0$, *the function* $\phi_s(t) = t + f(t)s^2 + g(t)s$ *is bijective and*

(*ii*) *for* $s < 0$, *the function* $\phi_s^a(t) = t + f(t)as^2 + g(t)s$ *is bijective.*

Proof. We need only to cover each point of the set of oval cone points by a unique plane π_t.

Since the points on

$$\langle (0,0,0,1), (1,s^2,s,0) \rangle$$

for $s > 0$ must be uniquely covered, this forces $\phi_s(t)$ to be bijective. Similarly, the points on

$$\langle (0,0,0,1), (1,as^2,s,0) \rangle$$

for $s < 0$ must be uniquely covered forcing $\phi_s^a(t)$ to be bijective. Q.E.D.

Hence, we have various flocks obtained from flocks of quadratic cones as follows:

Theorem 34 *Let* $\pi_t : tx_0 + f(t)x_1 + x_3 = 0$ *define a flock of a quadratic cone in* $PG(3,\mathbb{R})$. *Then the same planes define a flock of any* $\{a\}-$*semi-elliptic oval cone.*

Proof. Suppose $s > 0$ then $\phi_s(t)$ is bijective by assumption. If $s < 0$ we need only show that $\phi_s^a(t) = t + f(t)as^2$ is bijective. But, clearly, $\phi_s^a(t) = \phi_{\sqrt{a}s}(t)$ which is bijective by assumption. Q.E.D.

Theorem 35 *Let* $\pi_t : tx_0 + f(t)x_1 + g(t)x_2 + x_3$ *be a smooth flock of a quadratic cone in* $PG(3,\mathbb{R})$ *in the sense that* f *and* g *are smooth functions such that* $s^2 + g'(t)s + f'(t)$ *is never* 0, $f'(t) > (g'(t)/2)^2$ *for all* t *in* $\mathbb{R}$ *and, for each* s *in* $\mathbb{R}$, $\phi_s(t) \to \pm\infty$ *as* $t \to \pm\infty$ *respectively.*

If $a > 1$ *then the same planes define a flock of any* $\{a\}-$*semi-elliptic oval cone.*

Proof. Following the proof of Theorem 6, and using the above necessary and sufficient condition , we must show both $f'(t) > (g'(t)/2)^2$ and $f'(t)a > (g'(t)/2)^2$. Since $f'(t) > (g'(t)/2)^2$ holds by assumption and $f'(t)a > f'(t)$, the result follows. Q.E.D.

Thus, most of the flocks considered previously of $PG(3, \mathbb{R})$ will also produce flocks of $\{a\}$−semi-elliptic flocks. It is interesting to note that the corresponding spreads now produces flocks on oval flocks which are not quadratic.

We end this section with a question. Given a flock of an oval cone in $PG(3, \mathbb{K})$ for some field $\mathbb{K}$, is there a quadratic cone not necessarily in standard form such that the planes of the oval cone produce a corresponding flock of a quadratic cone?

References

[1] D. BETTEN, 4-dimensionale Translationsebenen mit 8-dimensionaler Kollineationsgruppe. *Geom. Dedicata* **2** (1973), 327-339.

[2] M. BILIOTTI AND N. L. JOHNSON, Bilinear flocks of quadratic cones. J. Geometry (to appear).

[3] M. BILIOTTI AND N. L. JOHNSON, Variations on a theme of Dembowski. Mostly Finite Geometries, Marcel Dekker (to appear).

[4] F. DE CLERCK AND H. VAN MALDEGHEM, On flocks of infinite quadratic cones. *Bull. Belgian Math. Soc., Simon Stevin* **1** (1994), 399-415.

[5] J. C. FISHER AND J. A. THAS, Flocks in $PG(3, q)$. *Math. Z.* **169** (1979), 1-11.

[6] V. JHA AND N. L. JOHNSON, Infinite flocks of a quadratic cone. Preprint.

[7] W. M. KANTOR, Some generalized quadrangles with parameters (q^2, q). *Math. Z.* **192** (1986), 45-50.

[8] D. E. KNUTH, Finite semifields and projective planes. *J. Algebra* **2** (1965), 182-217.

[9] S. E. PAYNE, A new infinite family of generalized quadrangles. *Congr. Numer.* **49** (1985), 115-128.

[10] T. PENTTILA AND L. STORME, Monomial flocks and ovals. Preprint.

[11] J. A. THAS, Generalized quadrangles and flocks of cones. *Europ. J. Combin.* **8** (1987), 441-452.

The Generalized Kantor–Knuth Flocks

NORMAN L. JOHNSON
and
XIAOBIN LIU
Department of Mathematics
The University of Iowa
Iowa City, IA 52242

Abstract

In this paper, we shall classify the generalized Kantor-Knuth flocks in $PG(3,\mathbb{K})$. Furthermore, the linear collineation group of a generalized Kantor-Knuth flock is completely determined over fields of characteristic not two.

1 Introduction

Recently, flocks of infinite quadratic cones in $PG(3,\mathbb{K})$ are studied by M. Biliotti, F. De Clerck, V. Jha, N. L. Johnson, X. Liu, H. Van Maldeghem and others. (see [1], [2], [3] and [5].)

Let $\mathbb{K}$ be a field. Let $\mathcal{Q}_C$: $x_0x_1 - x_2^2 = 0$ be the quadratic cone in $PG(3,\mathbb{K})$, whose vertex is $\mathbf{v} = (0,0,0,1)$. Let $\mathcal{P} = \{\pi_t : tx_0 + f(t)x_1 + g(t)x_2 + x_3 = 0 \mid t \in \mathbb{K}\}$ be a set of planes which do not contain $\mathbf{v}$. A *conical flock* $\mathcal{F}$ of $\mathcal{Q}_C$ in $PG(3,\mathbb{K})$ is a set of conics partitioning the points of $\mathcal{Q}_C - \mathbf{v}$. A flock $\mathcal{F}$ is called *linear* if the associated planes of $\mathcal{F}$ are collinear; a flock $\mathcal{F}$ is called *concurrent* if the associated planes of $\mathcal{F}$ are concurrent.

In [5], N. L. Johnson and X. Liu prove the following fundamental theorem of flocks of the quadratic cone in $PG(3,\mathbb{K})$ (the reader is also referred to De Clerck and Van Maldeghem where this result appears in slightly different form):

This article is dedicated to Professor T.G. Ostrom on his eightieth birthday.

Theorem 1 (Johnson-Liu [5]) *$\mathcal{P}$ is a set of associated planes of a flock of $\mathcal{Q}_C$ in $PG(3,\mathbb{K})$ iff $\phi_s(t) = ts^2 + g(t)s + f(t)$ is a bijective mapping from $\mathbb{K}$ to $\mathbb{K}$, for each $s \in \mathbb{K}$.*

If $\mathbb{K}$ is an algebraically closed field, any pair of the planes will intersect with the quadratic cone $\mathcal{Q}_C$ in $PG(3,\mathbb{K})$ nontrivially. Therefore, no partial flock of two conics can occur in this case. The Kantor-Knuth flock in $PG(3,q)$ has an infinite analogue, called the *generalized Kantor-Knuth flock.* We shall study this flock in this paper. In section 2, we shall classify the generalized Kantor-Knuth flocks in $PG(3,\mathbb{K})$. The linear collineation group of a generalized Kantor-Knuth flock will be determined in section 3.

2 The Classification Theorem

In [7], J. A. Thas studied the finite concurrent flocks. It turns out that the Kantor-Knuth flocks are the only finite nonlinear concurrent flocks. The associated planes of a Kantor-Knuth flock have equations:

$$tx_0 - mt^\sigma x_1 + x_3 = 0, \ t \in GF(q), \ q \text{ odd},$$

where m is a fixed nonsquare element of $GF(q)$. In general, a generalized Kantor-Knuth flocks, denoted by $\mathcal{F}(m,\sigma)$, has a set of associated planes:

$$tx_0 - mt^\sigma x_1 + x_3 = 0, \ t \in \mathbb{K}, \tag{1}$$

where m is a fixed element of $\mathbb{K}$ such that $\phi_s(t) = ts^2 - mt^\sigma$ is bijective for each $s \in \mathbb{K}$, and $1 \neq \sigma \in Aut(\mathbb{K})$. Notice that in $PG(3,\mathbb{R})$, such a flock does not exist, since $Aut(\mathbb{K})$ is trivial. When $\mathbb{K} = GF(q)$, if $q = 2^n$, then $ts^2 - mt^\sigma$ cannot be bijective, since $s^2 - mt^{\sigma-1} = 0$ has a solution of t. Hence there are no generalized Kantor-Knuth flocks in $PG(3,2^n)$. More generally, if $\mathbb{K}$ is a perfect field with characteristic 2, then there are no generalized Kantor-Knuth flocks in $PG(3,\mathbb{K})$. On the other hand, when $\mathbb{K} = GF(q)$, if $q = p^e$, p an odd prime, then m can be taken as a nonsquare and the flock is the Kantor-Knuth flock. Infinite generalized Kantor-Knuth flocks in $PG(3,\mathbb{K})$ may exist if $Aut(\mathbb{K})$ is nontrivial.

A classification of the generalized Kantor-Knuth flocks in $PG(3,\mathbb{K})$ will be given as follows.

Theorem 2 *Let $\mathbb{K}$ be a field such that $Aut(\mathbb{K})$ is nontrivial. Suppose that $m \in \mathbb{K}$ such that $(m^{\sigma+1})^{\frac{1}{2}} \in \mathbb{K}$, and $ts^2 - mt^\sigma$ is bijective for each $s \in \mathbb{K}$, and for all $\sigma \in Aut(\mathbb{K})$. Let $\sigma_1, \sigma_2 \in Aut(\mathbb{K})$ such that $\sigma_1 \neq 1 \neq \sigma_2$. Then $\mathcal{F}(m,\sigma_1) \cong \mathcal{F}(m,\sigma_2)$ iff $\sigma_2 = \sigma_1^{\pm 1}$.*

Proof. If $\sigma_2 = \sigma_1$, then $\mathcal{F}(m, \sigma_1) = \mathcal{F}(m, \sigma_2)$.

If $\sigma_2 = \sigma_1^{-1}$, then define $\theta \in GL(4, \mathbb{K})$ as follows:

$$\theta = \begin{bmatrix} 0 & -\frac{1}{m^{\sigma_2+1}} & 0 & 0 \\ -1 & 0 & 0 & 0 \\ 0 & 0 & \left(\frac{1}{m^{\sigma_2+1}}\right)^{\frac{1}{2}} & 0 \\ 0 & 0 & 0 & 1 \end{bmatrix}.$$

θ fixes $\mathcal{Q}_C$ and $\mathbf{v}$ and maps the planes $\pi_t^1 : tx_0 - mt^{\sigma_1}x_1 + x_3 = 0$ onto the planes $\pi_s^2 : sx_0 - ms^{\sigma_2}x_1 + x_3 = 0$, where $s = mt^{\sigma_1}$ runs over $\mathbb{K}$ as t does.

On the other hand, suppose that $\mathcal{F}(m, \sigma_1) \cong \mathcal{F}(m, \sigma_2)$. Then there is an element τ of $P\Gamma L(4, q)$ which fixes $\mathcal{Q}_C, \mathbf{v}$ and maps the associated planes of $\mathcal{F}(m, \sigma_1)$ to those of $\mathcal{F}(m, \sigma_2)$. Since τ fixes $(0, 0, 0, 1)$ and $(0, 0, 1, 0)$, τ can be represented as

$$(x_0, x_1, x_2, x_3)^\tau = (x_0^\alpha, x_1^\alpha, x_2^\alpha, x_3^\alpha)M,$$

where $\alpha \in Aut(\mathbb{K})$ and M is a nonsingular matrix with the form

$$\begin{bmatrix} a & b & c & d \\ a' & b' & c' & d' \\ 0 & 0 & c'' & 0 \\ 0 & 0 & 0 & 1 \end{bmatrix}.$$

Then

$$(x_0, x_1, x_2, x_3)^\tau = (ax_0^\alpha + a'x_1^\alpha,\ bx_0^\alpha + b'x_1^\alpha,\ cx_0^\alpha + c'x_1^\alpha + c''x_2^\alpha,\ dx_0^\alpha + d'x_1^\alpha + x_3^\alpha).$$

Suppose that τ maps $tx_0 - mt^{\sigma_1}x_1 + x_3 = 0$ to $sx_0 - ms^{\sigma_2}x_1 + x_3 = 0$. If (x_0, x_1, x_2, x_3) is on the plane $tx_0 - mt^{\sigma_1}x_1 + x_3 = 0$, then $x_3 = -tx_0 + mt^{\sigma_1}x_1$, and $s(ax_0^\alpha + a'x_1^\alpha) - ms^{\sigma_2}(bx_0^\alpha + b'x_1^\alpha) + (dx_0^\alpha + d'x_1^\alpha + x_3^\alpha) = 0$. Hence

$$(as - bms^{\sigma_2} + d - t^\alpha)x_0^\alpha + (a's - b'ms^{\sigma_2} + d' + m^\alpha t^{\alpha\sigma_1})x_1^\alpha = 0.$$

Consequently the following two equations hold:

(i) $as - bms^{\sigma_2} + d - t^\alpha = 0$;

(ii) $a's - b'ms^{\sigma_2} + d' + m^\alpha t^{\alpha\sigma_1} = 0$.

Since $\det(M) \neq 0$, $ab' - a'b \neq 0$. It follows from (i) and (ii) that

$$s = \frac{-b'd + bd' + b't^\alpha + bm^\alpha t^{\alpha\sigma_1}}{ab' - a'b},$$

and

$$s^{\sigma_2} = \frac{-a'd + ad' + a't^\alpha + am^\alpha t^{\alpha\sigma_1}}{m(ab' - a'b)}.$$

Let $t = 0$. Then

$$\left(\frac{-b'd + bd'}{ab' - a'b}\right)^{\sigma_2} = \frac{-a'd + ad'}{m(ab' - a'b)}. \quad (2)$$

Hence

$$\left(\frac{b't^\alpha + bm^\alpha t^{\alpha\sigma_1}}{ab' - a'b}\right)^{\sigma_2} = \frac{a't^\alpha + am^\alpha t^{\alpha\sigma_1}}{m(ab' - a'b)}. \quad (3)$$

Let $u = t^\alpha$. Then (3) becomes

$$\left(\frac{b'u + bm^\alpha u^{\sigma_1}}{ab' - a'b}\right)^{\sigma_2} = \frac{a'u + am^\alpha u^{\sigma_1}}{m(ab' - a'b)}.$$

Therefore, the equation

$$c_0 x + c_1 x^{\sigma_1} + c_2 x^{\sigma_2} + c_3 x^{\sigma_1\sigma_2} = 0,$$

holds for all $x \in \mathbb{K}$, where the coefficients are

$$\begin{aligned} c_0 &= \frac{-a'}{m(ab' - a'b)}, \\ c_1 &= \frac{-am^\alpha}{m(ab' - a'b)}, \\ c_2 &= \left(\frac{b'}{ab' - a'b}\right)^{\sigma_2}, \\ c_3 &= \left(\frac{bm^\alpha}{ab' - a'b}\right)^{\sigma_2}. \end{aligned}$$

Let $\sigma_0 = 1, \sigma_3 = \sigma_1\sigma_2$. Then the above equation can be represented as follows:

$$c_0 x^{\sigma_0} + c_1 x^{\sigma_1} + c_2 x^{\sigma_2} + c_3 x^{\sigma_3} = 0. \quad (4)$$

If the four σ_i's are distinct, then all c_i's must be zero. In fact, if not all c_i's are zero, then at least two c_i's are nonzero. Without loss of generality, let c_0 and c_3 be nonzero. Since $\sigma_0 \neq \sigma_3$, there exists $c \in \mathbb{K}$ such that $c^{\sigma_0} \neq c^{\sigma_3}$. As (4) holds for all $x \in \mathbb{K}$,

$$c_0(cx)^{\sigma_0} + c_1(cx)^{\sigma_1} + c_2(cx)^{\sigma_2} + c_3(cx)^{\sigma_3} = 0,$$

or

$$c_0 c^{\sigma_0} x^{\sigma_0} + c_1 c^{\sigma_1} x^{\sigma_1} + c_2 c^{\sigma_2} x^{\sigma_2} + c_3 c^{\sigma_3} x^{\sigma_3} = 0. \quad (5)$$

¿From (4) and (5), the x^{σ_0} term can be removed:

$$c_1(c^{\sigma_1} - c^{\sigma_0})x^{\sigma_1} + c_2(c^{\sigma_2} - c^{\sigma_0})x^{\sigma_2} + c_3(c^{\sigma_3} - c^{\sigma_0})x^{\sigma_3} = 0,$$

where $c^{\sigma_3} - c^{\sigma_0} \neq 0$.

Repeating this procedure at most twice, depending upon whether c_1 and c_2 are zero, the x^{σ_1} and x^{σ_2} terms will also be removed and (4) will become

$$c_3 d x^{\sigma_3} = 0,$$

where $0 \neq d \in \mathbb{K}$. Take $x = 1$, then $c_3 = 0$, a contradiction.

Hence all c_i's must be zero. But this would imply that $a = b = a' = b' = 0$, contradicting the fact that M is nonsingular. Therefore, either $\sigma_2 = \sigma_1$ or $\sigma_3 = 1$, that is, $\sigma_2 = \sigma_1^{\pm 1}$, as desired. Q.E.D.

Remark 3 *If* $\mathbb{K} = GF(q)$, q *is odd, then a generalized Kantor-Knuth flock is a Kantor-Knuth flock and the classification in the finite case is due to W. M. Kantor [6].*

3 The Linear Collineation Group of $\mathcal{F}(m, \sigma)$

We now consider the elements in $PGL(4, \mathbb{K})$. Then $\alpha = 1$, and we shall determine the linear collineation group G of the flock $\mathcal{F}(m, \sigma)$ in this section. Notice that in Theorem 2, if $\sigma_2 = \sigma_1 = \sigma$, then τ is an automorphism of $\mathcal{F}(m, \sigma)$. There are two possibilities:

(i) $\sigma_3 = \sigma^2 \neq 1$;

(ii) $\sigma_3 = \sigma^2 = 1$.

Theorem 4 *Let* $1 \neq \sigma \in Aut(\mathbb{K})$. *If* $\sigma^2 \neq 1$, *then the linear collineation group* G *of* $\mathcal{F}(m, \sigma)$ *in* $PG(3, \mathbb{K})$ *is* $G = H \cdot E$, *where*

$$H = \left\{ \begin{bmatrix} a & 0 & 0 & 0 \\ 0 & a^\sigma & 0 & 0 \\ 0 & 0 & (a^{\sigma+1})^{\frac{1}{2}} & 0 \\ 0 & 0 & 0 & 1 \end{bmatrix} \;\middle|\; a \in \mathbb{K}^*,\ \left(a^{\sigma+1}\right)^{\frac{1}{2}} \in \mathbb{K}^* \right\},$$

and

$$E = \left\{ \begin{bmatrix} 1 & 0 & 0 & d \\ 0 & 1 & 0 & -md^\sigma \\ 0 & 0 & 1 & 0 \\ 0 & 0 & 0 & 1 \end{bmatrix} \;\middle|\; d \in \mathbb{K} \right\}.$$

Proof. $\sigma^2 \neq 1$. (4) becomes

$$c_0 x + (c_1 + c_2)x^{\sigma} + c_3 x^{\sigma^2} = 0.$$

Hence $c_0 = c_3 = 0$ and $c_1 + c_2 = 0$. Then $a' = b = 0$ and

$$\left(\frac{b'}{ab' - a'b}\right)^{\sigma} = \frac{a}{(ab' - a'b)}. \tag{6}$$

Hence $\left(\frac{1}{a}\right)^{\sigma} = \frac{1}{b'}$, that is, $b' = a^{\sigma}$. Then we have

$$M = \begin{bmatrix} a & 0 & c & d \\ 0 & a^{\sigma} & c' & d' \\ 0 & 0 & c'' & 0 \\ 0 & 0 & 0 & 1 \end{bmatrix}.$$

Since M preserves the cone: $x_0 x_1 - x_2^2 = 0$,

$$(ax_0)(a^{\sigma}x_1) - (cx_0 + c'x_1 + c''x_2)^2 = 0.$$

Take three particular points $(1,0,0,0), (0,1,0,0)$ and $(1,1,1,0)$ on the cone. Then it is ready to see that $c = 0 = c'$, and $c''^2 = a^{\sigma+1}$. On the other hand, since $a' = 0 = b$, by (2), $d' = -\frac{mb'd^{\sigma}}{a^{\sigma}} = -md^{\sigma}$. Then

$$M = \begin{bmatrix} a & 0 & 0 & d \\ 0 & a^{\sigma} & 0 & -md^{\sigma} \\ 0 & 0 & (a^{\sigma+1})^{\frac{1}{2}} & 0 \\ 0 & 0 & 0 & 1 \end{bmatrix}. \tag{7}$$

We can write M as the product $\theta_a \cdot \epsilon_d$, where

$$\theta_a = \begin{bmatrix} a & 0 & 0 & 0 \\ 0 & a^{\sigma} & 0 & 0 \\ 0 & 0 & (a^{\sigma+1})^{\frac{1}{2}} & 0 \\ 0 & 0 & 0 & 1 \end{bmatrix},$$

and

$$\epsilon_d = \begin{bmatrix} 1 & 0 & 0 & d \\ 0 & 1 & 0 & -md^{\sigma} \\ 0 & 0 & 1 & 0 \\ 0 & 0 & 0 & 1 \end{bmatrix}.$$

Let $H = \{\theta_a \mid a \in \mathbb{K}^*,\ (a^{\sigma+1})^{\frac{1}{2}} \in \mathbb{K}^*\}$. Let $E = \{\epsilon_d \mid d \in \mathbb{K}\}$. Then $H \cdot E$ will form a group $G < PGL(4, \mathbb{K})$. Furthermore, H has a subgroup $H^* = \{\theta_a \in H \mid a^{\sigma} = a\} \cong Fix(\sigma)^*$. Q.E.D.

Theorem 5 *Let $1 \neq \sigma \in Aut(\mathbb{K})$. If $\sigma^2 = 1$, and the characteristic is not two then the linear collineation group G of $\mathcal{F}(m,\sigma)$ in $PG(3,\mathbb{K})$ is $G = (H \cup B) \cdot E$, where*

$$B = \left\{ \begin{bmatrix} 0 & b & 0 & 0 \\ b^\sigma m^{\sigma+1} & 0 & 0 & 0 \\ 0 & 0 & ((bm)^{\sigma+1})^{\frac{1}{2}} & 0 \\ 0 & 0 & 0 & 1 \end{bmatrix} \middle| \; b \in \mathbb{K}^*, \; \left(b^{\sigma+1}\right)^{\frac{1}{2}} \in \mathbb{K}^* \right\}.$$

Proof. $\sigma^2 = 1$. (4) becomes

$$(c_0 + c_3)x + (c_1 + c_2)x^\sigma = 0.$$

Hence $c_0 + c_3 = 0$ and $c_1 + c_2 = 0$, which implies that (6) and

$$\left(\frac{bm}{ab' - a'b}\right)^\sigma = \frac{a'}{m(ab' - a'b)} \tag{8}$$

It follows from (6) and (8) that

$$\left(\frac{b'}{bm}\right)^\sigma = \frac{am}{a'}. \tag{9}$$

Since $\sigma^2 = 1$, then

$$\left(\frac{am}{a'}\right)^\sigma = \frac{b'}{bm}. \tag{10}$$

If $b = 0 = a'$, then we have (7). Hence $H \cdot E$ is a subgroup of the linear collineation group G.

Suppose that $b \neq 0, a' \neq 0$. From (9) and (10) we have

$$\left(\frac{ab'}{a'b}\right)^\sigma = \frac{ab'}{a'b}.$$

We now consider the following three cases.

Case 1. $\frac{ab'}{a'b} = 1$. Then $ab' - a'b = 0$, a contradiction.

Case 2. $\frac{ab'}{a'b} = 0$. Then $a = 0 = b'$. By (8), $a'^\sigma = bm^{\sigma+1}$. Since $\sigma^2 = 1$, $a' = b^\sigma m^{\sigma+1}$. Hence

$$M = \begin{bmatrix} 0 & b & c & d \\ b^\sigma m^{\sigma+1} & 0 & c' & d' \\ 0 & 0 & c'' & 0 \\ 0 & 0 & 0 & 1 \end{bmatrix}.$$

Since M preserves the cone: $x_0x_1 - x_2^2 = 0$,

$$(bx_0)(b^\sigma m^{\sigma+1}x_1) - (cx_0 + c'x_1 + c''x_2)^2 = 0.$$

Again we take the three points on the cone, namely, $(1,0,0,0)$, $(0,1,0,0)$ and $(1,1,1,0)$. Then it is easy to see that $c = 0 = c'$, and $c''^2 = b^{\sigma+1}m^{\sigma+1}$. Since $a = 0 = b'$, by (2), $d'^\sigma = -\frac{a'^\sigma d}{mb} = -m^\sigma d$. Again since $\sigma^2 = 1$, $d' = -md^\sigma$. Therefore,

$$\begin{aligned} M &= \begin{bmatrix} 0 & b & 0 & d \\ b^\sigma m^{\sigma+1} & 0 & 0 & -md^\sigma \\ 0 & 0 & (b^{\sigma+1}m^{\sigma+1})^{\frac{1}{2}} & 0 \\ 0 & 0 & 0 & 1 \end{bmatrix} \\ &= \begin{bmatrix} 0 & b & 0 & 0 \\ b^\sigma m^{\sigma+1} & 0 & 0 & 0 \\ 0 & 0 & (b^{\sigma+1}m^{\sigma+1})^{\frac{1}{2}} & 0 \\ 0 & 0 & 0 & 1 \end{bmatrix} \begin{bmatrix} 1 & 0 & 0 & d_1 \\ 0 & 1 & 0 & -md_1^\sigma \\ 0 & 0 & 1 & 0 \\ 0 & 0 & 0 & 1 \end{bmatrix} \\ &= \beta_b \cdot \epsilon_{d_1}, \end{aligned}$$

where $d_1 = -(b^{-1}m^{-1}d)^\sigma$. Let $B = \{\beta_b \mid b \in \mathbb{K}^*,\ (b^{\sigma+1})^{\frac{1}{2}} \in \mathbb{K}^*\}$. Notice that

$$\beta_b^2 = \begin{bmatrix} b^{\sigma+1}m^{\sigma+1} & 0 & 0 & 0 \\ 0 & b^{\sigma+1}m^{\sigma+1} & 0 & 0 \\ 0 & 0 & b^{\sigma+1}m^{\sigma+1} & 0 \\ 0 & 0 & 0 & 1 \end{bmatrix},$$

which is in $H^* = \{\theta_a \in H \mid a^\sigma = a\}$, as $(b^{\sigma+1}m^{\sigma+1})^\sigma = b^{\sigma+1}m^{\sigma+1}$. Therefore, $B \cdot E$ is a subset of G.

Case 3. $\frac{ab'}{a'b} \neq 0, 1$. Then $ab' \neq 0$ and $ab' - a'b \neq 0$. To see that case 3 can never occur, we consider the corresponding group acting on the translation plane. The group of the translation plane must then permute the so-called base regulus nets. Since the corresponding translation plane is a semifield plane, it follows that the translation complement of the collineation group is the semidirect product of the the elation group by the stabilizer of two components.

Hence, we may assume that the the linear group fixes two components $x = 0$, and $y = 0$ and since this implies that the standard net is left invariant also, the group elements be written in the form $\begin{bmatrix} a & b & 0 & 0 \\ c & d & 0 & 0 \\ 0 & 0 & av & bv \\ 0 & 0 & cv & dv \end{bmatrix}$ for some

v in $K - \{0\}$. Using the calculations in N. L. Johnson [4] (p. 839 equations (1) – (4)), it follows that $y = x \begin{bmatrix} 0 & mt^\sigma \\ t & 0 \end{bmatrix}$ maps onto $y = x \begin{bmatrix} u^* & mt^{*\sigma} \\ t^* & u^* \end{bmatrix}$ where

(1) $t^* = (a^2tv - c^2mt^\sigma v)/\delta$ where $\delta = ad - bc$,
(2) $u^* = (abtv - cdvmt^\sigma)/\delta$,
(3) $u^* = (-abtv + dcvmt^\sigma)/\delta$,
(4) $mt^{*\sigma} = (-b^2tv + d^2vmt^\sigma)/\delta$.

First assume that the characteristic is not two.

Then using equations (2) and (3), and assuming that $\sigma \neq 1$, we obtain $ab = 0 = cd$.

Hence, either $a = d = 0$ or $a = c = 0$ since $ad - bc \neq 0$.

The indicated group element induces a group element acting on the flock such that, in the previous notation, we have $a = b' = 0$ or $a' = b = 0$. That is, case 3 does not occur when the characteristic is not two. Q.E.D.

Corollary 6 *Any generalized Kantor-Knuth flock admits a group which acts doubly transitively on the planes of the flock.*

Proof. The group E maps $x_3 = 0$ onto $-dx_0 + m(-d)^\sigma x_1 + x_3 = 0$ and H maps the plane given by $x_0 - x_1 + x_3 = 0$ onto $ax_0 - ma^\sigma x_1 + x_3 = 0$. Q.E.D.

References

[1] M. Biliotti and N. L. Johnson, Bilinear flocks of quadratic cones. J. Geometry (to appear).

[2] F. De Clerck and H. Van Maldeghem, On flocks of infinite quadratic cones. *Bull. Belgian Math. Soc., Simon Stevin* **1** (1994), 399-415.

[3] V. Jha and N. L. Johnson, Infinite flocks of a quadratic cone. J. Geometry (to appear).

[4] N.L.Johnson, Derivation of partial flocks of quadratic cones. *Rendiconti di Matematica* **12** (1992), 817-848.

[5] N. L. Johnson and X. Liu, Flocks of quadratic and semi-elliptic cones. Mostly Finite Geometries. Marcel Dekker.

[6] W. M. Kantor, Some generalized quadrangles with parameters (q^2, q). *Math. Z.* **192** (1986), 45-50.

[7] J. A. Thas, Generalized quadrangles and flocks of cones. *Europ. J. Combin.* **8** (1987), 441-452.

Spreads Corresponding to Flocks of Quadrics

Norman L. Johnson Mathematics Department, University of Iowa, Iowa City, Iowa 52242

Leo Storme* Department of Pure Mathematics and Department of Computer Algebra, University of Gent, Galglaan 2, Gent, Belgium

Abstract

In $PG(3,q)$, there exist three types of spreads which correspond via the Klein correspondence to flocks of quadrics. These spreads are characterized as being spreads consisting of $q, q \pm 1$ reguli sharing exactly one or two fixed lines. This article shows that any spread containing more than $3q/4$ reguli sharing exactly one or two fixed lines, is a spread corresponding to a flock of a quadric.

1 Introduction

A *spread* of $PG(3,q)$ is a set of q^2+1 lines which partition the point set of $PG(3,q)$. A *flock* of a quadratic cone, elliptic or hyperbolic quadric in $PG(3,q)$ is a set of respectively $q, q-1$, or $q+1$ disjoint nonsingular conics of this quadric.

These flocks correspond via the Klein correspondence to three types of spreads of $PG(3,q)$. A flock of a quadratic cone corresponds to a spread consisting of q reguli sharing exactly one line L; a flock of a hyperbolic quadric corresponds to a spread consisting of $q+1$ reguli all containing two fixed lines L and M; and a flock of an elliptic quadric corresponds to a spread containing $q-1$ reguli sharing two conjugate lines L and $L^{(q)}$ in a quadratic extension $PG(3,q^2)$.

The flocks of the elliptic and hyperbolic quadric all have been classified. A flock of an elliptic quadric is always linear [11, 14]. For q even, any flock of a hyperbolic quadric of $PG(3,q)$ is linear [15]. When q is odd, any flock

*Senior Research Assistant of the National Fund for Scientific Research (Belgium).

of the hyperbolic quadric of $PG(3,q)$ either is a linear flock, or a Thas flock, or an exceptional flock. [1, 16].

Since all these flocks have been classified, the corresponding spreads also have been determined, and consequently, also the corresponding translation planes have been determined. For the flock of the elliptic quadric, this is a Desarguesian plane. The planes corresponding to the flocks of the hyperbolic quadric respectively are the Desarguesian, regular nearfield plane, and the irregular nearfield planes of order $11^2, 23^2$, or 59^2.

The problem of classifying all flocks of the quadratic cone is still open. For a survey of the known flocks, we refer to De Clerck and Herssens [4], Thas [17], and Cherowitzo et al. [3].

Recently, the study of partial flocks of the quadratic cone has been initiated. In [12], Payne and Thas proved that any partial flock of deficiency one of a quadratic cone in $PG(3,q)$ can be uniquely extended to a flock. More recently, Storme and Thas [13] have shown that in $PG(3,2^r)$ there is a unique extension flock to any partial flock of k conics for r even and $k > 2^r - 2^{r/2} - 1$ or r odd and $k > 2^r - 2^{(r+1)/2}$.

This implies equivalent results on partial spreads consisting of k reguli sharing precisely one common line L, which can be uniquely extended to a spread.

Although the flocks of the hyperbolic and elliptic quadrics have been classified, the partial flocks of deficiency at least one are far from classified. In this case, a number of partial flocks of hyperbolic quadrics of deficiency one which cannot be extended to flocks are known in $PG(3,q)$ for small values of q [2, 8, 9], but no results for arbitrary q are known.

In [10], Johnson studied a related problem. He proved that the spreads in $PG(3,q)$ containing at least $q-2$ reguli that share a common line L contain exactly q reguli sharing L. Hence these spreads correspond to a flock of a quadratic cone in $PG(3,q)$.

This article improves this result and presents the corresponding result for spreads corresponding to flocks of the elliptic and hyperbolic quadric.

Our results show that a spread in $PG(3,q)$ containing more than $3q/4$ reguli sharing one or two fixed lines, is a spread corresponding to a flock of a quadric.

2 Spreads and flocks of quadrics

Theorem 1 *A partial flock of t conics of a quadric $\mathcal{Q}$ in $PG(3,q)$ is equivalent to a partial spread $\mathcal{S}$ of $PG(3,q)$ consisting of*
(a) t reguli which share precisely one line L if $\mathcal{Q}$ is a quadratic cone;
(b) t reguli which share precisely two lines L_1, L_2 if $\mathcal{Q}$ is a hyperbolic quadric;
(c) t reguli which share precisely two conjugate lines $L, L^{(q)}$ of $PG(3,q^2) \setminus PG(3,q)$ if $\mathcal{Q}$ is an elliptic quadric.

Proof: For (a), we refer to [6]. Cases (b) and (c) are proved similarly. ∎

Theorem 2 *(a) A partial spread $\mathcal{S}$ of $PG(3,q)$ consisting of t reguli $\mathcal{R}_i, i = 1,\ldots,t$, which share one line L is fixed by an elation group E with axis L acting regularly on the lines of $\mathcal{R}_i \setminus \{L\}$;*
(b) A partial spread $\mathcal{S}$ of $PG(3,q)$ consisting of t reguli $\mathcal{R}_i, i = 1,\ldots,t$, sharing two lines L and M is fixed by a homology group H with axis L and coaxis M acting regularly on the components of $\mathcal{R}_i \setminus \{L, M\}, i = 1,\ldots,t$.

Proof: [5, Theorem A]. ∎

Lemma 3 *Let $\mathcal{R}$ be a regulus of $PG(3,q)$ which extends to the regulus $\mathcal{R}^{(q^2)}$ of $PG(3,q^2)$ and let L and $L^{(q)}$ be two conjugate lines of $\mathcal{R}^{(q^2)}$ not defined over $PG(3,q)$. Then there exists a cyclic group C_{q+1} which fixes L and $L^{(q)}$ point by point, which acts transitively on the components of $\mathcal{R}$, and which fixes all regular spreads containing $\mathcal{R}$.*

Proof: Consider $\mathcal{R}$ and its complementary regulus $\mathcal{R}'$ defining the hyperbolic quadric $\mathcal{Q}$ in $PG(3,q)$.

As indicated in [7, p. 28], $\mathcal{Q}$ is stabilized by a group H isomorphic to $PGL(2,q)$ fixing the regulus $\mathcal{R}'$ line by line. This group H also fixes all conjugate lines U and $U^{(q)}$ of $\mathcal{R}'$ in $PG(3,q^2)\setminus PG(3,q)$. And H also contains a cyclic subgroup C_{q+1} of order $q+1$ fixing L and $L^{(q)}$.

This shows that there is a cyclic group C_{q+1} fixing $L, L^{(q)}$ point by point, acting transitively on the components of $\mathcal{R}$, and fixing all regular spreads containing $\mathcal{R}$. ∎

Lemma 4 *Let $\mathcal{R}_1$ and $\mathcal{R}_2$ be two distinct reguli of $PG(3,q)$ sharing two conjugate lines L and $L^{(q)}$ in $PG(3,q^2) \setminus PG(3,q)$.*

Assume that, as sets of points, $\mathcal{R}_1 \cap \mathcal{R}_2 = \emptyset$.

Then $\mathcal{R}_1$ and $\mathcal{R}_2$ are contained in one regular spread.

Proof: Using the Klein correspondence, the lines of $\mathcal{R}_1$ and $\mathcal{R}_2$ are mapped onto the points of conics C_1 and C_2, in planes Π_1 and Π_2, on the Klein quadric in $PG(5,q)$.

These planes Π_1 and Π_2 intersect in an external line to the Klein quadric because the two reguli have two lines in $PG(3,q^2) \setminus PG(3,q)$ in common.

Let $\langle \Pi_1, \Pi_2 \rangle$ be the 3-dimensional projective space generated by Π_1 and Π_2. Since the lines of $\mathcal{R}_1$ and $\mathcal{R}_2$ do not intersect, $\langle \Pi_1, \Pi_2 \rangle$ intersects the Klein quadric in an elliptic quadric. This defines a regular spread containing $\mathcal{R}_1$ and $\mathcal{R}_2$ in $PG(3,q)$. ∎

Lemma 5 *If two distinct reguli $\mathcal{R}_1, \mathcal{R}_2$ of $PG(3,q)$ share two conjugate lines L and $L^{(q)}$ of $PG(3,q^2) \setminus PG(3,q)$, and if, as sets of points, $\mathcal{R}_1 \cap \mathcal{R}_2 = \emptyset$, then there exists a cyclic group C_{q+1} of order $q+1$ acting regularly on the lines of $\mathcal{R}_1$ and $\mathcal{R}_2$.*

Proof: By the preceding lemma, $\mathcal{R}_1$ and $\mathcal{R}_2$ are contained in one regular spread $\mathcal{S}$.

This regular spread $\mathcal{S}$ defines an elliptic quadric $\mathcal{E}$ on the Klein quadric, which contains, in a quadratic extension, the points l and $l^{(q)}$ corresponding to L and $L^{(q)}$.

Let C_{q+1} be the cyclic group of order $q+1$ acting transitively on the components of $\mathcal{R}_1$, fixing $\mathcal{S}$, and also fixing L and $L^{(q)}$ (Lemma 4). Then C_{q+1} induces a cyclic group C'_{q+1} of projective transformations on $\mathcal{E}$. This group fixes the points l and $l^{(q)}$ on the Klein quadric corresponding to L and $L^{(q)}$. These two points define an external line $ll^{(q)}$ to $\mathcal{E}$ since l and $l^{(q)}$ are defined over $\mathbf{F}_{q^2}$, but not over $\mathbf{F}_q$.

There is a cyclic group C of order q^2-1 fixing the elliptic quadric $\mathcal{E}$, and the two points l and $l^{(q)}$. This group fixes the polar line M of $ll^{(q)}$ with respect to $\mathcal{E}$. This polar line intersects $\mathcal{E}$ in two points r_1 and r_2.

The cyclic group C fixes the polar planes $\langle r_1, ll^{(q)} \rangle$ and $\langle r_2, ll^{(q)} \rangle$, and acts transitively on the $q-1$ planes through $ll^{(q)}$ which intersect $\mathcal{E}$ in a conic. The cyclic subgroup C'_{q+1} of C fixes these planes.

Using the notation of the preceding lemma, C'_{q+1} fixes the planes Π_1 and Π_2 corresponding to the reguli $\mathcal{R}_1$ and $\mathcal{R}_2$.

So $\mathcal{R}_2$ is fixed by C_{q+1}. This proves the lemma. ∎

Theorem 6 *(a) Let S^t_E for $1 \le t \le q$ be a partial spread of cardinality $qt+1$ in $PG(3,q)$ consisting of t reguli which share exactly one line. Then either*

$\mathcal{S}_E^t$ is a maximal partial spread or $\mathcal{S}_E^t$ can be extended to a partial spread of type $\mathcal{S}_E^{t+1}$.
(b) Let $\mathcal{S}_H^t$ for $1 \leq t \leq q+1$ be a partial spread of cardinality $(q-1)t+2$ in $PG(3,q)$ consisting of t reguli that mutually share exactly two fixed lines. Then either $\mathcal{S}_H^t$ is a maximal partial spread or $\mathcal{S}_H^t$ can be extended to a partial spread of type $\mathcal{S}_H^{t+1}$.
(c) Let $\mathcal{S}^t$ for $1 \leq t \leq q-1$ be a partial spread of cardinality $t(q+1)$ in $PG(3,q)$ consisting of t reguli which share two conjugate lines L and $L^{(q)}$ in a quadratic extension.

Then either $\mathcal{S}^t$ is a maximal partial spread, or

(c.1) $\mathcal{S}^t$ can be extended to a partial spread consisting of $t(q+1)+1$ lines by a line intersecting L and $L^{(q)}$;

(c.2) $\mathcal{S}^t$ can be extended to a partial spread of type $\mathcal{S}^{t+1}$.

Proof: For (a) and (b), see [5].

For (c), let $\mathcal{S}^t = \mathcal{R}_1 \cup \cdots \cup \mathcal{R}_t$, where $\mathcal{R}_i, i = 1, \ldots, t$, are the reguli sharing the conjugate lines L and $L^{(q)}$.

Let C_{q+1} be the unique cyclic group of order $q+1$ fixing L and $L^{(q)}$, and fixing the complementary regulus $\mathcal{R}'_1$ of $\mathcal{R}_1$ line by line (Lemma 3).

This group also fixes L and $L^{(q)}$ point by point, and fixes $\mathcal{R}_2, \ldots, \mathcal{R}_t$ (Lemma 3).

Suppose M is a line extending $\mathcal{S}^t$ to a larger partial spread. Then there are two possibilities.

If M intersects L and $L^{(q)}$, then M is fixed by C_{q+1}. Then $\mathcal{S}^t$ can be extended to a partial spread with $t(q+1)+1$ lines.

If M does not intersect L and $L^{(q)}$, then $M, L, L^{(q)}$ define a regulus $\mathcal{R}_M$. This regulus $\mathcal{R}_M$ is also an orbit under C_{q+1} (Lemma 5), so $\mathcal{R}_M$ is skew to $\mathcal{R}_1, \ldots, \mathcal{R}_t$. Because if $\mathcal{R}_M$ and $\mathcal{R}_1$ have a point in common, then by using the cyclic group C_{q+1}, also M intersects a line of $\mathcal{R}_1$.

This is false.

So $\mathcal{R}_M$ extends $\mathcal{S}^t$ to a partial spread $\mathcal{S}^{t+1}$. ∎

3 Spreads corresponding to flocks of quadrics

Theorem 7 *Let S be a spread of $PG(3,q)$ containing at least t reguli passing through two conjugate lines L and $L^{(q)}$ of $PG(3,q^2) \setminus PG(3,q)$ where*

(1) $t \geq (3q-4)/4$ *if* q *is even;*
(2) $t \geq (3q-3)/4$ *if* $q \equiv 1 \pmod 4$*;*
(3) $t \geq (3q-5)/4$ *if* $q \equiv 3 \pmod 4$*;*

then $\mathcal{S}$ *is subregular; that is,* $\mathcal{S}$ *can be obtained from a regular spread by replacing a number of reguli by their complementary reguli.*

Proof: Consider the t reguli $\mathcal{R}_1, \ldots, \mathcal{R}_t$ containing L and $L^{(q)}$. Then the complementary reguli $\mathcal{R}'_1, \ldots, \mathcal{R}'_t$ contain $t(q+1)$ of the q^2+1 lines intersecting L and $L^{(q)}$.

So $q^2+1-t(q+1) = (q-1-t)(q+1)+2$ lines intersect L and $L^{(q)}$, and are disjoint from $\mathcal{R}_1, \ldots, \mathcal{R}_t$.

Case 1. If all these $(q-1-t)(q+1)+2$ lines lie in $\mathcal{S}$, then $\mathcal{R}'_1, \ldots, \mathcal{R}'_t$ and these $(q-1-t)(q+1)+2$ lines form the regular spread defined by L and $L^{(q)}$. Hence $\mathcal{S}$ is subregular.

Case 2. Assume now that one of those $(q-1-t)(q+1)+2$ lines does not lie in $\mathcal{S}$. Then there is a line M in $\mathcal{S}$, not lying in $\mathcal{R}_1, \ldots, \mathcal{R}_t$, and not intersecting L and $L^{(q)}$.

Let $\mathcal{R}_{t+1}$ be the regulus through $M, L, L^{(q)}$ defined by the cyclic group C_{q+1} (Lemma 5). This regulus is disjoint from $\mathcal{R}_1, \ldots, \mathcal{R}_t$.

Assume $\mathcal{R}_{t+1} \not\subset \mathcal{S}$. Let $N \in \mathcal{R}_{t+1} \setminus \mathcal{S}$ and let $\mathcal{R}'_{t+1}$ be the complementary regulus of $\mathcal{R}_{t+1}$.

Since $M \in \mathcal{S}$, no line of $\mathcal{R}'_{t+1}$ lies in $\mathcal{S}$. Hence, since the lines of $\mathcal{R}'_{t+1}$ are the only lines of $PG(3,q)$ intersecting L, $L^{(q)}$ and N, the lines of $\mathcal{S}$ intersecting N do not intersect L and $L^{(q)}$. So the $q+1$ lines of $\mathcal{S}$ intersecting N lie in orbits of size $q+1$ under C_{q+1}. These orbits are hyperbolic quadrics through L and $L^{(q)}$.

Let $\mathcal{H}$ be one of these hyperbolic quadrics. Then N intersects $\mathcal{H}$ in at most two points. Otherwise N is a line of $\mathcal{H}$. Then $\mathcal{H}$ would contain $N, L, L^{(q)}$, so $\mathcal{H}$ is the hyperbolic quadric defined by $\mathcal{R}'_{t+1}$ and $\mathcal{R}_{t+1}$. This is false.

So there are at least $(q+1)/2$ such hyperbolic quadrics $\mathcal{H}$. Each of these quadrics has a regulus of $q+1$ lines intersecting L and $L^{(q)}$. The $q+1$ lines of this regulus belong to the set of $(q-1-t)(q+1)+2$ lines not in $\mathcal{R}'_1, \ldots, \mathcal{R}'_t$.

Since two such hyperbolic quadrics have at most two lines intersecting L and $L^{(q)}$ in common, these hyperbolic quadrics contain at least $\sum_{i=0}^{(q-1)/2}(q+1-2i) = (q+1)(q+3)/4$ distinct lines intersecting L and $L^{(q)}$.

Hence $(q-1-t)(q+1)+2 \geq (q+1)(q+3)/4$. This contradicts the assumptions on t.

So $\mathcal{R}_{t+1} \subset \mathcal{S}$. Hence $\mathcal{S}$ has $t+1$ reguli passing through L and $L^{(q)}$. By induction on t, we eventually find that $\mathcal{S}$ turns out to be of Case 1. Hence $\mathcal{S}$ is subregular. ∎

Theorem 8 *Let $\mathcal{S}$ be a spread of $PG(3,q)$ containing at least t reguli passing through two lines L_1 and L_2 of $PG(3,q)$. If*
(1) $t \geq (3q+4)/4$ if q is even;
(2) $t \geq (3q+5)/4$ if $q \equiv 1 \pmod 4$;
(3) $t \geq (3q+3)/4$ if $q \equiv 3 \pmod 4$;
then $\mathcal{S}$ is a spread consisting of $q+1$ reguli containing L_1 and L_2. Hence $\mathcal{S}$ corresponds to a flock of a hyperbolic quadric.

Proof: This is proved by using similar arguments as in Theorem 3.1. ∎

Theorem 9 *Let $\mathcal{S}$ be a spread of $PG(3,q)$ containing t reguli sharing precisely one line L, where*
(1) $t > (3q-4)/4$ if q is even;
(2) $t > (3q-3)/4$ if $q \equiv 1 \pmod 4$;
(3) $t > (3q-5)/4$ if $q \equiv 3 \pmod 4$;
then $\mathcal{S}$ consists of q reguli sharing precisely one line L. Hence $\mathcal{S}$ corresponds to a flock of a quadratic cone.

Proof: Let $\mathcal{S}$ be a spread containing t reguli sharing one line L and let E be the elation group of order q fixing L point by point (Theorem 2.2).

Let L_1 be a line skew to L and let $\{L_1, \ldots, L_q\}$ be the orbit of the line L_1 under E.

Suppose that $\{M_1, \ldots, M_{q+1}\}$ is the complementary regulus of $\{L, L_1, \ldots, L_q\}$ defining the hyperbolic quadric $\mathcal{H}$.

Since L is fixed point by point by E, the orbit of a point $r \in \mathcal{H} \setminus L$ lies on a line M_i. This shows that the points of $PG(3,q) \setminus L$ partition into $q(q+1)$ line-orbits under E.

By assumption, there are precisely t reguli in $\mathcal{S}$ through L. Their complementary reguli consist of $t(q+1)$ such line-orbits under E. Hence there are $(q-t)(q+1)$ line-orbits left.

Let $M \in \mathcal{S}$ and assume that M does not belong to a regulus of $\mathcal{S}$ through L. Consider the regulus $EM \cup \{L\}$ and let $N \in EM \setminus \mathcal{S}$. Then N intersects

$q+1$ lines $T_1, \ldots, T_{q+1}$ of $\mathcal{S}$. These lines all determine a hyperbolic quadric $ET_i \cup \{L\}$ and the complementary reguli always consist of $q+1$ lines of the $(q-t)(q+1)$ remaining line-orbits.

If N intersects three lines of $ET_i \cup \{L\}$, then N belongs to the complementary regulus of $ET_i \cup \{L\}$. Then N intersects L. This is false.

So N contains at most two points of the quadric $ET_i \cup \{L\}$. Hence, we need at least $(q+1)/2$ hyperbolic quadrics having a complementary regulus consisting of $q+1$ lines of the $(q-t)(q+1)$ remaining line-orbits.

As before, this implies $(q-t)(q+1) \geq (q+1)(q+3)/4$. ∎

References

[1] L. Bader and G. Lunardon, On the flocks of $\mathcal{Q}^+(3,q)$. *Geom. Dedicata* **29** (1989), 177-183.

[2] M. Biliotti and N.L. Johnson, Maximal Baer groups in translation planes and compatibility with homology groups. To appear in *Geom. Dedicata.*

[3] W.E. Cherowitzo, T. Penttila, I. Pinneri, and G.F. Royle, Flocks and ovals. *Geom. Dedicata* **60** (1996), 17-37.

[4] F. De Clerck and C. Herssens, Flocks of the quadratic cone in $PG(3,q)$, for q small. *The Cage Reports* **8** (1992), 1-74.

[5] H. Gevaert and N.L. Johnson, On maximal partial spreads in $PG(3,q)$ of cardinalities q^2-q+1, q^2-q+2. *Ars Combin.* **26** (1988), 191-196.

[6] H. Gevaert, N.L. Johnson and J.A. Thas, Spreads covered by reguli. *Simon Stevin* **62** (1988), 51-62.

[7] J.W.P. Hirschfeld, Finite Projective Spaces of Three Dimensions. *Oxford University Press*, Oxford (1985).

[8] N.L. Johnson, Flocks of hyperbolic quadrics and translation planes admitting affine homologies. *J. Geom.* **34** (1989), 50-73.

[9] N.L. Johnson and R. Pomareda, A maximal partial flock of deficiency one of the hyperbolic quadric in $PG(3,9)$. *Simon Stevin* **64** (1990), 169-177.

[10] N.L. Johnson, Homology groups, nearfields and reguli. *J. Geom.* **42** (1991), 100-125.

[11] W.F. Orr, *The Miquelian Inversive Plane $IP(q)$ and the Associated Projective Planes.* PhD thesis, University of Wisconsin, Madison, WI 1973.

[12] S.E. Payne and J.A. Thas, Conical flocks, partial flocks, derivation and generalized quadrangles. *Geom. Dedicata* **38** (1991), 229-243.

[13] L. Storme and J.A. Thas, k-Arcs and partial flocks. *Linear Algebra Appl.* **226-228** (1995), 33-45.

[14] J.A. Thas, Flocks of finite egglike inversive planes. In *Finite Geometric Structures and their Applications* (Bressanone 1972), A. Barlotti, ed., pp. 189-191, Roma, Cremonese (1973).

[15] J.A. Thas, Flocks of nonsingular ruled quadrics in $PG(3, q)$. *Atti Accad. Naz. Lincei Rend. Cl. Sci. Fis. Mat. Natur.* (8) **59** (1975), 83-85.

[16] J.A. Thas, Flocks, maximal exterior sets and inversive planes. In *Finite Geometries and Combinatorial Designs. Contemp. Math.* **111** (1990), 187-218.

[17] J.A. Thas, Projective geometry over a finite field. In *Handbook of Incidence Geometry, Buildings and Foundations*, F. Buekenhout, ed., Chapter 7, pp. 295-348. Amsterdam: North-Holland.

N.L. Johnson, Mathematics Department, University of Iowa, Iowa City, Iowa 52242 (njohnson@math.uiowa.edu)

L. Storme, University of Gent, FCW, Galglaan 2, 9000 Gent, Belgium (ls@cage.rug.ac.be, http://cage.rug.ac.be/~ls)

A New Infinite Family of Extended Generalized Quadrangles

ANNA KASIKOVA Department of Mathematics, Kansas State University, Manhattan Kansas.

ERNEST SHULT Department of Mathematics, Kansas State University, Manhattan Kansas.

1. INTRODUCTION

A **generalized quadrangle** (or GQ) is a system of points and lines (each line being a subset of the set of points) such that

(GQ1) *Any two distinct points are on at most one line.*
(GQ2) *No point is collinear with all remaining points.*
(GQ3) *For each line L and point p not on L, p is collinear with exactly one point of L.*

A generalized quadrangle is said to have *order* (s,t) if and only if each line has $1+s$ points and each point lies on $1+t$ lines. All finite generalized quadrangles having at least three points per line and at least three lines per point, possess an order.

A special type of generalized quadrangle figures in this work:

PAYNE'S CONSTRUCTION OF $AS(q)$: Let $B : V \times V \to GF(q)$ be a non-degenerate symplectic form on a 4-space V over $GF(q)$ and let Δ be the associated polar space $(\mathcal{P}, \mathcal{L})$ of all 1-spaces as points and all isotropic 2-spaces as lines. Then Δ is the classical generalized quadrangle $W(3,q)$ of order (q,q) whose automorphism group contains $Sp(4,q)$. If x and y are 1-spaces which are not perpendicular with respect to B, they span a non-degenerate 2-space $\langle x, y \rangle_V$, which is a line of $\mathbf{P}(V) = PG(3,q)$ containing $1+q$ pairwise non-orthogonal 1-spaces of V – that is a coclique of $1+q$ points in the point- collinearity graph of Δ – with the property that each point of Δ is collinear to exactly one or all of its points. Such a set of points (arising from a non-degenerate 2-space of V) is called a **hyperbolic line** of Δ. Let $\mathcal{H}$ be the collection of all hyperbolic lines. Since every 2-space is either hyperbolic or isotropic, $(\mathcal{P}, \mathcal{H} \cup \mathcal{L})$ is the full $PG(3,q)$.

Fix a point p in $\mathcal{P}$. Let $p^{\perp}$ be all points of Δ which are collinear with p, including p itself. Set $X = \mathcal{P} - p^{\perp}$. Then for each point x in X, a line L on x meets $p^{\perp}$ in exactly one point so that $|X \cap L| = q$. Let $\mathcal{H}_p$ be the q^2 hyperpolic lines on the point p. Their intersections with X partition the q^3 points of X. Let $\mathcal{L}'$ be the union of these two families of q-subsets of X:

$$\mathcal{L}' = \{L \cap X | p \notin L \in \mathcal{L}\} \cup \{H \cap X | H \in \mathcal{H}_p\}.$$

Then

Theorem 1 *(Payne)* $(X, \mathcal{L}')$ *is a generalized quadrangle of order* $(q-1, q+1)$.

An **extended generalized quadrangle** or EGQ is a geometry with three sorts of objects: *points* $(\mathcal{P})$, *edges* $(\mathcal{E})$, which are certain 2-subsets of points, and *blocks*, $(\mathcal{B})$. which are certain larger sets of points. The basic axioms are:

(EGQ1) *Each block is a clique (that is a set of pairwise adjacent points) in the graph* $(\mathcal{P}, \mathcal{E})$.

(EGQ2) *For each point* p, *the set* $\mathcal{E}_p$ *of all edges on* p *and the set* $\mathcal{B}_p$ *of all blocks on* p *form the points and lines of a generalized quadrangle* $Q_p := (\mathcal{E}_p, \mathcal{B}_p)$.

Many extended generalized quadrangles exist and were surveyed eight years ago in Cameron, Hughes and Pasini ([**CHP**]). With one notable exception, all of these have point residues which are classical quadrangles. The exception is an extended generalized quadrangle constructed by J. Thas in 1982 ([**T**]). Its residual quadrangle has order $(q-1, q+1)$, q any prime power. These quadrangles are actually block designs, since every pair of points lies on a block. Much more recently, Pasini, found a class of "flat" extensions of $T_2^*(H)$ ([**Ps**]), and Yoshiara discovered three flag- transitive extensions of the uniqe generalized quadrangle of order (3,5) ([**Y**]).

This note gives a rough description of a family of extended generalized quadrangles whose residual quadrangles are the quadrangles $AS(q)$, for q odd[1] The full details are described in a paper submitted to the *Journal of Combinatorial Theory*.

---

[1] It is worth noting that as this work was written up, another family of extended generalized quadrangles with residual quadrangles of type $AS(q)$, q even, was announced by Del Fra, Pasechnik and Pasini ([**DPP**]). That construction depends upon the existence of hyperovals in projective planes of even order, and makes a nice complement to the construction described here.

2. DESCRIPTION OF THE CONSTRUCTION

Let $h : V \times V \to GF(q^2)$ be a non-degenerate Hermitian form on a 3-dimensional vector space over $GF(q^2)$. Let W be the associated 6-dimensional vector space over $GF(q)$: that is, W is V regarded as a $GF(q)$-module. Let $H : V \to GF(q)$ be the **Hermitian norm** defined by $H(u) := h(u,u)$ for all $u \in V$. A vector of V is **singular** if and only if H vanishes at it. A $GF(q^2)$-multiple of a singular vector is singular as $H(\alpha u) = N(\alpha)u$ where $N(\alpha) = \alpha^{q+1}$ is the (field) norm of α, always an element of $GF(q)$. There is a set of $1+q^3$ singular 1-spaces of V which, together with those subsets lying in the various non-degenerate 2-subpaces, form a 2-design (the Hermitian unital) whose points are doubly transitively permuted by the unitary group $GU(3,q^2)$.

Since the q-th power map induces a field automorphism of order two on $GF(q^2)$, there is an eigenvector θ for this automorphism for which $\theta^q = -\theta$. If q is odd, then $GF(q^2) = GF(q) \oplus \theta GF(q)$ as a 2-space over $GF(q)$. If q is even, one takes $\theta = 1$. One can then define a symplectic form $B : W \times W \to GF(q)$ by the formula

$$B(u,v) := \theta^{-1}(h(u,v) - h(v,u)). \tag{1}$$

We observe

$$B(u,v) = 0 \text{ if and only if } h(u,v) \in GF(q). \tag{2}$$

Let Δ denote the classical polar space of type $Sp(6,q)$ whose points are the 1-spaces of W (over $GF(q)$) and whose lines are the B-isotropic 2-subspaces of W.

The $1+q^3$ points of the unital $H(2,q^2)$ on $\mathbf{P}(V)$ correspond (as sets of vectors) to a family $\mathcal{O}$ of isotropic 2-subspaces of (W,B). Let $\{L_0, L_1, \ldots, L_{q^3}\}$ be the members of $\mathcal{O}$ and set $K = \cup L_i$, viewed as a set of points of Δ.

We define a graph $\Gamma = (K,E)$ where two vertices of K form an edge if and only if $B(x,y) = 0$ and x and y lie in distinct members of $\mathcal{O}$. When q is even, one may see that H is a quadratic form on W whose associated bilinear form is B. Thus K, being the set of H-singular points of $\mathbf{P}(W)$, is a geometric hyperplane of Δ which is a generalized quadrangle of type $Q^-(6,q)$ and $\mathcal{O}$ is a **spread** –that is, a set of pairwise opposite lines of the GQ partitioning the points. In this case the graph Γ is the collinearity graph of this quadrangle after a spread has been removed. So, from this point onward, we assume q is odd.

Let $\langle s_0\rangle$ and $\langle s_1\rangle$ be any two distinct points of the unital $H(2,q^2)$; then s_0 and s_1 are singular vectors of V which can be chosen so that $h(s_0,s_1) = 1$. There is then a vector n of Hermitian norm 1 generating an orthogonal complement to $\langle s_0, s_1\rangle_V$. (Such an n can be chosen $q+1$ ways.) Then as q is odd, $GF(q^2) = GF(q) \oplus \theta GF(q)$ and so W has an ordered basis $\mathcal{B} = (s_0, \theta s_0, s_1, \theta s_1, n, \theta n)$. Now one calculates that

$$B(s_0, \theta s_1) = \theta^{-1}(h(s_0,\theta s_1) - h(\theta s_1, s_0)) = \theta^{-1}(\theta^q - \theta) = -2,$$

since $\theta^q = -\theta$. In fact the Grammian of the symplectic form B with respect to the basis $\mathcal{B}$ is the matrix

$$\begin{pmatrix} 0 & 0 & 0 & -2 & 0 & 0 \\ 0 & 0 & 2 & 0 & 0 & 0 \\ 0 & -2 & 0 & 0 & 0 & 0 \\ 2 & 0 & 0 & 0 & 0 & 0 \\ 0 & 0 & 0 & 0 & 0 & -2 \\ 0 & 0 & 0 & 0 & 2 & 0 \end{pmatrix}.$$

Now $\langle s_0 \rangle$ and $\langle s_1 \rangle$ determine a non-degenerate 2-space of V whose singular 1-spaces form a line of the unital. This 2-subspace of V then determines a non-degenerate 4-space U of (W, B) with $GF(q)$-basis $(s_0, \theta s_0, s_1, \theta s_1)$. The $1+q$ points of this unital line determine $1+q$ members of $\mathcal{O}$ forming a partial spread $\mathcal{O}_U$ of the subpolar space Δ_U, a GQ isomorphic to $W(3,q)$. It follows from this alone, that all the lines of $\mathcal{O}$ are pairwise opposite in Δ, so that $\mathcal{O}$ is a 1-system in the sense of Shult and Thas ([**ST**]).

But it is clear from a study of Δ_U that this 1-system possesses

The BLT-property: *Any line of the polar space Δ which is not one of the lines of the 1-system $\mathcal{O} = \{L_0, \ldots L_{q^3}\}$, meets at most two members of $\mathcal{O}$, and meets exactly one member of $\mathcal{O}$ if and only if it is in the polar perp of that member.*

(This name arises in [**K**] for a dual property in [**BLT**].)

Finally, the group $SU(3,q^2)$ acts doubly transitively on the elements of $\mathcal{O}$ and transitively on the vertices of the graph $\Gamma(K, E)$. But a little more acts on the graph. Multiplication of V by a scalar matrix αI, in general is not a unitary transformation. But it does preserve the property among pairs of vectors of V, that their Hermitian inner product lies in $GF(q)$ and so induces an automorphism of the graph Γ. Multiplication by θI will play a special role below.

Now choose a vertex of Γ, say $v = \langle s_0 \rangle$ lying in L_0. Then L_0 is a special coclique of Γ containing v, and no polar point of $L_0^\perp - L_0$ belongs to the vertex set K. Thus every neighbor vertex in $\Gamma(v)$ determines a polar line on v not in $L_0^\perp$. By the BLT property, (1) the line so determined is unique, and (2) every line of Δ on v which is not in $L_0^\perp$, appears this way. Thus the subgraph $\Gamma(v)$ is isomorphic to the graph induced on all points of the $Sp(4,q)$-quadrangle $v^\perp/v$ not in the perp of the 'point' L_0/v. We now know the local structure of Γ.

All other sorts of vertices make their appearance with v in one of the subsgraphs Δ_U determined by a line of the unital on $GF(q^2)s_0$. There are

1. Two sorts of vertices at distance two from v, each with a distinctive "mu" value (recording the number of length-two paths connecting it with v).

2. Two sorts of vertices at distance three from v, one consisting of a unique vertex $v' = \langle \theta s_0 \rangle_{GF(q)}$, the others being a set of $q-1$ vertices.

As noted before, for each vertex x, the neighborhood graph $\Gamma(x)$ is isomorphic to the graph W^* induced on the q^3 vertices not collinear with a given point of the classical $W(3,q)$ quadrangle. The graph W^* is distance transitive with parameters (in the notation of [**BCN**])

$$(b_0, b_1, b_2; c_1, c_2, c_3) = (q^2 - 1, q^2 - q, 1; 1, q, q^2 - 1).$$

Now multiplication of V by θI induces a freely acting involutatory automorphism $\sigma : x \longrightarrow x'$. This transposes $v = \langle s_0 \rangle$ and $v' = \langle \theta s_0 \rangle$ and induces an isomorphism $\Gamma(v) \longrightarrow \Gamma(v')$. Now any (geodesic) path of length three connecting v' to v must pass through these two graphs. But the edges connecting $\Gamma(v')$ and $\Gamma(v)$ are defined by the rule:

- *For any pair of vertices (x, y) in $\Gamma(v') \times \Gamma(v)$, x is adjacent to y if and only if $\sigma(x) = x'$ and y are at distant three in the subgraph $\Gamma(v) \simeq W^*$.*

Thus when we pass to the homomorphic image

$$\bar{\Gamma} := \Gamma/\langle\sigma\rangle,$$

(whose vertices are the σ-orbits of vertices of Γ, two being adjacent in $\bar{\Gamma}$ if they induced a matching on their vertices, non-adjacent if the two orbits induced a 4-coclique $\bar{K}_4$), the neighbors of the orbit $\{v, v'\}$ form the graph of the distance 1-or-3 relation in W^*. The vertices and cliques in this graph clearly form the generalized quadrangle of type $AS(q)$ as constructed by Payne.

We are thus able to conclude

Theorem 2 *1. The graph $\bar{\Gamma}$ is distance regular with parameters*

$$(b_0, b_1, b_2; c_1, c_2, c_3) = (q^3, (q^2 - 1)(q - 1), 1; 1, 2(q^2 - 1), q^3)$$

but is not distance transitive. It is a $(q + 1)/2$-fold cover of the complete graph.

2. The vertices, edges and maximal cliques of $\bar{\Gamma}$ form an extended generalized quadrangle whose residual quadrangle is type $AS(q)$, for q odd.

REFERENCES

[BLT] Bader, L., Lunardon, G. and Thas J. A., A derivation of flocks of quadratic cones. *Forum Math.* **2** (1990), 163-174.

[B] Brouwer, A. E. Distance regular graphs of diameter three and strongly regular graphs. *Discrete Math.* **49** (1984), 101-103.

[BCN] Brouwer, A. E., Cohen, A. M., and Neumaier, A. *Distance Regular Graphs.* Springer-Verlag Berlin (1989).

[CHP] Cameron, P., Hughes, D. and Pasini, A. Extended generalized quadrangles. *Geom. Dedicata*, **35**, 193-228.

[DDP] Del Fra, A., Pasechnik, D. and Pasini, A. A new family of generalized quadrangles. Preprint.

[K] Kantor, W. A note on generalized quadrangles, flocks and BLT-sets. *J. Comb. Theory Ser. A* **68** (1991), 153-157.

[Ps] Pasini, A. *Diagram Geometries*, Oxford University Press, 1994.

[P] Payne, S. Quadrangles of order $(s-1, s+1)$. *J. Alg.* **22** (1972), 97-119.

[ST] Shult, E. E. and Thas, J. A. m-systems of polar spaces, *J. Comb Theory ser. A* **68** (1994), 184-204.

[T] Thas, J. A. Extensions of finite generalized quadrangles. *Symp. Math.* **28** , Academic Press (1986), 127-143.

[Y] Yoshiara, S. On some flag-transitive non-classical $c \cdot C_2$ geometries *European J. Comb.*, **14** (1993), 59-77.

Virtual Derivation

ROBERT A. LIEBLER Department of Mathematics, Colorado State University, Fort Collins, Colorado 80523

To T. G. Ostrom on his 80th birthday

1. Introduction

The purpose of this work is to present a new algebraic invariant for spreads in $PG(3,q)$ together with a hint of its geometric meaning. This work is also a return to some ideas first presented in Ravello in 1988 and in which my interest was rekindled by recent work of Deza and Grishukhin [3] on Steiner triple systems. The key idea is to study spreads as elements of the free $\mathbb{Z}$-module based on lines. Fundamental to such a study is the lattice Λ that is the $\mathbb{Z}$-span of the differences of pairs of opposite reguli (technical terms are explained below). If N denotes the incidence map from lines to points of $PG(3,q)$, then $\Lambda \leq kerN$ and the abelian group $(kerN)/\Lambda$ is the central object of our study.

The generalized Reed Muller codes on $M := Mat_{22}(GF(q))$ are $GF(q)$-vector spaces of functions from M to $GF(q)$. Any such function may be expressed as a polynomial function in 4 variables and (total) degree $\leq 4(q-1)$. The generalized Reed Muller code $\mathcal{R}_q(d,4)$ is just the set of such functions having degree less than or equal to d. Clearly, these codes (subspaces) are totally ordered by inclusion in the vector space of all functions from M to $GF(q)$.

Suppose the spread S of lines in $PG(3,q)$ has a set of slope matrices $\{m_s : s \in S\} \subseteq M$ with respect to lines $\ell, m \in S$ and $m_m = 0 \in M$. Consider the (characteristic) function $\overline{\sigma}(S)$ that takes value $1 \in GF(q)$ at x if $x = m_s$, for some $s \in S$ and value 0 if $x \in M \setminus \{m_s : s \in S\}$. We ask: what is the degree d of $\overline{\sigma}(S)$?

This question can be answered in two ways. The first is to simply find a polynomial function that takes exactly the same values as $\overline{\sigma}(S)$ and compute this polynomial's (total) degree. A second, more circuitous but also more effective for general S, way is to consider $\overline{\sigma}(S)^{\perp}$ the set of all (polynomial) functions f such that

$$0 = \sum_{m \in M} f(x)\sigma(S)(x) = \sum_{s \in S} f(m_s).$$

If m is the maximum degree of $f \in \overline{\sigma}(S)^{\perp}$, then (Theorem 1.4) $d = 4(q-1) - m$.

This research was partially supported by the National Security Agency.

This number d, which we call the **degree** of S with respect to l and m, measures the position of S in (a translate of) Λ. The degree does not depend on any of the choices leading to its construction other than the "vertical" line l. This new algebraic invariant most naturally takes values in the submodule lattice of a Hensel lift to characteristic q of $GF(q)M$ under $GF(q)AO^+(4,q)$, (see Section 5), but this module is unfortunately not now well understood.

Recall that a generalized or **virtual** character [2, p.208] of a finite group is an integral linear combination of characters. By analogy, say the spread S' is obtained from the spread S by **virtual derivation if** $S' - S \epsilon \Lambda$.

Our main result is the following relationship between degree and virtual derivation.

Theorem A. *Suppose S is a spread of lines in $PG(3,q)$. If S is virtually derivable from a regular spread, then S has degree $d \leq 4q-6$ with respect to any of its lines l. In case of equality, there is a unique term of highest degree. The case of equality occurs for Hall planes whenever l is not in the derivation set.*

Regular spreads have degree $2q-2$ and I wonder if any other spreads have lower degree. More importantly I wonder if the spread taxonomy based on the degree might actually be useful. Most importantly, I would dearly love to know of a spread with degree $4q-5$, but would settle for one that is not virtually derivable.

This last wish can be put in concrete terms. Suppose S is a spread of lines in $PG(3,q)$ with a set of slope matrices $\{m_s : s \epsilon S\}$ with respect to lines $l, m \epsilon S$. Let f be any quadratic function of the entries of the matrix $m = (m_{ij})$. If

$$\sum_{s \epsilon S} f(m_s) \neq 0$$

then S has degree $\geq 4q-5$. If this condition holds for some f in
$\mathcal{M} = \{m_{11}^2, m_{12}^2, m_{21}^2, m_{22}^2, m_{11}m_{12}, m_{11}m_{21}, m_{12}m_{22}, m_{21}m_{22}, m_{11}m_{22}+m_{12}m_{21}\}$
then S is not virtually derivable from a regular spread. **I conjecture a converse to Theorem A: Any spread S for which $\overline{\sigma}(S) \epsilon \mathcal{M}^\perp$ is virtually derivable.**

Our methods are mostly combinatorial and coding theoretic to start. But then turn algebraic culminating in a desperate endomorphism algebra computation in Lemma 3. Sections 2 and 3 contain preliminary material about generalized Reed Muller codes and the Klein correspondence respectively. Sections 4 and 5 contain a combinatorial analysis of Λ culminating in Proposition 2. Section 6 combines the coding theoretic and combinatorial preliminaries to prove Theorem A.

In order to be more precise more notation is required. The 1– and 2–dimensional subspaces of $V := GF(q^n)$ are called **points** and **lines** of the projective space $PG_{n-1}(q)$. Let $Q : V \to GF(q)$ be a quadratic form on V. A subspace $\alpha \leq V$ is called **(totally) singular** if $Q(a) = 0$ for all $a \epsilon \alpha$. The set of singular points of Q is called the **quadric** $\mathcal{Q}$ of Q. Two subspaces α, β are perpendicular (with respect to Q), written $\alpha \perp \beta$ or $\alpha \leq \beta^\perp$, if $Q(a+b) = Q(a) + Q(b)$ for all $0 \neq a \epsilon \alpha$, $0 \neq b \epsilon \beta$. The form Q is called **0** if its **radical** $V^\perp$ equals zero.

There is a classification of quadratic forms on V. Let $a = dimV/V^{\perp}$ and $b = dimV^{\perp}$. If a is even and not zero, there are two possibilities: **elliptic** denoted a^{-}/b, and **hyperbolic** denoted a^{+}/b. If a is odd there is only one possibility, denoted a/b. When $b = 0$, we avoid sacrilege by omitting /0 from this notation. For example, the form $Q(x_1, x_2, x_3, x_4, x_5) = x_1x_2 + x_3x_4$ on $GF(q)^5$ has type

$$4^{+}/1 \text{ or } (2^{+} \perp 2^{+})/1 \text{ or even } (2^{-} \perp 2^{-})/1.$$

Also abuse notation saying that a subspace U of an orthogonal geometry with quadric $\mathcal{Q}$ has the same **type** as $U \cap \mathcal{Q}$. Since this notation is intended to supplement not supplant, we also use the standard notation in the most important cases.

For example, the singular lines (subspaces of type 0/2) of the hyperbolic quadric (orthogonal geometry of type 4^{+}) fall into two **opposite reguli**, R and R' each of which is a set of mutually skew lines covering the quadric.

Let $\mathcal{P}$ and $\mathcal{L}$ denote the points and lines of $PG_3(q)$. A **spread** of $PG_3(q)$ is a subset $S \subseteq \mathcal{L}$ such that each point is on exactly one element of S. Let $\mathcal{S}$ denote the set of all spreads of $PG(3, q)$. Given a spread S and one of its lines ℓ, one can choose a basis $\{e_i\}$ of the underlying vector space so that $\ell =< e_1, e_2 >$. Then any line s skew to ℓ is the column space of a matrix $(I_2, m_s^t)^t$ whose transpose is in reduced row echelon form. There is no loss in taking $m_m = 0$ for some $m \in S$. The set $\{m_s : s \in S \setminus \{\ell\}\}$ is a called a set of **slope matrices** for S with respect to ℓ, m. This is explained more fully at the beginning of Section 5.

A spread S is **regular** if any three of its lines lie in a (unique) regulus R that is itself contained in S. Ostrom's concept of **derivation** [6] is an abstraction of the observation that by replacing the elements of R with the elements of its opposite regulus R' one obtains a new spread **derived** from R. A theorem of Orr [5] asserts that this process of derivation cannot be repeated in a nontrivial way. That is, a sequence of derivations can only lead to spreads that could have been obtained by derivation with respect to a disjoint set of reguli. But the Orr result presumes that each intermediate step is a spread. This study began by asking: what if one asks only that the final outcome be a spread? But perhaps it is not so simple, if the intermediate steps are not spreads, then what are they?

For a finite set X, define $\mathbb{Z}X$ to be the free module over the integers $\mathbb{Z}$ based on X and identify $Y \subseteq X$ with the sum $\sum_{y \in Y} y \in \mathbb{Z}X$ of its elements.

The assertion that the spread S' is obtained from the spread S by derivation with respect to the regulus R may now be expressed as $S' = S - R + R' \in \mathbb{Z}\mathcal{L}$. There arise several natural $\mathbb{Z}$-submodules, or **lattices** of $\mathbb{Z}\mathcal{L}$. Let $\Lambda := \langle R - R' \rangle \leq \mathbb{Z}\mathcal{L}$, where R and R' are opposite reguli as above. Since any regulus R may be embedded in a regular spread S, there is a spread S' (associated with a Hall plane) such that $R - R' = S - S' \in \mathbb{Z}\mathcal{L}$. Therefore $\Lambda \leq \Sigma := \langle S_1 - S_2 : S_1, S_2 \in \mathcal{S} \rangle$. Finally, since each spread covers the points of $PG(3, q)$, the difference of any two spreads is in the kernel of the natural incidence map $N : \mathbb{Z}\mathcal{L} \to Z\mathcal{P}$ arising from $PG_3(q)$.

Now $GL(4, q)$ acts as a rank 3 permutation group on $\mathcal{L}$, so $\mathbb{Q}\mathcal{L}$ has but three $\mathbb{Q}GL(4, q)$-constituents. Since $GL(4, q)$ acts nontrivially on $\mathcal{P}$ and since the rational incidence map $\mathbb{Q}\mathcal{L} \to \mathbb{Q}\mathcal{P}$ is surjective, its kernel is an irreducible $\mathbb{Q}GL(4, q)$-module W. It follows that each the three lattices enumerated in the last paragraph spans W and $dim\ kerN = |\mathcal{L}| - |\mathcal{P}| = q^4 + q^2$.

Again fix a line ℓ and let $A(\ell)$ be the set of lines meeting ℓ. Now let $\pi : \mathbb{Z}\mathcal{L} \to \mathbb{Z}(\mathcal{L} \setminus A(\ell))$ be the projection with kernel $\mathbb{Z}A(\ell)$. At the beginning of Section 5 we give the slope matrix map $\sigma : \mathbb{Z}(\mathcal{L} \setminus A(\ell)) \to \mathbb{Z}M$. Let $^{-}$ denote $\otimes GF(q)$ (reduction mod p and extension of coefficients to $GF(q)$). Then $\overline{\sigma\pi}$ has image $GF(q)M$, the free $GF(q)$ vector space based on M. This vector space is the ambient space of the generalized Reed Muller codes.

We actually find a polynomial function having degree $4q-2$ that models a pair of opposite reguli and thereby show that $\overline{\sigma\pi(\Lambda)}$ is not contained in $\mathcal{R}_q(2,4)^{\perp}$. We are then content to proceed modulo $\mathcal{R}_q(2,4)^{\perp}$. Since the quotient $GF(q)M/\mathcal{R}_q(2,4)^{\perp}$ is dual to the $GF(q)$-valued quadratic polynomial functions, this amounts looking for functions of degree ≤ 2 to annihilate $\lambda \in \Lambda$.

Denote reduction modulo $\mathcal{R}_q(2,4)^{\perp}$ with a second $^{-}$. The $GF(q)$-vector space $\overline{\overline{\mathbb{Z}\mathcal{L}}}$ has dimension $15 = 10+4+1$ with intermediate generalized Reed Muller codes appearing as a chain of subspaces. Our analysis falls into two parts, beginning with $\Lambda_\ell := \Lambda \cap \mathbb{Z}A(\ell) \leq kerN \cap \mathbb{Z}A(\ell)$ in Section 4 followed by $\sigma\pi(\Lambda) \leq \sigma\pi(kerN) \leq \mathbb{Z}M$ in sections 5 and 6. Here is a summary of our other results:

Summary

1. *Λ_ℓ is a pure submodule of $kerN$ having rank q^2 (Corollary 4.2).*
2. *$(kerN)/\Lambda$ has exponent dividing q (Proposition 2).*
3. *$\overline{kerN} \leq \mathcal{R}_q(0,4)^{\perp}$ has dimension ≤ 14 (Proposition 4.1).*
4. *$\overline{\overline{\Sigma}} \geq \overline{\;R_q(1,4)^{\perp}}$ has dimension ≥ 10 for every odd q for which there is a spread that is not virtually derivable from a regular spread (Proposition 4.2). (That extra space between the overline and $\mathcal{R}_q(1,4)^{\perp}$ denotes the second bar not the first.)*
5. *$\overline{\overline{\Lambda}}$ is 1-dimensional (Lemma 3).*

It must be pointed out that earlier work by G. James [4] implies that $kerN/\Lambda$ is a p-group where p divides q but his methods are highly non constructive, working in effect, with the combinatorially awkward coefficient ring $\mathbb{Z}[1/p, \zeta]$ where $\zeta \in \mathbb{C}$ is a primitive p-th root of unity. Indeed, James explicitly acknowledges [4, p 73-74] the difficulty of bringing his rather more general results into combinatorial focus.

Candid comments of W. M. Kantor, from which this exposition has greatly benefitted, are cheerfully acknowledged.

2. The Generalized Reed Muller codes

Let F denote $GF(q)$, let W be an F-vector space of dimension m and let G be $AGL(W, F)$. Unfortunately, despite its fundamental importance, the precise structure of the FG-permutation module $GF(p)W$ is still unclear. We give here a first approximation to this structure following [1, §5.4, §5.7].

The trick is to consider FW and to regard it as the vector space of polynomial functions from W to F. Let $X_i : W \to GF(q)$ denote the functions given by

$X_i((a_1 \quad \cdots \quad a_m)^t) = a_i$. Then $\{X_i : i = 1, \cdots, m\}$ spans the vector space $\epsilon F_1[]$ of linear functions from W to $GF(q)$ and generate the ring $\epsilon F[]$ of all polynomial functions from W to F.

More generally let $\epsilon F_d[], 0 \leq d \leq m(q-1)$, be the image of homogeneous polynomials $F_d[]$ of degree d under the evaluation map $\epsilon : F[] \to W^F$ from F-polynomials in m variables to functions from W to F. The kernel of ϵ is the ideal generated by $\{X_i^q - X_i\}$ and $\epsilon F[]$ is an F-algebra in a natural way. The F-submodules $\epsilon F_d[], 0 \leq d \leq m(q-1)$ represent that part of the natural grading of $F[]$ that is inherited by $\epsilon F[]$.

The (left) action of FG on $F[]$ is also inherited by $\epsilon F[]$ and is given by:

$$wg(f) : \mapsto w + f(g^t); \quad \in W; \ (w, g) \rtimes AGL(4, q),$$

where g^t denotes the transpose of the matrix g.

The following results appear in [1, Theorem 5.7.9, Theorem 5.4.2].

Theorem 1. *Suppose W has dimension 4 over F and $q = p^t$. Then*

1. *The F-vector space of all functions from W to F is $\epsilon F[]$.*
2. *For each $0 \leq d < 4(q-1)$, the* **generalized Reed Muller code**

$$\mathcal{R}_q(d, 4) := \bigoplus_{i \leq d} \epsilon F_i[]$$

   is FG-invariant.

3. *The image of the incidence map from lines to points of the affine geometry $AG(4, q)$ is $\mathcal{R}_q(3(q-1), 4)$.*
4. *$\mathcal{R}_q(d, 4)^\perp = \mathcal{R}_q(4q - 5 - d, 4)$*

3. The Klein Correspondence

Let $F = GF(q)$ and $V = F^4$ where $q = p^t$, and p is prime. Each $\ell \in \mathcal{L}$ may be described as the column space of a 4 by 2 matrix $M(\ell)$ whose transpose is in reduced row echelon form. For such a matrix M define $M_{ii'}, 1 \leq i < i' \leq 4$ to be the determinant of the submatrix consisting of its ith and i'th rows. The **Plücker coordinates** of such a matrix M are defined to be

$$\kappa(M) := (M_{12}, M_{13}, M_{14}, M_{23}, M_{24}, M_{34})^t \in GF(q)^6.$$

It is a simple matter to check that the point of $PG_5(q)$ given by $\kappa(\ell) := \langle \kappa(M) \rangle$ is unchanged if M is replaced by a column equivalent matrix and is a singular point of the hyperbolic quadric 6^+ associated with the

$$Q(x_{12}, x_{13}, x_{14}, x_{23}, x_{24}, x_{34}) = x_{12}x_{34} - x_{13}x_{24} + x_{14}x_{23}.$$

It is a remarkable fact that each point of this quadric $\mathcal{Q} \subset PG_5(q)$ is in the image of κ. There results the celebrated **Klein Correspondence** [7, Ch 12] between the lines of $PG_3(q)$ and singular points of a hyperbolic quadric $\mathcal{Q}(6^+)$ in $PG_5(q)$.

Figure 1 provides a brief summary of some of this correspondence's properties. The Klein correspondence together with Witt's Theorem imply all sorts of transitivity statements, for example:

$a \cap b \neq 0$	$\rightarrow$	$\kappa(a) \perp \kappa(b)$
lines on a point of $\mathcal{P}$	$\rightarrow$	$\mathcal{Q}(0/3)$ (+type t.s. plane)
lines in a plane	$\rightarrow$	$\mathcal{Q}(0/3)$ (-type t.s. plane)
lines on a point in a plane	$\rightarrow$	$\mathcal{Q}(0/2)$
All lines meeting a line	$\rightarrow$	$\mathcal{Q}(4^+/1)$
opposite reguli	$\rightarrow$	$\mathcal{Q}(3) \perp \mathcal{Q}(3)$
regular spread	$\rightarrow$	$\mathcal{Q}(4^-) \perp 2^-$
t.i. lines	$\rightarrow$	$\mathcal{Q}(5) \perp 1^-$
t.i. lines meeting a line	$\rightarrow$	$\mathcal{Q}(3/1)$

(The term t.i. lines denotes totally isotropic lines with respect to some 0 alternate form.)

Figure 3.1: The Klein correspondence $\mathcal{L} \xrightarrow{\kappa} \mathcal{Q}(6^+)$

Lemma 1. *Fix a line $\ell \in \mathcal{L}$ and a regulus R disjoint from ℓ. Then there is a unique regular spread containing both ℓ and R. Moreover, the subgroup of $PGL(V, F)$ fixing ℓ acts transitively on such reguli.*

Proof: Let R be such a regulus. Then $\kappa(\ell)$ is a singular point, $\kappa(R)$ is of type 3 and the span of $\kappa(\ell)$ and any singular point in $\kappa(R)$ is of type 2+ by the Klein correspondence. It follows that the span $\langle \kappa(\ell), \kappa(R) \rangle \leq 6^+$ is of type 4^-. The transitivity statement follows from Witt's Theorem [7, Theorem 7.4].

Suppose a set $\mathcal{R}$ of reguli is given. The Klein correspondence and the $\perp$ relation in $PG_5(q)$ allows us to quickly decide the nature of multi-set $\mathcal{R}'$ of lines in the reguli $\{R' : R \in \mathcal{R}\}$. For example, the left half Figure 2 is a Hasse diagram for the types of subspaces of $PG(5, q)$ that correspond under κ to a partition of the lines (other than ℓ) of a regular spread S on the line ℓ into reguli on ℓ. The right half of this figure is a Hasse diagram for the types of subspaces arising as perps of those on the left hand side, and therefore determines the lines that correspond under κ to the reguli opposite to those in the partition. (The terms **flock, pencil** and **bundle** are used to describe partitions of a quadric with plane sections that are disjoint, share one point and share two points, respectively. Such a family of plane sections is called **linear** if the associated planes all fall on a line of projective space.)

The linear case is the context of our next

Lemma 2. *The perps of a pencil in a 4^- on $\kappa(\ell)$ forms a linear flock of a 3/1 with vertex $\kappa(\ell)$ and covers the singular points apart from $\kappa(\ell)$ in a 3/1. The difference in $\mathbb{Z}\mathcal{Q}(6^+)$ of two such pencils in the same 4^- has as its support the difference in $\mathbb{Z}\mathcal{Q}(6^+)$ of the singular points in two 3/1 of the $4^+/1$ given by $\kappa(\ell)^\perp$. These 3/1s intersect in a $2^-/1$.*

This implies

Corollary 3. *Let Λ_ℓ denote the sub-lattice of Λ having support in the set of lines $m \in \mathcal{L}$ that meet ℓ. Let A and B be any two 3/1's in the $4^+/1$ given by $(\kappa(\ell))^\perp$. Then $A - B \in \Lambda_\ell$.*

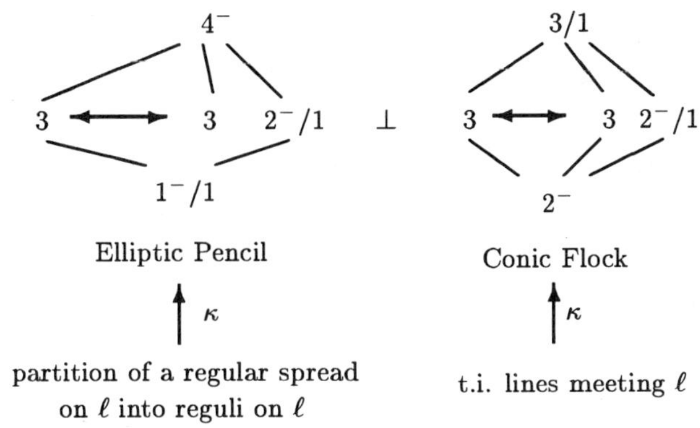

Figure 3.2: Lemma 3.2

Proof: If $A \cap B$ is a $2^-/1$ this follows directly from the lemma. Otherwise pick C to be the span of intersecting $2^-/1$s in A and in B. Apply Lemma 2.1 twice, first with $A - C$ then $C - B$.

4. The structure of Λ_ℓ

Continue the notation of the introduction and take F, V, q, p, t as in the first sentence of Section 3, but now fix a line $\ell \in \mathcal{L}$ and let G be the F-linear transformations of V leaving ℓ invariant. Also let $A(\ell) \subset \mathcal{L}$ denote the set of lines meeting ℓ. $\Lambda_\ell := \Lambda \cap \mathbb{Z}A(\ell)$ denote the sub-lattice of Λ having support in $A(\ell)$. Note that there is no loss in ignoring the multiplicity m_ℓ when considering elements of Λ_ℓ, since elements of Λ all have the form $\sum_{x \in \mathcal{L}} m_x x$ where $\sum_{x \in \mathcal{L}} m_x = 0$.

This and the following section contain a combinatorial analysis of Λ. In it we give a number of explicit constructions eventually showing that $kerN/\Lambda$ has exponent dividing q in Proposition 2. In this section we treat Λ_ℓ by constructing a submodule $\mathcal{N}$ of Λ_ℓ that is both pure [2, p 84] in $kerN$ and of rank equal to $kerN \cap \mathbb{Z}A(\ell)$. That is, we show $(kerN\mathbb{Z} \cap A(\ell))/\mathcal{N}$ is both torsion free and not free as a $\mathbb{Z}$-module, from which it of course follows that $kerN \cap \mathbb{Z}A(\ell) = \mathcal{N}$.

For points p_1, p_2 of ℓ and planes π_1, π_2 on ℓ, let $\mathcal{L}_{ij}$ be the set of lines on p_i in π_j. Define $n(p_1, p_2, \pi_1, \pi_2) := \mathcal{L}_{11} - \mathcal{L}_{12} - \mathcal{L}_{21} + \mathcal{L}_{22} \in \mathbb{Z}\mathcal{L}$. Note that $n(p_1, p_2, \pi_1, \pi_2)$ has a support of only $4q$ lines and these fall in four planar pencils all in $A(\ell)$.

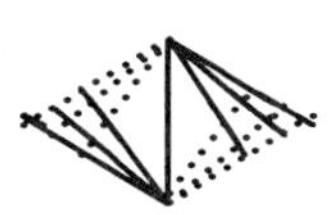

Note also that $n(p_1, p_2, \pi_1, \pi_2) \in \ker N$. Define

$$\mathcal{N} = \langle n(p_1, p_2, \pi_1, \pi_2) : p_1, p_2 \text{ points of } \ell;\ \pi_1, \pi_2 \text{ planes on } \ell \rangle.$$

Lemma 1. $\mathcal{N} \leq \Lambda_\ell$.

Proof: Use Corollary 3 and work in the quotient space $H := \kappa(\ell)^{\perp}/\kappa(\ell)$ which has type 4^+. A singular point of H is the image under κ of a planar pencil on ℓ (this is the fourth entry in Figure 1). Let L_{ij} denote the singular point $\kappa(\mathcal{L}_{ij})$ of H and let $P \in \mathcal{Q}(H) \cap (\langle L_{21}, L_{22}\rangle \setminus \{L_{21}, L_{22}\})$. Let $\mathcal{A}_i$ be the planes of type 3 in a linear pencil of $\mathcal{Q}(H)$ on $L_{1,i} \in H$, for $i = 1, 2$. Finally, let $\mathcal{B}_i$ be the planes of type 3 in the linear bundle of $\mathcal{Q}(H)$ on $\langle P, L_{1i}\rangle$ for $i = 1, 2$.

Let $\mathbb{Z}\mathcal{Q}(4^+)$ denote the free $\mathbb{Z}$-module based on the singular points of H. By Lemma 3, it is sufficient to show that when $L_{11} - L_{12} - L_{21} + L_{22}$ is regarded as an element of $\mathbb{Z}\mathcal{Q}(4^+)$ it can be expressed in the form $\sum m_A \mathcal{Q}(A)$ where $\sum m_A = 0$, and in each case A ranges through the subspaces of H that are of type 3. This is accomplished by

$$\sum_{A\in\mathcal{A}_1} \mathcal{Q}(A) - \sum_{A\in\mathcal{A}_2} \mathcal{Q}(A) + \sum_{B\in\mathcal{B}_1} \mathcal{Q}(B) - \sum_{B\in\mathcal{B}_2} \mathcal{Q}(B).$$

The following figure may be helpful in this verification. In it each component is a graphical representation of an element of $\mathbb{Z}\mathcal{Q}(4^+)$ as a labelled $(q+1) \times (q+1)$ grid. Multisets of singular points appear as the grid with each position labeled by its multiplicity.

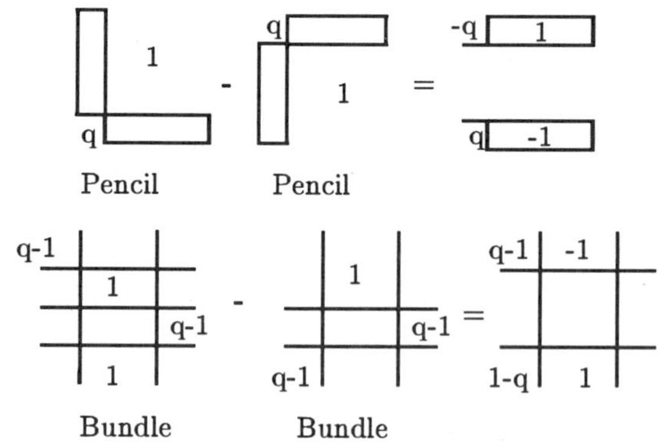

(Blank regions are weighted 0.)

Corollary 2. *If Λ/Λ_ℓ has rank q^4, then $\mathcal{N} = \Lambda_\ell$.*

Proof: Since $kerN$ has rank $q^4 + q^2$, it suffices to show that $\mathcal{N}$ contains a pure submodule of $\mathbb{Z}\ell$ of rank q^2. Fix p_1 in ℓ and π_1 on ℓ. Then $\{n(p_1, p_2, \pi_1, \pi_2) : p_2 \neq p_1,\ \pi_2 \neq \pi_1\}$ is a set of cardinality q^2. It is a basis for a $\mathbb{Z}$-pure submodule of $\mathbb{Z}^\ell$ because the lines of corresponding to $\mathcal{L}_{22}$ of the preceding proof fall in the support of exactly one element of this set.

5. The Structure of Λ/Λ_ℓ

We now turn to the structure of Λ/Λ_ℓ. In other words we no longer keep track of the multiplicities of lines meeting ℓ and focus instead on $\mathcal{L} \setminus A(\ell)$, the q^4 lines skew to ℓ. This study seems to best be accomplished in a setting that leads directly to the generalized Reed Muller codes of Section 2.

More formally, this is accomplished with the **natural projection** π from $\mathbb{Z}\mathcal{L}$ onto $\mathbb{Z}(\mathcal{L} \setminus A(\ell))$ having $\mathbb{Z}A(\ell)$ as its kernel. Continue the notation of the preceding section and let $m \in \mathcal{L}$ be a line of $V = F^4$ that is skew to ℓ. Change bases so that

$$\ell = \left\langle \begin{smallmatrix} 0 \\ I_2 \end{smallmatrix} \right\rangle \quad \text{and} \quad m = \left\langle \begin{smallmatrix} I_2 \\ 0 \end{smallmatrix} \right\rangle,$$

where $\langle\ \rangle$ is used to denote the column space of a matrix. Let A be the subgroup of G that acts trivially on ℓ and trivially on V/ℓ. Then A is elementary abelian of order q^4 and has elements of the form:

$$g(a) = \begin{pmatrix} I_2 & 0 \\ a & I_2 \end{pmatrix}, \text{ where } a \in M := Mat_{22}(GF(q)).$$

Also A is a normal subgroup of G, $G/A \cong GL(2,q) \times GL(2,q)$, and G acts on A inducing the natural action of $GL(2,q) \times GL(2,q)$ on M given by left and right matrix multiplication

$$(A, B) : m \mapsto AmB^{-1}. \tag{5.1}$$

Now A acts regularly on the lines skew to ℓ. Use it to construct a bijection with the set of possible slope matrices

$$\sigma : \mathcal{L} \setminus A(\ell) \to M \text{ via } (g(a)m \mapsto a).$$

Abuse notation and use σ to denote the associated $\mathbb{Z}$-linear map $\mathbb{Z}(\mathcal{L} \setminus A(\ell)) \to ZM$. Now σ and $\sigma\pi$ are G-isomorphisms. Moreover G may now be regarded as an affine permutation group on $\mathcal{L} \setminus A(\ell)$ with the associated permutation representation being realized by $\mathbb{Z}M$.

Lemma 1. *Regard M as an F-vector space in the natural way. Then $\sigma\pi(\Lambda)$ contains the characteristic function of each* **affine line** *of M.*

Proof: Since $\sigma\pi(\Lambda)$ is A-invariant, it suffices to show that each 1-dimensional F-subspace of M has its characteristic function in $\sigma\pi(\Lambda)$. Under the $GL(2,q) \times GL(2,q)$ action on M there are but two orbits of 1-dimensional F-subspaces, namely those corresponding to a nonsingular and those corresponding to a singular.

To exhibit the characteristic function of an nonsingular 1-space in $\sigma\pi(\Lambda)$, observe that the F-scalar matrices in M are the image under σ of a regulus R containing ℓ and m. Moreover R is invariant under the group $\{g(a) \mid a \text{ is scalar }\} \leq A$, which maps under σ to an F-subspace of M.

To exhibit the characteristic function of a singular 1-space in $\sigma\pi(\Lambda)$, consider the configuration arising in Lemma 1 where ℓ is taken to in π_1 on p_1 but **off** p_2. The only lines of this configuration not in $A(\ell)$ are those in π_2 on p_2. The result follows from the fourth line of Figure 1.

Proposition 2. *The abelian group $\mathbb{Z}M/\sigma\pi(\Lambda)$ is of exponent dividing q.*

Proof: Fix a 2-dimensional F-subspace $N \leq M$ and consider the sum Σ in $\mathbb{Z}M$ of all 1-spaces in N. Then Σ counts m to multiplicity $q+1$ and each of the remaining elements of N to multiplicity 1. By Lemma 1, $\Sigma \in \sigma\pi(\Lambda)$ as is N, since it is a union of any parallel class of affine lines. It follows that $qm \in \sigma\pi(\Lambda)$.

Now the G-invariance of Λ and the fact that $\sigma\pi$ is a G-homomorphism implies that $qx \in \sigma\pi(\Lambda)$ for all $x \in \mathcal{L} \setminus A(\ell)$. From this it follows that $\pi(\Lambda)$ has rank q^4 as a $\mathbb{Z}$-module and so Corollary 2 implies that $kerN/\Lambda$ is torsion. The result follows.

Let $\bar{\ }$ denote $\otimes F$ (reduction mod p and extension of coefficients to $GF(q)$). Then $\overline{\sigma\pi}$ has range FM, the free $GF(q)$-vector space based on M. Recall that FM is the ambient space of the generalized Reed Muller codes. In order to study the size of Im $\overline{\sigma\pi}$, we recall Theorem 1.4 and seek polynomial functions of low degree in its perp. Since quadratic functions suffice, we may as well reduce modulo $\mathcal{R}_q(4q-7,4) = \mathcal{R}_q(2,4)^\perp$.

Denote reduction modulo $\mathcal{R}_q(2,4)^\perp$ by a second bar: $\bar{\bar{\ }}$.

6. The Structure of $Im\ \bar{\bar{\ }}$

By Lemma 1 and Theorem 1.3, $\overline{\Lambda} \geq \mathcal{R}_q(3(q-1),4)$ but actually $\overline{\Lambda}$ is much bigger. It turns out that $\bar{\bar{\ }}$ is sufficient for our purposes, but to see this we need even more machinery.

Lemma 1. *Let $K := GF(q^2)$ and $\bar{\ }$ denote its nontrivial F-automorphism. Let*

$$\mathcal{S} = \{\ s(a,b) := \begin{pmatrix} a & b \\ \bar{b} & \bar{a} \end{pmatrix} \ :\ a,b \in K\ \}.$$

1. *Then $\mathcal{S}$ is an F-algebra isomorphic to $Mat_{22}(F)$.*
2. *Let $\rho : \mathcal{S} \to M$ be such an ring isomorphism. Define $\mathcal{S}$ as a G-module by:*

   $$gs = \rho^{-1}(A)s\rho^{-1}(B^{-1}) \text{ where } g\rho(s) = A\rho(s)B^{-1}.$$

   Then ρ induces a G-module isomorphism from $\mathcal{S}$ to M.
3. *$R := \{\rho(s(a,0)) : a\bar{a} = 1\}$ and $R' := \{\rho(s(0,b)) : b\bar{b} = 1\}$ are opposite reguli.*

Proof: The set of matrices of the indicated form is clearly closed under addition multiplication and scalar multiplication by F. Thus, they form a 4-dimensional noncommutative F-algebra. After checking that $\mathcal{A}$ has no nontrivial 2-sided ideals, apply the Wedderburn structure theorem [2, 3.22] to obtain part 1.

Part 2 is immediate from the action of G on M given in Equation(5.1).

For part 3, it is enough to check that each line in $\sigma^{-1}(R)$ meets each line in $\sigma^{-1}(R')$. This is equivalent to showing that $\rho(s(a,0)) - \rho(s(0,b))$ in not invertible whenever $a\bar{a} = b\bar{b} = 1$. The result follows from the choice of ρ as a ring isomorphism.

The representation given in Lemma 1 has the great advantage that it presents $Mat_{22}(F)$ as a K-module while avoiding explicit irreducible polynomials. Unfortunately, it is defined over the wrong field to use directly with the Reed Muller codes as formulated in Section 2. The natural functions X, Y defined by by $X(s(a,b)) = a$ and $Y(s(a,b)) = b$ are K-valued.

We are rescued by the functions X^{q+1} and Y^{q+1} that are "accidently" F-valued and of degree 2 **over** F.

Note that the function $(1-(X^{q+1}-1)^{q-1})(1-Y^{q^2-1})$ takes value 1 on $s(a,b)$ whenever $a\bar{a}=1$ and $b=0$, and takes the value 0 elsewhere. Thus for R, R' as in Lemma 1.3, $R-R' \in GF(p)S$ has characteristic function

$$
\begin{aligned}
&(1-(X^{q+1}-1)^{q-1})(1-Y^{q^2-1})-(1-X^{q^2-1})(1-(Y^{q+1}-1)^{q-1})\\
&= \left[1-\sum_{i=0}^{q-1}(-1)^{q-1-i}\binom{q-1}{i}(X^{q+1})^i\right](1-(Y^{q+1})^{q-1})\\
&- (1-(X^{q+1})^{q-1})\left[1-\sum_{i=0}^{q-1}(-1)^{q-1-i}\binom{q-1}{i}(Y^{q+1})^i\right]\\
&= (q-1)[X^{q+1}-Y^{q+1})](X^{q+1}Y^{q+1})^{q-2}+\text{terms of lower total } F \text{ degree.}
\end{aligned}
$$

It follows that $R-R'$ has characteristic function of total degree $2+4(q-2)$ **over** F and in view of Theorem 1.4. This establishes the first part of:

Lemma 2. $0 \neq \overline{\overline{\Lambda}} \leq \overline{R_q(1,4)^{\perp}}$.

Proof: Let R be as in Lemma 1.3. Note that $\sum_{\{a\in K:a\bar{a}=1\}} a = 0$, since the index set for this summation is a subgroup of the multiplicative group of K. Note also that the function $\pi_{i,j}$ taking $s(a,b)$ to its i,j-entry is F-linear, for $i,j \in \{1,2\}$ and that these functions span all F-linear functions on $\mathcal{S}$. It follows that

$$\sum_{x\in\rho^{-1}(R)} q(x) = \sum_{x\in\rho^{-1}(R')} q(x)$$

for all F-linear functions q on $\mathcal{S}$.

Fix a regulus U with opposite regulus U' and let $f_U \in FM$ be the function that takes values $+1$ on the slopes of U, -1 on the slopes of U' and 0 elsewhere. In case U contains ℓ, f_U appears in $\overline{\Lambda}$ as the characteristic function of an affine line in M, as shown in the second paragraph of the proof of Lemma 1. Such a function is already in $\mathcal{R}_q(3(q-1),4)$ by Theorem 1.3. Lemma 1 implies that $\overline{\Lambda}$ is generated by $\mathcal{R}_q(3(q-1),4)$ and the set of functions f_U, $U \in R^G$. Since ρ is a FG-module isomorphism, it is sufficient to work with f_R. The result follows from the last displayed equation and the fact that ρ is a ring homomorphism.

Lemma 3. *Assume* $q \neq 2$. *Then* $\overline{\overline{\Lambda}}$ *is the unique* FG*-submodule of* $\overline{R_q(1,4)^{\perp}}$ *of dimension* 1.

We remark that $\overline{\mathrm{R}_q(1,4)^{\perp}}$ is the direct sum of 1- and 9-dimensional irreducible FG-modules in odd characteristic but it seems to have the more complicated structure $1/(2+4+2)/1$ in even characteristic.

Proof: By Theorem 1.4, the FG-module $\overline{\mathrm{R}_q(1,4)^{\perp}}$ is contragredient to the FG-module N of quadratic forms on M. The module N is 10-dimensional. Direct matrix computations (using Maple) show that the ring $End_{FG}(N)$ is 2-dimensional. It is a local ring if $2|q$ and it is F^2 in odd characteristic.

The same computations show that $\overline{\overline{R-R'}}$ spans the unique FG-submodule of $\overline{\mathrm{R}_q(1,4)^{\perp}}$ of dimension 1 and that any of the quadratic forms listed in $\mathcal{M}$ annihilates the characteristic function of $R-R'$. The result follows.

Proposition 4. *1.* $\overline{\overline{kerN}} \leq \overline{R_q(0,4)^{\perp}}$ *has dimension at most* 14.

2. $\overline{\overline{\Sigma}} \geq \overline{R_q(1,4)^{\perp}}$ *and has dimension at least* 10 *in every odd characteristic where there is a spread that is not virtually derivable from a regular spread.*

Proof: Let $[N]$ denote an incidence matrix for N and let $[x]$ denote the vector of coordinates associated with $x \in \mathbb{Z}\mathcal{P}$ or in $\mathbb{Z}\mathcal{L}$ as the case may be. Suppose $u \in kerN$ and let $a = \{\alpha \in \mathcal{P} \subset \mathbb{Z}\mathcal{P} : \alpha \text{ is off } \ell\}$. Then

$$\begin{aligned} 0 &= [a]^t([N][u]) = ([a]^t[N])[u] = (q[A(\ell) \setminus \{\ell\}] + (q+1)[\mathcal{L} \setminus A(\ell)])[u] \\ &\equiv [\mathcal{L} \setminus A(\ell)][u] \qquad (\text{mod } q\ ). \end{aligned}$$

It follows that $\overline{kerN} \leq \mathcal{R}_q(0,4)^{\perp}$.

Let S be such a spread and let R be a regular spread both containing the line ℓ. Then $\overline{\overline{S-R}} = \overline{\overline{S}} \in \overline{\overline{\Sigma}} \setminus \overline{\overline{\Lambda}}$. Thus the contragredient of $\overline{\overline{\Sigma}}$ must be an affine invariant FG-homomorphic image of $\mathcal{R}_q(2,4)$. The result follows from the fact that the 9-dimensional section of $\mathcal{R}_q(2,4)/\mathcal{R}_q(1,4)$ does not split over $\mathcal{R}_q(1,4)$ since any affine invariant code is closed under differentiation.

Proof of Theorem A: Let S be a spread that is virtually derivable from a regular spread. Then, as above, $\overline{\sigma\pi(S)} \in \overline{\Lambda}$ and Lemma 3 implies that S has dual degree at most $4q-6$. The sentence preceding Lemma 6.2 establishes the assertion about Hall planes.

References

[1] E. F. Assmus Jr. and J. D. Key, Designs and their Codes, Cambridge U. Press, Cambridge, 1992.

[2] C. W. Curtis and I, Reiner. Methods of Representation Theory I, Wiley-Interscience, New York, 1981.

[3] M. Deza and V. Grishukhin, "Once more about 80 Steiner triple systems." 1995 (preprint).

[4] G. D. James, Representations of General Linear Groups, London Math. Soc. Lecture Notes Series **94**, Cambridge 0 Press, 1984.

[5] W.F. Orr. The Miquelian inversive plane IP(q) and the associated projective planes. Ph.D. Thesis (Madison, Wisconsin, May 1973).

[6] T. G. Ostrom, Derivable Nets, Canad. Math. Bull. **8** (1965), 601-613.

[7] D. Taylor. The Geometry of the Classical Groups, Heldermann, Berlin, 1992.

Some Generalized Quadrangles in Characteristic 2

GEOFFREY MASON Mathematics Department, University of California, Santa Cruz, CA

Dedicated to Ted Ostrom, Octogenarian

1. INTRODUCTION

One of the themes of Ted Ostrom's work on translation planes (TP's) is the possibility of using representation theory (of the collineation group) to restrict the nature of the translation plane (cf. [O1], [O2], [O3], [O4]). In some of my own work, often joint with others (e.g., [M1], [M2], [BM], [EM], [FMW]) I have attempted to lay the foundations of such a theory and take the first few steps toward it beyond those of Ted himself. In this paper I want to go beyond the theme of TP's and initiate an analogous theory for certain kinds of *generalized quadrangles* (GQ's) in characteristic 2 (see Section 2). One may think of our GQ's as tapestries of interwoven TP's, so that one expects in principle that theorems about TP's should have strong analogues for GQ's. That is indeed the case, as we shall see. The main reason is that elements of order 3 behave particularly well in their action on GQ's; their behavior is analogous to the action of involutions on TP's. There is also a feature in our GQ-theory not present in that of TP's, namely the possibility of certain indecomposable group representations playing a rôle.

In order to keep the paper reasonably short we have included only what is necessary to prove the main theorems (stated below). The attentive reader will doubtless be able to extend some of our results. The main problem left open from our work is the question of existence of GQ's of *Ott-Schaefer type* (see Section 5). It should be possible to complete our discussion of this case computationally.

The paper is organized as follows: in Section 2 we discuss the GQ's we have in mind, together with related TP's. Section 3 considers automorphisms, in particular those of order 2 or 3. In Section 4 we prove some results of the kind that say that, under suitable circumstances, a simple group acting on one of our GQ's is necessarily $SL(2, q)$ for some even q. In Section 5 we prove the main theorem; it is

Supported by grants from the NSF and the Committee on Research, UC Santa Cruz.

a finiteness theorem saying essentially that a GQ of the kind under consideration which admits an irreducible action of $\mathrm{SL}(2,q)$ is of type (q^n, q^n) with $n \le 2$, i.e., it has dimension 3 or 6. We discuss the possibility that the case $n = 2$ occurs, but at present it remains an open question.

Thanks to Norman Johnson for organizing the Geometry Sessions in Iowa, and to Bruce Cooperstein and Bill Kantor for discussions about GQ's. Finally, congratulations to Ted Ostrom.

2. GQ's AND TP's

Fix q a power of 2, with $F = \mathrm{GF}(q)$. The GQ's of interest to us may be described via axioms similar to those for a TP, namely spreads. Thus we take a vector space V of dimension $3n$ over F, and we assume that the following hold.

(1) There is a family $\mathcal{C}$ of $q^n + 1$ subspaces of V, each of dimension n, and such that

$$V = C_1 \oplus C_2 \oplus C_3 \tag{2.1}$$

for any 3 pairwise distinct members $C_1, C_2, C_3 \in \mathcal{C}$.

(2) For each $C \in \mathcal{C}$ there is a $2n$-dimensional subspace $T = T(C)$ of V such that $C \subseteq T$ and such that each $x \in V \setminus C$ lies in *exactly one* of the differences $T \setminus C$, $(C \oplus D) \setminus C$ $(D \in \mathcal{C}, D \neq C)$.

We call the members of $\mathcal{C}$ *components.* We let $\mathcal{T}$ be the set of subspaces $T(C)$, $C \in \mathcal{C}$; they are called *tangents.* We refer to [PT] for details about GQ's in general, and this type in particular, which are there called TGQ's of type (q^n, q^n). One knows (loc. cit.) that the subspace

$$N = \bigcap_{C \in \mathcal{C}} T(C) \tag{2.2}$$

is n-dimensional (this depends on the fact that F has characteristic 2). We call N the *nucleus* of the GQ.

Clearly we have

$$T(C) = C \oplus N \tag{2.3}$$

for each component C. Moreover

$$V = C_1 \oplus C_2 \oplus N \tag{2.4}$$

for distinct components C_1, C_2. In particular we see that the following holds.

LEMMA 2.1. *Set $\bar{V} = V/N$. Then the family of subspaces $T(C)/N$, $C \in \mathcal{C}$, is a spread set in $\bar{V}$.*

When we speak of the "spread on $\bar{V}$," it will always mean the one described in Lemma 2.1.

There are many other spreads lying around:

LEMMA 2.2. *Let A, B be distinct components. Then the family of subspaces $(A \oplus C) \cap B \oplus N$, $C \in \mathcal{C} \setminus \{A\}$, together with N constitute a spread set in $B \oplus N$.*

PROOF. Easy consequence of (2.1)–(2.4). ∎

We denote this spread on $B \oplus N$ as $\pi(A, B)$.

3. LINEAR AUTOMORPHISMS

We retain the notation of Section 2. The group of linear automorphisms is

$$G = G(\mathcal{C}) = \{g \in \mathrm{GL}(V) \mid g \text{ preserves } \mathcal{C}\}.$$

Obviously, G is a group which preserves the family of tangents $\mathcal{T}$. It also acts on the nucleus because of (2.2), so that we have a short exact sequence of FG-modules

$$O \to N \to V \to \bar{V} \to O \tag{3.1}$$

From these comments together with Lemma 2.1 we see that G preserves the spread on $\bar{V}$. Moreover we have

LEMMA 3.1. *G acts faithfully on $\bar{V}$.*

PROOF. If $g \in G$ acts trivially on $\bar{V}$ then g leaves each tangent $T(C)$, $C \in \mathcal{C}$, invariant. By (2.3) we see then that $[g, C] \subseteq N \cap C = 0$. Thus g acts trivially on each $C \in \mathcal{C}$, and hence on V. So $g = 1$ as required. ∎

Because of this lemma, we may use the theory of TP's and their automorphisms to our advantage. For example, if $t \in G$ is an involution one knows that t is either an elation or Baer involution when regarded as a collineation of the spread on $\bar{V}$. For $g \in G$ we set

$$V^g = \{v \in V \mid gv = v\}.$$

LEMMA 3.2. *Let $t \in G$ have order 2. Exactly one of the following holds:*

(a) *t leaves invariant a unique component C. In this case the space V^t is precisely the tangent $T(C)$.*

(b) *t leaves invariant exactly $q^{n/2} + 1$ components. In this case n is even, the space V^t has dimension $3n/2$, and the t-invariant components induce a generalized quadrangle on V^t.*

PROOF. By Lemma 3.1, t acts faithfully on $\bar{V}$. If t induces an elation on $\bar{V}$ then it fixes just one component $T(C)/N$ of $\bar{V}$, i.e., t fixes just one tangent $T = T(C)$, and hence just one component $C \in \mathcal{C}$. Now (a) will follow if we can show that $\dim V^t = 2n$, for certainly $V^t \subseteq T$. To see this, note that if $D \in \mathcal{C} \setminus \{C\}$ then $t(D)$ is a component distinct from either C or D. So $V = C \oplus D \oplus t(D)$, from which we visibly see that $V^t = C \oplus \{d + t(d) \mid d \in D\}$. So $\dim V^t = 2n$, as required.

Now assume that t acts as a Baer involution on $\bar{V}$. So n is even, $\dim(\bar{V})^t = n$, and t fixes exactly $q^{n/2} + 1$ components of the spread on $\bar{V}$. So t fixes the same number of tangents and components of $\mathcal{C}$, namely $T \in \mathcal{T}$ is t-invariant if, and only if, $T = T(C)$ and $C \in \mathcal{C}$ is t-invariant, if, and only if, $T(C)/N$ is a t-invariant component of the spread on $\bar{V}$. Moreover the space C^t of t-invariants for any $C \in \mathcal{C}$ fixed by t has dimension $n/2$, i.e., C is a free $\langle t \rangle$-module. It is now immediate that the spaces $\{C_i^t \mid C_i \in \mathcal{C}, t(C_i) = C_i\}$ are the components of a generalized quadrangle on V^t with $T(C_i^t) = C_i^t \oplus N^t$. This completes the proof of (b). ∎

We say that t is an *elation* of V if (a) of Lemma 3.2 holds; we call t a *Baer involution* if (b) holds. So t is elation/Baer on V if, and only if, t is elation/Baer on $\bar{V}$. We must, of course, always be clear which of these two $\langle t \rangle$-modules we have in mind.

LEMMA 3.3. *Let $u \in G$ have odd order. Exactly one of the following holds:*

(a) $V^u = 0$.
(b) $V^u \neq 0$ *and the u-invariant components of $\mathcal{C}$ induce a generalized quadrangle on* V^u.
(c) $0 \neq V^u \subseteq N$ *and there are either* 0 *or* 2 *u-invariant components in* $\mathcal{C}$.
(d) $V^u \neq 0 = N^u$ *and u induces a (generalized) homology on* $\bar{V}$.

PROOF. We may assume $V^u \neq 0$ and try to show that one of (b)–(d) holds. If there are at least two components $C_1, C_2 \in \mathcal{C}$ satisfying $C_i^u \neq 0$ then u induces a planar automorphism of $\bar{V}$, in which case (b) holds. So we may assume that u does not induce a planar automorphism of $\bar{V}$.

Next we claim that u leaves no more than two components of $\mathcal{C}$ variant. Otherwise it fixes at least 3, say C_1, C_2, C_3, as well as N. Moreover one of these spaces may be assumed to contain non-zero u-invariant vectors. From (2.1), (2.4) we see that there are isomorphisms of $\langle u\rangle$-modules $C_i \cong C_j \cong N$, which tells us that each of the C_i, $1 \leq i \leq 3$, satisfies $C_i^u \neq 0$. Thus we are again in case (b).

To complete the proof, it suffices to show that if $T = T(C) = C \oplus N$ is a u-invariant tangent then we cannot have $C^u \neq 0 \neq N^u$. Suppose this is false. As there are $q^n + 1$ components in $\mathcal{C}$, and since q is even while u has odd order, u must fix a second component, say B. Then u induces an automorphism of the spread $\pi(B, C)$ (cf. Lemma 2.2). Since u has fixed points on at least two of the components of the spread $\pi(B, C)$, namely C and N, then u must be a planar automorphism of $\pi(B, C)$ and hence fix at least 3 components of $\pi(B, C)$. But then u fixes the corresponding components of $\mathcal{C}$, which contradicts a previous assertion. ∎

If we take u in Lemma 3.3 to have order 3 then even stronger results accrue:

PROPOSITION 3.4. *Let $u \in G$ have order* 3. *Exactly one of the following holds:*

(a) *u acts on V as a scalar.*
(b) *n is divisible by* 3, $\dim V^u = n$, *and the u-invariant components of $\mathcal{C}$ induce a (cube root) generalized quadrangle on* V^u.
(c) $V^u = N$.
(d) $V^u \in \mathcal{C}$ *and u induces a homology on* $\bar{V}$.

PROOF. First note that if u acts as a scalar on $\bar{V}$ then the same is true for the action of u on V, so (a) holds in this case. If u is not a scalar on $\bar{V}$ then it cannot leave invariant all components of $\mathcal{C}$, so there are $C_0, C_1, C_2 \in \mathcal{C}$ with $u(C_i) = C_{i+1}$ (indices taken $\pmod 3$). Now (2.1) tells us that $\dim V^u = n$, i.e., V is a free $\langle u\rangle$-module. Now (b)–(d) hold if, and only if, the corresponding (b)–(d) of Lemma 3.3 hold. ∎

The combined effects of Lemmas 3.2 and 3.3 and Proposition 3.4 place severe restrictions on the possible actions of G on V. We give some illustrative examples in Section 4.

We complete this section with a different kind of result. It says that in certain circumstances the extension (3.1) is a *non-split* extension of FG-modules, so that various homological considerations become relevant. This possibility occurred to me during a conversation with Bill Kantor in Sussex several years ago.

PROPOSITION 3.5. *Let $Z_2 \times Z_2 \cong U \leq G$ and assume that there is a tangent T in $\mathcal{T}$ such that each $u \in U^\sharp$ is an elation with fixed points T (cf. Lemma 3.2(a)). Then*

$$[U, V] = T$$

and in particular (2.1) *is a non-split extension of FU-modules.*

PROOF. Let $T = C \oplus N$ with $C \in \mathcal{C}$ and pick $B \in \mathcal{C}\backslash\{C\}$. As C is the unique component in $\mathcal{C}$ which is u-invariant for *any* $u \in U^\sharp$ then the U-images of B give 4 distinct components $B_0 = B$, B_1, B_2, B_3 in $\mathcal{C}$. Let $u_i \in U$ satisfy $u_i(B) = B_i$, $1 \leq i \leq 3$.

Since $\dim V^{u}i = 2n$ then $\dim([u_i, V]) = n$. As u_i exchanges B and B_i we see that $[V, u_i] = \{b + u_i b \mid b \in B\} = (B \oplus B_i) \cap T$. Thus the 3 spaces $[V, u_i]$, $1 \leq i \leq 3$, are 3 distinct components of the spread $\pi(B, C)$ (cf. Lemma 2.2). In particular $[U, V]$, being the sum of the spaces $[V, u_i]$, $1 \leq i \leq 3$ is equal to T. This completes the proof of the proposition. ∎

LEMMA 3.6. *Let $Z_2 \times Z_2 \cong U \leq G$. Then U cannot consist solely of Baer involutions with the same fixed-point subspace.*

REMARK. This is a very useful result, and is another illustration of how stronger results accrue in the case of quadrangles versus planes. In the Hall planes, for example, one finds four-groups such that all involutions are Baer involutions with the same fixed points. The lemma says that we cannot "lift" this situation to a generalized quadrangle.

PROOF OF THE LEMMA. Assume the lemma is false. Then we can certainly find a component B which is not fixed by any involution in U. So we may adopt the notation $B = B_0$, B_1, B_2, of the last lemma. Now for $i = 1, 2$, u_i fixes the subspace consisting of all $b + u_i b$ where b lies in B. Call these two subspaces D_1, D_2 respectively. By assumption we know that u_1 fixes D_2 pointwise, so that in fact u_1 fixes $D_1 + D_2$ pointwise. As $V = B_0 + B_1 + B_2$ then $D_1 \cap D_2 = \emptyset$, whence u_1 fixes a subspace of dimension $2n$ (V having dimension $3n$). But this is impossible since we are assuming that u_1 is a Baer involution and hence has fixed points of dimension $3n/2$. ∎

We can now prove

THEOREM 3.7. *In order for the extension* (3.1) *to be a split extension of G-modules, it is both necessary and sufficient that $\bar{V}$ be a projective G-module.*

PROOF. Suppose first that $\bar{V}$ is a projective G-module. Then by the very definition of projective, the sequence (3.1) necessarily splits. As for the converse, we will assume that $\bar{V}$ is not a projective G-module and show that (3.1) is then a non-split extension of G-modules. Let G be a minimal counterexample to the theorem. If H is a proper subgroup of G then $\bar{V}$ is necessarily a projective H-module: for otherwise the minimality of G implies that (3.1) is non-split as an s.e.s. of H-modules, and *a fortiori* it is non-split as an s.e.s. of G-modules.

By Chouinard's theorem [C] we see that G is an elementary abelian 2-group of order 2^k for some integer k. Observe that, as a consequence of Lemma 3.2, we must have $k > 1$. Let H be a subgroup of G of maximal order subject to the following conditions: (i) the H-fixed subspace $W = V^H$ is a subquadrangle of V; (ii) V is a free (i.e., projective) H-module. Note that the trivial subgroup 1 of G satisfies this condition, so H certainly exists. On the other hand, since $\bar{V}$ is not a free G-module then H is not equal to G.

Suppose to begin with that $|G : H| > 2$. If G/H does not act faithfully on W then there is some g in $G \setminus H$ acting trivially on W, in which case $K = \langle H, g \rangle$ is a proper subgroup of G such that V is not a free K-module. This is impossible, so G/H acts faithfully on W. By Lemma 3.2 and the maximality of H, each involution of G/H must induce an elation on W. But then the non-splitting of (3.1) as G-module follows by applying Proposition 3.5 to the action of G/H on W. This contradiction thus shows that in fact we have $|G : H| = 2$.

Since $k > 1$ we can choose a subgroup K of H such that $|H : K| = 2$. In this case we let $X = V^K$ be the space of K-fixed points on V, and consider the action of G/K on X as before. Again the action is faithful. By Lemma 3.6, G/K cannot consist of Baer involutions with the same fixed points; on the other hand any element in $H \setminus K$ induces a Baer involution on X. Using Lemma 3.2, we conclude that in fact V is a projective G-module, which is the desired contradiction. ∎

4. ACTIONS OF SIMPLE GROUPS

We now take a non-abelian simple group $X \subseteq G(\mathcal{C})$.

THEOREM 4.1. *Suppose that N is a trivial FX-module. Then $X \cong \mathrm{SL}_2(q_0)$ with q_0 some power of* 2.

PROOF (SKETCH). Let A be a maximal elementary abelian 2-subgroup of X, so that $|A| \geq 4$. By Lemma 3.2, A is a group of elations with axis some tangent $T \in \mathcal{T}$. Now one checks easily that A is strongly closed, alternatively that A is weakly closed in $C_X(a)$ for each $1 \neq a \in A$. In any case, any number of classification theorems tell us that X is a Bender group, i.e., $X \cong \mathrm{SL}_2(q_0)$, $(\mathrm{S})\mathrm{U}_3(q_0)$ or $\mathrm{Sz}(q_0)$ with q_0 some power of 2.

Before proceeding we interpolate a general result:

LEMMA 4.2. *Let $H \leq G(\mathcal{C})$ act trivially on N with H of odd order. Then H is a Frobenius complement; in particular, Sylow subgroups of H are cyclic.*

PROOF. From Lemma 3.3 we see that $N = V^h$ for all $1 \neq h \in H$. So H acts on $\bar{V}$ in a "fixed-point-free" manner. That is, $\bar{V} \rtimes H$ is a Frobenius group with complement H. ∎

Returning to the proof of Theorem 4.1, since $(\mathrm{S})\mathrm{U}_3(q_0)$ contains non-cyclic odd p-groups then $X \not\cong (\mathrm{S})\mathrm{U}_3(q_0)$ by Lemma 4.2. Finally, we show that $X \not\cong \mathrm{Sz}(q_0)$, so assume false. Lemma 3.3 tells us that an element of X of order 5 acts fixed-point-free on $\bar{V}$. A theorem of Martineau [1] says then that $\bar{V}$ is a direct sum of standard 4-dimensional $\mathrm{Sz}(q_0)$ modules (possibly twisted by Galois automorphisms and allowing for field extensions if necessary). In any case, $\bar{V}$ is a sum of $\frac{n}{2}$ such modules over a splitting field. As N is a trivial FX-module of dimension n, one can deduce from a theorem of Peter Sin [S] that V has a trivial summand as FX-module. However, the involutions of X are elations, so that there is no such summand as a consequence of Proposition 3.5. This completes the proof of Theorem 4.1. ∎

THEOREM 4.3. *Suppose that X has order divisible by 3 and that every element of order 3 induces a homology on $\bar{V}$ (cf. Proposition 3.4). Then $X \cong \mathrm{SL}_2(q_0)$ for q_0 some odd power of 2.*

PROOF (SKETCH). Let $\langle x \rangle = R$ be a subgroup of X of order 3. Since $N^x = 0$ then X acts faithfully on N. In fact, since $N^x = 0$ for each $x \in X$ of order 3 then a Sylow 3-subgroup of X must be cyclic.

Next we claim that there are no subgroups in X isomorphic to A_4 or $\mathrm{SL}_2(3)$. For otherwise, there is $Q \leq X$ with $Q \trianglelefteq QR$ and $Q \cong Z_2 \times Z_2$ or Q_8, respectively. In the latter case, the involution $t \in Q_8$ acts freely on $\bar{V}$, so $\bar{V}$ is a projective QR-module by the theorem of Chouinard [C]. Now among the projective indecomposable modules P for QR (characteristic 2) only the projective cover $P(1)$ of the trivial module satisfies the inequality $\dim P \geq 2 \dim P^x$ with equality. For the other two projective indecomposables the inequality is strict. So $\bar{V}$ is a sum of copies of $P(1)$ as FQR-modules, in particular QR has non-trivial points on $\bar{V}$. Such points must come from the component $C \in \mathcal{C}$ satisfying $C = V^x$, so $[Q, C] = [Q, R, C] = [Q, C, R] + [R, Q, C] \subseteq [C, R] = 0$. But then t is an elation with axis $C \oplus N$, whence t and hence X acts trivially on N, contradiction.

If $Q \cong Z_2 \times Z_2$ then the above argument still works if $\bar{V}$ is a free Q-module, so assume that this is not the case. Since any $u \in Q^\sharp$ acts as a Baer involution it must be the case that V^Q is a square root generalized quadrangle. In any case, applying Proposition 3.4 to the action of x on V^Q shows that Q leaves invariant the R-invariant component C. This leads to the contradiction $[Q, C] = 0$ as before.

Finally, prove Theorem 4.3 using induction on $|X|$. The above reductions allow us to readily conclude (with the classification of finite simple groups [G]) that X is a Bender group once more. Since $(\mathrm{S})\mathrm{U}_3(q_0)$ has no F-module on which x acts without fixed points, we must have $X \cong \mathrm{SL}_2(q_0)$, as required. ∎

COROLLARY 4.4. *Suppose that n is not divisible by 6. Then $X \cong \mathrm{SL}_2(q_0)$ or $\mathrm{Sz}(q_0)$.*

PROOF. If n is odd then involutions are elations by Lemma 3.2. Then an involution in X acts trivially on N, whence so too does X. Now we are done in this case by Theorem 4.1.

Suppose that n is not divisible by 3. If X contains elements of order 3 then they either act trivially on N or induce homologies on $\bar{V}$ by Proposition 3.4. Now apply Theorems 4.1 and 4.3 to complete the proof of the corollary if $3 \mid |X|$. If $|X|$ is not divisible by 3 then $X \cong \mathrm{Sz}(q_0)$. ∎

5. IRREDUCIBLE ACTION OF $\mathrm{SL}(2, q)$

In this section we assume that V is of dimension $3n$ over $F = \mathrm{GF}(q)$, q a power of 2, with generalized quadrangle $\mathcal{C}$ as before. We take $X \leq G(\mathcal{C})$ with $X \cong \mathrm{SL}(2, q)$, $q \geq 4$.

THEOREM 5.1. *Assume that $\bar{V} = V/N$ is an irreducible FX-module. Then $n \leq 2$.*

We proceed in a sequence of lemmas which will establish much more than is stated in the theorem.

LEMMA 5.2. *Suppose that involutions of X are elations. Then $n = 1$ and V is the standard GQ of type (q, q) associated with $\mathrm{Sp}(4, q)$.*

PROOF. In this case involutions act trivially on N, whence so too does X itself. By Proposition 3.4, an element $u \in X$ of order 3 satisfies $V^u = N$, in particular $(\bar{V})^u = 0$.

The only simple FX-module with this property is the natural module (or a Galois twist). In any case $n = 1$, and the lemma follows. ∎

From now on we assume that involutions of X are *not* elations. They are therefore Baer elements in their action on both V and $\bar{V}$ (Lemma 3.2).

We fix a Sylow 2-subgroup U of X and choose a subgroup $A \leq U$ of maximal order subject to the following conditions:

(a) A is planar in its action on $\bar{V}$.
(b) $\bar{V}$ is a free FA-module.

The use of such subgroups was instigated in [M2], and led to the notion of "dispersive" modules (loc. cit.). Alperin and I now called such modules "subfree" (cf. [AM1], [AM2]). We do not strictly need to use the following lemma, but include it to show how one can make use of these ideas when studying TGQ's.

LEMMA 5.3. *N is an irreducible FX-module.*

PROOF. Since A is planar on $\bar{V}$ it fixes at least 3 components C_1, C_2, C_3 on $\mathcal{C}$. Since $\bar{V}$ is a free FA-module, the same is true of each C_i by (2.1). Then by (2.4), $N \cong C_i$ (as FA-module), so also N is a free FA-module.

Now if $x \in U \setminus A$ then $\langle A, x\rangle$ cannot be both planar and free on $\bar{V}$ by maximality of A. So either $[x, \bar{V}^A] = 0$ or else x induces an elation on $\bar{V}^A$. It follows that $\dim \bar{V}^U \geq \frac{1}{2} \dim \bar{V}^A$. But $\bar{V}$ is an irreducible FX-module, so that $\dim \bar{V}^U = 1$. We conclude that $\dim \bar{V}^A = 2$.

Now consider the FX-module N: it is a free A-module and $\dim N^A = \frac{1}{2} \dim \bar{V}^A = 1$. By the main theorem of [M3], N is an irreducible FX-module. ∎

LEMMA 5.4. $n = 2$.

PROOF. We know that N is a simple FX-module with an element of order 3 in X acting without non-trivial fixed points on N. This forces N to be a natural FX-module (possibly Galois twisted). In any case, $|N| = q^2$ and the lemma holds. ∎

This completes the proof of Theorem 5.1. The reader is invited to find a simpler proof which avoids the use of Lemma 5.3.

LEMMA 5.5. *q is an odd power of 2.*

PROOF. Apply Corollary 3.7 of [BM]. ∎

LEMMA 5.6. (3.1) *is a non-split extension of FX-modules.*

PROOF. If A is as above then $|A| = 2$. So by Lemma 5.5 we have $|U : A| = \frac{1}{2}q \geq 4$. Thus there is a 4-group of elations induced by the action of U/A on $\bar{V}^A$. By Proposition 3.5, the extension $1 \to N^A \to V^A \to \bar{V}^A \to 1$ is non-split as FU/A-module. The lemma follows. ∎

Now recall some facts about the Ext groups $\mathrm{Ext}_{FX}(L, M)$ for simple FX-modules L, M [A]. Taking L to be the natural module N, and taking M to have dimension 4, we know (loc. cit., Theorem 3) that $\mathrm{Ext}_{FX}(N, M) = 0$ unless $M \cong N \otimes N^\alpha$ for non-trivial Galois automorphism α. Moreover $\mathrm{Ext}_{FX}(N, N \otimes N^{\alpha}) = F$, in other words there is a *unique* non-split extension V of the shape

$$O \to N \to V \to N \otimes N^{\alpha} \to O. \qquad (5.1)$$

Thus by Lemma 5.6 our space V must have the general shape of (5.1). Note also that the spread on $\bar{V}$ must be of Ott-Schaefer type and that α must necessarily generate $\mathrm{Aut}(F)$ [BM].

LEMMA 5.7. *Let φ be the Frobenius automorphism of F, let $\beta = \varphi^{-1}\alpha$, and assume that $\beta \neq 1$. Then V is (isomorphic to) the unique maximal proper FX-submodule of $N^{\beta} \otimes N^{\beta} \otimes N$.*

PROOF. Apply Lemma 6 of [A] to see that $N^{\beta} \otimes N^{\beta} \otimes N$ is uniserial as FX-module if $\beta \neq 1$. Its composition factors are (in order, starting from the socle and proceeding "upwards"): N, $N \otimes N^{\beta\varphi}$, N. So we get an FX-module of type (5.1), and it must be isomorphic to V by the uniqueness statement. ∎

This gives us an explicit model of V with which we can calculate (at least if $\alpha \neq \varphi$). For from the structure of the Ott-Schaefer spreads on $\bar{V}$ (cf. [BM], [L]) one knows that the elements of order 3 are indeed homologies (in agrement with Proposition 3.4). Thus we can write down explicitly what the components of $\mathcal{C}$ have to be, in terms of a suitable co-ordinate system of $N^{\beta} \otimes N^{\beta} \otimes N$.

I finish with two comments:

(i) I conjecture that the case $n = 2$ does not actually occur.
(ii) I hope someone can prove me wrong.

REFERENCES

[A] J.L. Alperin, Projective modules for $\mathrm{SL}(2, 2^n)$, *J. Pure Appl. Algebra* **15** (1979), 219–234.

[AM1] J.L. Alperin and G. Mason, On simple modules for $\mathrm{SL}(2, q)$, *Bull. London Math. Soc.* **25** (1993), 17–22.

[AM2] ———, Geometry, Steinberg representations and complexity, London Math. Soc. Lecture Note Ser., no. 211, Cambridge Univ. Press, 1995, pp. 1–5.

[BM] O.E. Barriga and G. Mason, Spread-invariant representations of $L_2(2^n)$ and a characterization of some Ott-Schaefer planes, *J. Algebra* **137** (1991), 233–251.

[C] L.G. Chouinard, Projectivity and relative projectivity over group rings, *J. Pure Appl. Algebra* **7** (1976), 287–302.

[G] D. Gorenstein, *Finite Simple Groups, an Introduction to Their Classification,* Plenum Publishing Corp., New York, 1982.

[EM] M. Eastman and G. Mason, Spread-invariant representations of symmetric groups, *J. Algebra* **128** (1990), 86–100.

[FMW] D. Foulser, G. Mason, and M. Walker, On irreducible translation planes in odd characteristic, I, Lecture Notes in Math., No. 969, Springer, Berlin-New York, 1982, pp. 167–175.

[L] H. Lüneburg, *Translation planes*, Springer-Verlag, Berlin-New York, 1980.

[Ma] R.P. Martineau, On 2-modular representations of the Suzuki groups, *Amer. J. Math.* **94** (1972), 55–72.

[M1] G. Mason, Irreducible translation planes and representations of Chevalley groups in characteristic 2, in *Finite geometries* (Pullman, Wash., 1981) (C.M. Campbell *et al.*, eds.), Lecture Notes in Pure and Appl. Math., No. 82, Marcel Dekker, New York, 1983, pp. 333–346.

[M2] ———, Representation theory and translation planes, in *The Arcata Conference on Representations of Finite Groups* (Arcata, Calif., 1986) (P. Fong, ed.), Proc. Sympos. Pure Math., No. 47, Part 1, Amer. Math. Soc., Providence, 1987, pp. 211–222.

[M3] ———, Varieties attached to an $SL_2(2^k)$-module, *Proc. Amer. Math. Soc.* **116** (1992), 343–350.

[O1] T.G. Ostrom, *Lectures on Finite Projective Planes,* Notas de Curso Dept. de Mat., No. 27, Universidade Federal de Pernambuco, Departamento de Matematica, Recife, 1989.

[O2] ———, Lectures on Finite Translation Planes, *Confer. Sem. Mat. Univ. Bari (1983),* No. 191, 1984.

[O3] ———, Elementary abelian 2-groups on the line at infinity of translation planes, *J. Geom.* **17** (1981), 128–139.

[O4] ———, Reducibility in finite translation planes, *Algebras Groups Geom.* **2** (1985), 455–477.

[PT] S.E. Payne and J.A. Thas, *Finite generalized quadrangles,* Research Notes in Mathematics, no. 110, Pitman, Boston, 1984.

[S] P. Sin, Extensions of simple modules for $Sp_4(2^n)$ and $Suz(2^m)$, *Bull. London Math. Soc.* **24** (1992), 159–164.

Some p-Ranks Related to Geometric Structures

G. ERIC MOORHOUSE Dept. of Mathematics, University of Wyoming, Laramie WY 82071–3036, U.S.A. moorhous@UWyo.edu

Dedicated to Professor T. G. Ostrom

Abstract The p-rank of the point-hyperplane incidence matrix A of $PG(n, p^e)$ is well-known. Let A_S be the submatrix formed by the rows of A indexed by an arbitrary subset S of the points. We show that the p-rank of A_S is related to the Hilbert function (or a modification thereof) for $\mathcal{I}(S)$, the ideal of $F[X_0, X_1, \ldots, X_n]$ generated by all homogeneous polynomials vanishing on S. This leads to a determination of $\text{rank}_p(A_S)$ in case S is a naturally embedded Grassmann variety. The cases when S is a quadric or a Hermitian variety have been treated by Blokhuis and the author [2] and the author [10] respectively.

1 HILBERT FUNCTIONS AND p-RANKS

Let $F = GF(q)$, $q = p^e$, and let A be the incidence matrix of points versus hyperplanes of $PG(n, F)$. Thus A is a square matrix of size $N = (q^{n+1} - 1)/(q - 1)$ having entries 1 and 0 corresponding to incident and non-incident point-hyperplane pairs. Now let A_S be an $s \times N$ submatrix of A, whose rows are indexed by an s-subset S of the points of $PG(n, F)$. The intent of Theorem 1 is to describe a general approach to finding the p-rank of A_S. This approach makes use of a modification of the Hilbert function of $\mathcal{I}(S)$, the ideal in the polynomial ring $R := F[X_0, X_1, \ldots, X_n]$ generated by all homogeneous polynomials which vanish on S. Much of our terminology and background results are standard in algebraic geometry; see eg. [7].

For each integer $k \geq 0$, let R_k be the F-subspace of R consisting of all homogeneous polynomials of degree k. The natural action of $G := GL(n + 1, F)$ on

The author is grateful to David Wagner and Andries Brouwer for useful discussions, and to the Fields Institute for Research in Mathematical Sciences where part of this research was conducted.

$R_1 \cong F^{n+1}$ extends uniquely to an action of G on the algebra $R = \bigoplus_{k \geq 0} R_k$, and each R_k is an FG-submodule. Let $\mathfrak{I} \subseteq R$ be a homogeneous ideal, i.e. the ideal $\mathfrak{I}$ is generated by homogeneous polynomials. The *Hilbert function* of $\mathfrak{I}$ is defined by

$$h_{\mathfrak{I}}(k) = \dim(R_k \,/\, \mathfrak{I} \cap R_k) = \binom{n+k}{n} - \dim(\mathfrak{I} \cap R_k)\,.$$

It is known that for all sufficiently large k, the function $h_{\mathfrak{I}}(k)$ agrees with a polynomial $P_{\mathfrak{I}}(k)$, called the *Hilbert polynomial of* $\mathfrak{I}$. Over algebraically closed fields, Hilbert polynomials provide a precise algebraic definition of the degree and dimension of projective varieties.

Denote by $R_k^\dagger$ the subspace of R_k spanned by all monomials $X_0^{\alpha_0} X_1^{\alpha_1} \cdots X_n^{\alpha_n}$ where $\alpha_0, \alpha_1, \ldots, \alpha_n$ are non-negative integers summing to k such that the multinomial coefficient $\binom{k}{\alpha_0,\, \alpha_1,\, \cdots,\, \alpha_n}$ is not divisible by p. This is more than an F-subspace; as shown in [2], it is an FG-submodule of R_k. Note that the inclusion $R_k^\dagger \subseteq R_k$ is proper iff $k \geq p$. By Lucas' Theorem (see Section 2) we have

$$dim(R_k^\dagger) = \prod_{\ell=0}^{e-1} \binom{n+k_\ell}{n}$$

where the p-adic expansion of k is given by $k = \sum_{\ell=0}^{e-1} k_\ell p^\ell$, $0 \leq k_\ell \leq p-1$. In particular, $\dim(R_{q-1}^\dagger) = \binom{p+n-1}{n}^e = (\dim R_{p-1})^e$. We modify the Hilbert function of $\mathfrak{I}$ by defining

$$h_{\mathfrak{I}}^\dagger(k) = dim(R_k^\dagger \,/\, \mathfrak{I} \cap R_k^\dagger)\,.$$

In general, the values of $h_{\mathfrak{I}}^\dagger(k)$ for $k \gg 0$ are not given by any polynomial.

Let G_S be the subgroup of $G = GL(n+1, F)$ preserving the point set S in $PG(n, F)$. As explained in Section 5, the row and column spaces of A_S over F are naturally contragredient FG_S-modules. Similarly, the row and column spaces of $J - A_S$ are naturally contragredient FG_S-modules, where J is an $s \times N$ matrix of 1's. In Section 3 we prove

THEOREM 1

(i) $\text{rank}_p(J - A_S) = h_{\mathcal{I}(S)}^\dagger(q-1)$. Moreover, the column space $Col(J - A_S)$ is isomorphic to $R_{q-1}^\dagger \,/\, \mathcal{I}(S) \cap R_{q-1}^\dagger$ as an FG_S-module.

(ii) $\text{rank}_p A_S$ differs from $\text{rank}_p(J - A_S)$ by at most one. If $|H \cap S| \equiv 1 \bmod p$ for every hyperplane H of $PG(n, F)$, then $\text{rank}_p A_S = 1 + \text{rank}_p(J - A_S)$, and $Col(A_S) \cong \langle \mathbf{1} \rangle \oplus Col(J - A_S)$ where $\langle \mathbf{1} \rangle$ is a trivial FG_S-module of dimension one.

We are most interested in the case S is a 'discrete variety' of the form $\mathcal{Z}(\mathfrak{I})$, i.e. the set of points of $PG(n, F)$ where a given homogeneous ideal $\mathfrak{I}$ vanishes. Although the Hilbert function of $\mathfrak{I}$ is often readily available, care is required in applying Theorem 1 since in general, the inclusion $\mathcal{I}(S) \supseteq \mathfrak{I}$ may be proper. The *radical* of $\mathfrak{I}$ is the ideal $\sqrt{\mathfrak{I}} := \{f \in R : f^k \in \mathfrak{I} \text{ for some positive integer } k\} \subseteq R$. Clearly

$$\mathcal{I}(S) = \mathcal{I}(\mathcal{Z}(\mathfrak{I})) \supseteq \sqrt{\mathfrak{I}} \supseteq \mathfrak{I}\,.$$

Since F is not algebraically closed, $\mathcal{I}(\mathcal{Z}(\mathfrak{I}))$ is typically larger than $\sqrt{\mathfrak{I}}$ (considering that for all $\mathfrak{I}$, $X_i^q X_j - X_i X_j^q$ vanishes on $PG(n, F)$ for all $i \neq j$).

DEFINITION $\mathfrak{I}$ satisfies FFN(k) (Finite Field Nullstellensatz of degree k) if every $f \in R_k$ which vanishes on $\mathcal{Z}(\mathfrak{I})$, belongs to $\mathfrak{I}$.

In many cases $\mathfrak{I}$ satisfies FFN$(q-1)$, which implies that $h^{\dagger}_{\mathcal{I}(\mathcal{Z}(\mathfrak{I}))}(q-1) = h^{\dagger}_{\mathfrak{I}}(q-1)$ and Theorem 1 applies. Moreover, $h^{\dagger}_{\mathfrak{I}}(q-1) \le (h_{\mathfrak{I}}(p-1))^e$ (see Lemma 2) with equality in many cases, including Examples 1.1, 1.2, 1.4 below.

1.1 Example: Projective Spaces

Let $\mathfrak{I} = (0)$, $S = \mathcal{Z}(0) = PG(n, p^e)$. Then

$$\operatorname{rank}_p A_S = \operatorname{rank}_p A = h^{\dagger}_{(0)}(q-1) + 1 = (h_{(0)}(p-1))^e + 1 = \binom{p+n-1}{n}^e + 1.$$

This is the well-known result of Goethals and Delsarte [6], MacWilliams and Mann [9], and Smith [11]; see also [3]. In this case, $Col(A) \cong R^{\dagger}_{q-1} \oplus \langle \mathbf{1} \rangle$ as FG-modules.

1.2 Example: Quadrics

Let $Q(X) = Q(X_0, X_1, \ldots, X_n)$ be a quadratic form, i.e. $Q(X) \in R_2$. We will assume that the corresponding bilinear form $f(X,Y) := Q(X+Y) - Q(X) - Q(Y)$ is nondegenerate. Thus the quadric $S := \mathcal{Z}(Q)$ is nondegenerate, and if $p = 2$ then n is odd. Homogeneous polynomials of degree $k \ge 2$ belonging to the principal ideal $\mathfrak{I} = (Q)$ are precisely those polynomials of the form $Q(X)f(X)$ where $f \in R_{k-2}$, so

$$h_{(Q)}(k) = \dim(R_k) - \dim(R_{k-2}) = \binom{n+k}{n} - \binom{n+k-2}{n}.$$

In [2] it is also shown that (Q) satisfies FFN(q) for $n \ge 4$, and FFN$(q-1)$ for $n = 3$, and that consequently

$$\operatorname{rank}_p A_{\mathcal{Z}(Q)} = \left[\binom{n+p-1}{n} - \binom{n+p-3}{n}\right]^e + 1$$

for $n \ge 3$. This statement fails for $n \le 2$, as does FFN$(q-1)$.

1.3 Example: Hermitian Varieties

Suppose $q = p^{2d}$ and let $U(X) = \sum_{i=0}^{n} \sum_{j=0}^{n} a_{ij} X_i X_j^{p^d}$ be a nondegenerate unitary form; thus $a_{ji} = a_{ij}^{p^d}$ and $\det[a_{ij}] \ne 0$. Then $S := \mathcal{Z}(U)$ is a nondegenerate Hermitian variety. In [10] it is shown that the principal ideal (U) satisfies FFN(q) for $n \ne 3$, and

$$\operatorname{rank}_p A_{\mathcal{Z}(U)} = h^{\dagger}_{(U)}(q-1) + 1 = \left[\binom{n+p-1}{n}^2 - \binom{n+p-2}{n}^2\right]^d + 1$$

for all $n \ge 1$.

1.4 Example: Grassmann Varieties

Let $V = F^{\nu+1}$. Every subspace $U \le V$ of dimension $r+1$ gives rise to a one-dimensional subspace $\bigwedge^{r+1} U \le \bigwedge^{r+1} V$. This defines an injective mapping from the collection of projective r-subspaces of $PG(\nu, F)$ to points of $PG(\bigwedge^{r+1} V) = PG(n, F)$ where $n = \binom{\nu+1}{r+1} - 1$. The image of this mapping is the *Grassmann variety* $S = G_{r,\nu}(F)$, a discrete variety of the form $S = \mathcal{Z}(\mathfrak{I})$ where $\mathfrak{I}$ is the ideal generated by a certain collection of homogeneous quadratic polynomials known as van der Waerden syzygies. In Section 4, we show that $\mathfrak{I}$ satisfies FFN$(q-1)$, and consequently

THEOREM 2 If $S = \mathcal{Z}(\mathfrak{I})$ is the Grassmann variety $G_{r,\nu}(p^e)$ naturally embedded in $PG(n, p^e)$ where $n = \binom{\nu+1}{r+1} - 1$, then $\text{rank}_p A_S = \delta^e + 1$ where $\delta = h_{\mathfrak{I}}(p-1)$ is given by the formula

$$h_{\mathfrak{I}}(k) = \prod_{j=0}^{r} \frac{(\nu + k - r + j)!\, j!}{(\nu - r + j)!\,(k+j)!}\,.$$

For example, the Grassmann variety $G_{0,n}(p^e)$ coincides with $PG(n, p^e)$, and the p-rank value from Theorem 2 agrees with that obtained in Example 1.1. Also, the Grassmann variety $G_{1,3}(p^e)$ coincides with the Klein quadric in $PG(5, p^e)$, and the value $\text{rank}_p A_S = [p(p+1)^2(p+2)/12]^e + 1$ from Theorem 2 agrees with the value given in Example 1.2.

2 STANDARD MONOMIALS

The goal of this section is to provide combinatorial interpretations of the Hilbert function values $h_{\mathfrak{I}}(k)$ and $h^{\dagger}_{\mathfrak{I}}(k)$ as the cardinalities of certain sets of monomials. In many cases, including Examples 1.2–1.4 above, this leads to explicit computation of $h^{\dagger}_{\mathfrak{I}}(q-1)$ and hence the desired p-rank values; in other cases, they may at least provide useful bounds for p-ranks. We require only a few notions from the theory of Gröbner bases, as introduced in [12], [13]; several broader texts on the subject are now available, including [5].

Let $X = (X_0, X_1, \ldots, X_n)$ be an $(n+1)$-tuple of indeterminates, and let $R = K[X] = K[X_0, X_1, \ldots, X_n]$ be the polynomial ring over a fixed arbitrary field K. A *monomial* is a polynomial of the form $X_0^{\alpha_0} X_1^{\alpha_1} \cdots X_n^{\alpha_n} \in R$ for some non-negative integers $\alpha_0, \alpha_1, \ldots, \alpha_n$ (note that we require coefficient 1). The set of monomials is a multiplicative submonoid $\mathcal{M} \subset R$. A *monomial ordering* is a total ordering $<$ on $\mathcal{M}$ such that

(i) $(\mathcal{M}, <)$ is well-ordered with least element 1, and

(ii) for all $m, m', m'' \in \mathcal{M}$, $m < m' \Rightarrow m''m < m''m'$.

The most well-known monomial ordering, and the only one we shall require, is the *lexicographical ordering* defined by

$$X_0^{\alpha_0} X_1^{\alpha_1} \cdots X_n^{\alpha_n} < X_0^{\beta_0} X_1^{\beta_1} \cdots X_n^{\beta_n} \iff \begin{array}{r}\text{there exists } i_0 \text{ such that } \alpha_{i_0} < \beta_{i_0} \\ \text{and } \alpha_i = \beta_i \text{ whenever } i < i_0\,.\end{array}$$

For any nonzero polynomial $f \in R$, let $\text{Init}(f)$ denote the *initial monomial* of f, i.e. the largest monomial with respect to $<$ which appears in f. For any subset $B \subseteq R$, define $\text{Init}(B) := \{\text{Init}(f) : 0 \neq f \in B\}$.

Now let $\mathfrak{I} \subseteq R$ be an ideal. A monomial $m \in \mathcal{M}$ is *standard* (with respect to $\mathfrak{I}$) if $m \notin \text{Init}(\mathfrak{I})$. Let $k \geq 0$, and let R_k be the k-homogeneous component of R. We immediately have

LEMMA 1 $R_k = (\mathfrak{I} \cap R_k) \oplus \langle \mathcal{SM}_{\mathfrak{I}}(k) \rangle$ where $\mathcal{SM}_{\mathfrak{I}}(k)$ is the set of standard monomials of degree k. In particular, $h_{\mathfrak{I}}(k) = |\mathcal{SM}_{\mathfrak{I}}(k)|$.

Now suppose the field K (not necessarily finite) has prime characteristic p. Given an integer $k \geq 0$, choose $e \geq 0$ such that $p^e > k$. Then k has a unique p-adic expansion of the form $k = k_0 + k_1 p + k_2 p^2 + \cdots + k_{e-1} p^{e-1}$ where $0 \leq k_\ell \leq p-1$. As before, let $R_k^\dagger$ be the subspace of R_k spanned by all monomials $m(X) = X_0^{\alpha_0} X_1^{\alpha_1} \cdots X_n^{\alpha_n}$ such that the multinomial coefficient

$$\binom{k}{\alpha_0,\ \alpha_1,\ \cdots,\ \alpha_n} := \begin{cases} \dfrac{k!}{\alpha_0!\alpha_1!\cdots\alpha_n!}, & \alpha_\ell \geq 0,\ \sum \alpha_\ell = k; \\ 0, & \text{otherwise} \end{cases}$$

is not divisible by p (and so in particular, $\deg(m) = \alpha_0 + \alpha_1 + \cdots + \alpha_n = k$). Lucas' Theorem (see [3], [2]) states that

$$\binom{k}{\alpha_0,\ \alpha_1,\ \cdots,\ \alpha_n} \equiv \prod_{\ell=0}^{e-1} \binom{k_\ell}{\alpha_{0\ell},\ \alpha_{1\ell},\ \cdots,\ \alpha_{n\ell}} \mod p$$

where $\alpha_i = \sum_\ell \alpha_{i\ell} p^\ell$, $0 \leq \alpha_{i,\ell} \leq p-1$. It follows that the monomial $m(X)$ belongs to $R_k^\dagger$ if and only if it is expressible in the form

$$m(X) = \textstyle\prod_{\ell=0}^{e-1} m_\ell(X)^{p^\ell}$$

where $m_\ell(X)$ is a monomial of degree k_ℓ. In particular,

$$dim(R_k^\dagger) = \prod_{\ell=0}^{e-1} \binom{n+k_\ell}{n}.$$

We say the monomial $m(X)$ is *p-standard* if each of the monomials m_ℓ (as above) is standard.

LEMMA 2 Let $k \geq 0$ be an integer, with p-adic coefficients k_ℓ as above. Then
(i) $R_k^\dagger = (\mathfrak{I} \cap R_k^\dagger) + \langle \mathcal{SM}_{\mathfrak{I}}^{(p)}(k) \rangle$ where $\mathcal{SM}_{\mathfrak{I}}^{(p)}(k)$ is the set of p-standard monomials of degree k.
(ii) $h_{\mathfrak{I}}^\dagger(k) \leq |\mathcal{SM}_{\mathfrak{I}}^{(p)}(k)| = \prod_{\ell=0}^{e-1} h_{\mathfrak{I}}(k_\ell)$.
(iii) $h_{\mathfrak{I}}^\dagger(q-1) \leq (h_{\mathfrak{I}}(p-1))^e$.

Proof: We must show that every monomial in $R_k^\dagger$, but not in $\mathfrak{I}$, is p-standard. Accordingly, suppose $m(X) = \prod_{\ell=0}^{e-1} m_\ell(X)^{p^\ell}$ for some monomials $m_\ell \in R_{k_\ell}$. If m is not p-standard, then for some ℓ_0 and some $f \in R_{k_{\ell_0}}$ we have $m_{\ell_0} = \text{Init}(f)$. But then since $\text{char}(K) = p$, the polynomial

$$f(X)^{p^{\ell_0}} \textstyle\prod_{\ell \neq \ell_0} m_\ell(X)^{p^\ell}$$

lies in $\mathfrak{I} \cap R_k^\dagger$, with initial monomial $m \in \text{Init}(\mathfrak{I} \cap R_k^\dagger)$. This proves (i), and the remaining conclusions clearly follow as well. □

LEMMA 3 Let $k \geq 0$. Then
(i) $h_{\mathcal{I}(\mathcal{Z}(\mathfrak{I}))}(k) \leq h_{\mathfrak{I}}(k)$ and $h_{\mathcal{I}(\mathcal{Z}(\mathfrak{I}))}^\dagger(k) \leq h_{\mathfrak{I}}^\dagger(k)$.
(ii) If $\mathfrak{I}$ satisfies FFN(k), then equality holds in (i).
(iii) Suppose the ideal $\mathfrak{I}$ is prime. Then $\mathfrak{I}$ satisfies FFN(k) iff $\mathfrak{I}$ satisfies FFN(ℓ) for all $\ell \leq k$, iff $h_{\mathcal{I}(\mathcal{Z}(\mathfrak{I}))}(k) = h_{\mathfrak{I}}(k)$, iff $h_{\mathcal{I}(\mathcal{Z}(\mathfrak{I}))}(\ell) = h_{\mathfrak{I}}(\ell)$ for all $\ell \leq k$.

Proof: (i) and (ii) are clear. Suppose $\mathfrak{I}$ is a prime ideal satisfying FFN(k), and let $0 \le \ell < k$. For $i = 0, 1, 2, \ldots, n$, the polynomial $X_i^{k-\ell} f(X_0, \ldots, X_n) \in R_k$ vanishes on $\mathcal{Z}(\mathfrak{I})$. If $f \notin \mathfrak{I}$ then $X_0, X_1, \ldots, X_n \in \mathfrak{I}$ and $\mathfrak{I} \supset R_k \ni f$, a contradiction. Hence $\mathfrak{I}$ satisfies FFN(ℓ), and (iii) follows. □

The following is well-known.

LEMMA 4 Suppose $f \in R$ has degree $\le q-1$ in each of $X_0, X_1, \ldots, X_n$. If f vanishes on F^{n+1} then $f = 0$.

Proof: If $\sum_\alpha a_{\alpha_0\alpha_1\cdots\alpha_n} x_0^{\alpha_0} x_1^{\alpha_1} \cdots x_n^{\alpha_n} = 0$ (sum over $\alpha = (\alpha_0, \alpha_1, \ldots, \alpha_n) \in \{0, 1, 2, \ldots, q-1\}^{n+1}$) for all $x = (x_0, x_1, \ldots, x_n) \in F^{n+1}$, then the vector $(a_\alpha)_\alpha$ lies in the left null space of the $q^{n+1} \times q^{n+1}$ matrix with (α, x)-entry equal to $x_0^{\alpha_0} x_1^{\alpha_1} \cdots x_n^{\alpha_n}$. By [2, Lemma 2.3], the latter matrix is nonsingular, and so $(a_\alpha) = 0$. □

3 INCIDENCE MATRICES

Most of this section can be found in [2] and [10], either directly or implicitly. In particular, the following is shown in [2]. Let $F^{k\times\ell}$ denote the vector space of $k \times \ell$ matrices over $F = GF(q)$, and let A be the incidence matrix of $PG(n, F)$ as in Section 1.

LEMMA 5 $Col(J_{N,N} - A) \cong R_{q-1}^\dagger$ as FG-modules, where $G = GL(n+1, F)$.

We sketch the proof of Lemma 5, describing the natural isomorphism $R_{q-1}^\dagger \to Col(J_{N,N} - A)$, since this is useful in verifying Theorem 1. For each $f(X) = f(X_0, X_1, \ldots, X_n) \in R_{q-1}^\dagger$ and $x \in F^{n+1}$, the value of $f(x) \in F$ clearly depends only on the subspace $\langle x \rangle$. So we have a well-defined FG-homomorphism $\varphi : R_{q-1}^\dagger \to F^{N\times 1}$ which maps f to the column vector having values $f(x)$ as $\langle x \rangle$ ranges over the one-dimensional subspaces of F^{n+1} (i.e. the points of $PG(n, F)$). For any nonzero $\ell(X) \in R_1$, $\varphi(\ell^{q-1})$ is the column of $J_{N,N} - A$ corresponding to the hyperplane $\mathcal{Z}(\ell)$. However, $Col(J_{N,N} - A)$ and $R_{q-1}^\dagger$ both have dimension $\binom{p+n-1}{n}^e$. Therefore $R_{q-1}^\dagger = \langle \ell(X)^{q-1} : \ell \in R_1 \rangle$ and $\varphi : R_{q-1}^\dagger \to Col(J_{N,N} - A)$ is the required isomorphism. □

Now let $\pi_S : F^{N\times 1} \to F^{s\times 1}$ be the projection onto the coordinates corresponding to the points of S. This gives rise to an exact sequence of FG_S-modules:

$$0 \longrightarrow \mathcal{I}(S) \cap R_{q-1}^\dagger \longrightarrow R_{q-1}^\dagger \overset{\pi_S \circ \varphi}{\longrightarrow} Col(J_{s,N} - A_S) \longrightarrow 0\,.$$

This proves Theorem 1(i).

Since $\mathrm{rank}_p(J_{s,N}) = 1$, it is clear that $\mathrm{rank}_p A_S$ differs from $\mathrm{rank}_p(J_{s,N} - A_S)$ by at most 1. Now suppose that $|H \cap S| \equiv 1 \bmod p$ for every hyperplane H of $PG(n, F)$. Then the sum of the rows of A_S (modulo p) equals $\mathbf{1} = (1, 1, \ldots, 1) \in F^{1\times N}$, so that $Row(A_S) = Row(J_{s,N} - A_S) + \langle \mathbf{1} \rangle$. To see that the latter sum is direct, observe that every row of $J_{s,N} - A_S$ has sum $q^n \equiv 0 \bmod p$, whereas every row of A_S has sum $\equiv 1 \mod p$. The proof of Theorem 1 follows by taking contragredients (Section 5).

4 GRASSMANN VARIETIES

For terminology and basic properties on Grassmann varieties, we follow [12], [13]; see also [4] for a more general approach.

We develop further the description of Example 1.4, this time over an arbitrary field K, and using explicit coordinates. Consider the vector space $V = K^{\nu+1}$, with standard basis $\{e_0, e_1, \ldots, e_\nu\}$. Fix an integer r, $0 \leq r \leq \nu$. By $\bigwedge^{r+1} V$, we mean the K-vector space of dimension $n+1 := \binom{\nu+1}{r+1}$ with basis consisting of the symbols

$$e_{\tau_0\tau_1\cdots\tau_r} := e_{\tau_0} \wedge e_{\tau_1} \wedge \cdots \wedge e_{\tau_r}, \quad 0 \leq \tau_0 < \tau_1 < \cdots < \tau_r \leq \nu .$$

Given an $(r+1)$-dimensional subspace $U \leq V$, let $x = [x_{ij}]$ be an $(r+1) \times (\nu+1)$ matrix whose rows form a basis for U. Then $\bigwedge^{r+1} U$ is the point (one-dimensional subspace) of $\bigwedge^{r+1} V$ spanned by the vector

$$\Phi(x) := \sum_{[\tau_0\tau_1\cdots\tau_r]\in\Lambda} \det \begin{bmatrix} x_{0,\tau_0} & x_{0,\tau_1} & \cdots & x_{0,\tau_r} \\ x_{1,\tau_0} & x_{1,\tau_1} & \cdots & x_{1,\tau_r} \\ \vdots & \vdots & \ddots & \vdots \\ x_{r,\tau_0} & x_{r,\tau_1} & \cdots & x_{r,\tau_r} \end{bmatrix} e_{\tau_0\tau_1\cdots\tau_r} .$$

The Grassmann variety $G_{r,\nu}(K) \subseteq PG(\bigwedge^{r+1} V) = PG(n, K)$ is the set of all such points. To justify the term 'variety', we must find an ideal $\mathfrak{I}$ whose zero set coincides with $G_{r,\nu}(K)$.

Let $X = [X_{ij}]_{0\leq i\leq r,\, 0\leq j\leq \nu}$ be an $(r+1) \times (\nu+1)$ matrix of indeterminates, and let $K[X] = K[X_{00}, X_{01}, \ldots, X_{r\nu}]$ be the corresponding polynomial algebra over K. As coordinate functions for $\bigwedge^{r+1} V$, we require $\binom{\nu+1}{r+1}$ additional indeterminates, for which purpose we adopt the set Λ consisting of the $\binom{\nu+1}{r+1}$ *bracket symbols*

$$[\tau_0\,\tau_1\,\cdots\,\tau_r], \qquad 0 \leq \tau_0 < \tau_1 < \cdots < \tau_r \leq \nu.$$

Let $K[\Lambda]$ denote the polynomial algebra in these new indeterminates. Let $\varphi : K[\Lambda] \to K[X]$ be the unique algebra homomorphism such that

$$\varphi([\tau_0\,\tau_1\,\cdots\,\tau_r]) = \det \begin{bmatrix} X_{0\tau_0} & X_{0\tau_1} & \cdots & X_{0\tau_r} \\ X_{1\tau_0} & X_{1\tau_1} & \cdots & X_{1\tau_r} \\ \vdots & \vdots & \ddots & \vdots \\ X_{r\tau_0} & X_{r\tau_1} & \cdots & X_{r\tau_r} \end{bmatrix} .$$

As described above, the Grassmann variety $G_{r,n}(K) \subseteq PG(\bigwedge^{r+1} V) = PG(n, K)$ is obtained by evaluating the expression

$$\Phi(X) := \sum_{[\tau_0\tau_1\cdots\tau_r]\in\Lambda} \varphi([\tau_0\,\tau_1\cdots\tau_r])\, e_{\tau_0\tau_1\cdots\tau_r}$$

at constant matrices $[x_{ij}]$ of full rank. Moreover, $G_{r,n}(K) = \mathcal{Z}(\mathfrak{I})$ where $\mathfrak{I} \subseteq K[\Lambda]$ is the *syzygy ideal* generated by certain homogeneous quadratic polynomials known as *van der Waerden syzygies,* as described in [13], [12]; the exact form of these generators will not be required here.

Let $<$ denote the lexicographical order on monomials for each of our sets X, Λ of indeterminates; thus

$$X_{00} < X_{01} < \cdots < X_{0\nu} < X_{10} < \cdots < X_{1\nu} < \cdots < X_{r0} < \cdots < X_{r\nu}\,;$$

$$\textstyle\prod_{i,j} X_{ij}^{\alpha_{ij}} < \prod_{i,j} X_{ij}^{\beta_{ij}} \iff \begin{array}{r}\text{there exists } (i_0, j_0) \text{ such that } \alpha_{i_0 j_0} < \beta_{i_0 j_0} \\ \text{and } \alpha_{ij} = \beta_{ij} \text{ whenever } X_{ij} < X_{i_0 j_0}\,;\end{array}$$

$$[\tau_0\,\tau_1\ \cdots\ \tau_r] < [\rho_0\,\rho_1\ \cdots\ \rho_r] \iff \begin{array}{r}\text{there exists } j_0 \text{ such that } \tau_{j_0} < \rho_{j_0} \\ \text{and } \tau_j = \rho_j \text{ for all } j < j_0.\end{array}$$

Products of indeterminates in Λ are represented by *tableaux*, which are arrays having the indeterminates as rows: thus the monomial $T = \prod_{i=1}^{k}[\tau_{i0}\,\tau_{i1}\ \cdots\ \tau_{ir}]$ of degree k may be expressed as

$$T = \begin{bmatrix} \tau_{10} & \tau_{11} & \cdots & \tau_{1r} \\ \tau_{20} & \tau_{21} & \cdots & \tau_{2r} \\ \vdots & \vdots & & \vdots \\ \tau_{k0} & \tau_{k1} & \cdots & \tau_{kr} \end{bmatrix}.$$

The *degree* of this tableau is k, the number of rows. We may assume that the rows have been listed in weakly increasing order:

$$[\tau_{10}\,\tau_{11}\ \cdots\ \tau_{1r}] \le [\tau_{20}\,\tau_{21}\ \cdots\ \tau_{2r}] \le \cdots \le [\tau_{k0}\,\tau_{k1}\ \cdots\ \tau_{kr}].$$

Let

$$T = [\tau_{ij} : 1 \le i \le k,\ 0 \le j \le r], \qquad T' = [\tau'_{ij} : 1 \le i \le k',\ 0 \le j \le r]$$

be two tableaux. Then $T < T'$ iff either $k < k'$, or $k = k'$ and there exists $i \le k$ such that $[\tau_{i0}\,\tau_{i1}\ \cdots\ \tau_{ir}] < [\tau'_{i0}\,\tau'_{i1}\ \cdots\ \tau'_{ir}]$ and the first $i-1$ rows of T' coincide with those of T. By [12], [13] we have

LEMMA 6 Let T be an arbitrary tableau, as above. Then T is standard ($T \notin \mathrm{Init}(\mathfrak{I})$) if and only if every column of T is weakly increasing, i.e. $\tau_{1j} \le \tau_{2j} \le \cdots \le \tau_{kj}$.

The number of standard tableaux of degree k is given by [8, p.387] in slightly different language:

$$h_{\mathfrak{I}}(k) = \det \begin{bmatrix} \binom{\nu+k}{k} & \binom{\nu+k-1}{k-1} & \cdots & \binom{\nu+k-r}{k-r} \\ \binom{\nu+k}{k+1} & \binom{\nu+k-1}{k} & \cdots & \binom{\nu+k-r}{k-r+1} \\ \vdots & \vdots & \ddots & \vdots \\ \binom{\nu+k}{k+r} & \binom{\nu+k-1}{k+r-1} & \cdots & \binom{\nu+k-r}{k} \end{bmatrix}.$$

Although it was assumed in [8] that $\mathrm{char}(K) = 0$, it is clear that the number of standard tableaux with k rows cannot depend on the choice of K. By [1, p.95], the $(r+1) \times (r+1)$ determinant above can be evaluated in closed form, whereby we obtain

LEMMA 7 $h_{\mathfrak{I}}(k) = \displaystyle\prod_{j=0}^{r} \frac{(\nu+k-r+j)!\,j!}{(\nu-r+j)!\,(k+j)!}$, independent of the choice of field K.

Case (i) of the following is found in [12, pp.81–82].

LEMMA 8 Let T, T' be tableaux such that either
(i) T, T' are both standard; or
(ii) T, T' are both p-standard where $p = \text{char}(K)$ is prime.
If $\varphi(T)$ and $\varphi(T')$ have the same initial monomial, then $T = T'$.

Proof: We may assume that T, T' are of the same degree k. First suppose that the monomial $T = [\tau_{ij}]_{1\le i\le k,\, 0\le j\le r}$ is standard. Clearly $\text{Init}(\varphi([\tau_{i0}\,\tau_{i1}\cdots\tau_{ir}])) = X_{0,\tau_{i0}}X_{1,\tau_{i1}}\cdots X_{r,\tau_{ir}}$, and so

$$\text{Init}(\varphi(T)) = \prod_{i=1}^{k} X_{0,\tau_{i0}}X_{1,\tau_{i1}}\cdots X_{r,\tau_{ir}}\,.$$

The number of times an integer τ appears in column j of T equals the exponent of $X_{j,\tau}$ in $\text{Init}(\varphi(T))$, and since the jth column of T is weakly increasing, this means we can recover the jth column of T from $\text{Init}(\varphi(T))$, for each $j = 0, 1, 2, \ldots, r$. This verifies the Lemma in Case (i).

Now suppose $T = \prod_{\ell=0}^{e-1} T_\ell^{p^\ell}$ is p-standard, i.e. T_ℓ is a standard tableau of degree $k_\ell \le p-1$. Then

$$\text{Init}(\varphi(T)) = \prod_{\ell=0}^{e-1} (\text{Init}\,\varphi(T_\ell))^{p^\ell}.$$

The degree of this polynomial with respect to X_{ij} is given by

$$\deg_{X_{ij}}(\text{Init}\,\varphi(T)) = \sum_{\ell=0}^{e-1} \deg_{X_{ij}}(\text{Init}\,\varphi(T_\ell))p^\ell$$

where $\deg_{X_{ij}}(\text{Init}\,\varphi(T_\ell)) \le k_\ell \le p-1$. By the uniqueness of p-adic expansions, the monomial $\text{Init}(\varphi(T))$ uniquely determines each of the monomials $\text{Init}(\varphi(T_\ell))$ ($\ell = 0, 1, 2, \ldots, e-1$), which in turn (by case (i)) uniquely determines each of the tableaux $T_0, T_1, \ldots, T_{e-1}$ and hence also T. This proves the Lemma in Case (ii). □

For the remainder of Section 4, we replace K by the finite field F of order $q = p^e$.

LEMMA 9 (i) The syzygy ideal $\mathfrak{I} \subseteq F[\Lambda]$ satisfies FFN$(q-1)$.
(ii) $h_{\mathfrak{I}}^{\dagger}(q-1) = (h_{\mathfrak{I}}(p-1))^e$.

Proof: Let $R = F[\Lambda]$, and suppose $f(\Lambda) \in R_{q-1}$ vanishes on $G_{r,\nu}(F)$. Since $R_{q-1} = (\mathfrak{I}\cap R_{q-1}) \oplus \langle \mathcal{T}\rangle$ where $\langle\mathcal{T}\rangle$ is the F-span of $\mathcal{T}$, the set of standard tableaux of degree $q-1$, we may assume that $f \in \langle\mathcal{T}\rangle$ vanishes on $G_{r,\nu}(F)$. We must show that $f = 0$. By hypothesis, the polynomial $\widehat{f}(X) := f(\Phi(X)) \in F[X]$ vanishes on $F^{(r+1)\times(\nu+1)}$. Clearly $\deg_{X_{ij}}(\widehat{f}) \le \deg(f) = q-1$, and so by Lemma 4, $\widehat{f}(X)$ is the zero polynomial. If $f(\Lambda) \ne 0$, then let $T = \text{Init}(f) \in \mathcal{T}$; then by Lemma 8(i) we have $\text{Init}(\widehat{f}) = \text{Init}\,\varphi(T) \ne 0$, a contradiction. Thus $f(\Lambda) = 0$ as required, and (i) follows.

By Lemma 2, we have $R_{q-1}^{\dagger} = (\mathfrak{I}\cap R_{q-1}^{\dagger}) + \langle\mathcal{T}'\rangle$ where $\mathcal{T}'$ is the set of all p-standard tableaux of degree $q-1$. We must show that this sum is direct. Accordingly, suppose $0 \ne f(\Lambda) \in \mathfrak{I}\cap\langle\mathcal{T}'\rangle$. By hypothesis, the polynomial

$\widehat{f}(X) := f(\Phi(X)) \in F[X]$ vanishes on $F^{(r+1)\times(\nu+1)}$. As above, this implies that $\widehat{f}(X) = 0$ and $f(\Lambda) = 0$, a contradiction. Thus $R^{\dagger}_{q-1} = (\mathfrak{I} \cap R^{\dagger}_{q-1}) \oplus \langle T' \rangle$ and conclusion (ii) follows from Lemma 2. □

LEMMA 10 Let $S = G_{r,\nu}(F) \subseteq PG(n, F)$ and let H be any hyperplane of $PG(n, F)$, $F = GF(q)$. Then $|H \cap S| \equiv 1 \mod q$.

Proof: (Due to A. E. Brouwer.) Let N_H be the number of incident point-line pairs (P, ℓ) such that $P \in S \smallsetminus H$ and $\ell \subseteq S$. For every point $P \in S$, the number of lines contained in S passing through P, equals $[(q^{r+1}-1)/(q-1)][(q^{\nu-r}-1)/(q-1)] \equiv 1 \mod q$. Thus $N_H \equiv |S \smallsetminus H| \mod q$. However, every line $\ell \subseteq S$ not contained in H, contains exactly q points of $S \smallsetminus H$, so that $N_H \equiv 0 \mod q$. Furthermore, $|S| = \prod_{i=0}^{r}[(q^{\nu-i+1}-1)/(q^{r-i+1}-1)] \equiv 1 \mod q$, so that $|S \cap H| = |S| - |S \smallsetminus H| \equiv 1 \mod q$. □

Now Theorem 2 follows from Theorem 1 and Lemmas 3, 7, 8 and 10.

5 AUTOMORPHISMS OF MATRICES

Let $B \in K^{k\times\ell}$, the vector space of all $k \times \ell$ matrices over a field K. Let G be a group, and let $Perm(k)$ denote the group of all $k \times k$ permutation matrices. Consider a *permutation action* of G on B, i.e. a homomorphism $G \to Perm(k) \times Perm(\ell)$, $g \mapsto (\Pi_1(g), \Pi_2(g))$ such that

$$\Pi_1(g)B = B\Pi_2(g)$$

for all $g \in G$. In the special case that B is square and invertible, it is well known that Π_1 and Π_2 are equivalent linear representations (although not necessarily equivalent permutation representations). In this section we prove a natural generalisation of this fact, which was alluded to in [2], but not proven there.

Our intent is to show that the row and column spaces of B over K are naturally contragredient KG-modules (in general not isomorphic, as this author erroneously stated in [2]). In order to deal just with left KG-modules, we consider instead the column spaces

$$Col(B) = \{Bx : x \in K^{\ell\times 1}\} \le K^{k\times 1}, \qquad Col(B^{\top}) = \{B^{\top}y : y \in K^{k\times 1}\} \le K^{\ell\times 1}.$$

Since $\Pi_1(g)(Bx) = B(\Pi_2(g)x) \in Col(B)$, we see that $Col(B)$ is indeed a left KG-module via $v \mapsto \Pi_1(g)v$. Similarly, since $B^{\top}\Pi_1(g) = \Pi_2(g)B^{\top}$, $Col(B^{\top})$ is a left KG-module via $w \mapsto \Pi_2(g)w$.

LEMMA 11 $Col(B^{\mathsf{T}}) \cong Col(B)^*$ as left KG-modules.

Proof: Choose $M_1 \in GL(k, K)$, $M_2 \in GL(\ell, K)$ such that

$$M_1 B M_2 = \begin{bmatrix} I_r & O_{r,\ell-r} \\ O_{k-r,r} & O_{k-r,\ell-r} \end{bmatrix}$$

where I_r is an identity matrix of size $r = \operatorname{rank}_K B$ and the O's consist of zeroes. Now

$$(M_1 \Pi_1(g) M_1^{-1})(M_1 B M_2) = (M_1 B M_2)(M_2^{-1} \Pi_2(g) M_2)$$

for all $g \in G$. From this it is easy to see that

$$M_1 \Pi_1(g) M_1^{-1} = \begin{bmatrix} Q(g) & * \\ O & * \end{bmatrix}, \qquad M_2^{-1} \Pi_2(g) M_2 = \begin{bmatrix} Q(g) & O \\ * & * \end{bmatrix}$$

where $Q : G \to GL(r, K)$ is a homomorphism. Via $v \mapsto (M_1 \Pi_1(g) M_1^{-1})v$, we have a left KG-module $Col(M_1 B)$ isomorphic to $Col(B)$, and since

$$Col(M_1 B) = Col(M_1 B M_2) = \{(\underbrace{*, *, \cdots, *}_{r \text{ times}}, \underbrace{0, 0, \cdots, 0}_{k-r \text{ times}})^{\mathsf{T}}\},$$

the associated matrix representation of degree r is explicitly given by Q. Similarly, via

$$w \mapsto (M_2^{-\mathsf{T}} \Pi_2(g) M_2^{\mathsf{T}})w = (M_2 \Pi_2(g) M_2^{-1})^{-\mathsf{T}} w = \begin{bmatrix} Q(g)^{-\mathsf{T}} & * \\ O & * \end{bmatrix} w,$$

$Col(M_2^{\mathsf{T}} B^{\mathsf{T}})$ becomes a left KG-module isomorphic to $Col(B^{\mathsf{T}})$. Again, $Col(M_2^{\mathsf{T}} B^{\mathsf{T}}) = Col(M_2^{\mathsf{T}} B^{\mathsf{T}} M_1^{\mathsf{T}}) = \{(*, *, \ldots, *, 0, 0, \ldots, 0)^{\mathsf{T}}\}$ and so the associated matrix representation of degree r is given by $Q^{-\mathsf{T}}$. □

REFERENCES

1. E. Arbarello, M. Cornalba, P. A. Griffiths and J. Harris, *Geometry of Algebraic Curves, Volume I,* Springer-Verlag, New York (1985).
2. A. Blokhuis and G. E. Moorhouse, Some *p*-ranks related to orthogonal spaces, *J. Algeb. Comb., 4:* 295–316 (1995).
3. A. E. Brouwer and H. A. Wilbrink, Block Designs, in *Handbook of Incidence Geometry, Buildings and Foundations,* ed. F. Buekenhout, pp.349–382, North-Holland, Amsterdam and New York (1995).
4. A. M. Cohen and R. H. Cushman, Gröbner bases and standard monomial theory, in *Computational Algebraic Geometry,* ed. F. Eyssette and A. Galligo, pp.41–60, Birkhäuser Boston (1993).
5. D. Cox, J. Little and D. O'Shea, *Ideals, Varieties, and Algorithms,* Springer-Verlag, New York (1992).
6. J. M. Goethals and P. Delsarte, On a class of majority-logic decodable cyclic codes, *IEEE Trans. Inform. Theory, 14:* 182–188 (1968).
7. R. Hartshorne, *Algebraic Geometry,* Springer-Verlag, New York (1977).

8. W. V. D. Hodge and D. Pedoe, *Methods of Algebraic Geometry, Volume II*, Camb. Univ. Press, Cambridge (1952).
9. F. J. MacWilliams and H. B. Mann, On the p-rank of the design matrix of a difference set, *Inform. and Control, 12:* 474–489 (1968).
10. G. E. Moorhouse, Some p-ranks related to Hermitian varieties, to appear in *J. Stat. Plan. Inf.*
11. K. J. C. Smith, On the p-rank of the incidence matrix of points and hyperplanes in a finite projective geometry, *J. Comb. Theory, 1:* 122–129 (1969).
12. B. Sturmfels, *Algorithms in Invariant Theory*, Springer-Verlag, Vienna (1993).
13. B. Sturmfels and N. White, Gröbner bases and invariant theory, *Adv. Math., 76:* 245–259 (1989).

Building a Cyclic q-Clan

S. E. PAYNE* Department of Mathematics, University of Colorado at Denver, Campus Box 170, Post Office Box 173364, Denver, Colorado 80217-3364 U.S.A., spayne@carbon.cudenver.colorado.edu

TIM PENTTILA Department of Mathematics, University of Western Australia, Nedlands 6907, Western Australia, Australia, penttila@maths.uwa.edu.au

GORDON F. ROYLE Department of Computer Science, University of Western Australia, Nedlands 6907, Western Australia, Australia, gordon@cs.uwa.edu.au

ABSTRACT

For $q = 2^e$ let $\mathcal{C}$ be a q-clan whose associated generalized quadrangle $GQ(\mathcal{C})$ of order (q^2, q) admits a collineation θ that acts as a cycle of length $q+1$ on the lines through the special point (∞). With some natural additional assumptions on the action of θ on certain subquadrangles of order q, i.e., on the associated ovals, the q-clan $\mathcal{C}$ can be constructed from one carefully chosen matrix in $\mathcal{C}$ and the action of θ. With the help of computer calculations this approach is used to construct new examples with $q = 4^k$ for $k = 5$, $k = 6$, $k = 7$, and $k = 8$. We conjecture that these examples belong to a new infinite family.

1. INTRODUCTION AND REVIEW

This essay is a sequel to [4], but we recall sufficiently many definitions and results to render this work intelligible to a reader already familiar with the construction of a generalized quadrangle $GQ(\mathcal{C})$ of order (q^2, q) from a q-clan $\mathcal{C}$ (cf. [3, 2]).

Let $q = 2^e$, $F = GF(q)$, $\tilde{F} = F \cup \{\infty\}$. We assume that each q-clan $\mathcal{C} = \{A_t : t \in F\}$ is normalized so that $A_0 = \begin{pmatrix} 0 & 0 \\ 0 & 0 \end{pmatrix}$, and for $t \in F$, $A_t = \begin{pmatrix} x_t & t^{1/2} \\ 0 & z_t \end{pmatrix}$. As in

*The research for this paper was performed while the first author enjoyed the warm hospitality of the Combinatorial Computing Research Group of the University of Western Australia and the financial support of the Australian Research Council.

[4] we also write $A_\infty = \left(\begin{smallmatrix} 0 & 0 \\ 0 & 0 \end{smallmatrix}\right)$. Let $P = \left(\begin{smallmatrix} 0 & 1 \\ 1 & 0 \end{smallmatrix}\right)$, and for $\alpha, \beta \in F^2$ put $\alpha \circ \beta = \alpha P \beta^T$. Then the elation group $G^\otimes$ of $GQ(\mathcal{C})$ is defined as follows:

$$G^\otimes = \left\{ ((\alpha, \beta), c) \in F^2 \times F^2 \times F : \alpha, \beta \in F^2,\ c \in F \right\}$$

with binary operation given by

$$((\alpha, \beta), c) \circ ((\alpha', \beta'), c') = ((\alpha + \alpha', \beta + \beta'), c + c' + \beta \circ \alpha'). \tag{1}$$

There are two families of subgroups of $G^\otimes$, each subgroup having order q^3 and being elementary abelian, that play important roles in q-clan geometry.

$$\text{For } \vec{0} \neq \gamma \in F^2, \quad \mathcal{L}_\gamma = \left\{ (\gamma \otimes \alpha, c) \in G^\otimes : \alpha \in F^2,\ c \in F \right\}. \tag{2}$$

$$\text{For } \vec{0} \neq \alpha \in F^2, \quad \mathcal{R}_\alpha = \left\{ (\gamma \otimes \alpha, c) \in G^\otimes : \gamma \in F^2,\ c \in F \right\}. \tag{3}$$

For nonzero $\alpha, \gamma \in F^2$, $\mathcal{L}_\gamma = \mathcal{L}_\alpha$ (resp., $\mathcal{R}_\gamma = \mathcal{R}_\alpha$) if and only if $\{\alpha, \gamma\}$ is F-dependent. Hence we may think of the $\mathcal{L}_\gamma$ (resp., $\mathcal{R}_\alpha$) as being indexed by the points of $PG(1, q)$. Identify the elements of $\tilde{F}$ with the points of $PG(1, q)$ in the following way. For $t \in F$, write $\gamma_t = (1, t)$, and $\gamma_\infty = (0, 1)$. So for $\vec{0} \neq \gamma = (a, b)$, $\gamma \equiv \gamma_{b/a}$. For $t \in \tilde{F}$ we then write $\mathcal{L}_t = \mathcal{L}_{\gamma_t}$ and $\mathcal{R}_t = \mathcal{R}_{\gamma_t}$.

Now let $\mathcal{C} = \{A_t : t \in \tilde{F}\}$ be a normalized q-clan, so for distinct $t, s \in F$, $A_t - A_s$ is anisotropic. For $t \in \tilde{F}$ and $\alpha \in F^2$, write $g_t(\alpha) = \alpha A_t \alpha^T$. Then we may use $\mathcal{C}$ to define subgroups of $G^\otimes$ as follows.

$$\text{For } t \in \tilde{F}, \quad A(t) = \left\{ \left(\gamma_{t^{\frac{1}{2}}} \otimes \alpha, g_t(\alpha)\right) \in G^\otimes : \alpha \in F^2 \right\}. \tag{4}$$

It follows that for each $t \in \tilde{F}$, $A(t)$ is a subgroup of $\mathcal{L}_{t^{\frac{1}{2}}}$ where $\infty^{\frac{1}{2}} = \infty$ by convention. And then $\mathcal{J}(\mathcal{C}) = \{A(t) : t \in \tilde{F}\}$ is a 4-gonal family for $G^\otimes$, and $\mathcal{L}_{t^{\frac{1}{2}}}$ is the tangent space with respect to $\mathcal{J}(\mathcal{C})$ at $A(t)$. This notation for $G^\otimes$ and $\mathcal{J}(\mathcal{C})$ was introduced in [4]. By the Fundamental Theorem of q-Clan Geometry (cf. [5, 1]) and its corollaries, each collineation of $GQ(\mathcal{C})$ that fixes the two points (∞) and $((\vec{0}, \vec{0}), 0)$ is induced by an automorphism θ of the elation group $G^\otimes$. Moreover, each such automorphism is defined in terms of a field automorphism $\sigma \in \mathrm{Aut}(F)$ and two 4×4 matrices over F, H and S, with H invertible. From Eq. (10) of [4] and the material that follows it, if we write $\theta = \theta(\sigma, H, S)$, then we have

$$\begin{aligned} &\theta(\sigma_1, H_1, S_1) \cdot \theta(\sigma_2, H_2, S_2) \\ &\quad = \theta\left(\sigma_1 \cdot \sigma_2,\ {H_1}^{\sigma_1} H_2,\ [\det(H_2)]^{\frac{1}{2}}\ {S_1}^{\sigma_2} + {H_1}^{\sigma_2} S_2 \left({H_1}^{\sigma_2}\right)^T\right). \end{aligned} \tag{5}$$

By putting $\tau = 2$ in Theorem 3.1 of [4] (since $y_t = t^{\frac{1}{2}}$ here), we see that if $\theta(\sigma, H, S)$ stabilizes the 4-gonal family $\mathcal{J}(\mathcal{C})$, then $H = A \otimes B$, where $A, B \in GL(2, q)$. We may assume $\Delta = \det(B) = 1$ and write $A = \left(\begin{smallmatrix} a_4 & a_2 \\ a_3 & a_1 \end{smallmatrix}\right)$. Then

$$S = (I \otimes B) \begin{pmatrix} a_4^2 A_{(a_2/a_4)^2} & a_2 a_3 P \\ 0 & a_3^2 A_{(a_1/a_3)^2} \end{pmatrix} (I \otimes B)^T. \tag{6}$$

And it follows (cf. part (ii) of Theorem 3.1 of [4]) that if $\mu = \det(A)$, $1 = \det(B)$, then

$$\theta = \theta(\sigma, A \otimes B, S):$$
$$((\alpha,\beta),c) \mapsto \Big((\alpha^\sigma,\beta^\sigma)(A\otimes B),\ \mu c^\sigma + a_4^2\alpha^\sigma\left(BA_{(a_2/a_4)^2}B^T\right)(\alpha^\sigma)^T$$
$$+\ a_2a_3\left(\alpha^\sigma \circ \beta^\sigma\right) + a_3^2\beta^\sigma\left(BA_{(a_1/a_3)^2}B^T\right)(\beta^\sigma)^T\Big). \quad (7)$$

Until further notice we assume that $\theta(\sigma, H, S)$ is *linear*, i.e., $\sigma = \mathrm{id}$. And since S is determined by $A \otimes B$, we may write $\theta(\mathrm{id}, A\otimes B, S) = \theta(A \otimes B)$. With A as above and $B = \left(\begin{smallmatrix} b_4 & b_2 \\ b_3 & b_1\end{smallmatrix}\right)$, we now have (cf. Eqs. (22) through (25) of [4])

$$\theta(A\otimes B): \mathcal{L}_\gamma \mapsto \mathcal{L}_{\gamma A}, \quad \text{so that } \mathcal{L}_t \mapsto \mathcal{L}_{\bar t},$$
$$\text{where } \bar t = \frac{a_1t + a_2}{a_3t + a_4},\ t \in \tilde F, \quad (8a)$$
$$\theta(A\otimes B): \mathcal{R}_\alpha \mapsto \mathcal{R}_{\alpha B}, \quad \text{so that } \mathcal{R}_t \mapsto \mathcal{R}_{\bar t},$$
$$\text{where } \bar t = \frac{b_1t + b_2}{b_3t + b_4},\ t \in \tilde F. \quad (8b)$$

From [4] we recall that for $\alpha \in PG(1,q)$, $A(t) \cap \mathcal{R}_\alpha = \left\{d\left(\gamma_{t^{\frac12}} \otimes \alpha, g_t(\alpha)\right) = \left(\gamma_{t^{\frac12}} \otimes d\alpha, g_t(d\alpha)\right) : d \in F\right\}$. So $A(t) \cap \mathcal{R}_\alpha$ is the point $\bar p_\alpha(t) = \left(\gamma_{t^{\frac12}} \otimes \alpha, g_t(\alpha)\right)$ in the projective plane $\bar{\mathcal{R}}_\alpha$. And $\bar{\mathcal{O}}_\alpha = \left\{\bar p_\alpha(t) : t \in \tilde F\right\}$ is an oval in $\bar{\mathcal{R}}_\alpha$. The map (use the notation $(a,b)^{(2)} = (a^2, b^2)$)

$$\pi_\alpha : \bar{\mathcal{R}}_\alpha \to PG(2,q) : (\gamma \otimes \alpha, c) \mapsto \left(\gamma^{(2)}, c\right) \quad (9)$$

is an isomorphism mapping $\bar{\mathcal{O}}_\alpha$ to the oval

$$\mathcal{O}_\alpha = \left\{p_\alpha(t) = (\gamma_t, g_t(\alpha)) : t \in \tilde F\right\} \text{ in } PG(2,q). \quad (10)$$

It follows that in $\theta = \theta(A \otimes B)$ as given above, the matrix A determines the action of θ on the subgroups $\mathcal{L}_t$, and hence on the lines of $GQ(\mathcal{C})$ through the point (∞). Similarly, the matrix B determines the action of θ on the subgroups $\mathcal{R}_\alpha$, and hence on the ovals $\mathcal{O}_\alpha$. In [6] it was shown that for the Subiaco q-clans $\mathcal{C}$ there was a certain collineation $\theta = \theta(A \otimes B)$, where $A = M$ and $B = M^{-5}$ for a carefully chosen M, and where θ permutes the lines through (∞) in one cycle of length $q+1$.

Motivated by the computer-generated examples of q-clans for $q = 64$ and $q = 256$, we wish to mimic this situation and make the assumption that a collineation of the form $\theta(M \otimes M^{-p})$ exists for some other value of p. The general theory is worked out in Section 2, where it is shown how to use $\theta(M \otimes M^{-p})$ to generate all the matrices of $\mathcal{C}$ starting with a special member. Then in Section 3, again motivated by the existing examples, we consider an additional assumption that the generalized quadrangle $GQ(\mathcal{C})$ admit a certain involution. In Section 4 we report on some computer results based on these ideas, whereby a search loops over all the possibilities for p and a starting matrix and attempts to generate a complete q-clan. This search succeeds in producing a q-clan for $q = 1024$, $q = 4096$, $q = 16384$, and $q = 65536$. In each case the value of p is $(q-1)/3$, and hence we conjecture the existence of an infinite family of this form.

2. A CYCLIC q-CLAN

Let $F \subset E = GF(q^2)$, and let ζ be a primitive element for E. Put $\lambda = \zeta^{q-1}$ and $\delta = \lambda + \lambda^{-1}$. For any rational number a with denominator prime to $q+1$, put $[a] = \lambda^a + \lambda^{-a} \in F$. Then we have the following basic arithmetic properties for the symbol $[a]$:

$$[a] = \lambda^a + \lambda^{-a}; \quad [1] = \delta; \quad [0] = 0. \tag{11a}$$

$$[a] = [b] \text{ if and only if } a \equiv \pm b \pmod{q+1}. \tag{11b}$$

$$[a]\,[b] = [a+b] + [a-b]\,. \tag{11c}$$

$$\text{For } \sigma = 2^i \in \mathrm{Aut}\,(F), \quad [a]^\sigma = [a\sigma]\,. \tag{11d}$$

$$\left[\frac{a+c}{2}\right]\left[\frac{a}{2}\right]\left[\frac{c}{2}\right] = [a+c] + [a] + [c]\,. \tag{11e}$$

Put $M = \begin{pmatrix} 0 & 1 \\ 1 & \delta^{\frac{1}{2}} \end{pmatrix}$. Then M has eigenvalues $\lambda^{\frac{1}{2}}$ and $\lambda^{-\frac{1}{2}}$. By diagonalizing M over E, with a bit of routine computation we may compute the powers of M:

$$M^j = \frac{1}{\delta^{\frac{1}{2}}}\begin{pmatrix} \left[\frac{j-1}{2}\right] & \left[\frac{j}{2}\right] \\ \left[\frac{j}{2}\right] & \left[\frac{j+1}{2}\right] \end{pmatrix}. \tag{12}$$

We note that M is symmetric with determinant 1 and that M has multiplicative order $q+1$. Also observe that $MPM = P$ and hence $M^j P M^j = P$ for any j.

We are now ready to make the basic assumption of this work.

BASIC ASSUMPTION. There is a normalized q-clan $\mathcal{C} = \{A_t : t \in F\}$ with $A_\delta = \begin{pmatrix} x_\delta & \delta^{\frac{1}{2}} \\ 0 & z_\delta \end{pmatrix}$ given, and there is an integer p modulo $q+1$ for which $\theta = \theta\,(M \otimes M^{-p})$ is an automorphism of $G^{\otimes}$ leaving $\mathcal{J}\,(\mathcal{C})$ invariant.

Since M has eigenvalues $\lambda^{\pm\frac{1}{2}}$, $\theta\,(M \otimes M^{-p})$ will permute the members of $\mathcal{J}\,(\mathcal{C})$ (and hence the members of $\mathcal{C}$) in a cycle of length $q+1$.

BASIC GOAL. Starting with $A_0 = \begin{pmatrix} 0 & 0 \\ 0 & 0 \end{pmatrix} = A_\infty$, and $A_\delta = \begin{pmatrix} x_\delta & \delta^{\frac{1}{2}} \\ 0 & z_\delta \end{pmatrix}$, and the collineation $\theta\,(M \otimes M^{-p})$, we want to reconstruct the entire q-clan $\mathcal{C}$.

To acheve this goal we will make use of Equations 5, 6 and 12. For linear automorphisms we have

$$\theta\,(H_1, S_1)\,\theta\,(H_2, S_2) = \theta\,(H_1 H_2, S_3)\,,$$

where we can compute S_3 either by using Equation 5 or by using Equation 6 to go directly from $H_1 H_2$ to S_3.

We illustrate the procedure and the computational techniques by computing θ^2 before proceeding to the general case.

By using Equation 6 on θ and on θ^2 we see that

$$\theta = \theta\left(M \otimes M^{-p}\right) = \theta\left(M \otimes M^{-p}, \begin{pmatrix} 0 & P \\ 0 & M^{-p} A_\delta M^{-p} \end{pmatrix}\right), \tag{13}$$

and that

$$\theta^2 = \theta\left(M^2 \otimes M^{-2p}\right) = \theta\left(M^2 \otimes M^{-2p}, \begin{pmatrix} M^{-2p}A_\delta M^{-2p} & \delta P \\ 0 & M^{-2p}\left(\delta A_{\left(\frac{[3]}{[2]}\right)}\right)M^{-2p}\end{pmatrix}\right). \tag{14}$$

Now we want to use Eq. 5 to compute θ^2 as $\theta \cdot \theta$. So

$$\theta\cdot\theta = \theta\left(M^2 \otimes M^{-2p}, \begin{pmatrix} 0 & P \\ 0 & M^{-p}A_\delta M^{-p}\end{pmatrix} + \left(\begin{pmatrix} 0 & 1 \\ 1 & \delta^{\frac{1}{2}}\end{pmatrix}\otimes M^{-p}\right)\begin{pmatrix} 0 & P \\ 0 & M^{-p}A_\delta M^{-p}\end{pmatrix}\left(\begin{pmatrix} 0 & 1 \\ 1 & \delta^{\frac{1}{2}}\end{pmatrix}\otimes M^{-p}\right)\right). \tag{15}$$

To simplify the bottom line of Eq. 15 we observe that

$$\begin{pmatrix} 0 & M^{-p} \\ M^{-p} & \delta^{\frac{1}{2}}M^{-p}\end{pmatrix}\begin{pmatrix} 0 & P \\ 0 & M^{-p}A_\delta M^{-p}\end{pmatrix}\begin{pmatrix} 0 & M^{-p} \\ M^{-p} & \delta^{\frac{1}{2}}M^{-p}\end{pmatrix} = \begin{pmatrix} M^{-2p}A_\delta M^{-2p} & \delta^{\frac{1}{2}}M^{-2p}A_\delta M^{-2p} \\ M^{-p}PM^{-p} + \delta^{\frac{1}{2}}M^{-2p}A_\delta M^{-2p} & \delta^{\frac{1}{2}}M^{-p}PM^{-p} + \delta M^{-2p}A_\delta M^{-2p}\end{pmatrix}.$$

We now mention two computational simplifications that will be used without special mention from now on. First recall that if $A = \left(\begin{smallmatrix} x & y \\ w & z\end{smallmatrix}\right)$ and $A' = \left(\begin{smallmatrix} a & b \\ c & d\end{smallmatrix}\right)$, then $A \equiv A'$ means that $x = a$, $z = d$, and $y + w = b + c$.

For $S = \left(\begin{smallmatrix} C & D_1 \\ D_2 & E\end{smallmatrix}\right)$, $\alpha S\beta^T = \alpha C\alpha^T + \alpha\left(D_1 + {D_2}^T\right)\beta^T + \beta E\beta^T$. Hence in $\theta(H, S)$, we may replace C with any $C' \equiv C$, and E with any $E' \equiv E$, where we usually choose C' and E' to be upper triangular. And we may replace D_1 with $D_1 + {D_2}^T$ and D_2 with 0. Also, $M^jP\left(M^j\right)^T = M^jPM^j = P$. Then in the lower right hand block of the last displayed equation above, $M^{-p}PM^{-p} \equiv 0$. And after the indicated simplifications, the upper right hand block is $M^{-p}PM^{-p} + \delta^{\frac{1}{2}}M^{-2p}\left(A_\delta + {A_\delta}^T\right)M^{-2p} \equiv P + \delta M^{-2p}PM^{-2p} \equiv (1+\delta)P$. Now we may complete the computation of $\theta\cdot\theta$.

$$\theta^2 = \theta\cdot\theta = \theta\left(M^2 \otimes M^{-2p}, \begin{pmatrix} M^{-2p}A_\delta M^{-2p} & \delta P \\ 0 & M^{-p}A_\delta M^{-p} + \delta M^{-2p}A_\delta M^{-2p}\end{pmatrix}\right). \tag{16}$$

Comparing the last two equations for θ^2 we find

$$\delta A_{\left(\frac{[3]}{[2]}\right)} \equiv \frac{[2]}{\delta}A_{\left(\frac{[3]}{[2]}\right)} \equiv M^pA_\delta M^p + \delta A_\delta. \tag{17}$$

Now we shall perform a similar computation, but in the general case. Since $\theta^j = \theta\left(M^j \otimes M^{-pj}, S_j\right)$ for some S_j, we may use Eqs. 6 and 12 to write out the following:

$$\theta^j = \theta\left(M^j \otimes M^{-pj}, \begin{pmatrix} M^{-pj}C_jM^{-pj} & D_j \\ 0 & M^{-pj}E_jM^{-pj} \end{pmatrix}\right), \quad \text{where} \tag{18}$$

$$C_j = \frac{[j-1]}{\delta} A_{\left(\frac{[j]}{[j-1]}\right)}, \tag{18a}$$

$$D_j = \frac{[j]}{\delta} P, \tag{18b}$$

$$E_j = \frac{[j]}{\delta} A_{\left(\frac{[j+1]}{[j]}\right)} = C_{j+1}. \tag{18c}$$

REMARK 2.1. $\left\{\infty = \frac{[1]}{[0]}, \delta = \frac{[2]}{[1]}, \ldots, \frac{[q+1]}{[q]} = 0\right\} = \tilde{F}$.

Proof. We have $\frac{[j+1]}{[j]} = \frac{[k+1]}{[k]}$ iff $[j+1]\,[k] = [j]\,[k+1]$ iff $[j+1+k] + [j+1-k] = [j+k+1] + [j-k-1]$ iff $j-k+1 \equiv \pm(j-k-1) \pmod{q+1}$ iff $2(j-k) \equiv 0 \pmod{q+1}$. ∎

THEOREM 2.2. *There is a recurrence formula for* A_t:

(i) $A_{\left(\frac{[j+1]}{[j]}\right)} \equiv A_\delta + \frac{[j-1]}{[j]} M^p A_{\left(\frac{[j]}{[j-1]}\right)} M^p$, *which is equivalent to*
(ii) $E_j \equiv \frac{[j]}{\delta} A_\delta + M^p E_{j-1} M^p$ *(for E_j as in Eq. 17).*

Proof. The idea is to use the fact that θ^{j+1} as given by Eq. 17 must be the same as that computed according to $\theta^{j+1} = \theta^j \cdot \theta$. So

$$\begin{aligned}
\theta^j \cdot \theta &= \theta\left(M^j \otimes M^{-pj}, \begin{pmatrix} M^{-pj}C_jM^{-pj} & \frac{[j]}{\delta}P \\ 0 & M^{-pj}E_jM^{-pj} \end{pmatrix}\right) \\
&\quad \cdot \theta\left(M \otimes M^{-p}, \begin{pmatrix} 0 & P \\ 0 & M^{-p}A_\delta M^{-p} \end{pmatrix}\right) \\
&= \theta\left(M^{j+1} \otimes M^{-p(j+1)}, \begin{pmatrix} M^{-pj}C_jM^{-pj} & \frac{[j]}{\delta}P \\ 0 & M^{-pj}E_jM^{-pj} \end{pmatrix}\right. \\
&\quad + \frac{1}{\delta} \begin{pmatrix} \left[\frac{j-1}{2}\right] M^{-pj} & \left[\frac{j}{2}\right] M^{-pj} \\ \left[\frac{j}{2}\right] M^{-pj} & \left[\frac{j+1}{2}\right] M^{-pj} \end{pmatrix} \begin{pmatrix} 0 & P \\ 0 & M^{-p}A_\delta M^{-p} \end{pmatrix} \\
&\quad \left. \cdot \begin{pmatrix} \left[\frac{j-1}{2}\right] M^{-pj} & \left[\frac{j}{2}\right] M^{-pj} \\ \left[\frac{j}{2}\right] M^{-pj} & \left[\frac{j+1}{2}\right] M^{-pj} \end{pmatrix}\right) \\
&= \theta\left(M^{j+1} \otimes M^{-pj}, S\right),
\end{aligned}$$

where we need to work a bit to find S.

$$S = \begin{pmatrix} M^{-pj}C_jM^{-pj} & \frac{[j]}{\delta}P \\ 0 & M^{-pj}E_jM^{-pj} \end{pmatrix}$$
$$+\frac{1}{\delta}\begin{pmatrix} 0 & \left[\frac{j-1}{2}\right]M^{-pj}P + \left[\frac{j}{2}\right]M^{-p(j+1)}A_\delta M^{-p} \\ 0 & \left[\frac{j}{2}\right]M^{-pj}P + \left[\frac{j+1}{2}\right]M^{-p(j+1)}A_\delta M^{-p} \end{pmatrix}\begin{pmatrix} \left[\frac{j-1}{2}\right]M^{-pj} & \left[\frac{j}{2}\right]M^{-pj} \\ \left[\frac{j}{2}\right]M^{-pj} & \left[\frac{j+1}{2}\right]M^{-pj} \end{pmatrix}$$
$$\equiv \begin{pmatrix} M^{-pj}C_jM^{-pj} & \frac{[j]}{\delta}P \\ 0 & M^{-pj}E_jM^{-pj} \end{pmatrix}$$
$$+\frac{1}{\delta}\begin{pmatrix} [j]\,M^{-p(j+1)}A_\delta M^{-p(j+1)} & \left(\left[\frac{j-1}{2}\right]\left[\frac{j+1}{2}\right]P + \left[\frac{j}{2}\right]\left[\frac{j+1}{2}\right]M^{-p(j+1)}A_\delta M^{-p(j+1)}\right) \\ \left([j]\,P + \left[\frac{j+1}{2}\right]\left[\frac{j}{2}\right]M^{-p(j+1)}A_\delta M^{-p(j+1)}\right) & [j+1]\,M^{-p(j+1)}A_\delta M^{-p(j+1)} \end{pmatrix}$$
$$\equiv \begin{pmatrix} \left(M^{-pj}C_jM^{-pj} + \frac{[j]}{\delta}M^{-p(j+1)}A_\delta M^{-p(j+1)}\right) & \left(\frac{\left[\frac{j-1}{2}\right]\left[\frac{j+1}{2}\right]}{\delta}P + \frac{\left[\frac{j+1}{2}\right]\left[\frac{j}{2}\right]}{\delta}M^{-p(j+1)}\left(\left[\frac{1}{2}\right]P\right)M^{-p(j+1)}\right) \\ 0 & \left(\frac{[j+1]}{\delta}M^{-p(j+1)}A_\delta M^{-p(j+1)} + M^{-pj}E_jM^{-pj}\right) \end{pmatrix}$$
$$\equiv \begin{pmatrix} \left(M^{-pj}C_jM^{-pj} + \frac{[j]}{\delta}M^{-p(j+1)}A_\delta M^{-p(j+1)}\right) & \frac{[j+1]}{\delta}P \\ 0 & \left(M^{-pj}E_jM^{-pj} + \frac{[j+1]}{\delta}M^{-p(j+1)}A_\delta M^{-p(j+1)}\right) \end{pmatrix}.$$

If we now use Eq. 18 to see what the above matrix should be, both the upper left and lower right blocks yield the desired recurrence. ∎

COROLLARY 2.3. *The q-clan being generated by powers of θ is automatically indexed so that if $A_t \equiv \begin{pmatrix} x_t & y_t \\ 0 & z_t \end{pmatrix}$, then $y_t = t^{\frac{1}{2}}$.*

Proof. The "y" entry of $A_{\left(\frac{[j+1]}{[j]}\right)}$ should be $\frac{\left[\frac{j+1}{2}\right]}{\left[\frac{j}{2}\right]}$. By Theorem 2.2 (since M^p is symmetric with determinant 1) it is $\left[\frac{1}{2}\right] + \frac{[j-1]\left[\frac{j}{2}\right]}{[j]\left[\frac{j-1}{2}\right]} = \left[\frac{1}{2}\right] + \frac{\left[\frac{j-1}{2}\right]}{\left[\frac{j}{2}\right]} = \frac{\left[\frac{j+1}{2}\right]}{\left[\frac{j}{2}\right]}$. ∎

COROLLARY 2.4.

$$A_{\left(\frac{[j+1]}{[j]}\right)} \equiv \sum_{i=0}^{j-1}\frac{[i+1]}{[j]}M^{p(j-1-i)}A_\delta M^{p(j-1-i)}$$
$$\equiv \sum_{k=0}^{j-1}\frac{[j-k]}{[j]}M^{pk}A_\delta M^{pk}, \quad j \geq 1.$$

Proof. Straightforward induction starting with part (i) of Theorem 2.2. ∎

Recall that

$$\begin{pmatrix} a & b \\ c & d \end{pmatrix} \begin{pmatrix} x & y \\ 0 & z \end{pmatrix} \begin{pmatrix} a & c \\ b & d \end{pmatrix} \equiv \begin{pmatrix} a^2x + aby + b^2z & (ad+bc)y \\ 0 & c^2x + dcy + d^2z \end{pmatrix}. \tag{19}$$

Then use $M^{pk} = \frac{1}{\delta^{\frac{1}{2}}} \begin{pmatrix} \left[\frac{pk-1}{2}\right] & \left[\frac{pk}{2}\right] \\ \left[\frac{pk}{2}\right] & \left[\frac{pk+1}{2}\right] \end{pmatrix}$ and $A_\delta \equiv \begin{pmatrix} x_\delta & \delta^{\frac{1}{2}} \\ 0 & z_\delta \end{pmatrix}$ to write out $M^{pk} A_\delta M^{pk}$ and use Cor. 2.4 to obtain the following.

COROLLARY 2.5.

$$A_{\left(\frac{[j+1]}{[j]}\right)} \equiv \sum_{k=0}^{j-1} \frac{[j-k]}{[j]} \cdot \begin{pmatrix} \frac{[pk-1]x_\delta + ([pk]+[pk-1]+[1]) + [pk]z_\delta}{\delta} & \delta^{\frac{1}{2}} \\ 0 & \frac{[pk]x_\delta + ([pk]+[pk+1]+[1]) + [pk+1]z_\delta}{\delta} \end{pmatrix}.$$

REMARK 2.6. Corollary 2.5 gives each member of the q-clan as a concrete function of A_δ.

An easy computation shows that

$$\theta^{-1} = \theta\left(M^{-1} \otimes M^p, \begin{pmatrix} A_\delta & P \\ 0 & 0 \end{pmatrix}\right). \tag{20}$$

Now put $j = -1 \equiv q \pmod{q+1}$ in Eq. 18 and compare with Eq. 20 to obtain

$$A_\delta \equiv \frac{[2]}{\delta} M^p A_{\left(\frac{[1]}{[2]}\right)} M^p, \quad \text{i.e., } \delta A_{\delta^{-1}} \equiv M^{-p} A_\delta M^{-p}. \tag{21}$$

THEOREM 2.7. *Replacing the exponent $-p$ with p is equivalent to interchanging x_δ and z_δ in A_δ.*

Proof. $\theta\,(\mathrm{id}, I \otimes P, (\begin{smallmatrix} 0 & 0 \\ 0 & 0 \end{smallmatrix}))$ is an involutory automorphism of $G^\otimes$ mapping $(\gamma \otimes \alpha, c)$ to $(\gamma \otimes \alpha P, c)$, i.e., it leaves invariant each $\mathcal{L}_\gamma$. Also, it maps $\left(\gamma_{t^{\frac{1}{2}}} \otimes \alpha, \alpha A_t \alpha^T\right)$ to $\left(\gamma_{t^{\frac{1}{2}}} \otimes \alpha P, \alpha P\,(PA_tP)\,(\alpha P)^T\right)$. So it replaces the q-clan $\mathcal{C} = \{A_t : t \in F\}$ with the q-clan $\mathcal{C}' = \{A'_t = PA_tP : t \in F\}$.

If $\phi = \theta\,(\mathrm{id}, I \otimes P, (\begin{smallmatrix} 0 & 0 \\ 0 & 0 \end{smallmatrix}))$ and $\theta = \theta\left(\mathrm{id}, M \otimes M^{-p}, \left(\begin{smallmatrix} 0 & P \\ 0 & M^{-p}A_\delta M^{-p} \end{smallmatrix}\right)\right)$, so θ is the putative isomorphism of $G^\otimes$ permuting the members of $\mathcal{J}(\mathcal{C})$ in a cycle of length $q+1$, then $\phi \cdot \theta \cdot \phi$ is an automorphism of $G^\otimes$ permuting the members of $\mathcal{J}(\mathcal{C}')$ in a cycle of length $q+1$.

First note

$$\begin{aligned} (I \otimes P) \begin{pmatrix} 0 & P \\ 0 & M^{-p}A_\delta M^{-p} \end{pmatrix} (I \otimes P)^T &= \begin{pmatrix} 0 & P \\ 0 & PM^{-p}A_\delta M^{-p}P \end{pmatrix} \\ &= \begin{pmatrix} 0 & P \\ 0 & M^p\,(PA_\delta P)\,M^p \end{pmatrix} \\ &= \begin{pmatrix} 0 & P \\ 0 & M^p A'_\delta M^p \end{pmatrix}. \end{aligned}$$

Then

$$\phi\cdot\theta\cdot\phi=\theta\left(\mathrm{id},M\otimes PM^{-p},(I\otimes P)\begin{pmatrix}0 & P\\ 0 & M^{-p}A_\delta M^{-p}\end{pmatrix}(I\otimes P)^T\right)$$
$$\cdot\,\theta\left(\mathrm{id},I\otimes P,\begin{pmatrix}0&0\\0&0\end{pmatrix}\right)$$
$$=\theta\left(\mathrm{id},M\otimes PM^{-p}P,\begin{pmatrix}0 & P\\ 0 & M^{p}A'_\delta M^{p}\end{pmatrix}\right)$$
$$=\theta\left(\mathrm{id},M\otimes M^{p},\begin{pmatrix}0 & P\\ 0 & M^{p}A'_\delta M^{p}\end{pmatrix}\right). \quad \blacksquare$$

3. CYCLIC q-CLANS WITH INVOLUTIONS

Continue with exactly the same setup as in the previous section, but now in addition consider when the automorphism $\phi=\theta(\sigma,A\otimes B)$ of $G^{\otimes}$ would leave invariant the 4-gonal family $\mathcal{J}(\mathcal{C})$. If $A=\left(\begin{smallmatrix}a_4 & a_2\\ a_3 & a_1\end{smallmatrix}\right)$, such an automorphism must have the form (see Eq. 20 and Theorem 3.1 of [4])

$$\phi=\theta\left(\sigma,A\otimes B,(I\otimes B)\begin{pmatrix}a_4^2A_{(a_2/a_4)^2} & a_2a_3P\\ 0 & a_3^2A_{(a_1/a_3)^2}\end{pmatrix}(I\otimes B)^T\right), \tag{22}$$

where $\phi: A(t)\mapsto A(t')$ satisfies

$$t'=\frac{a_1^2t^\sigma+a_2^2}{a_3^2t^\sigma+a_4^2},\quad t\in\tilde{F}. \tag{23}$$

Now suppose ϕ is an involution interchanging $A(\infty)$ and $A(0)$, so that $a_1=a_4=0$, and $A=\left(\begin{smallmatrix}0 & a\\ a^{-1} & 0\end{smallmatrix}\right)$. If we also assume that ϕ normalizes the group $\langle\theta\rangle$ of order $q+1$ studied in the previous section, then A conjugates M to a power of M. But $A^{-1}MA=\left(\begin{smallmatrix}\delta^{\frac{1}{2}} & a^2\\ a^{-2} & 0\end{smallmatrix}\right)$ is a power of M if and only if it is equal to $M^{-1}=\left(\begin{smallmatrix}\delta^{\frac{1}{2}} & 1\\ 1 & 0\end{smallmatrix}\right)$ and then $a=1$, i.e., $A=P$.

But then writing out that $\phi^2=\mathrm{id}$ shows that B must be an involutory matrix for which $BM^{-p}B=M^p$. If p is prime to $q+1$ (as it is for $p=-2/3$, $q=4^h$), then $BM^{-1}B=M$, and we must have $B=P$ also. So: *the basic additional assumption of this section* is that

$$\phi=\theta\left(\mathrm{id},P\otimes P,\begin{pmatrix}0&P\\0&0\end{pmatrix}\right)$$

is an automorphism of $G^{\otimes}$ leaving $\mathcal{J}(\mathcal{C})$ invariant. (24)

For $t\in\tilde{F}$,

$$A\left(\frac{[j+1]}{[j]}\right)=\left\{\left(\gamma_{\left(\frac{[j+1]}{[j]}\right)^{\frac{1}{2}}}\otimes\alpha,\alpha A_{\left(\frac{[j+1]}{[j]}\right)}\alpha^T\right):\alpha\in F^2\right\}.$$

Since for all $j \not\equiv 0, -1 \pmod{q+1}$

$$\begin{aligned}\left(\gamma_{\left(\frac{[j+1]}{[j]}\right)^{\frac{1}{2}}} \otimes \alpha\right)(P \otimes P) &= \left(([j], [j+1])^{\frac{1}{2}} \otimes \frac{1}{\left[\frac{j}{2}\right]}\alpha\right)(P \otimes P) \\ &= \left(([j+1], [j])^{\frac{1}{2}} \otimes \frac{1}{\left[\frac{j}{2}\right]}\alpha P\right) \\ &= \left(\gamma_{\left(\frac{[j]}{[j+1]}\right)^{\frac{1}{2}}} \otimes \frac{\left[\frac{j+1}{2}\right]}{\left[\frac{j}{2}\right]}\alpha P\right),\end{aligned}$$

ϕ maps the element $\left(\gamma_{\left(\frac{[j+1]}{[j]}\right)^{\frac{1}{2}}} \otimes \alpha, \alpha A_{\left(\frac{[j+1]}{[j]}\right)}\alpha^T\right)$ of $A\left(\frac{[j+1]}{[j]}\right)$ to the element

$$\left(\gamma_{\left(\frac{[j]}{[j+1]}\right)^{\frac{1}{2}}} \otimes \frac{\left[\frac{j+1}{2}\right]}{\left[\frac{j}{2}\right]}\alpha P, \frac{\left[\frac{j+1}{2}\right]}{\left[\frac{j}{2}\right]}\alpha P\left(\frac{[j]}{[j+1]}PA_{\left(\frac{[j+1]}{[j]}\right)}P\right)\left(\frac{\left[\frac{j+1}{2}\right]}{\left[\frac{j}{2}\right]}\alpha P\right)^T\right),$$

which is supposed to be in $A\left(\frac{[j+1]}{[j]}\right)$. And this is the case if and only if

$$\frac{[j]}{[j+1]}PA_{\left(\frac{[j+1]}{[j]}\right)}P \equiv A_{\left(\frac{[j]}{[j+1]}\right)}, \quad j \not\equiv 0, -1 \pmod{q+1}.$$

Clearly ϕ interchanges $A(0)$ and $A(\infty)$, so we may restate this result as follows:

$$\phi = \theta\left(\mathrm{id}, P \otimes P, \begin{pmatrix} 0 & P \\ 0 & 0 \end{pmatrix}\right) \text{ is a collineation of } GQ(\mathcal{C}) \text{ iff}$$
$$[j]\, PA_{\left(\frac{[j+1]}{[j]}\right)}P \equiv [j+1]\, A_{\left(\frac{[j]}{[j+1]}\right)} \quad \text{for all } j \mod q+1. \tag{25}$$

REMARK 3.1. Note that Eq. 25 implies that $PA_tP \equiv tA_{t^{-1}}$ for all nonzero $t \in F$ if ϕ is a collineation.

THEOREM 3.2. *The following are equivalent:*

(i) *ϕ is a collineation of $GQ(\mathcal{C})$.*
(ii) $A_\delta \equiv M^p P A_\delta P M^p$.
(iii) $([p] + [1])\, x_\delta + [p-1]\, z_\delta = [p] + [p-1] + [1]$, *where* $A_\delta = \begin{pmatrix} x_\delta & \delta^{\frac{1}{2}} \\ 0 & z_\delta \end{pmatrix}$.

Proof. We know that (i) is equivalent to Eq. 25. In Eq. 25 put $j = 1$, to get the fact that (i) implies

$$PA_\delta P \equiv \delta A_{\delta^{-1}}. \tag{26}$$

By Eq. 21, Eq. 26 implies condition (ii). Hence now we must show that condition (ii) implies Eq. 25. So suppose that $A_\delta \equiv M^p P A_\delta P M^p$, which by Eq. 21 implies

$PA_\delta P \equiv \delta A_{\delta^{-1}}$. Then for all $j \geq 1$,

$$\begin{aligned}
P\left([j]\,A_{\left(\frac{[j+1]}{[j]}\right)}\right)P &\equiv \sum_{k=0}^{j-1}[j-k]\,PM^{pk}P\,(PA_\delta P)\,PM^{pk}P \\
&\equiv \sum_{k=0}^{j-1}[j-k]\,M^{-pk}\,(\delta M_{\delta^{-1}})\,M^{-pk} \\
&\equiv \sum_{k=0}^{j-1}[j-k]\,M^{-p(k+1)}A_\delta M^{-p(k+1)} \\
&\equiv \sum_{k=1}^{j}[j-k+1]\,M^{-pk}A_\delta M^{-pk} \\
&\equiv \sum_{k=1}^{j}[q-j+k]\,M^{-pk}A_\delta M^{-pk} \\
&\equiv \sum_{i=0}^{j-1}[q-i]\,M^{-p(j-i)}A_\delta M^{-p(j-i)}.
\end{aligned}$$

So at this point we know that Eq. 26 implies

$$P\left([j]\,A_{\left(\frac{[j+1]}{[j]}\right)}\right)P \equiv \sum_{i=0}^{j-1}[q-i]\,M^{-p(j-i)}A_\delta M^{-p(j-i)}, \quad j \geq 1. \qquad \bullet \ (27)$$

Put $j = q$ in Corollary 2.4 to get $0 \equiv \sum_{k=0}^{q-1}[q-k]\,M^{pk}A_\delta M^{pk}$. Fix j, $0 < j \leq q-1$. Then $\sum_{k=0}^{j-1}[q-k]\,M^{pk}A_\delta M^{pk} \equiv \sum_{k=j}^{q-1}[q-k]\,M^{pk}A_\delta M^{pk}$. Multiply through on the left and on the right by M^{-pk} to get $\sum_{k=0}^{j-1}[q-k]\,M^{p(k-j)}A_\delta M^{p(k-j)} \equiv \sum_{k-j}^{q-1}[q-k]\,M^{p(k-j)}A_\delta M^{p(k-j)}$, i.e.,

$$\begin{aligned}
\sum_{i=0}^{j-1}[q-i]\,M^{-p(j-i)}A_\delta M^{-p(j-i)} &\equiv \sum_{i=j}^{q-1}[q-i]\,M^{p(i-j)}A_\delta M^{p(i-j)} \\
&\equiv \sum_{k=0}^{q-j-1}[q-j-k]\,M^{pk}A_\delta M^{pk} \\
&\equiv [q-j]\,A_{\left(\frac{[q-j+1]}{[q-j]}\right)} \quad (\text{put } i-j=k) \\
&\equiv [-j-1]\,A_{\left(\frac{[-j]}{[-j-1]}\right)} \\
&\equiv [j+1]\,A_{\left(\frac{[j]}{[j+1]}\right)} \quad (\text{by Cor. 2.4}). \qquad (28)
\end{aligned}$$

So by Eqs. 27 and 28 together we see that condition (ii) does imply Eq. 25. This completes a proof that conditions (i) and (ii) are equivalent.

$$M^pP = \frac{1}{\delta^{\frac{1}{2}}}\begin{pmatrix} \left[\frac{p-1}{2}\right] & \left[\frac{p}{2}\right] \\ \left[\frac{p}{2}\right] & \left[\frac{p+1}{2}\right] \end{pmatrix} P = \begin{pmatrix} \frac{\left[\frac{p}{2}\right]}{\left[\frac{1}{2}\right]} & \frac{\left[\frac{p-1}{2}\right]}{\left[\frac{1}{2}\right]} \\ \frac{\left[\frac{p+1}{2}\right]}{\left[\frac{1}{2}\right]} & \frac{\left[\frac{p}{2}\right]}{\left[\frac{1}{2}\right]} \end{pmatrix}. \qquad (29)$$

Recall that

$$\begin{pmatrix} a & b \\ c & d \end{pmatrix} \begin{pmatrix} x_\delta & \delta^{\frac{1}{2}} \\ 0 & z_\delta \end{pmatrix} \begin{pmatrix} a & c \\ b & d \end{pmatrix} \equiv \begin{pmatrix} a^2 x_\delta + ab\delta^{\frac{1}{2}} + b^2 z_\delta & (ad+bc)\,\delta^{\frac{1}{2}} \\ 0 & c^2 x_\delta + cd\delta^{\frac{1}{2}} + d^2 z_\delta \end{pmatrix}.$$

Hence condition (ii) is equivalent to

$$\begin{aligned} \begin{pmatrix} x_\delta & \delta^{\frac{1}{2}} \\ 0 & z_\delta \end{pmatrix} &\equiv (M^p P)\, A_\delta\, (M^p P)^T \\ &\equiv \begin{pmatrix} \frac{[p]x_\delta + \left[\frac{p}{2}\right]\left[\frac{p-1}{2}\right]\left[\frac{1}{2}\right] + [p-1]z_\delta}{[1]} & \delta^{\frac{1}{2}} \\ 0 & \frac{[p+1]x_\delta + \left[\frac{p+1}{2}\right]\left[\frac{p}{2}\right]\left[\frac{1}{2}\right] + [p]z_\delta}{[1]} \end{pmatrix}. \end{aligned}$$

Comparing first the upper left hand entries, then lower right hand entries, we have

$$\begin{aligned} ([p]+[1])\,x_\delta + [p-1]\,z_\delta = [p]+[p-1]+[1] = \left[\frac{p}{2}\right]\left[\frac{p-1}{2}\right]\left[\frac{1}{2}\right] \quad &\text{iff} \\ \left[\frac{p+1}{2}\right]\left[\frac{p-1}{2}\right] x_\delta + \left[\frac{p-1}{2}\right]\left[\frac{p-1}{2}\right] z_\delta = \left[\frac{p}{2}\right]\left[\frac{p-1}{2}\right]\left[\frac{1}{2}\right] \quad &\text{iff} \\ \left[\frac{p+1}{2}\right] x_\delta + \left[\frac{p-1}{2}\right] z_\delta = \left[\frac{p}{2}\right]\left[\frac{1}{2}\right] \quad &\text{iff} \\ \left[\frac{p+1}{2}\right]\left[\frac{p+1}{2}\right] x_\delta + \left[\frac{p-1}{2}\right]\left[\frac{p+1}{2}\right] z_\delta = \left[\frac{p+1}{2}\right]\left[\frac{p}{2}\right]\left[\frac{1}{2}\right] \quad &\text{iff} \\ [p+1]\,x_\delta + ([p]+[1])\,z_\delta = [p+1]+[p]+[1]. \end{aligned}$$

Hence conditions (ii) and (iii) are equivalent. ∎

Warning: In the above proof we divided by $\left[\frac{p-1}{2}\right]$, which is not allowed if $p = 1$.

4. COMPUTER RESULTS

A computer search was performed guided by the results of the previous sections. The basic idea of the search is simply to loop over all the possibilities for the value p and starting matrix A_δ, and then to construct a candidate q-clan $\mathcal{C}$ as outlined in Section 2. The resulting candidate can then be checked directly to verify that it is indeed a proper q-clan.

The first task is simply to choose a single fixed value for δ and compute the function $[a]$. Rather than working in the large field $GF(q^2)$ we first select a suitable $\delta \in GF(q)$ and then precompute the function $[a]$ according to the following recurrence:

$$\begin{aligned} [0] &= 0, \\ [1] &= \delta, \\ [a+1] &= [a-1] + [a]\,[1]. \end{aligned}$$

Then we loop through values of p. For each value of p we first compute and store the matrix M^p, which will be repeatedly used. Then an inner loop runs through all the possible pairs of values for x_δ and z_δ. For each such pair we form the matrix

$$A_\delta = \begin{pmatrix} x_\delta & \delta^{\frac{1}{2}} \\ 0 & z_\delta \end{pmatrix},$$

and then use the recurrence

$$A_{\left(\frac{[j+1]}{[j]}\right)} \equiv A_\delta + \frac{[j-1]}{[j]} M^p A_{\left(\frac{[j]}{[j-1]}\right)} M^p$$

to compute the $q-1$ matrices $A_{\frac{[j+1]}{[j]}}$ as j ranges from 1 to $q-1$.

The list of matrices thus produced is then checked firstly to see if each matrix is itself anisotropic and then to see if the pairwise sums are anisotropic.

This procedure is implemented in the computer algebra system Magma.

For small values of $q \leq 1024$, it is feasible to carry out the process exactly as described above. As expected, for each value of q the procedure finds a q-clan when $p = 5$. However, it also finds additional q-clans where $p = 21$, $p = 85$, and $p = 341$ for $q = 64$, $q = 256$, and $q = 1024$ respectively. There are also many choices for x_δ and z_δ that yield a q-clan—in these cases there are $q/2$ choices for x_δ and two choices for z_δ for each x_δ. However, amongst this large number of possibilities there is a unique pair satisfying the condition

$$[p+1]\, x_\delta + ([p] + [1])\, z_\delta = [p+1] + [p] + [1]$$

equivalent to the existence of the special involution.

This evidence led us to perform a more restricted search for the values $q = 4096$, $q = 16384$, and $q = 65536$. In these cases the search was restricted to $p = (q-1)/3$ and (x_δ, z_δ) pairs that satisfy the involution equation above. In each case a q-clan was successfully constructed. The details of these q-clans are given in the table below.

q	Defining Polynomial	p	δ	x_δ	z_δ
64	$\omega^6 + \omega^4 + \omega^3 + \omega + 1$	21	ω	ω^{18}	ω^{12}
256	$\omega^8 + \omega^4 + \omega^3 + \omega^2 + 1$	85	ω^3	ω^{49}	ω^{12}
1024	$\omega^{10} + \omega^6 + \omega^5 + \omega^3 + \omega^2 + \omega + 1$	341	ω^3	ω^{1007}	ω^{663}
4096	$\omega^{12} + \omega^7 + \omega^6 + \omega^5 + \omega^3 + \omega + 1$	1365	ω^5	ω^{1356}	ω^{1039}
16384	$\omega^{14} + \omega^7 + \omega^5 + \omega^3 + 1$	5461	ω^3	ω^{4705}	ω^{1909}
65536	$\omega^{16} + \omega^5 + \omega^3 + \omega^2 + 1$	21845	ω^3	ω^{12933}	ω^{1880}

From the determination of the groups of the Subiaco q-clans in [6], it can be seen that these q-clans are not isomorphic to the Subiaco q-clans, and are therefore new.

This seems to be very compelling evidence for the existence of a new infinite family of q-clans; indeed the process above can be viewed as a recipe for finding a q-clan that has succeeded for every attainable value of q. However, finding a "canonical" form for this q-clan amenable to a formal proof seems a very difficult problem.

REFERENCES

[1] L. Bader, G. Lunardon, and S. Payne, On q-clan geometry, $q = 2^e$, *Bull. Belg. Math. Soc. Simon Stevin* **1** (1994), 301–328.

[2] F. Buekenhout, editor, *Handbook of Incidence Geometry: Buildings and Foundations,* North-Holland, Amsterdam, 1995.

[3] S.E. Payne, A new infinite family of generalized quadrangles, *Congr. Numer.* **49** (1985), 115–128.

[4] ———, A tensor product action on q-clan generalized quadrangles with $q = 2^e$, *Linear Algebra Appl.* **226/228** (1995), 115–137.

[5] ———, The fundamental theorem of q-clan geometry, *Des. Codes Cryptogr.* **8** (1996), 181–201.

[6] S.E. Payne, T. Penttila, and I. Pinneri, Isomorphisms between Subiaco q-clan geometries, *Bull. Belg. Math. Soc. Simon Stevin* **2** (1995), 197–222.

Hyper-Reguli in Projective Space of Dimension 5

Rolando Pomareda
Department of Mathematics, Universidad de Chile
Casilla 653, Santiago-Chile.

Dedicated to Professor T.G. Ostrom on his 80 th birthday

Abstract

The concept of a regulus in three dimensional projective space was generalized to a "hyper-regulus" in projective space of dimension 2d-1 by T. G. Ostrom. We prove that there are only two opposite hyper-regulis to an André hyper-regulus in an affine Desarguesian plane of order q^3.

1 Introduction

In his paper [2], Ostrom defines a hyper-regulus and answer the question as to whether there are hyper-reguli not isomorphic to an André hyper-regulus, but also raises the question whether it is possible that an André hyper-regulus has an opposite hyper-regulus other than the canonical one. We prove that the answer is negative for the affine Desarguesian plane of order q^3, where the largest power of 3 that divides $q-1$ is 1.

2 André Hyper-Reguli.

Let π be an affine Desarguesian plane of order q^3 coordinatized by the field K isomorphic to $GF(q^3)$, where $q-1$, is divisible by 3 but not by 9, and q is a power of an odd prime p. Throughout the paper K, will denote the field of 3×3 matrices

$$A_{x,y,z} = \begin{pmatrix} x & y & z \\ \theta z & x & y \\ \theta y & \theta z & x \end{pmatrix}$$

with entries in $F = GF(q)$, and θ is a different from 1 cubic root of unity in F.

Let W denote the diagonal matrix $diag(1,\theta,\theta^2)$, then, for each 3×3 matrix T, with entries in F, there exists a set $\{A,B,C\}$ of three matrices in K such that

Partially supported by Fondecyt 1940575

$$T = A + WB + W^2C.$$

Let L be the extension of F obtained by adjoining an element γ which is the cubic root of θ so that each element of L may be written uniquely in the form $x_0 + x_1\gamma + x_2\gamma^2, x_0, x_1, x_2 \in F$. Let $\bar{z}$ and $\bar{\bar{z}}$ denote z^q and z^{q^2}, respectively, for each $z \in L$. Hence $\bar{\gamma} = \gamma\theta$ or $\gamma\theta^2$, according to $(q-1)/3$ is congruent to 1 or 2 mod 3.

The map $z = a + b\gamma + c\gamma^2 \rightarrow A_{a,b,c}$ defines a field isomorphism. Also $detA_{a,b,c} = N(z)$. Here $N(z) = z\bar{z}\bar{\bar{z}}$ for $z \in L$.

Let G denote the multiplicative group of matrices in K consisting of matrices of determinant 1. For each $g \in G$, both $W-g$ and W^2-g are singular matrices. Also, if $T = A+WB+W^2C$, A, B, C in K then $T_g = A+gB+Wg^{-1}W^2C$ is such that $T-T_g$ is a singular matrix for each $g \in G$. Moreover, if $detB \neq detC$ then $T_g \neq T_h$ for each $g \neq h$ in G. Also

$$detT_g = detA + detB + detC + Tr(A\bar{C}\bar{\bar{B}}) + Tr(f(g))$$

$$detT_g = detA + detB + detc + Tr(A\bar{C}\bar{\bar{B}}) + Tr(fg)$$

where

$$\begin{aligned} f(g) &= g\{B\bar{A}\bar{\bar{A}} + C\bar{C}\bar{\bar{A}} + B\bar{C}\bar{\bar{B}}\} \\ &+ g^{-1}\{A\bar{A}\bar{\bar{C}} + A\bar{B}\bar{\bar{B}} + C\bar{B}\bar{\bar{C}}\} \\ &+ g^{q-1}A\bar{B}\bar{\bar{C}} \end{aligned}$$

here $T_rM = M + \bar{M} + \bar{\bar{M}}$.

The points (x, y) of π may be regarded as elements of a vector space V of dimension 6 over F.; we may take $V = V_0 \oplus V_0$ where the vector space V_0 is essentially the set of triples over F, and the set of lines passing through the origin is $\mathcal{L} = \{L(M)|M \in K\} \cup \{L(\infty)\}$, where $L(M) = \{(x_1, x_2, x_3, (x_1, x_2, x_3)M)|x_i \in F\}$, and $L(\infty) = \{(0, 0, 0, x_1, x_2, x_3)|x_i \in F\}$.

If $T = WB$ or W^2C then $detT_g$ is constant for each $g \in G$. However, it is possible that other T have the same property.

The partial spread $\mathcal{A}$ whose components are given by the N equations $y = xm$, where $m^N = 1$, $N = q^2 + q + 1$, is an André hyper-regulus. A hyper-regulus is said to be an André hyper-regulus if it is isomorphic under collineations of the projective space of dimension 5, whose points are the 1-spaces of V, to the system $\mathcal{A}$.

3 Hyper-Reguli and its Admissible Subspaces

We recall that a 3-dimensional subspace S of V is said to be an admissible space with respect to π if, for each line ℓ through $(0, 0), dim(S \cap \ell) = 0$ or 1.

THEOREM 1 *Let S be an admissible subspace. Let $\mathcal{A}$ be the partial spread consisting of the N lines throug (0,0) such that $dim(\ell \cap S) = 1$. If $\mathcal{A}$ is the André hyper-regulus defined by the N equations $y = xm$, wehre $m^N = 1$, then S is defined by a matrix of the form WB or W^2C, where B, C are in G.*

Proof. Let $T = A + WB + W^2C$ be the nonsingular matrix with A, B, C in K that defines the admissible subspace S. Then $\mathcal{A}$ consists of the components defined by the matrices $T_g = A + gB + Wg^{-1}W^2C$. Hence $detT_g$ is constant for each g in G. Without loss of generality, we may assume that $Tr(f_g) = 0$ for each g in G.

Let

$$\alpha = B\bar{A}\bar{\bar{A}} + C\bar{C}\bar{\bar{A}} + B\bar{C}\bar{\bar{B}}$$

$$\beta = A\bar{A}\bar{\bar{C}} + A\bar{B}\bar{\bar{B}} + C\bar{B}\bar{\bar{C}}$$

and $\gamma = A\bar{B}\bar{\bar{C}}$. Hence $f(g) = g\alpha + g^{-1}\beta + g^{q-1}\gamma$. Since the two scalar matrices $diag(\theta, \theta, \theta)$ and $diag(\theta^2, \theta^2, \theta^2)$ are in G we also obtain

$$\theta Tr(g\alpha) + \theta^2 Tr(g^{-1}\beta) + Tr(g^{q-1}\gamma) = 0$$

and

$$\theta^2 Tr(g\alpha) + \theta Tr(g^{-1}\beta) + Tr(g^{q-1}\gamma) = 0$$

for each $g \in G$. Therefore, $Tr(g\alpha) = Tr(g^{-1}\beta) = Tr(g^{q-1}\gamma) = 0$ for each $g \in G$. Hence $\alpha = \beta = \gamma = 0$, and it follows that two of the three matrices A, B, C must be the zero matrix. Since $B = C = 0$ implies T_g is constant for each g in G we must have that $T = WB$ or W^2C. Also, the fact that S must intersect the components in $\mathcal{A}$ implies that both B and C are in G.

COROLLARY 1 *Let S and S_1 be admissible subspaces with respect to π that intersect the same André hyper-regulus. If $dim(S \cap S_1) = 0$ then both are of the form WB or both are of the form W^2C, B, C in G.*

Proof. Clearly $WB - W^2C$ is a singular matrix, when B and C are in G.

COROLLARY 2 *The opposite hyper-regulus to the hyper-regulus $\mathcal{A}$ consists of the subspaces $y = x^{q^i} m$ $m^N = 1$; equivalently they are defined by the matrices of the form Wg or by the matrices W^2g, $g \in G$.*

REFERENCES

1. J. André, Über nich-Desarguesher Ebenen mit transitiver Translations gruppe. Math.Z. 60 (1954) 156-186.

2. T. G. Ostrom, Hyper-Reguli. Journal of Geometry 48 (1993) 157-166.

Parallelisms of $PG(3,3)$ Invariant Under a Collineation of Order 5

Alan R. Prince Department of Mathematics, Heriot-Watt University, Edinburgh, Scotland EH14 4AS

Abstract

All parallelisms of $PG(3,3)$, which are invariant under a collineation of order 5, are classified.

1 Introduction

We denote by $PG(3,q)$ the 3-dimensional projective space over the field of q elements. A *spread* of $PG(3,q)$ is a set of q^2+1 lines which partition the points of the space. A *parallelism* or *packing* of $PG(3,q)$ is a collection of q^2+q+1 spreads which partition the lines of the space. A *regular* spread is one which contains the regulus determined by any three of its lines. A *regular* parallelism is one consisting only of regular spreads.

For $q=2$, the parallelisms are completely known: there are two up to projective equivalence (see [7, Theorem 17.5.6]). For $q>2$, $PG(3,q)$ has at least two projectively inequivalent parallelisms consisting of one regular spread and q^2+q subregular spreads of index one (see [1],[3] and [7, Theorem 17.4.2]; a *subregular* spread of *index* 1 is a spread obtained from a regular spread by replacing a single regulus in the spread by its opposite). Apart from the examples given by this theorem, the only other known example is a regular parallelism of $PG(3,8)$ discovered by Denniston [5]. This example is *cyclic*, which means that the spreads involved are permuted in a single cycle by some collineation of the space. Denniston [4] has shown that there are no cyclic parallelisms in $PG(3,3)$ and $PG(3,4)$.

A regular spread of $PG(3,q)$ gives rise to a translation plane of order q^4 via a construction due to Michael Walker (unpublished notes, see also [6],[8],[9]). The only known regular spreads occur in $PG(3,2)$ and $PG(3,8)$. It is an open problem whether or not there are regular parallelisms of $PG(3,q)$ for q odd (Lunardon has confirmed to the author that there is an error in [9]). In a paper [11] in preparation, the author shows that there are no regular parallelisms of $PG(3,3)$.

R.H. Bruck [2] has considered a possible construction of a projective plane of order q^2+q containing a subplane of order q, starting from a certain partial plane embedded in $PG(3,q)$. This construction depends on the existence of a family of spreads with certain properties. If such a family exists, then it would be possible to construct many parallelisms in $PG(3,q)$ from the spreads involved and, in particular, *orthogonal* parallelisms (that is two parallelisms with the property that any spread from one and any spread from the other have at most one line in common). For $q>2$, there are no known examples of orthogonal parallelisms (but, for examples in higher dimensions, see [12]).

In this paper, we construct some new parallelisms of $PG(3,3)$. By the result of Denniston mentioned above, there are no cyclic parallelisms of $PG(3,3)$, that is parallelisms invariant under a collineation of order 13. The collineation group $PGL(4,3)$ of the space $PG(3,3)$ has order $2^8 \cdot 3^6 \cdot 5 \cdot 13$ so that, following Denniston's result, it is natural to consider parallelisms invariant under a collineation of order 5. In this case, there are examples and we give a complete classification. We prove the following theorem (representatives of the projective equivalence classes are listed later in the paper).

THEOREM. There are precisely seven parallelisms of $PG(3,3)$, up to projective equivalence, which are invariant under a collineation of order 5.

2 Preliminary Results

The concrete model we take for $PG(3,3)$ is as follows. We consider the field $\mathbf{F}_{81}$ generated over $\mathbf{F}_3$ by a primitive element θ satisfying $\theta^4 = \theta + 1$. Then, $\mathbf{F}_{81}$ is a 4-dimensional vector space over $\mathbf{F}_3$ and we take its 1-spaces, 2-spaces and 3-spaces to be the points, lines and planes of $PG(3,3)$. The 40 points of $PG(3,3)$ may be represented by θ^i for $i = 0, 1, 2, \cdots, 39$. Since $\theta^4 = \theta + 1$ we have:

$\theta^4 = \theta + 1$	$\theta^5 = \theta^2 + \theta$	$\theta^6 = \theta^3 + \theta^2$
$\theta^7 = \theta^3 + \theta + 1$	$\theta^8 = \theta^2 - \theta + 1$	$\theta^9 = \theta^3 - \theta^2 + \theta$
$\theta^{10} = -\theta^3 + \theta^2 + \theta + 1$	$\theta^{11} = \theta^3 + \theta^2 - 1$	$\theta^{12} = \theta^3 + 1$
$\theta^{13} = -\theta + 1$	$\theta^{14} = -\theta^2 + \theta$	$\theta^{15} = -\theta^3 + \theta^2$
$\theta^{16} = \theta^3 - \theta - 1$	$\theta^{17} = -\theta^2 + 1$	$\theta^{18} = -\theta^3 + \theta$
$\theta^{19} = \theta^2 - \theta - 1$	$\theta^{20} = \theta^3 - \theta^2 - \theta$	$\theta^{21} = -\theta^3 - \theta^2 + \theta + 1$
$\theta^{22} = -\theta^3 + \theta^2 - 1$	$\theta^{23} = \theta^3 + \theta - 1$	$\theta^{24} = \theta^2 + 1$
$\theta^{25} = \theta^3 + \theta$	$\theta^{26} = \theta^2 + \theta + 1$	$\theta^{27} = \theta^3 + \theta^2 + \theta$
$\theta^{28} = \theta^3 + \theta^2 + \theta + 1$	$\theta^{29} = \theta^3 + \theta^2 - \theta + 1$	$\theta^{30} = \theta^3 - \theta^2 - \theta + 1$
$\theta^{31} = -\theta^3 - \theta^2 - \theta + 1$	$\theta^{32} = -\theta^3 - \theta^2 - 1$	$\theta^{33} = -\theta^3 + \theta - 1$
$\theta^{34} = \theta^2 + \theta - 1$	$\theta^{35} = \theta^3 + \theta^2 - \theta$	$\theta^{36} = \theta^3 - \theta^2 + \theta + 1$
$\theta^{37} = -\theta^3 + \theta^2 - \theta + 1$	$\theta^{38} = \theta^3 - \theta^2 - 1$	$\theta^{39} = -\theta^3 + 1$

It is straightforward to calculate the 130 lines of $PG(3,3)$ and we list them in the following table. We denote the point represented by θ^i simply by i.

We list the lines in ten groups of size 13 to show the action of the collineation of $PG(3,3)$ of order 13 arising from the linear map of the vector space $\mathbf{F}_{81}$ whose action on the the basis elements $1, \theta, \theta^2, \theta^3$ is as follows:

$$1 \mapsto 1, \theta \mapsto \theta^2, \theta^2 \mapsto \theta^3, \theta^3 \mapsto \theta + \theta^2 = \theta^5$$

Line	Points	Line	Points	Line	Points	Line	Points	Line	Points
$L1$	0 1 4 13	$L14$	1 2 5 14	$L27$	1 7 12 33	$L40$	1 8 24 26	$L53$	1 10 37 38
$L2$	0 2 24 17	$L15$	2 3 6 15	$L28$	2 19 26 4	$L41$	2 22 12 32	$L54$	2 33 29 13
$L3$	0 3 12 39	$L16$	3 5 27 20	$L29$	3 38 32 24	$L42$	3 10 26 28	$L55$	3 4 7 16
$L4$	0 5 26 34	$L17$	5 6 9 18	$L30$	5 30 28 12	$L43$	5 33 32 36	$L56$	5 24 19 13
$L5$	0 6 32 11	$L18$	6 27 35 1	$L31$	6 16 36 26	$L44$	6 4 28 21	$L57$	6 12 38 17
$L6$	0 27 28 31	$L19$	27 9 25 2	$L32$	27 13 21 32	$L45$	27 24 36 23	$L58$	27 26 30 39
$L7$	0 9 36 37	$L20$	9 35 14 3	$L33$	9 17 23 28	$L46$	9 12 21 8	$L59$	9 32 16 34
$L8$	0 35 21 29	$L21$	35 25 15 5	$L34$	35 39 8 36	$L47$	35 26 23 22	$L60$	35 28 13 11
$L9$	0 25 23 7	$L22$	25 14 20 6	$L35$	25 34 22 21	$L48$	25 32 8 10	$L61$	25 36 17 31
$L10$	0 14 8 19	$L23$	14 15 20 6	$L36$	14 11 10 23	$L49$	14 28 22 33	$L62$	14 21 39 37
$L11$	0 15 22 38	$L24$	15 20 1 9	$L37$	15 31 33 8	$L50$	15 36 10 4	$L63$	15 23 34 29
$L12$	0 20 10 30	$L25$	20 18 2 35	$L38$	20 37 4 22	$L51$	20 21 33 24	$L64$	20 8 11 7
$L13$	0 18 33 16	$L26$	18 1 3 25	$L39$	18 29 24 10	$L52$	18 23 4 12	$L65$	18 22 31 19

Line	Points	Line	Points	Line	Points	Line	Points	Line	Points
$L66$	1 11 21 31	$L79$	1 16 23 39	$L92$	1 17 19 34	$L105$	1 22 30 36	$L118$	1 28 29 32
$L67$	2 31 23 37	$L80$	2 13 8 34	$L93$	2 39 38 11	$L106$	2 10 16 21	$L119$	2 36 7 28
$L68$	3 37 8 29	$L81$	3 17 22 11	$L94$	3 34 30 31	$L107$	3 33 13 23	$L120$	3 21 19 36
$L69$	5 29 22 7	$L82$	5 39 10 31	$L95$	5 11 16 37	$L108$	5 4 17 8	$L121$	5 23 38 21
$L70$	6 7 10 19	$L83$	6 34 33 37	$L96$	6 31 13 29	$L109$	6 24 39 22	$L122$	6 8 30 23
$L71$	27 19 33 38	$L84$	27 11 4 29	$L97$	27 37 17 7	$L110$	27 12 34 10	$L123$	27 22 16 8
$L72$	9 38 4 30	$L85$	9 31 24 7	$L98$	9 29 39 19	$L111$	9 26 11 33	$L124$	9 10 13 22
$L73$	35 30 24 16	$L86$	35 37 12 19	$L99$	35 7 34 38	$L112$	35 32 31 4	$L125$	35 33 17 10
$L74$	25 16 12 13	$L87$	25 29 26 38	$L100$	25 19 11 30	$L113$	25 28 37 24	$L126$	25 4 39 33
$L75$	14 13 26 17	$L88$	14 7 32 30	$L101$	14 38 31 16	$L114$	14 36 29 12	$L127$	14 24 34 4
$L76$	15 17 32 39	$L89$	15 19 28 16	$L102$	15 30 37 13	$L115$	15 21 7 26	$L128$	15 12 11 24
$L77$	20 39 28 34	$L90$	20 38 36 13	$L103$	20 16 29 17	$L116$	20 23 19 32	$L129$	20 26 31 12
$L78$	18 34 36 11	$L91$	18 30 21 17	$L104$	18 13 7 39	$L117$	18 8 38 28	$L130$	18 32 37 26

A cyclic parallelism is just a parallelism with a base spread containing one line from each of these ten groups which is then acted on by the collineation of order 13. A computer search shows that there is no spread containing one line from each of the ten groups. This confirms Denniston's result given in the Introduction.

It is straightforward to generate by computer all the spreads and the total number of them is 8424. This number can be checked theoretically. Up to projective equivalence, it is known that there are only two types of spread in $PG(3,3)$ corresponding to the two translation planes of order 9: the Desarguesian plane and the Hall plane. We shall call the corresponding spreads type D or type H, respectively, the type D spreads being just the regular spreads. The linear translation complements in $GL(4,3)$ of the Desarguesian and Hall planes (defined via spreads in a 4-dimensional vector space over $\mathbf{F}_3$) are the group $\Gamma L(2,9)$ of order $2 \cdot 80 \cdot 72$ and a group of order $2^5.5!$ respectively (see, for example, [10, Sections 8 and 48]). In the projective context, the stabilisers of the type D and type H spreads in $PGL(4,3)$ have orders $80 \cdot 72$ and $2^4 \cdot 5!$ (since we factor out a centre of order 2). Taking the index of these stabilisers in $PGL(4,3)$ yields that there are precisely 2106 type D spreads and 6318 type H spreads. Thus, the number 8424 of spreads generated by the computer search is in agreement with this (and, incidentally, confirms that the Desarguesian and Hall planes are the only translation planes of order 9).

3 Parallelisms invariant under a collineation of order 5

We turn our attention to the problem of determining all parallelisms which are invariant under a collineation of order 5 of $PG(3,3)$. The collineation group $PGL(4,3)$ has order $2^8 \cdot 3^6 \cdot 5 \cdot 13$ so that a Sylow 5-subgroup has order 5. Denote by τ the Singer cycle of $PG(3,3)$, that is the collineation induced by multiplication by θ on $\mathbf{F}_{81}$. Then, τ^8 is a collineation of order 5. Any parallelism invariant under a collineation of order 5 is projectively equivalent to one which is invariant under τ^8.

Consider a τ^8-invariant parallelism. Since a parallelism consists of 13 spreads, at least 3 of the spreads involved must be τ^8-invariant. This turns out to be a very restrictive condition on a spread. By searching through the 8424 spreads it is found that only 9 of them are τ^8-invariant. These are:

S1	1	15	36	43	46	58	65	99	103	113
S2	1	23	43	46	47	70	93	94	103	113
S3	3	14	45	64	83	87	89	91	112	124
S4	3	17	35	45	53	54	64	75	89	112
S5	6	14	34	55	83	87	91	116	124	128
S6	6	17	34	35	53	54	55	75	116	128
S7	7	15	36	51	58	65	74	99	108	118
S8	7	23	47	51	70	74	93	94	108	118
S9	12	21	31	41	66	97	98	107	117	127

Here, the numbers denote the lines listed earlier: we write i instead of Li. We remark that the first 8 of these spreads are type H while the last is type D. In fact, the last spread is the standard regular spread obtained inside $\mathbf{F}_{81}$ by taking the subspace $\mathbf{F}_9$ and its translates. Since $\mathbf{F}_9$ is generated by θ^{10}, this 2-space is the line with points 0,10,20,30 which is $L12$ in the list of lines. The translates are the lines with points $i, 10+i, 20+i, 30+i$ and it is easily checked from the list of lines that this gives the last spread listed above (for example $L21$ is 5,15,25,35). We say that two spreads are compatible if they are disjoint (as sets of lines). The next step is to determine all triples of pairwise compatible spreads, amongst the 9 spreads which are τ^8-invariant. A computer search shows that there are precisely 36 such compatible triples and, with the numbers referring to the nine spreads listed above, they are:

1 3 6	1 3 8	1 3 9	1 4 5	1 4 8	1 4 9
1 5 8	1 5 9	1 6 8	1 6 9	1 8 9	2 3 6
2 3 7	2 3 9	2 4 5	2 4 7	2 4 9	2 5 7
2 5 9	2 6 7	2 6 9	2 7 9	3 6 7	3 6 8
3 6 9	3 7 9	3 8 9	4 5 7	4 5 8	4 5 9
4 7 9	4 8 9	5 7 9	5 8 9	6 7 9	6 8 9

Let σ denote the collineation of $PG(3,3)$ induced by the Frobenius automorphism $x \mapsto x^3$ of $\mathbf{F}_{81}$. The group $<\tau, \sigma>$ acts on the set of nine τ^8-invariant spreads and on the set of 36 compatible triples. The permutation on the nine invariant spreads induced by τ is $(1,5,7,6,8,4,2,3)$ and the permutation induced by σ is $(1,4,8,5)(2,3,7,6)$. It is easily verified that in the action of the group $<\tau, \sigma>$ on the 36 triples, there are three orbits of sizes 16,16,4 with representative triples: 1 3 6, 1 3 9 and 1 8 9.

Any τ^8-invariant parallelism must involve at least three τ^8-invariant spreads, and hence must involve exactly three such from the list of compatible triples (the number of invariant spreads in a parallelism is 3,8,13 and there is no set of eight τ^8-invariant spreads which are pairwise compatible). Thus, any τ^8-invariant parallelism consists of one of the invariant triples together with two cycles of length 5 in the action of τ^8 on the set of all spreads. The cycles of length 5 must consist of 5 spreads which are pairwise compatible. Testing each of the 8424 spreads yields that there are 417 cycles of length 5 consisting of pairwise compatible spreads. The next step is to test for each of the invariant triples whether we can add two cycles of length 5 consisting of pairwise compatible spreads to obtain a parallelism. Since we are determining parallelisms up to projective equivalence, we may assume the invariant triple is one of the three representatives for the orbits under the group $< \tau, \sigma >$. It turns out that the triple 1 3 6 cannot be extended to a parallelism, that the triple 1 3 9 can be extended in 10 ways to give a parallelism and that the triple 1 8 9 can be extended in 4 ways.

4 The parallelisms involving the triple 1 8 9

A complete search by computer yields only the following four parallelisms extending the triple 1 8 9 by the addition of two cycles of length 5 under the action of τ^8. We list the 13 spreads involved. The numbers refer to the 130 lines and we have ordered the last ten spreads in each parallelism to show the cycles of length 5. Each of the lines 1-130 occurs precisely once in each array.

PARALLELISM 1									
1	15	36	43	46	58	65	99	103	113
7	23	47	51	70	74	93	94	108	118
12	21	31	41	66	97	98	107	117	127
2	17	27	32	34	38	42	63	100	101
48	75	24	8	55	30	78	67	71	109
13	35	33	68	128	90	28	82	18	88
40	54	61	95	116	44	110	104	20	11
59	53	126	56	6	114	25	115	81	122
3	22	37	73	84	92	119	121	124	130
64	49	79	29	86	19	50	96	91	4
89	105	85	5	16	125	52	62	87	80
45	72	76	10	60	26	129	69	83	106
112	57	9	123	120	111	77	102	14	39

PARALLELISM 2									
1	15	36	43	46	58	65	99	103	113
7	23	47	51	70	74	93	94	108	118
12	21	31	41	66	97	98	107	117	127
2	18	30	68	78	101	115	116	124	126
48	20	90	95	28	109	63	6	91	27
13	61	53	28	20	32	63	64	30	109
40	100	114	32	25	11	82	55	83	33
59	71	38	8	42	122	104	128	14	61
3	22	50	54	60	85	92	121	123	130
64	49	52	53	120	76	19	96	73	4
89	105	129	17	84	9	125	62	29	80
45	72	77	75	86	37	26	69	5	106
112	57	119	35	16	79	111	102	10	39

PARALLELISM 3									
1	15	36	43	46	58	65	99	103	113
7	23	47	51	70	74	93	94	108	118
12	21	31	41	66	97	98	107	117	127
2	20	27	32	38	78	82	87	89	122
48	81	24	8	30	28	104	83	45	101
13	100	33	68	90	110	115	14	112	109
40	71	61	95	44	25	63	124	3	88
59	18	126	56	114	42	67	91	64	11
4	26	50	54	57	60	62	85	116	123
80	111	52	53	22	120	69	76	6	73
106	92	129	17	49	84	102	9	34	29
39	19	77	75	105	86	121	37	55	5
130	125	119	35	72	16	96	79	128	10

PARALLELISM 4									
1	15	36	43	46	58	65	99	103	113
7	23	47	51	70	74	93	94	108	118
12	21	31	41	66	97	98	107	117	127
2	24	35	42	71	95	104	112	114	122
48	33	54	78	18	56	115	3	38	101
13	61	53	28	20	32	63	64	30	109
40	126	17	110	81	8	67	89	90	88
59	27	75	25	100	68	82	45	44	11
4	26	37	57	62	73	84	116	119	124
80	111	79	22	69	29	86	6	50	91
106	92	85	49	102	5	16	34	52	87
39	19	76	105	121	10	60	55	129	83
130	125	9	72	96	123	120	128	77	14

In the group $< \tau, \sigma >$ of order 160, the stabiliser of the triple 1 8 9 is the subgroup $< \tau^4, \tau\sigma >$ of order 40. Note that $(\tau\sigma)^2 = \tau^{28}\sigma^2$ and $(\tau\sigma)^4 = 1$. The subgroup $< \tau^4, \tau\sigma >$ acts on the set of parallelisms extending the triple 1 8 9. It fact, both τ^4 and σ^2 interchange the first and third parallelisms as well as interchanging the second and fourth, while $\tau\sigma$ interchanges the first parallelism and the second. Thus:

- all four of the parallelisms listed above are projectively equivalent.

Of the two $< \tau^8 >$-orbits of length five which complete the triple 1 8 9 to a parallelism the first orbit listed (in each parallelism) consists of type D spreads and the second orbit consists of type H spreads. Thus, the unique parallelism (up to projective equivalence) extending the triple 1 8 9 consists of 6 type D orbits and 7 type H orbits. We say it is of type (6,7).

5 The parallelisms involving the triple 1 3 9

A complete search by computer yields only ten parallelisms extending the triple 1 3 9 by the addition of two cycles of length 5 under the action of τ^8. In the group $< \tau, \sigma >$, the stabiliser of the triple 1 3 9 is the subgroup $< \tau^8, \tau^4\sigma^2 >$ of order 10. This subgroup acts on the ten parallelisms extending the triple. There are 6 orbits of this action and the parallelisms listed below are representatives for these orbits. The full set of ten extensions is obtained by the addition of the images under $\tau^4\sigma^2$ of the second, third, fourth and sixth parallelisms.

PARALLELISM 1									
1	15	36	43	46	58	65	99	103	113
3	14	45	64	83	87	89	91	112	124
12	21	31	41	66	97	98	107	117	127
2	17	27	32	34	38	42	63	100	101
48	75	24	8	55	30	78	67	71	109
13	35	33	68	128	90	28	82	18	88
40	54	61	95	116	44	110	104	20	11
59	53	126	56	6	114	25	115	81	122
4	22	52	85	93	102	118	120	123	125
80	49	129	76	70	121	7	84	73	26
106	105	77	9	23	96	108	86	29	111
39	72	119	37	47	62	74	16	5	92
130	57	50	79	94	69	51	60	10	19

PARALLELISM 2									
1	15	36	43	46	58	65	99	103	113
3	14	45	64	83	87	89	91	112	124
12	21	31	41	66	97	98	107	117	127
2	17	60	68	101	105	110	115	116	126
48	75	120	95	109	72	25	63	6	27
13	35	84	56	88	57	42	67	34	24
40	54	86	32	11	22	78	82	55	33
59	53	16	8	122	49	28	104	128	61
4	19	37	52	62	70	73	81	90	118
80	125	79	129	69	23	29	100	44	7
106	26	85	77	102	47	5	71	114	108
39	111	76	119	121	94	10	18	38	74
130	92	9	50	96	93	123	20	30	51

PARALLELISM 3									
1	15	36	43	46	58	65	99	103	113
3	14	45	64	83	87	89	91	112	124
12	21	31	41	66	97	98	107	117	127
2	18	30	35	50	68	101	104	111	116
48	20	90	54	52	95	109	115	92	6
13	81	44	53	129	56	88	63	19	34
40	100	114	17	77	32	11	67	125	55
59	71	38	75	119	8	122	82	26	128
4	25	29	33	37	62	70	74	84	105
80	42	5	61	79	69	23	51	86	72
106	78	10	126	85	102	47	118	16	57
39	28	123	27	76	121	94	7	60	22
130	110	73	24	9	96	93	108	120	49

PARALLELISM 4									
1	15	36	43	46	58	65	99	103	113
3	14	45	64	83	87	89	91	112	124
12	21	31	41	66	97	98	107	117	127
2	18	35	37	42	72	95	104	114	116
48	20	54	79	78	57	56	115	38	6
13	81	53	85	28	22	32	63	30	34
40	100	17	76	110	49	8	67	90	55
59	71	75	9	25	105	68	82	44	128
4	26	33	50	51	86	88	93	96	123
80	111	61	52	118	16	11	70	62	73
106	92	126	129	7	60	122	23	69	29
39	19	27	77	108	120	101	47	102	5
130	125	24	119	74	84	109	94	121	10

PARALLELISM 5									
1	15	36	43	46	58	65	99	103	113
3	14	45	64	83	87	89	91	112	124
12	21	31	41	66	97	98	107	117	127
2	20	37	70	74	77	84	105	121	130
48	81	79	23	51	119	86	72	96	4
13	100	85	47	118	50	16	57	62	80
40	71	76	94	7	52	60	22	69	106
59	18	9	93	108	129	120	49	102	39
5	26	33	34	38	54	56	101	110	115
10	111	61	55	30	53	32	109	25	63
123	92	126	128	90	17	8	88	42	67
73	19	27	116	44	75	68	11	78	82
29	125	24	6	114	35	95	122	28	104

PARALLELISM 6									
1	15	36	43	46	58	65	99	103	113
3	14	45	64	83	87	89	91	112	124
12	21	31	41	66	97	98	107	117	127
2	24	44	47	68	71	74	78	82	88
48	33	114	94	95	18	51	28	104	11
13	61	38	93	56	20	118	110	115	122
40	126	30	70	32	81	7	25	63	101
59	27	90	23	8	100	108	42	67	109
4	26	50	54	57	60	62	85	116	123
80	111	52	53	22	120	69	76	6	73
106	92	129	17	49	84	102	9	34	29
39	19	77	75	105	86	121	37	55	5
130	125	119	35	72	16	96	79	128	10

In the first parallelism, all the spreads extending the triple 1 3 9 are of type D, whereas in the other five parallelisms the additional spreads are of type H. Thus, the first parallelism is of type (11,2) whereas the other five are of type (1,12).

For any parallelism, we define a parameter λ to be the number of spreads meeting each constituent spread of the parallelism in at most one line (this parameter is of interest in the context of searching for an orthogonal mate to the given parallelism). Clearly, projectively equivalent parallelisms have the same value of λ. The value of λ in the six cases are respectively: $\lambda = 30, 30, 30, 20, 25, 360$.

These invariants (*type* and λ) show that the only possible projective equivalence amongst these six parallelisms is between the second and third. We now give an argument to show that these two parallelisms are not equivalent.

Suppose that β is a collineation of $PG(3,3)$ mapping the second parallelism to the third. Then, β fixes the unique type D spread (the third one, S9) in each parallelism. It is straightforward to determine, by computer, the orbit of the spread S1 under the action of the stabiliser of S9 (we already know that the collineations τ and σ fix S9 and it is not difficult to find an extra collineation which, together with τ and σ, generates the full stabiliser of S9). It turns out that this orbit has length 48 and contains S1 and S3 but none of the other spreads in the second and third parallelisms. Thus, β must leave the pair $\{S1, S3\}$ invariant. It is also easy to generate, by computer, the orbit of the ordered pair (S1,S3) under the action of the stabiliser of S9. It turns out that this orbit has length 576. Since the stabiliser of S9 has order 5760, the subgroup fixing each of S1,S3 has order 10 and so must be $< \tau^8, \tau^4\sigma^2 >$. Moreover, the ordered pair (S3,S1) is not one of the 576 pairs in the orbit of (S1,S3). This shows that $\beta \in < \tau^8, \tau^4\sigma^2 >$ and, since no element of $< \tau^8, \tau^4\sigma^2 >$ maps the second parallelism to the third, this is a contradiction.

We have established:

- the above six parallelisms are projectively inequivalent.

Together with the results of the previous section, this completes the proof of the theorem stated in the Introduction. Note that the unique parallelism (up to projective equivalence) obtained in the previous section, being of type (6,7), is not projectively equivalent to any of the six parallelisms of this section.

6 Concluding Remarks

The first parallelism of Section 5 is of particular interest since it is of type (11,2), so that all but two of the spreads involved are regular. The author is investigating whether this parallelism could give rise to a translation plane of order 3^4 via a modification of the construction due to Walker, which applies only to regular parallelisms (see the Introduction).

Although there are no regular parallelisms of $PG(3,3)$, in other words no parallelisms of type (13,0), at the other extreme, there are parallelisms of type (0,13), consisting completely of type H spreads (see [11]).

References

[1] A. Beutelspacher, On parallelisms in finite projective spaces, *Geom. Dedicata* **3** (1974), 35-40.

[2] R.H.Bruck, Construction problems of finite projective planes, *Combinatorial Mathematics and its Applications* (Univ. of North Carolina at Chapell Hill, 1967), University of North Carolina Press, 1969, 426-514.

[3] R.H.F. Denniston, Some packings of projective spaces, *Rend. Accad. Naz. Lincei* **52** (1972) 36-40.

[4] R.H.F. Denniston, Packings of $PG(3,q)$, *Finite Geometric Structures and their Applications* (Bressanone, 1972), Cremonese, 1973, 195-199.

[5] R.H.F. Denniston, Cyclic packings of the projective space of order 8, *Rend. Accad. Naz. Lincei* **54** (1973) 373-377.

[6] J.W. Freeman, Reguli and pseudo-reguli of $PG(3,q)$, *Geom. Dedicata* **9** (1980) 267-280.

[7] J.W.P. Hirschfeld, *Finite Projective Spaces of Three Dimensions*, Oxford University Press, Oxford,1985.

[8] V. Jha and N.L. Johnson, Regular parallelisms from translation planes, *Discrete Math.* **59** (1986) 91-97.

[9] G. Lunardon, On regular parallelisms in $PG(3,q)$, *Discrete Math.* **51** (1984) 229-235.

[10] H.Lüneburg, *Translation Planes*, Springer-Verlag, 1980.

[11] A.R. Prince, Uniform parallelisms of $PG(3,3)$, in preparation.

[12] D.R. Stinson and S.A. Vanstone, Orthogonal packings in PG(5,2) *Proc. Winnipeg Conference* (1984).

Classical Spaces That Interact with Spreads

F. A. SHERK Department of Mathematics, University of Toronto, Toronto, Canada M5S 1A1

To T. G. Ostrom on his Eightieth Birthday

1 INTRODUCTION

For the past thirty years, a finite translation plane of dimension $t + 1$ over its kernel has usually been described in terms of the t -spread in $PG(2t+1, q)$ that characterizes it [**9**, pp 219-226]. The t - spread - a set of q^{t+1} t - dimensional subspaces that partition $PG(2t+1, q)$ - is best presented in terms of a *spread set* $\{M_i\}$ of q^{t+1} $(t+1)\times(t+1)$ matrices over $GF(q)$, with the property that $M_j - M_i$ is non-singular for all $i, j \neq i$.

The non-Desarguesian translation plane is therefore intimately linked with a Desarguesian, or *classical* space (for ease of reference, we shall define a *classical* space to be any geometrical space that can be coordinatized by a field). A spread in $PG(2t+1, q)$ induces (and in some cases is induced by) other incidence structures in other classical spaces. Study of these incidence structures often reveals properties of spreads that are not easily seen when attention is confined to spreads alone, or to the translation planes that arise from the spreads.

In this paper, we confine our attention to 1- spreads, hereafter simply called spreads. A *spread* is a set of q^2+1 lines partitioning the projective 3-space $\Sigma \cong PG(3,q)$, where q is any prime or prime power.

Given any spread S in Σ, we concentrate on its interaction with a regular spread $\mathfrak{R}$ which has the largest intersection with S. Spread S induces an indicator set I in a plane $\pi \cong AG(2,q^2)$, and finally I induces a design W in $E \cong EG(2,q)$, whose blocks are circles, called a *weighted configuration*.

What is proposed here is a simultaneous study of S in Σ, I in π, and W in E. We refer to work in the literature that deals with special cases of this general study, and prove a few new results. We also propose (in Section 3) a classification of spreads $\{S\}$ that involves structural properties of I and W.

In Section 2 we connect the designs S, I, and W by embedding all in $PG(3,q^2)$. Since the ideas are all in the literature, we often curtail or suppress proofs. In Section 4 we prove a few results about the structure of S, I, and W. Finally, in Section 5, we illustrate the new classification in the case $q=5$.

Throughout, we let F be a field $\cong GF(q)$. The field $K \cong GF(q^2)$ is a quadratic extension of F.

The definitions of $PG(3,q)$ and $AG(2,q^2)$ are well-known. $EG(2,q)$ is an incidence structure whose points $\{z\}$ are the elements of K. The set G of mappings $\{z' = rz + c \mid r, c \in K,\ r \neq 0\}$ is a group of point transformations. The *line* $\ell_{c,r}$ of $EG(2,q)$ is the set $\{r\lambda + c \mid \lambda \in F\}$; with this definition of line, it is clear that $EG(2,q) \cong AG(2,q)$. The *circle* $c(\rho)$ of $EG(2,q)$ is the set $\{ru^i + c \mid u^{i(q+1)} = 1\}$, where $\rho = r^{q+1}$; $c(\rho)$ has *centre* c and *radius* ρ.

If we add an ideal point, ∞, to K, and extend G to the group of linear fractional transformations $\{z' = (az+b)/(cz+d) \mid a,b,c,d \in K,\ ad \neq bc\}$, we get the (Miquelian) *inversive plane* $M(q)$ [**6**, p 449, where $M(q)$ is identified as $IP(q)$].

Models of $EG(2,q)$ and $M(q)$ that we use are respectively the affine and projective lines over K.

2 THE CONTEXT OF S, I, AND W

Let S be any spread in $\Sigma \cong PG(3,q)$. Any three mutually skew lines a, b, c in Σ generate a *regulus* $R(a,b,c)$, whose *opposite regulus* $R'(a,b,c)$ is the set of $q+1$

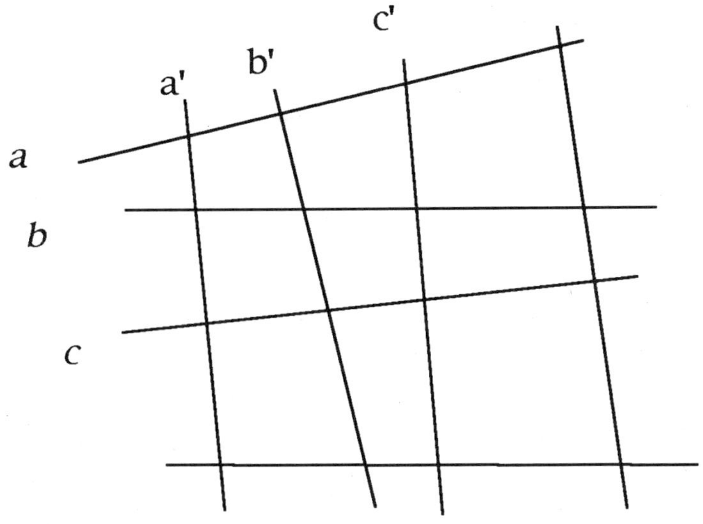

Figure 2.1 The regulus - opposite regulus configuration.

lines meeting a, b and c. $R(a, b, c)$ is the set of $q+1$ lines, each of which meets every line of $R'(a, b, c)$ [**6**, p 435].

A *regular spread* in Σ is a spread which contains the entire regulus $R(a, b, c)$ whenever it contains a, b and c.

LEMMA 2.1 [**6**, p 436] Any four mutually skew lines which do not all belong to one regulus generate a unique regular spread.

We are interested in the intersection (as a set of lines) of S with any regular spread in Σ. Let $\{\mathfrak{R}_i\}$ be the complete set of regular spreads in Σ, and let $m = \max_{\forall i} |S \cap \mathfrak{R}_i|$.

DEFINITION 2.1 The *deficiency* $d = d(S)$ of S is $q^2 + 1 - m$.

Deficiency provides a rough classification of spreads. The following facts are well-known:

LEMMA 2.2 a) $d = 0$ if and only if S is regular.

b) $d = q + 1$ if and only if S is subregular of index 1 [**6**, p 442].

c) $d \leq q^2 - 3$ (immediate corollary of Lemma 2.1).

d) If $d > 0$, then $d \geq q+1$.

Thus S is classified according to d, and the potential classes are

$$\{d\} = \{0,\ q+1,\ q+2,\ \ldots,\ q^2-3\}.$$

Analysis of I and W sharpens this classification. See also [**13**, p 41].

Now given S with deficiency d, fix a regular spread $\mathfrak{R}$ in Σ, whose intersection with S is maximum ($|S \cap \mathfrak{R}| = q^2+1-d$). Fix three lines in $S \cap \mathfrak{R}$, labelling them $\ell_\infty, \ell_0, \ell_1$. Introduce coordinates to Σ in such a way that $\ell_\infty = \{(0\ Y)\}$, $\ell_0 = \{(X\ 0)\}$, and $\ell_1 = \{(X\ X)\}$, where $0, X, Y$ are the 1×2 matrices $(0\,0)$, $(x_1\ x_2)$, $(y_1\ y_2)$ $(x_1, x_2, y_1, y_2 \in F)$. [13, p 49]. Then any line of Σ which is skew to ℓ_∞ is $\{(X\ XM)\}$, where M is a 2×2 matrix over F. For example, S contains ℓ_∞ and $\{(X\ XM_i)\}$ $(0 \leq i < q^2)$, where $\{M_i\}$ is the spread set for S. The spread set for $\mathfrak{R}$ is $\{0, A, A^2, \ldots, A^{q^2-1} = I\}$, a representation for the field K.

When F is extended to K, Σ is extended to $\Sigma^* \cong PG(3, q^2)$. The lines of Σ^* are partitioned into three classes:

a) extended lines of Σ, henceforth called simply lines of Σ.

b) lines containing exactly one point of Σ.

c) lines containing no point of Σ.

An example of a line of class a) is the line ℓ_∞ (Figure 2.2). A relevant classical theorem, recognized by Bruck as important in this context, states:

THEOREM 2.1 [**6**. p 439] Through each point of a line of class c) there passes exactly one line of class a); the set of q^2+1 lines of Σ thus defined form a regular spread. Conversely, any regular spread in Σ corresponds to two lines of class c) in Σ^* which define it.

Let a be one of the lines that define $\mathfrak{R}$. Then a meets all q^2+1-d lines of $S \cap \mathfrak{R}$, including ℓ_∞. Let π' be the plane of Σ^* spanned by a and ℓ_∞, and let π be the affine plane $\pi' \setminus \ell_\infty$. The indicator set I is the set of q^2 points of intersection of S with π. It partitions into two subsets: I' (the points of I on a) and I'' (the points of I not on a); $|I''| = d$, the deficiency of S.

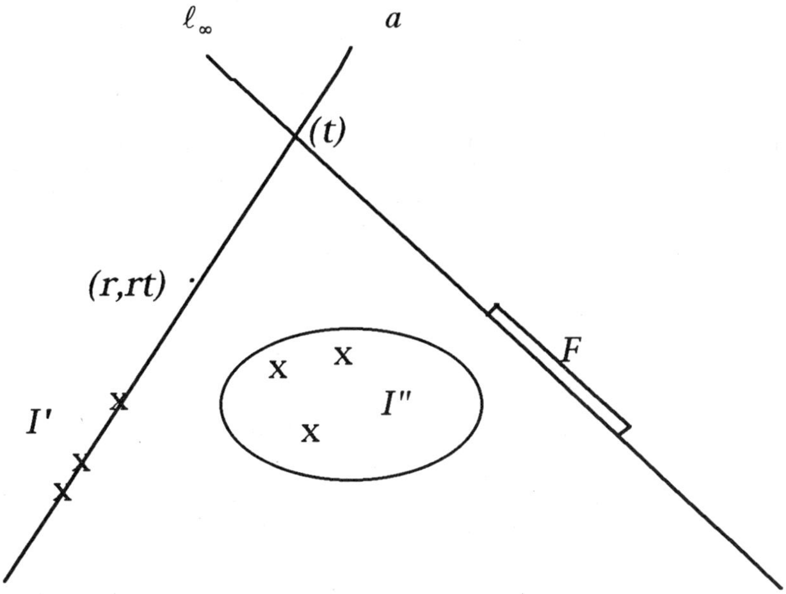

Figure 2.2 The indicator plane

Now the coordinates chosen for Σ and extended to Σ^* impose coordinates (x, y) $(x, y \in K)$ on each point of π. Line $x = 0$ meets ℓ_∞ in a point which we label (∞), and any other line, $y = xk$, of slope k, meets ℓ_∞ in point (k). In parrticular, a contains $(0, 0)$, and has equation $y = xt$ for some $t \in K \setminus F$.

The characteristic property of an indicator set is that the join of any two of its points has slope in $K \setminus F$ [**16**, pp 133, 134]. Using standard vector geometry, it is not hard to prove

THEOREM 2.2 Any regular spread of Σ which contains ℓ_∞ is indicated in π either as a line with slope in $K \setminus F$, or as a subplane $\cong AG(2, q)$, all of whose slopes are in $K \setminus F$.

Let $P \in I''$, and let C_P be the set of $q+1$ points of a which, when joined to P, yield a line whose slope is in $\{\infty\} \cup F$. Since no point of C_P can be in I, we refer to C_P as the set of points of a which are *annihilated* by P. In the Bruck approach [**6**, p 492], two lemmas now readily follow:

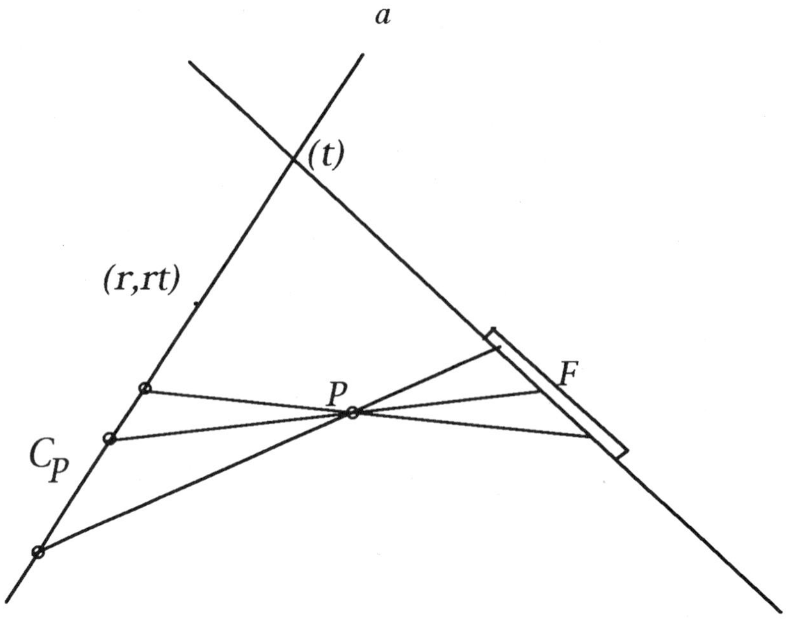

Figure 2.3 Points C_P annihilated by P

LEMMA 2.3 C_P indicates a regulus in the regular spread $\mathfrak{R}$, and P is a point of the set C'_P which indicates the opposite regulus.

LEMMA 2.4 Interpreting line a as E, a model of $EG(2, q)$, C_P is a circle $c(\rho)$ in E. Also, C'_P lies on the line of slope t^q through (c, ct).

The proofs are straightforward applications of the analytic geometry of π.

We conclude from Lemma 2.4 that I'', the set of d points of I which are not on a, induces a design W in E whose blocks are the circles annihilated by points of I''. The same circle C_P of W is annihilated by points P *and* Q of I'' if and only if Q lies in C'_P.

DEFINITION 2.2 The *weight* of a circle $C \in W$, denoted by w or $w(C)$, is the integer $|C' \cap I''|$, where C' indicates the opposite regulus to that indicated by C. Thus each circle of W has a weight w, and $1 \leq w \leq q+1$.

DEFINITION 2.3 The design W is called a *configuration of weighted circles*, or simply a *weighted configuration*.

Let X be any point of W. Then X is a common point of $v = v(X)$ circles

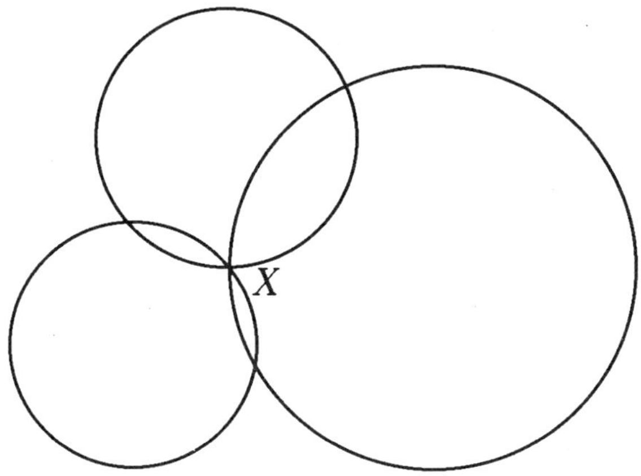

Figure 2.4 A point of valency 3

$C_1, \ldots, C_v$ of W. The integer v is called the *valency* of X. It is readily seen that $1 \le v \le q+1$, that $\sum_{i=1}^{v} w(C_i) = q+1$, and that the total number of vertices in W is d, the deficiency of S.

Counting involving weights and valencies establishes

THEOREM 2.3 Let W be the weighted configuration induced in E by the indicator set I. Let d_v (b_w) be the number of vertices (circles) in W of valency v (weight w). Let d (b) be the total number of vertices (circles) in W. Then

$$a)\sum_{v=1}^{q+1} d_v = d \qquad b)\sum_{w=1}^{q+1} b_w = b$$

$$c)\sum_{w=1}^{q+1} w b_w = d \qquad d)\sum_{v=1}^{q+1} v d_v = b(q+1)$$

(note the preceding remark that d is both the number of vertices and the deficiency of S).

Theorem 2.3, in a more general context, is Theorem 2.1 of [**15**, p 161].

Although weighted configurations appear as such explicitly only in [**15**], there are examples elsewhere in the literature. The sets of disjoint circles in $M(q)$ $(IP(q))$ that define subregular spreads [**6**, p 492] are weighted configurations in which every vertex has valency 1 (and every block has weight q+1). The Bruen chain [**7**], and the more general t - nests of Baker and Ebert [**1**, **2**, **3**, **4**, **5**, **10**, **16**], are essentially weighted configurations in which every vertex has valency 2 and every block has weight $(q+1)/2$. Put another way, we may claim that the approach to spreads *via* weighted configurations is a generalization of all the above work. We would claim also that the intricacies of t - nests are explained more naturally in the context of circles in E rather than of reguli in $\mathfrak{R}$.

3 CLASSIFICATION OF SPREADS BY WEIGHTED CIRCLES

We have defined W in E as a configuration induced by I in π, which in turn is induced by S in Σ. Using the characteristic property of indicator sets in π (above, Section 2), it is easy to define I independently of S, and to induce S from I. Also, W can be defined independently of I. However, the most general definition of a weighted configuration [**15**] involves only the parameters of weights and valencies; while such a design satisfies Theorem 2.3, it may not be realizable in $E \cong EG(2,q)$, or in $M(q)$ [**15**, pp 161, 162]. Even if it is realized in E, it may not be induced by an indicator set [**15**, pp 169 - 171]. Nevertheless, a classification of S (or of I) based on an independent classification of weighted configurations is worthwhile, if only because the number of vertices in W is less than the number of lines in S.

Given a weighted configuration W in E, one must look for an indicator set I in π which induces W. As noted above, I may not exist. Also, if I does exist, it may not be unique (in terms of defining just one translation plane). The process of finding I from W is a replacement process, generalizing the process of derivation [**9**, pp 223 - 6, **12**, **13**, pp 8 - 21, **14**].

REPLACEMENT IN S *The d lines of $\mathfrak{R}$ indicated by the vertices of W are replaced by d other lines. If W contains a circle C of weight w, then w of the replacement lines lie in the regulus which is opposite to the regulus indicated by C.*

The formula given above is unambiguous only when $w = q + 1$. In other cases, one must experiment with the set C' of $q + 1$ points of π indicating the opposite regulus (Lemma 2.3). The choice of w points of C' then must be consistent with other choices required by other circles which are blocks of W.

Finally, we note that, as in any other classification scheme, there is the isomorphism problem. Two equivalent spreads (spreads which define isomorphic translation planes) may induce different weighted configurations, and it may even be that two non-equivalent spreads would induce the same weighted configuration. We consistently narrow the choices here by tying the induced indicator sets and weighted configurations to the deficiency of S. Nevertheless, a great deal of freedom remains. In any case, when we do address the isomorphism problem it is easier, because we seek isomorphisms among spreads with a common deficiency.

4. PROPERTIES OF S, I, AND W

In this section we prove some new results by simultaneously considering S in Σ, I in π, and W in E.

DEFINITION 4.1 A *regular indicator set* in π is the indicator set of a regular spread containing ℓ_∞.

Theorem 2.2 states that a regular indicator set is either a line with slope in $K \setminus F$ or a subplane, all of whose slopes are in $K \setminus F$. Either type is generated by three of its points: $\{A, B, C\}$, with the property that the vectors $B - A$, $C - A$ are linearly independent. $B - A$ and $C - A$ are linearly dependent if and only if A, B, C belong to a subline in π.

LEMMA 4.1 Any three points of I lie in at least one regular indicator set. If the three points are not in one subline, then they lie in exactly one regular indicator set.

Proof: Let A, B, C be any three points of I, indicating lines a, b, c of Σ. By Lemma 2.1, a, b, c and ℓ_∞ all lie in a regular spread.

Another way to state Lemma 4.1 is that if A, B, C are three non-collinear points of I, then the subplane of π generated by A, B, C has no slope from the set $\{\infty\} \cup F$.

LEMMA 4.2 There are $q^2(q-1)^2/2$ regular indicator sets containing point $O = (0,0)$.

Proof: π contains $q^2 - q = q(q-1)$ lines through O with slope in $K \setminus F$. As for subplanes, when ℓ_∞ is interpreted as $M(q)$, the set of slopes of any subplane forms a circle. Therefore the set of slopes of any subplane which is a regular indicator set forms a circle that is disjoint from the circle $\{\infty\} \cup F$. We calculate that the number of such circles is $q(q-1)(q-2)/2$ (whether q is even or odd). There are $q+1$ subplanes with a given slope set and containing O, so the total number of regular indicator sets containing O is $q(q-1) + (q+1)q(q-1)(q-2)/2 = q^2(q-1)^2/2$.

THEOREM 4.1 Let d be the deficiency of S. Further to the properties listed in Lemma 2.2, we also have

$$d \le q^2 - 4.$$

Proof: From Lemma 2.2 we have $d \le q^2 - 3$. Suppose that $d = q^2 - 3$. Then I intersects every regular indicator set of π in 3 points. Consider any triad $\{O, P, Q\}$ of points in I that contains $O = (0,0)$. If the two points P, Q, as vectors, were linearly dependent, then $\{O, P, Q\}$ would be a subline, and yet another point R of I could be added to some regular indicator set containing $\{O, P, Q\}$, contradicting the assumption $d = q^2 - 3$. It follows from Lemma 4.1 that the triad $\{O, P, Q\}$ generates a unique regular indicator set. There are $\binom{q^2}{2} = q^2(q^2-1)/2$ different triads $\{O, P, Q\}$. Since $q^2(q^2-1)/2 > q^2(q-1)^2/2$ (cf. Lemma 4.2), at least two different triads must generate the same regular indicator set. Once again, this contradicts the assumption that $d = q^2 - 3$.

THEOREM 4.2 Suppose that W is induced by I, and that W contains a vertex X of valency 2. Let $C_1 = c_1(\rho_1)$ and $C_2 = c_2(\rho_2)$ be the two blocks of W meeting at X, and let $w(C_1) = w$. If C_1 and C_2 intersect (rather than touch at X), then

$$1 < (w, q+1) < q+1,$$

where $(w, q+1)$ is the g.c.d. of w and $q+1$.

Proof: We must examine the interaction between C'_1 and C'_2, indicating the opposite reguli (cf. Lemma 2.3), and containing w and $q+1-w$ points of I'' respectively. Since C_j $(j = 1, 2)$ is the set of points

$$\left\{\left(c_j + d_j u^i,\ c_j t + d_j u^i t\right)\right\} \qquad (u^{i(1+q)} = 1,\ \rho_j = d_j^{1+q}),$$

we calculate, using Lemma 2.4, that C'_j is the set

$$\left\{P_{j,i}\right\} = \left\{\left(c_j + d_j u^i,\ c_j t + d_j u^i t^q\right)\right\}.$$

Vertex X, being a common point to C_1 and C_2, has the property that each line joining X to a point of C'_j has slope from $\{\infty\} \cup F$.

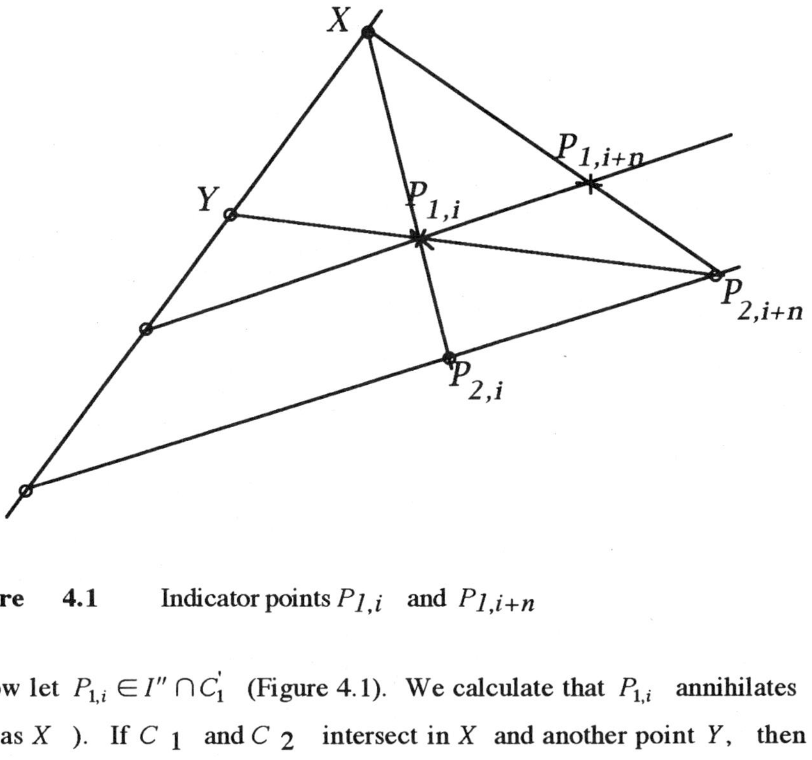

Figure 4.1 Indicator points $P_{1,i}$ and $P_{1,i+n}$

Now let $P_{1,i} \in I'' \cap C'_1$ (Figure 4.1). We calculate that $P_{1,i}$ annihilates $P_{2,i}$ (as well as X). If C_1 and C_2 intersect in X and another point Y, then $P_{1,i}$ also annihilates Y and a point $P_{2,i+n}$ of C'_2. Thus the choice of $P_{1,i}$ in I'' eliminates $P_{2,i}$ and $P_{2,i+n}$ as candidates for points of I'', and forces $P_{1,i+n}$ to be in I''.

By repetition, we produce k points $\left\{P_{1,i}, P_{1,i+n}, P_{1,i+2n}, \ldots\right\}$ in $I'' \cap C'_1$, where $P_{1,i+kn} = P_{1,i}$ and $q+1 = |C'_1| = rk$ for some integer r. Also, $w = sk$ for some $s < r$. Since C'_2 must contain at least one point of I, $1 < w < q+1$, and k is a proper divisor of $q+1$; hence $1 < (w, q+1) < q+1$.

COROLLARY 4.2 If two circles C_1 and C_2 of W, with weights w and $q+1-w$, meet in a point X of valency 2, and if $(w, q+1) = 1$, then C_1 and C_2 touch at X.

Theorem 4.2 illustrates one property that a weighted configuration W must have in order to be induced by an indicator set. In contrast to the condition imposed by Theorem 4.1, it is a geometrical, rather than combinatorial, condition. For the case q odd, at least, it is equivalent to a restriction on the angle at which C_1 and C_2 meet, and also on the radii ρ_1 and ρ_2. It is clear that a much fuller study of the geometrical properties of W, and its interaction with I, will enlarge our understanding of spreads.

5. SPREADS IN $PG(3,5)$

To illustrate the preceding, we restrict our attention to the case $q = 5$. Czerwinski and Oakden **[8]** have listed all translation planes of order 25, finding 22 isomorphism classes. For each plane, a spread set is given, and it is instructive to examine indicator sets and weighted configurations that are induced by these spread sets. This author has spent much time in this exercise, finding several interesting weighted configurations which have supported or destroyed conjectures. But it is difficult, without machine calculations, to find the deficiency of any given spread. As a result, some of the examples of W found by this method are not "canonical", in the sense that $|W|$ is the deficiency of S.

Previous to examining weighted configurations induced by spreads in **[8]**, this author constructed several as designs in $E \cong EG(2,5)$, and found most of these to be induced by indicator sets in $\pi \cong AG(2,25)$. Besides the empty set (induced by the regular spread) and sets of disjoint circles (induced by subregular spreads), some W with $d < 16$ were:

W_1	$(d = d_2 = 12,\ \ b = b_3 = 4)$	(Figure 5.1)
W_2	$(d = 14, d_2 = 12, d_3 = 2,\ \ b = 5,\ b_2 = 3,\ b_4 = 2)$	(Figure 5.2)
W_3	$(d = d_2 = 15,\ \ b = b_3 = 5)$	(Figure 5.3)

W_1 and W_3, being weighted configurations with $d_2 = d$, are equivalent to t-nests **[1, 2, 3, 4, 5, 7, 10, 16]**. The design W_1 is particularly interesting, since, even though it is certainly induced by a spread, I was unable to identify the spread in [**8**] that induces it. Assuming that the list in [**8**] is complete, my failure illustrates the difficulty in determining the deficiency of a given spread by hand calculation.

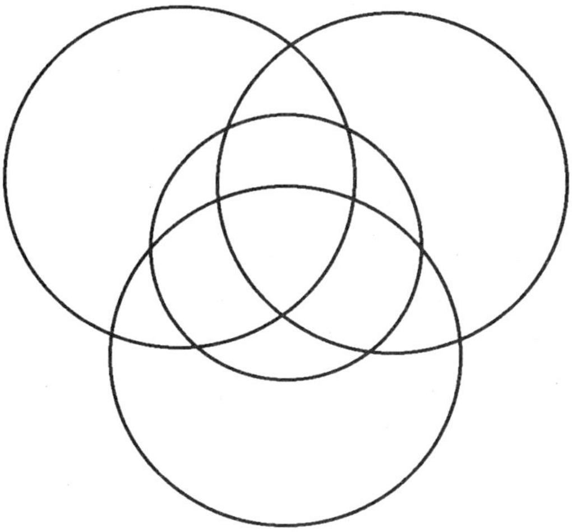

Figure 5.1 W_1

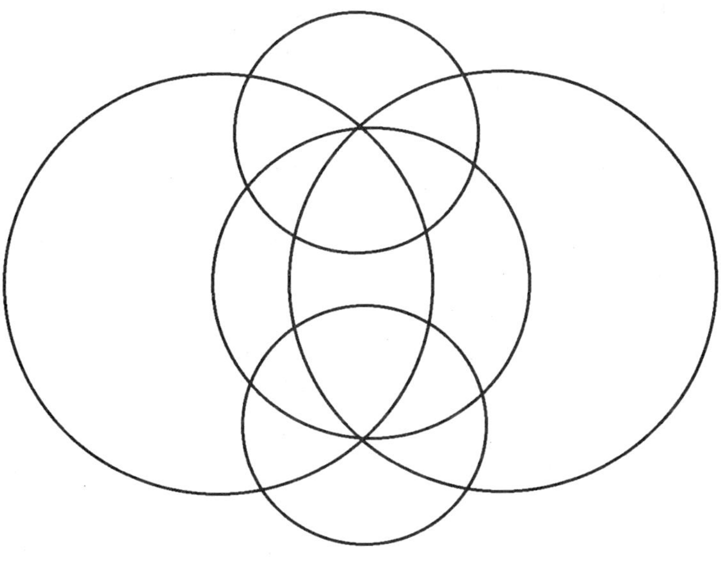

Figure 5.2 W_2

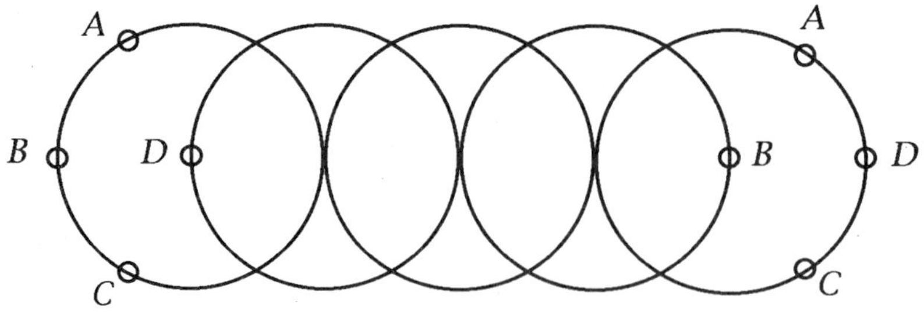

Figure 5.3 W_3

It may well be better, if one wishes to construct a complete list of spreads in $\Sigma \cong PG(3,q)$, to use the classification suggested here. Thus one would first classify the spreads according to their deficiencies, and then try to identify subclasses with the same deficiency. The approach would be through weighted configurations.

For $q = 5$, we have from Theorem 4.1 that $0 \le d \le 21$. From Lemma 2.2 and an easy extension thereof, we have that if $0 < d \le 12$, then $d = 6$ or 12. Trial and error shows that $d \ne 13$. Thus we have the potential classes

$$\{d\} = \{0, 6, 12, 14, 15, 16, 17, 18, 19, 20, 21\},$$

which we consider one by one. Theorem 2.3 takes care of the combinatorial properties that W muat satisfy, and there are geometrical conditions in E and π that W and I must conform to. We also know that if S contains a regulus, then $d \le 18$, since, by Lemma 2.1, $|S \cap \mathfrak{R}| \ge 7$ for some regular spread $\mathfrak{R}$.

Analysis of [**8**] suggests that there are more spreads in the class $d = 18$ than in any other, and that classes $d = 17$ and $d = 21$ are empty. Since $d < 21$ when $q = 5$, it is possible that the inequality of Theorem 4.1 can be sharpened.

REFERENCES

1. R. D. Baker and G. L. Ebert, Nests of size $q+1$ and another family of translation planes, *J. London Math. Soc.* (2) 38 (1988) 341 - 355.

2. ________________________, A new class of translation planes, *Annals of Discrete Math.* 37 (1988) 7 - 20.

3. ____________________, A Bruen chain for $q = 19$, *Designs, Codes and Cryptography* 4 (1994) 307 -312.

4. ____________________, Filling the nest gaps, *Finite Fields and their Applications* 2 (1996) 42 - 61.

5. R. D. Baker, G. L. Ebert and R. Weida, Another look at Bruen chains, *J. Comb. Theory, Ser. A* 48 (1988) 77 - 90.

6. R. H. Bruck, Construction problems of finite projective planes, Chapter 27 in *Combinatorial Mathematics and its Applications* (T. A. Dowling, Ed.) Univ. North Carolina Press, Chapel Hill (1969) 426 - 514.

7. A. A. Bruen, Inversive geometry and some translation planes I, *Geom. Ded.* 7 (1977) 81 - 98.

8. Terry Czerwinski and David Oakden, The translation planes of order twenty-five, *J. Comb. Theory, Ser. A* 59 (1992) 193 - 217.

9. P. Dembowski, Finite Geometries, Springer - Verlag, Berlin, 1968.

10. G. L. Ebert, Spreads admitting regular elliptic covers, *Europ. J. Combinatorics* (1989) 10 319 - 330.

11. Rudolph Mathon and Gordon F. Royle, The translation planes of order 49, *Designs, Codes and Cryptography* 5 (1995) 57 - 72.

12. T. G. Ostrom, Vector spaces and construction of finite projective planes, *Arch. Math.* 19 (1968) 1 - 25.

13. __________, Finite translation planes, *Lecture Notes in Mathematics* 158 Springer - Verlag, Berlin 1968.

14. __________, Classification of finite translation planes, *Proc. Conf. on Projective Planes, Washington State University*, Wash. State Univ. Press 1973 195 - 213.

15. F. A. Sherk, Configurations of weighted circles in finite inversive planes, *J. of Geometry* 45 (1992) 158 -173.

16. F. A. Sherk and Günther Pabst, Indicator sets, reguli, and a new class of spreads, *Canadian J. Math.* 29 (1977) 132 -154.

Coset Representation of Homogeneous Geometrical Structures

THOMAS Q. SIBLEY Mathematics Department, St. John's University, Collegeville, MN

A congruence relation provides a minimal geometric structure for describing metrical, linear and other relationships. The homogeneous congruence spaces in this paper include a variety of metric spaces, affine spaces and all vertex transitive polyhedra. We show how to obtain all homogeneous congruence spaces using cosets of groups derived from these spaces. Chen and Teh [1] also use cosets of groups to construct related graphs, but their class of structures is different.

DEFINITIONS. A *congruence space* $(S,\equiv)$ is a set S together with a *congruence* (or *equidistance*) relation $\equiv$ on ordered pairs of points. That is, I) $\equiv$ is an equivalence relation and ii) for all $a,b,c \in S$, $ab \equiv ba$ and ($aa \equiv bc$ iff $b = c$). A non-empty subset T of an congruence space $(S,\equiv)$ is a *subspace* iff for $a, b, c \in T$ and $d \in S$ if $ab \equiv cd$, then $d \in T$.

EXAMPLE 1. Consider Figures 1 and 2. (As in all figures, only the vertices are in the space; the different types of edges indicate the congruence relation.) We can partition Figure 1 into subspaces, for example {0,2,4} and {1,3,5} or {0,3}, {1,4} and {2,5}. However, the only subspaces of Figure 2 are the trivial ones: the whole space and individual elements.

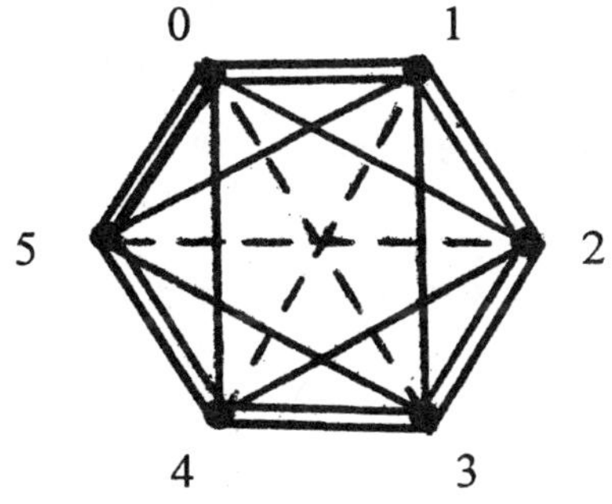

Figure 1.

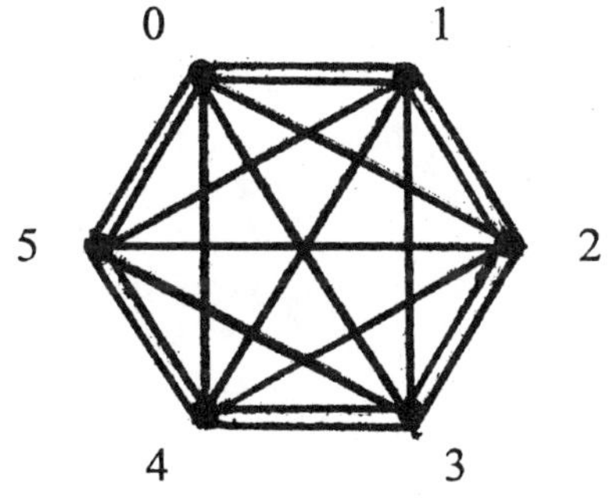

Figure 2.

DEFINITIONS. A bijection α on a congruence space $(S,\equiv)$ is an *isometry* iff for all $x, y \in S$, $xy \equiv \alpha(x)\alpha(y)$. A congruence space $(S,\equiv)$ is *one-point homogeneous* iff for every $x, z \in S$, there is an isometry α with $\alpha(x) = z$. A congruence space $(S,\equiv)$ is *edge homogeneous* iff it is one-point homogeneous and whenever $xy \equiv zw$, there is an isometry α with either $\alpha(x) = z$ and $\alpha(y) = w$ or $\alpha(x)= w$ and $\alpha(y) = z$. The *order* of a finite set S is the number of elements in S.

EXAMPLE 2. The spaces of Figures 1 and 2 are both one-point homogeneous and the first one is edge homogeneous. Their groups of isometries are both D_6.

THEOREM 1. (Generalized LaGrange's Theorem) Let $(S,\equiv)$ be a finite one-point homogeneous congruence space and T a subspace. Then the order of T divides the order of S.

Proof. See Sibley [3].■

The definitions below generalize congruence on the real numbers using an absolute value. For example, the group Z_6 gives the design in Figure 1. For our purposes, $|a|$ must satisfy that two elements have the same absolute value iff they are equal or inverses of each other.

DEFINITIONS. On a group G define $|a| = \{a,a^{-1}\}$ and $ab \equiv cd$ iff $|a^{-1}b| = |c^{-1}d|$.

THEOREM 2. If G is a group, then $(G,\equiv)$ is an edge homogeneous congruence space. Furthermore, if H is a subgroup of G, then all the left cosets of H are isometric subspaces of G.

Proof. See Sibley [3] which generalizes this result to a class of loops.■

Free Congruence Spaces

DEFINITIONS. A function α from one congruence space $(S,\equiv)$ to another $(T,\equiv')$ is a *homomorphism* iff whenever $ab \equiv cd$ in S, $\alpha(a)\alpha(b) \equiv' \alpha(c)\alpha(d)$. A non-empty subset R generates the smallest subspace of $(S,\equiv)$ which contains R. A congruence space $(F,\equiv)$ generated by $\{a_i : I \in I\}$ is *free* iff for any congruence space $(S,\equiv')$ and any subset $\{s_i : I \in I\}$ of S, there is a unique homomorphism of $(F,\equiv)$ into $(S,\equiv')$ which takes each a_i to the corresponding s_i.

EXAMPLE 3. The only free congruence spaces are the following minimalist ones. On any set S define $\equiv$ by $ab \equiv cd$ iff $\{a,b\} = \{c,d\}$. The subset of generators is all of S.

The congruence spaces derived from free groups are almost free among one-point homogeneous congruence spaces in the following sense. For a free group F generated by $\{a_i : I \in I\}$ and with identity e, one can show that $(F,\equiv)$ is generated by $\{a_i : I \in I\} \cup \{e\}$. Let $(S,\equiv')$ be any one-point homogeneous space and $\{s_i : I \in I\} \cup \{o\}$ any subset of S. It is easy to define a homomorphism of $(F,\equiv)$ into $(S,\equiv')$ from the obvious mapping of the generators of $(F,\equiv)$ to $\{s_i : I \in I\} \cup \{o\}$. However, the extension to all of F is not unique. For example, map the generators 0 and 1 of the space $(Z,\equiv)$, derived from the integers $(Z,+)$, to the points 0 and 1 in Figure 1. This can be extended to a homomorphism by either of the functions $f(x) = x(\text{mod } 6)$ or $f(x) = x(\text{mod } 2)$.

Coset Representations

DEFINITIONS. Let $(S,\equiv)$ be any one-point homogeneous congruence space and o a fixed element of S. Then I is the group of isometries of S and I_o is the subgroup of isometries of S fixing o.

From Theorem 2, we can make I into an edge homogeneous congruence space. How does this space relate to S? Figure 3 shows the situation with S the four vertices of a square and $I = D_4$, together with the natural intermediate space of cosets I/I_o. Theorem 3 shows that in general we can define an edge homogeneous congruence space on the cosets of any subgroup. Furthermore Theorem 3 ensures there is a homomorphism from the space of the group onto the space of the cosets even when the subgroup is not normal. Theorem 4 generalizes the process illustrated in Figure 3 to any one-point homogeneous space S by showing that I/I_o can be "inserted" onto S, that is, there is always a one-to-one homomorphism from the coset space I/I_o onto the original space. When S is edge homogeneous, as in Figure 3, then the space of cosets is isomorphic to S. If we start with Figure 2, the space of cosets is isomorphic to Figure 1. Theorems 3 and 4 imply that the coset spaces of groups give all possible edge homogeneous spaces and, up to an insertion, all one-point homogeneous congruence spaces. (Sibley [2] provides a more general solution to the problem of defining a congruence relation on equivalence classes of subspaces.)

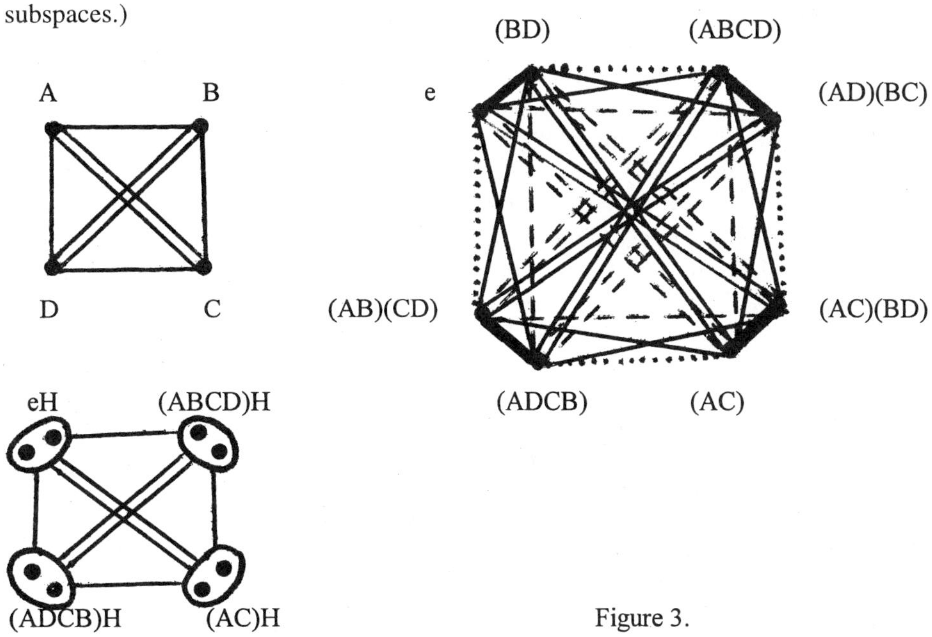

Figure 3.

DEFINITIONS. Given a group G with a congruence relation $\equiv$ as defined above and H a subgroup of G, define $\equiv_H$ on G/H, the left cosets of H, by $a_1Ha_2H \equiv_H a_3Ha_4H$ iff there are $b_i \in a_iH$ such that $b_1b_2 \equiv b_3b_4$. A congruence space $(S,\equiv)$ is *inserted* onto a congruence space $(T,\equiv')$ iff there is a bijection $\beta:S \rightarrow T$ which is a homomorphism from $(S,\equiv)$ onto $(T,\equiv')$. If β^{-1} is also a homomorphism, then β is an *isomorphism*.

THEOREM 3. Let G be any group with its congruence relation $\equiv$ and H be any subgroup of G. Then $(G/H, \equiv_H)$ is an edge homogeneous congruence space and the mapping $\alpha(a) = aH$ is a homomorphism from G onto G/H.

Proof. Given a group G and a subgroup H, $\equiv_H$ is clearly reflexive, symmetric and satisfies eHfH $\equiv_H$ fHeH because the corresponding relation in G satisfies these properties. Consider the collection of all "edges" ah to bh* between the cosets aH and bH. The terms $(ah)^{-1}bh^* = h^{-1}a^{-1}bh^*$ form the double coset $H(a^{-1}b)H$. Two double cosets HxH and HyH are either equal or disjoint. Suppose, to show transitivity, that aHbH $\equiv_H$ cHdH and cHdH $\equiv_H$ eHfH. The first relation implies either $H(a^{-1}b)H = H(c^{-1}d)H$ or $H(a^{-1}b)H = H(d^{-1}c)H$ and similarly for the second relation. Hence either $H(a^{-1}b)H = H(e^{-1}f)H$ or $H(a^{-1}b)H = H(f^{-1}e)H$. If $H(a^{-1}b)H = H(e^{-1}f)H$, then aHbH $\equiv_H$ eHfH. Otherwise, we combine aHbH $\equiv_H$ fHeH with the previously shown eHfH $\equiv_H$ fHeH to show transitivity. Thus $\equiv_H$ is a congruence relation on G/H.

Next $\sigma_g(xH) = (gx)H$ is an isometry of G/H because $(gx)^{-1}(gy) = x^{-1}g^{-1}gy = x^{-1}y$ and so (gx)H(gy)H $\equiv_H$ xHyH for g, x, y $\in$ G. Further, $\sigma_b\sigma_a^{-1}$ takes aH to bH and so G/H is one-point homogeneous. Suppose aHbH $\equiv_H$ cHdH. Then without loss of generality either $a^{-1}b = c^{-1}d$ or $a^{-1}b = d^{-1}c$. In the first case the isometry $\sigma_c\sigma_a^{-1}$ takes aH to cH and bH to dH. In the second case we use $\sigma_d\sigma_a^{-1}$. This shows G/H is edge homogeneous. Finally the natural mapping $\alpha(x) = xH$ from G onto G/H is a homomorphism.■

THEOREM 4. For $(S,\equiv)$ a one-point homogeneous space with o an element of S and I the group of isometries, the space $(I/Io, \equiv_{Io})$ can be inserted onto $(S,\equiv)$. This insertion is an isomorphism iff $(S,\equiv)$ is edge homogeneous.

Proof. Note that $\mu(\sigma I_o) = \sigma(o)$ is a well-defined bijection from I/I_o onto S because every isometry $\sigma\alpha$ in σI_o maps o to $\sigma(o)$. For $\rho I_o \tau I_o \equiv_{Io} \pi I_o \kappa I_o$ we can assume $\rho\tau \equiv \pi\kappa$ and $\rho^{-1}\tau = \pi^{-1}\kappa$. Consider $\sigma_\pi\sigma_\rho^{-1}$ on I, which maps ρ to π and τ to κ. Let $\rho(o) = r$, $\tau(o) = t$, $\pi(o) = p$ and $\kappa(o)=k$. Then rt $\equiv$ pk because $\pi\rho^{-1}$ takes r to p and t to k. Thus μ is an insertion of I/I_o onto S. Now suppose that S is edge homogeneous and rt $\equiv$ pk. So either there is an isometry ψ taking r to p and t to k or an isometry taking r to k and t to p. Without loss of generality, suppose the former. For ρI_o, τI_o, πI_o and κI_o as above, ψ must take ρI_o to πI_o and τI_o to κI_o. This shows $\rho I_o \tau I_o \equiv_{Io} \pi I_o \kappa I_o$ and that μ is an isomorphism in this case. If S is not edge homogeneous, then for some rt $\equiv$ pk, there is no corresponding isometry ψ. Then μ^{-1} would not be a homomorphism.■

We need cosets and insertions to construct all one-point homogeneous congruence spaces from groups. To see this, form a congruence space S using the Petersen graph and its complement as the types of edges. An elementary check shows that neither congruence space derived from Z_{10} or D_5 can be inserted into S. Equivalently, the smallest transitive group of automorphisms of the Petersen graph has twenty elements, rather than ten.

We can readily convert a metric space into a congruence space by defining ab $\equiv$ cd iff d(a,b) = d(c,d). The vertices of vertex transitive polyhedra or polytopes thus form one-point homogeneous congruence spaces. Similarly, Euclidean, hyperbolic, spherical and elliptical geometries of any number of dimensions form one-point homogeneous congruence spaces.

A more unusual interpretation of ≡ converts affine planes (and affine spaces) into congruence spaces. Suppose ⟨ab⟩ is the line on a and b for an affine plane S. For a ≠ b and c ≠ d we define ab ≡ cd iff ⟨ab⟩ is parallel to ⟨cd⟩. The group of homotheties of the affine plane is the group of the isometries of the congruence space. Affine planes with the first Desargues property give one-point homogeneous congruence spaces. (See Sibley [3].)

REFERENCES

1. C. C. Chen and H. H. The, Constructions of point-colour-symmetric graphs, Journal of Combinatorial Theory Series B 27 (1979) 160-167.

2. Thomas Q. Sibley, Homomorphisms for equidistance relations, Cuttington Research Journal, 2 (1983), 1-11.

3. Thomas Q. Sibley, Sylow-like theorems in geometry and algebra, Journal of Geometry, 30 (1987), 1-11.

Division Algebras Three-Dimensional Over Algebraic Number Fields

G. P. Wene Department of Mathematics, University of Texas at San Antonio, San Antonio, Texas

ABSTRACT. Proofs by Dickson [2] and Kaplansky [8] imply that if the polynomial $f(X) = X^3 - BX^2 - \beta X - b$ is irreducible over the rational numbers or any algebraic number field F that contains no subfield three-dimensional over the rational numbers then the algebra **E** over F with basis $1, i, j$ and multiplication given by

$$i^2 = j, ij = ji = b + \beta i + Bj, j^2 = 4bB - \beta^2 - 8bi - 2\beta j$$

is a division algebra. The commutative division algebras constructed here are never associative; the automorphism groups are computed.

1. INTRODUCTION

The study of three-dimensional commutative nonassociative division algebras began with Dickson [4] in1905 . He continued this study in [5], [6] and [7] and gave a capsule summary on page 69 of his monograph [8]. I repeat Dickson's description, keeping his notation. Let F be any field of characteristic not equal to 2. Let B, β, b be elements of F such that $x^3 - Bx^2 - \beta x - b$ is irreducible over F. Define an algebra with basis $1, i, j$ by

$$i^2 = j, \qquad ij = ji = b + \beta i + Bj, \qquad j^2 = 4bB - \beta^2 - 8bi - 2\beta j.$$

Then this is a division algebra, nonassociative except in the purely inseparable case (characteristic 3, $B = \beta = 0$). He further showed that any three-dimensional commutative division algebra over a field of characteristic not 2 is a field or one of these algebras. Following Dickson, we will insist that each division algebra has a unity element which we will denote by 1.

Dickson was primarily concerned with finite division algebras.

There are no semifields two-dimensional over any field F that are not fields. There are, however, left near-fields two-dimensional over both finite and infinite fields; the interested reader is directed to the articles by Sabharwal[17][18] concerning infinite Hughes planes.

The study of three-dimensional division algebras was continued by Zelinsky [20], Menichetti [14] [15]; Kaplansky [10] and [11] applied Dickson's construction over algebrically closed fields and Harrison [9] studies commutative algebras three-dimensional over an algebrically closed field of characteristic zero.

Cohen and Ganley [3] investigate commutative division algebras quadratic over their (finite) middle nuclei and give as an example the "Dickson Commutative Semifields" that are exhaustedly treated in the two paper by Burmester [1] and [2].

Zemmer [21] studied commutative division algebras **D** of order 2n over a finite field F with $q^k = 2ns+1$ elements and their automorphism groups. Knuth[13] gives a construction for commutative division algebras over finite fields of characteristic two.

Oehmke [16] showd that a finite division algebra of characteristic not equal to two and satisfying the *flexible* relation: $(xy)\,x = x\,(yx)$ must be commutative.

The classical reference concerning finite division algebras (also called "semifields") and their role in gometry is Knuth [12].

It appears that no one has noticed that the proof that Dickson's construction resulting in a division algebra does not rely on the fact that the field F has finite characteristic. I will investigate those Dickson commutative algebras three-dimensional over the algebraic number fields and their automorphism groups.

Since the name Dickson has been applied to many different algebras, I will use the name Eisenstein to refer to the algebras constructed in this paper.

2. Prelimanaries

The first result is a general result concerning algebras and automorphisms of the base field.

Theorem 1. *Let A be an algebra with basis $\{u_i \mid i = 0, 1, \cdots, n\}$ over the field F and multiplication given by*

$$u_i u_j = \sum_{k=0}^{n} C_{ijk} u_k.$$

where $C_{ijk} \in F$ for all $i, j, k = 0, 1, \cdots, n$. If * is an automorphism of F, then * induces an automorphims $^\sigma$ in A defined by

$$\left(\sum_{i=0}^{n} \alpha_i u_i\right)^{\sigma} = \left(\sum_{i=0}^{n} \alpha_i^* u_i\right)$$

iff $C^*_{ijk} = C_{ijk}$ for all $i, j, k = 0, 1, \cdots, n$.

Proof. Let $a = \sum_{i=0}^{n} \alpha_i u_i$ and $b = \sum_{i=0}^{n} \beta_i u_i$ be elements of A. Clearly $(a+b)^\sigma = a^\sigma + b^\sigma$. Furthermore, $(ab)^\sigma = \sum_{i,j,k=0}^{n} (\alpha_i \beta_j C_{ijk} u_k)^\sigma = \sum_{i,j,k=0}^{n} \left(\alpha_i^* \beta_j^* C_{ijk}^* u_k\right)$ and $a^\sigma b^\sigma = \left(\sum_{i=0}^{n} \alpha_i u_i\right)^\sigma \left(\sum_{i=0}^{n} \beta_i u_i\right)^\sigma = \left(\sum_{i=0}^{n} \alpha_i^* u_i\right)\left(\sum_{i=0}^{n} \beta_i^* u_i\right) = \sum_{i,j,k=0}^{n} \left(\alpha_i^* \beta_j^* C_{ijk} u_k\right)$. These two expressions are equal iff $C^*_{ijk} = C_{ijk}$ for all $i, j, k = 0, 1, \cdots, n$. ∎

The next series of results are well known; for proofs see [20].

Let D be a division algebra 3-dimensional over the field F. If $0 \neq a \in D$, we will call the mapping $R_a : D \to D$ defined by $xR_a = xa$ right multiplication by a; we have the obvious inductive definition: $xR_a^n = xR_a^{n-1}R_a$. Left multiplication is defined similarly. Then $a^2a + \alpha a^2 + \beta a + \delta = 0$ for $\alpha, \beta, \delta \in F$ and $1R_a^3 + \alpha R_a^2 + \beta R_a + \delta = 0$, and we say that the minimal polynomial of a is $X^3 + \alpha X^2 + \beta X + \delta$.

Theorem 2. *Let D be a division algebra three-dimensional over the field F then the right and left minimum polynomials of each element of D are irreducible.*

Thus we see that there are no division algebras 3-dimensional over the real numbers or the complex numbers.

Theorem 3. *A division algebra D of dimension $n \geq 3$ over the field F has no subalgebra of dimension $n - 1$.*

A division algebra D 3-dimensional over a field F is not associative iff each of the nuclei N_l, N_m and N_r of D is isomorphic to F.

Theorem 4. *A division algebra D 3-dimensional over a field F is generated by any one of its nonscalar elements. That is, if a is not a scalar, then $1, a, a^2$ is a basis for D.*

Theorem 5. *There does not exist an automorphism over F of D of period 2.*

Proof. Suppose otherwise; let $^\sigma$ be an automorphism of D of period 2. Let $S \subset D$ such that $s \in S$ if and only if $s^\sigma = s$. Then S is a subdivision ring of D and by Theorem 4, $S = D$ or $S = Q1$. In the former case $^\sigma$ is the identity automorphism. In the latter case, $i + i^\sigma = \alpha$ and $ii^\sigma = \beta$ for scalars α and β. Since $i\,(i + i^\sigma) = j + \beta = \alpha i$, we see that $\{1, i, j\}$ is not linearly independent. ■

Corollary 6. *the algebra* **D** *has no automorphism over F of period 2n, $n = 1, 2, 3 \cdots$.*

Proof. If $^\sigma$ is an automorphism of period 2n, then $^{\sigma^n}$ is an automorphims of period 2. ■

3. The Construction

We now construct a large class of commutative division algebras 3-dimensional over those algebraic number fields F that contain no subfield of degree three over the rational numbers. Let B, β, b be elements of F such that $x^3 - Bx^2 - \beta x - b$ is irreducible over F. Define an algebra with basis $1, i, j$ by

$$i^2 = j, \qquad ij = ji = b + \beta i + Bj, \qquad j^2 = 4bB - \beta^2 - 8bi - 2\beta j.$$

We will call these algebras three-dimensional Dickson commutative algebras.

Theorem 7. *All three-dimensional Dickson commutative algebras are division algebras.*

Proof. See Kaplansky[10]

Theorem 8. *A three-dimensional Dickson commutative algebra* $\mathbf{D}$ *is never associative.*

Proof. By Schafer[19], page 130, the algebra $\mathbf{D}$ is associative iff

$$(ij)\,i = i^2 i^2 = jj$$

$$(ij)\,i = bi$$

$$j^2 = -8bi$$

Equating components gives $b = -8b$. But b cannot be zero. ■

Example 9. *The polynomial* $X^3 - 2$ *is irreducible over the rational numbers. Define an algebra* $\mathbf{D}$ *with basis* $\{1, i, j\}$ *by*

$$i^2 = j, \qquad ij = ji = 2, \qquad j^2 = -16i.$$

Example 10. *Choose a new basis for the algebra in Example 1 by setting* $\overline{1} = 1$ *and* $I = 1 + i$. *The new basis will be* $\left\{\overline{1}, I, J\,(= I^2)\right\}$. *The element* J *will be* $J = 1 + 2i + j$.

$$IJ = 3 + 3i + 3j = 3 - 3I + 3J.$$

We can compute J^2 *directly or use Dickson's formula to get*

$$J^2 = 27 - 24I + 6J.$$

This system is a division algebra isomorphic to that of Example 1.

Example 11. *Let* F *be the field* $Q(\sqrt[3]{2})$ *where* $\sqrt[3]{2}$ *is the real cube root of two. Let* $\mathbf{A}$ *be an algebra over* F *with basis* $\{1, i, j\}$ *and the multiplication:*

$$i^2 = j, \qquad ij = ji = 2, \qquad j^2 = -16i.$$

The algebra $\mathbf{A}$ *has divisors of zero:* $\left(j + i\sqrt[3]{2} + \sqrt[3]{4}\right)\left(i - \sqrt[3]{2}\right) = 0$. *We can easily compute nilpotent elements of* $\mathbf{A}$: *A simple computation shows that if* $(x + yi + zj)^2 = 0$ *then*

$$x^3 = 2y^3 = -32z^3.$$

The elements $u = -4\sqrt[3]{4} - 4\sqrt[3]{2}i + 2j$ *is nilpotent.*

4. The Automorphism Groups

Let **D** be the algebra constructed in Example 7. If F is an algebraic number field such that 2 is a prime and F contains no subfield of degree 3 over the rational numbers then the algebra obtained by extension of the scalar field to F, denoted by $\mathbf{D}\otimes_Q F$ (tensored over Q), is also a division algebra. This algebra contains a copy of the division algebra **D**.

Lemma 12. *Let* $^\sigma : \mathbf{D}\otimes_Q F \to \mathbf{D}\otimes_Q F$ *be an automorphism that is linear over* Q *but not necessarily linear over* F. *Then* $^\sigma$ *restricted to* **D** *is the identity map.*

Proof. We must have $1^\sigma = 1$ and $^\sigma$ restricted to **D** is completely determined by its action on i. Let $i^\sigma = x + yi + zj$. Then $j^\sigma = (x^2 + 2yzb) + (2xy - 8z^2b)\,i + (y^2 + 2xz)\,j$. Since $^\sigma$ is an automorphism we must have $(ij)^\sigma = b^\sigma = b = i^\sigma j^\sigma$.

$$i^\sigma j^\sigma = x^3 + 6bxyz + by^3 - 8b^2z^3 + \left(3x^2y - 6by^2z - 24bxz^2\right)i + \left(3xy^2 + 3x^2z - 6byz^2\right)j$$

Equating components gives

$$\begin{aligned} x^3 + 6bxyz + by^3 - 8b^2z^3 &= b \\ 3x^2y - 6by^2z - 24bxz^2 &= 0 \\ 3xy^2 + 3x^2z - 6byz^2 &= 0 \end{aligned}$$

Noting that $y\,(3xy^2 + 3x^2z - 6byz^2) - z\,(3x^2y - 6by^2z - 24bxz^2) = 3xy^3 + 24bxz^3 = 0$, we get $xy^3 + 8bxz^3 = 0$. If $x \neq 0$ this gives $y^3 + 8bz^3 = 0$. Since the field F contains no subfield of degree three over Q this last is impossible. Hence $x = 0$ and either $y = 0$ or $z = 0$. If $x = y = 0$, the first equation yeilds $z = -\frac{\sqrt[3]{b}}{2}$; the only remaining possibility is $x = z = 0$ and then $y = 1$. ∎

Theorem 13. *The automorphism group of* $\mathbf{D}\otimes_Q F$ *is isomorphic to the group of automorphisms of* F.

Proof. Let $x + yi + zj \in \mathbf{D}\otimes_Q F$ then $(x + yi + zj)^\sigma = x^* + y^*i + z^*j$ where * is the automorphism of F induced by $^\sigma$. Any automorphism * of F induces an automorphism of $\mathbf{D}\otimes_Q F$ by Theorem 2. ∎

Example 14. *The polynomial* $X^3 - 2$ *is irreducible over the rational numbers. Define an algebra* **D** *with basis* $\{1, i, j\}$ *over* $Q(\sqrt{2})$ *and mutltiplication given by*

$$i^2 = j, \qquad ij = ji = 2, \qquad j^2 = -16i.$$

There is no element of $Q(\sqrt{2})$ *that is a zero of* $X^3 - 2$. *The automorphism* * : $Q(\sqrt{2}) \to Q(\sqrt{2})$ *given by* $\left(s + t\sqrt{2}\right)^* = s - t\sqrt{2}$ *induces an automophism of* **D** *over* Q *of period two.*

Theorem 15. *There exists a commutative nonassociative division algebra* **D** *such that the group of automophism of* **D**, *Aut*(**D**), *contains a group isomorphic to the finite abelian group* $Z_2 \otimes Z_2 \otimes \cdots \otimes Z_2$, *that is n copies of* Z_2.

Proof. The polynomial $X^3 - 2$ is irreducible over $F = Q\left(\sqrt{2}, \sqrt{3}, \cdots, \sqrt{p_n}\right)$ where p_n is the nth prime. The division algebra **D** has a basis $\{1, i, j\}$ over F and mutltiplication given by

$$i^2 = j, \qquad ij = ji = 2, \qquad j^2 = -16i.$$

$Aut(F) \cong Z_2 \otimes Z_2 \otimes \cdots \otimes Z_2$, n copies of Z_2. Each automorphism of F induces an automorphism of **D** by Theorem 4. The result now follows. ■

Corollary 16. *There exists a commutative nonassociative division algebra* **D** *such that the group of automophism of* **D** , *Aut*(**D**), *contains a subgroup isomorphic to the infinite abelian group in which each element has order 2.*

Proof. Let **D** be an algebra constructed over the field Q according to the construction in Section 3. Apply Zorn's lemma to the chain of algebras $\mathbf{D} \subset \mathbf{D} \otimes_Q F_1 \subset \mathbf{D} \otimes_Q F_2 \subset \cdots$ fields $F_1 \subset F_2 \subset F_3 \cdots \subset F_n \subset \cdots$ where $F_1 = Q$ and $F_k = F_{k-1}[\sqrt{p_k}]$ and p_k is the kth prime. ■

Example 17. $Q\left(\sqrt{2}, \sqrt{3}\right)$ *is of degree 4 over* Q *and has Galois group isomorhic to the Klein 4-group.*

Example 18. *The splitting field of* $X^8 - 2$ *is of degree 16 over* Q *and has a 16 element Galois group isomorphic to* $< \sigma, \tau \mid \sigma^8 = \tau^2 = 1, \sigma\tau = \tau\sigma^3 >$.

5. The Autotopism Groups

Any triple $\alpha(P, Q, S)$ of non-singular additive maps of a division ring **D** onto itself satisfying$(xP)(yQ) = (xy)S$ for all x, y in **D** is called an *autotopism* of **D**. We will use the following well-known fact:

Lemma 19. *The autotopisms of a division ring* **D** *form a group under the composition* $\alpha(P, Q, S) \circ \alpha(P_1, Q_1, S_1) = \alpha(PP_1, QQ_1, SS_1)$.

Example 20. *Any automorphism* σ *of a division ring* **D** *will be an autotopism. Let* **D** *be a commutative division algebra 3-dimensional over the field* F *and* $m, n \in N_r$. *Then each of* $\alpha(R_m, R_n, R_{mn})$ *and* $\alpha(\sigma R_m, \sigma R_n, \sigma R_{mn})$ *is an autotopism of* **D**.

Lemma 21. *Let* **D** *be a commutative division algebra 3-dimensional over the field* F. *Any autotopism* $\alpha(P, Q, S)$*of* **D** *for which one of* P, Q *and* S *is the identity map,* I, *is of the form* $\alpha(R_m, R_n, R_{mn})$ *where* m, n *are elements of the nucleus.*

Proof. Case 1. $S = I$. We have $x^P y^Q = xy$. There exist elements a and $b \in \mathbf{D}$ such that $x^P y^Q = (xb)(ay) = xy$. Substituting $y = a^{-1}$into this gives us $(xb) = xa^{-1}$. Since $\mathbf{D}$ is a semifield, $b = a^{-1}$. Now $(xa^{-1})(ay) = xy$ and y any nonzero element of the nucleus gives $(xa^{-1})a = x$.The element a must be in the nucleus. Suppose otherwise; for some y and z, $z(ay) - (za)y \neq 0$. Since $\mathbf{D}$ is a semifield, there is an unique $x \in \mathbf{D}$ such that $xa^{-1} = z$. We now have $z(ay) - (za)y = (xa^{-1})(ay) - ((xa^{-1})a)y = xy - xy \neq 0$.

Case 2. $P = I$. We have $x\left(y^Q\right) = (xy)^S$. We must have $y^Q = y^S$; hence, $x\left(y^S\right) = (xy)^S$. Setting $y = 1$, gives $x\left(1^S\right) = x^S$. This last implies $x\left(y1^S\right) = (xy)1^S$. Thus 1^S is in the nucleus.

Case 3. $Q = I$. The argument will not be given.

Lemma 22. *Let $\alpha(P, Q, S)$be an autotopisms of a commutative division algebra* $\mathbf{D}$ *3-dimensional over the field F. If $P = R_a$ and $Q = R_a$ then $a \in N$ and $S = R_{a^2}$.*

Proof. Let 1 denote the identity of $\mathbf{D}$; $(1a)(1a) = a^2 = 1^S$. Furthermore, $(xa)(a^2) = (xa)^S$; since R_a is a bijection for all $a \neq 0$, $x^S = xa^2$.

Since is commutative, $a^2a = aa^2$. Let A_n be a product of n a's under any association of the a's; we will show that $A_n = aR_a^{n-1}$ for all $n = 1, 2, 3, \cdots$. This is clearly true for $n = 1, 2$ and 3. Suppose now that $A_i = aR_a^{i-1}$ for all positive integers $i \leq k$. We can write $A_{k+1} = A_{k+1-i}A_i$ for some $i \neq 0$. By the induction hypothesis, $A_{k+1-i}A_i = (A_{k-i}A_{i-1})a^2 = (A_{k-1})a^2 = (A_k)a = aR_a^k$. Thus the element a generates an assoicative subalgebra of $\mathbf{D}$. ∎

References

[1] M. V. D. Burmester, On the commutative non-associative division algebras of even order of L. E. Dickson, *Rend. Mat. e Appl., V. Ser.* **21** (1962), 143-166.

[2] M. V. D. Burmester, On the non-unicity of translation planes over division algebras, *Archiv der Mat.* **15** (1964), 364-370.

[3] S. D. Cohen and M. J. Ganley, Commutative semifields, two dimensional over their middle nuclei, *J. Algebra* **75** (1982), 373-385.

[4] L. E. Dickson, On finite algebras, *Nachr. Ges. Wiss. Göttingen* (1905), 358-393.

[5] L. E. Dickson, Linear algebras in which division is always uniquely possible, *Trans. Amer. Math. Soc.* **7** (1906), 370-390.

[6] L. E. Dickson, On linear algebras, *Amer. Math. Monthly* **13** (1906), 201-205.

[7] L. E. Dickson, On triple algebras and ternary forms, *Bull. Amer. Math. Soc.* **14** (1908), 160-168.

[8] L. E. Dickson, "Linear Algebras," Cambridge Univ. Press, 1914.

[9] D. K. Harrison, Commutative nonassociative algebras and cubic forms, *J. Algebra* **32** (1973),518-528.

[10] Irving Kaplansky, Three-dimensional division algebras, *J. Algebra* **40** (1974), 384-391.

[11] Irving Kaplansky, Three-dimensional division algebras II, *Houston J. Math.* **1** (1975), 63-79.

[12] D. Knuth, Finite semifields and projective planes, *J. Algebra* **2** (1965), 182-217.

[13] D. E. Knuth, A class of projective planes, *Trans. Amer. Math. Soc.* 115 (1965), 541-459.

[14] G. Menchetti, Algebre tridimensionali su un campo di Galois, *Ann. Mat. Pura Appl.* **97** (1973), 283-301.

[15] G. Menchetti, On a Kaplansky conjecture concerning three-dimensional division algebras over a finite field, *J. Algebra* **47** (1977), 400-410.

[16] R. H. Oehmke, On finite division rings, *Proc. Amer. Math. Soc.* 79 (1980), 174-176.

[17] R. S. Sabharwal, Inifinite Hughes planes, *Math. Student*, XLI (1973), 307-310.

[18] R. S. Sabharwal, Inifinite planar left near-fields, *Math. Student*, XLI (1973), 322-324.

[19] R. D. Schafer, "An Introduction to Nonassociative Alegras", Academic Press, New York, 1966.

[20] D. Zelinsky, "Nonassociative Algebras of Order Three," unpublished Master's thesis, University of Chicago, 1943.

[21] J. L. Zemmer, Jr., On the subalgebras of finite division algebras, *Canadian J. Math.* 4 (1932), 491-503.

Index